MW01005843

THE SOIL UNDERFOOT

Infinite Possibilities for a Finite Resource

THE SOIL UNDERFOOT

Infinite Possibilities for a Finite Resource

Edited by
G. Jock Churchman
University of Adelaide, Australia
Edward R. Landa
University of Maryland, College Park, USA

Cover illustration by Jay Stratton Noller

CRC Press is an imprint of the
Taylor & Francis Group, an **informa** business

CRC Press
Taylor & Francis Group
6000 Broken Sound Parkway NW, Suite 300
Boca Raton, FL 33487-2742

Printed on acid-free paper
Version Date: 20140131

International Standard Book Number-13: 978-1-4665-7156-3 (Hardback)

Library of Congress Cataloging-in-Publication Data

The soil underfoot : infinite possibilities for a finite resource / editors: Gordon John Churchman, Edward R. Landa.
pages cm
Includes bibliographical references and index.
ISBN 978-1-4665-7156-3 (hardcover : alk. paper) 1. Soils. 2. Soil science. 3. Soil and civilization. I. Churchman, G. J. II. Landa, Edward.

S591.S7563 2014
631.4--dc23 2014001312

Visit the Taylor & Francis Web site at
http://www.taylorandfrancis.com

and the CRC Press Web site at
http://www.crcpress.com

To Patrick, Harriet, Alice, Audrey, and Reuben, and every child of their generation worldwide, including those yet to be born.

To Judith, for the sky above and earth below.

Contents

SECTION IV Technologies and Uses

SECTION V Future Strategies

Foreword

Vanishing Soils: The World's Dirty Secret

Soil scientists do not make the front pages. Never. Meanwhile major media outlets commonly write about "empty skies" (declines of neotropical migratory birds), "empty oceans" (overfishing and depleted marine systems), mass extinction, deforestation, and, of course, climate disruption. Each of these conservation and environmental crises deserves the spotlight. But what could be more fundamental than the soil that grows the plants from which 99% of humankind's calorie intake is derived? This is also the soil that can either store or emit vast amounts of carbon, that can hold on to nutrients or let those nutrients flow into our rivers and estuaries to the detriment of fisheries and people, that can erode into rivers where silting harms fish and reduces hydropower generation. Indeed, the entire arc of human history can be viewed as a series of civilizations bankrupting their soils. This unique book provides a cultural, religious, historical, and scientific tour-de-force regarding the role of soil in the human epic. If you have gardened and felt the comfort and seduction of warm, fertile soil in your hands, you know how primal is the link between people and soil. When someone back in the recesses of time coined the term "mother earth" I have to believe she or he was thinking of warm soil.

Today, around the world the mean rate of soil loss is roughly 10 times the rate at which soil is replenished (Montgomery 2007). Unless we fundamentally change our agriculture practices, current rates of soil loss and erosion will pose severe challenges for agricultural productivity, as well as the need for massive clearing of lands as yet undisturbed. Fortunately, the world is slowly waking up to the soil crisis. In 2009, the globalsoilmap.net project was initiated with support from the Bill and Melinda Gates Foundation. The United Nations Environment Programme (UNEP), the Food and Agriculture Organization (FAO), and the World Bank have all recognized how crucial soil is to their mission of sustainable development. But we still lack global- national-level soil monitoring. We lack clear site-specific indicators that can give us an early warning signal that we might be on the brink of irreversible soil loss. And we have not developed policy incentives to reward land owners, farmers, or local communities who treat their soils well. The loss of any species is tragic and sad. But the loss of one's fertile soil is catastrophic. The conservation community, the environmental community, the agricultural community, and the development community need to unite around the issue of protecting and restoring our soils.

Just as greenhouse gas emissions should be on everyone's mind, so too should soil loss and soil health. My hope is that this book will inspire new science, new awareness, and new policy aimed at addressing the soil crisis. The national, religious, historical, and cultural stories told in this book provide a compelling picture of how humanity is linked to the earth's soil. The chapters on climate change, soil carbon, nutrients and ecosystem services begin to frame the sorts of new research we need if we are to have the tools that could inform a modern soil conservation policy. The science of ecosystem services is taking hold around the world as a tool for conservation and environmental policy—but this science has not adequately captured the value of supporting services such as soil formation. The value of water supply or fisheries, or even crop pollinators is straightforward to capture. This is in part because incremental changes in water supply can be monetized, as can incremental changes in fish supply. In contrast, models that link gradual changes in soil quality, or the risk of crossing a soil threshold to changes in human welfare in real time (as opposed to historical stories) need to be developed. In much of the world we have built our cities on the most fertile soils, thereby squandering a valuable resource when buildings and settlements could be much more wisely placed on unproductive soils (Nizeyimana et al. 2001). It is now routine to inform plans for infrastructure or development with maps of where biodiversity is concentrated and especially valuable. We need to adopt a similar approach for our world's soils. This book is a starting point for what I hope will be a twenty-first century of soil science and conservation that transforms land use, land-planning, and agriculture around the world.

REFERENCES

Montgomery, D. 2007. Soil erosion and agricultural sustainability. *Proceedings of the National Academy of Sciences of the United States* 104: 13268–13272.

Nizeyimana, E.L., G.W. Petersen, M.L. Imhoff, H.R. Sinclair, S.W. Waltman, D.S. Reed-Margetan, E.R. Levine, and J.M. Russo 2001. Assessing the impact of land conversion to urban use on soils of different productivity levels in the USA. *Soil Science Society of America Journal* 65: 391–402.

Peter Kareiva
The Nature Conservancy

Introduction

G. Jock Churchman and Edward R. Landa

> If all of the elephants in Africa were shot, we would barely notice it, but if the nitrogen fixing bacteria in the soil, or the nitrifiers, were eliminated, most of us would not survive for long because the soil could no longer support us.
>
> **Hans Jenny (*in* Stuart 1984)**

By general consensus, and well within the lifetimes of most of us alive today, the world's population will ascend from the current 7 billion to at least 9 billion. There are already about a billion people who are chronically undernourished (FAO 2012). Even so, it is often said (e.g., Lenné and Wood 2011) that the world actually produces more than enough food to provide adequate nourishment for everyone. Patently, it is very poorly distributed, so that obesity and early deaths from related diseases occur at the same time as chronic undernourishment and premature deaths from hunger—an unjust and obscene juxtaposition. However, proclaiming that there is enough food for all, "is rather like saying there is enough money for all: there certainly is, but will it ever be distributed equally?" (Lenné and Wood 2011, 213).

The stark truth is that 2 million more tiny mouths cry for food every week, and, unachievable redistribution aside, it is becoming incrementally harder to find the extra food to sustain them all. Among other reasons, the world—its environment and its people's tastes and dietary expectations—are all changing rapidly. "Simply put," writes Lenné (2011, 14) in a review of food security requirements, "the world must produce 50% more food, on less land, with less fresh water, using less energy, fertilizer and pesticide—by 2030—a daunting challenge that *must* be met."

The largest part of the world's food comes from its soils, either directly from plants, or via animals fed on pastures and crops. Thus, these finite soil resources will be called on to grow much of the extra food. Much evidence, including some presented in this book, argues persuasively that hardly any more land can be brought into agricultural production; so it is the productivity of soils on the land already providing most of our food today, that will need to be raised—drastically—if the "coming famine," as Julian Cribb (2010) calls it, is to be averted.

This is not to claim that improving the productivity of soils will alone ensure global food security. When it comes to feeding the world, improving and maintaining soils may constitute a necessary, but not necessarily sufficient, condition. Other technical solutions to world hunger have been proposed, including improvements in the productivity of plants, the storage of crops and food products, and the efficiency of water use; the latter one is integral to the productive use of soils, and is much discussed in this book. Other possible improvements are in the realm of human interactions, and include items such as terms of trade, setting of market prices for food, minimization of food wastes, form of ownership of land, and cost and use of energy (see, e.g., Weis 2007; Cribb 2010; Rosin et al. 2012). These same authors and others also warn of the effects of alternative uses for productive soils, such as for growing biofuels.

Short-term gains in food production may occur at the expense of the long-term productivity of the soils. For soil is not just an inert, unforgiving medium for the growth of plants. It is a living body that can suffer ill health if stressed. The quality of soil, and indeed its very presence as the sustainable blanket of the earth, may be at risk as a result. This book was planned with the idea in mind that it will be necessary to maintain, and if possible, improve the quality—and hence good health—of soils, while enabling them to provide crops and pastures for food to support more people in a changing environment.

In another edited volume, Lal and Stewart (2010) concentrate on the nexus between food security and soil quality from a technical viewpoint, mainly through case studies from around the world that, *en masse*, present a compelling picture of the challenges ahead. While likewise underpinned by science, our intent in this volume was somewhat different. We have looked to diverse cultural and geographical sources for descriptions and analyses of past attitudes and approaches to soils and their management, including philosophical and ethical frameworks that have given them value. We aimed to arm the book's readers with the "wisdom of our elders," together with current ideas and approaches to the wise—as well as the unwise—management of soils. Then readers may be able to draw their own inferences as to how the world's soils can be sustained, while continuing to provide the ultimate support for the survival of an expanding human population. Clearly, mere survival is not our ultimate goal as human beings; so in this volume, we also look at aspects of the soil that enhance the "good life," once our "daily bread" has been provided.

Richard Ruggiero, a wildlife biologist with the U.S. Fish and Wildlife Service, recently spoke (National Public Radio {U.S.} 2013) about the illegal ivory trade in central Africa and the threat of extinction of the forest elephants. In his words, these noble creatures should not be "reduced to numbers in a balance book of a business that trades in their teeth"—a rich and moving statement about an animal for which we all seem to have an innate attraction and fascination.

But what about soils? They often lay (literally and figuratively) below the radar screen. Underfoot... Unseen... Ignored. For their stewardship, perhaps an approach drawn somewhat counterintuitively from this metaphor is needed. In soil science, we often separate to reveal. We sift the soil to expose its skeleton of sand, silt, and clay. We extract the soil with dilute acids to assess plant-available nutrients. We isolate, culture, and enumerate microorganisms from the soil to understand the rates and pathways by which elements cycle. In these ways, we scientists try for a more-whole view of the soil from a look at its parts in our balance book.

Nonetheless, getting to know soil through a close look at each of its parts may indeed be like the old story—a version of which comes from Indian, Chinese, and also African cultures—of the group of blind men who each described an elephant after touching a part of it, a different part for each man. The result was that none of them could describe the whole animal and, indeed, were inclined to dispute each other's description.

And thinking more about elephants may teach us something else about soils. The opening quotation from Hans Jenny—a towering figure in soil science—does not diminish elephants. Indeed, it celebrates them, and puts a rare and equally bright spotlight on the soil beneath their massive feet. And Jenny was not the only person in the past who recognized both the high value of soils and their general neglect. Leonardo da Vinci said, "We know more about the movement of celestial bodies than about the soil underfoot." His circa 1500 statement remains true more than half a millennium later, and inspired the title for this book.

With all this talk of elephants, we do not wish to create the impression that soils are only a grand-scale feature of the landscape. Indeed, their complexity and fascination is rooted (literally!) in the network of macropores and micropores that govern water retention, aeration, plant growth, and that form the habitat for microorganisms, insects, earthworms, and a myriad of other creatures. This book sprang from a previous volume—*Soil and Culture* (Landa and Feller 2010). A focal point of the film chapter there (Landa 2010) was the movie *Dune*, and its iconic image of the warrior riding the huge sandworm—pure fiction, and yet glorious, in that it resonates with reality. In the unlit subsurface, there exist scaled-down analogs—tiny nematodes with adhering fungal cells hitching rides through the soil's interconnected pores (Figure I.1). And in addition to living creatures moving through these pores, there is a flux of clay minerals, organic matter, and other dissolved and particulate materials carried by infiltrating water (Ugolini 2005). The slow removal of such materials from one zone and deposition in another, over long periods of time, results in a vertical stratification of the soil. The field presence of the resultant "profiles" underpins our thinking as soil scientists, and its imagery (see, e.g., http://soils.usda.gov/gallery/photos/profiles/) is iconic. For those of us

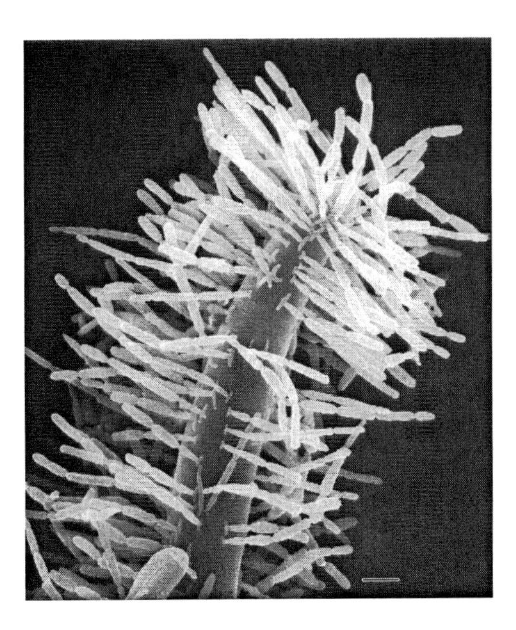

FIGURE I.1 Nematodes are microscopic worms of a different phylum (*Nematoda*) from the common earthworm (*Annelida*). They live in water films in soil pores, and a teaspoonful of soil may contain several hundred individuals, most less than a millimeter in length. Yeast cells of the genus *Botryozyma* have been shown to attach to the outer skin (cuticle) of nematodes, providing a means for their dispersal in the environment traversed by the nematode. This scanning electron microscope image shows cells of *Botryozyma nematodophila* on the anterior end of a *Panagrellus dubius* nematode. The scale bar = 10 μm. (Adapted from Kerrigan, J. and J.D. Rogers. 2013. Biology, ecology and ultrastructure of *Ascobotryozyma* and *Botryozyma*, unique commensal nematode-associated yeasts. *Mycologia* 105: 34–51. Photo courtesy of Julia Kerrigan, Clemson University.)

who study soils, there remains a wonderment, an unending supply of questions, and a joy in the unravelling of mysteries unseen.

This book is about soils—their value, their management, and their myriad uses viewed within human and societal contexts in the past, present, and supposed futures. With the view that the tally of a balance book of entries from a net cast wide can indeed reveal, we have embarked on our path.

The volume begins with five chapters that describe major challenges that we have to face, now and in the future, and some possible responses to them. Kevin Tate and Benny Theng show that soils and their management can be intimately linked with processes of climate change, both causally and for mitigation. Their analysis places us firmly within the "Anthropocene," the era within which "human activities have become the major driver of environmental change." Sharon Hall then gives a comprehensive overview of the problems involved in trying to feed the coming 9–10 billion people, and proposes some possible ways to meet this massive challenge. Nikolaus Kuhn discusses soil loss to remind us that the continued existence of soils cannot be taken for granted. Indeed, soils can even disappear if treated badly for long enough. Kuhn reminds us further that soils may lose quality and functions, as well as suffer complete loss, by erosion. Zeng-Yei Hseu and Zueng-Sang Chen emphasize this point for chemical pollution and other forms of soil degradation by describing soils in Taiwan, one of the Earth's most heavily populated countries, as a case study. With the future inevitably placing us more into the "Technosphere," Peter Haff then looks to some possible futures for soils, as scenarios within a world that will be shaped strongly by technology. Among his messages is that, if you can't beat the technological trends, you can join them.

In the following six chapters, the volume discusses a number of different ways in which soils are, or have been, valued. This section of the book can be seen as a continuation of some of the themes covered in the 2010 volume *Soil and Culture* (Landa and Feller, 2010). Filmmaker

Deborah Garcia writes about some of the ways we see soil, the story of how her film *Symphony of the Soil* came about, and its role as a celebration of soil. Alexandra Toland and Gerd Wessolek examine how the visual arts, in their depictions of soils, can assist in the protection of their vulnerable subject. There follow three chapters that describe how soils have been seen in different religious and/or spiritual philosophies. Norman Habel instances the Bible to show how soil has been accorded a sacred status in the Abrahamic religious traditions. Garth Harmsworth and Nick Roskruge describe the beliefs and concepts of soils, the environment, and the land within the rich Māori tradition to which they belong. From China, Renfang Shen and Xinhua Peng integrate soil into the ancient philosophy of Confucianism. Finally in this section, Brent Clothier and Mary Beth Kirkham present a quite recent scientific way of valuing soils as natural capital that provide valuable ecosystem services.

Eleven chapters then provide some stories about soil in a variety of cultures, both ancient and historic. While not aiming to cover the whole world, these glimpses enable reflection on how soil has been seen and treated in different times and places, by diverse groups and by some influential individuals. Bruce James, Winfried Blum, and Carmelo Dazzi begin the section with an analysis of the relationship of the highly influential Roman Empire to their soils, as mediated through the requirements of its people for their staple diet of wheat, grapes, and olives. Concentrating on a 200-year span at the height of the Empire around 2000 years ago, they describe a resilient agricultural system that could inspire the possibility of sustainable practices even now and into the future. Erhan Akça and Selim Kapur describe how several successive civilizations—including the Romans—have occupied and modified the varied but often unstable soils of the pivotal region of Anatolia. They present archaeological evidence for the possibility that this region includes the location of the beginnings of the agricultural revolution. From within the adjacent territory of Greece, Richard Doyle and Mary Savina describe how evidence from soils, combined with that from geomorphology and archaeology, has been used to decipher the landscape history of a small catchment. This history is shown to reflect changes in geology, climate, and human occupation and use of the land that have occurred in the whole region over about the last million years.

Interpreting the history of the landscape in order to tease out the nature of soil management by earlier peoples is one approach to "ethnopedology," or the study of the indigenous knowledge of soils. Another approach to this topic is through old texts, and this is exemplified in Pichu Rengasamy's chapter on the awareness of ancient farmers in India of the relation between soil properties and crop production, and the wisdom they have encapsulated in stories and sayings. In Japan, Masanori Okazaki and Koyo Yonebaysahi's historical analysis of cropping, especially of rice, in both paddy fields and uplands, shows how taxation of crop yields was used until quite recently as a major source of revenue for both the government and feudal owners. The area of land under agriculture has increased greatly over this time, although agricultural production is in recent decline in this highly urbanized and industrialized country. Rog-Young Kim, Su-Jung Kim, E. Jin Kim, and Jae Yang give examples of how the ancient wisdom concerning agricultural land use and soil management in Korea has been handed down in folk sayings. While the legacy of ancient soil management can be negative, through degradation and loss, or neutral, through conservation, it can also be locally positive, as is shown by Antoinette WinklerPrins' descriptions of the dark soils in Amazonia that apparently have resulted from burning piles of organic wastes on the land surface. Their discovery has led to many current studies of "biochar," as burnt organic materials specifically designed as soil conditioners. Again in South America, but on hillslopes, the Incas increased agricultural productivity by devising a number of ingenious methods of manipulating soils that helped control water, temperature, and erosion, as described by Francisco Mamani-Pati, David Clay, and Hugh Smeltekop. These sustainable practices from about eight centuries ago constitute a form of "precision farming," and continue today in the Andean region. In Aoteoroa-New Zealand, the established tribal Māori culture and its attitudes to land, including soils, were challenged by the conquest of the country by Europeans almost two centuries ago, and Garth Harmsworth and Nick Roskruge describe how the newly dominant culture affected Māori attitudes and practices.

We are then taken into modern historical times in two chapters, set respectively in the United States and France. In the first, Edward Landa shows how science can be a very human pursuit, with the clash of strong personalities inside an institutional hierarchy sometimes preventing the free dissemination and discussion of matters under investigation—in this case, explanations of the maintenance of soil fertility. The last chapter in this group sees Christian Feller and Jean-Paul Aeschlimann examining written sources and noting the changing meaning of key words to assess whether a famous Frenchman from the sixteenth century could also be credited with the discovery of the fundamental idea that plant fertility originated in chemicals derived from the soil. This landmark discovery is usually attributed to Justus von Liebig, from Germany, some two centuries later.

A number of different technologies for the modification of soils, and examples of important uses for soils outside of commercial agriculture, are described in the following five chapters. One of the simplest and cheapest materials for improving soils for plant growth originates within each of us—human excreta—according to Rebecca Lines-Kelly, who argues for its return to the soil fertility cycle. For Alvin Smucker and Bruno Basso, the key to improving plant production in arid and semi-arid soils is found in improving their retention of water using polymer membrane barriers. This particular technology can be seen as illustrative of many innovative proposals for raising productivity by modifying soils or enhancing the efficiency of supply of water or nutrients from soils to plants. Richard Stirzaker is another who puts prime importance for plant production on the efficient supply of water via the soil. His chapter recounts how he has achieved increasingly efficient irrigation solutions by experiments he has carried out in his home garden. This chapter reminds us that the home garden is the place where many of us encounter the problems, challenges, and delights of soils, while the remaining two chapters in the group draw attention more generally to the importance of soils for recreation. John Gordon, Patricia Bruneau, and Vanessa Brazier, using examples from Scotland, describe how soils form an integral part of "humanity's natural and cultural heritage, linking the geosphere, biosphere and human culture." Areas are reserved for this purpose and these serve as venues for both passive and active forms of recreation. Active recreation also takes place on sports grounds, and Richard Gibbs outlines the "purposeful manipulation of soils" that is required for this purpose.

Three final chapters are aimed at finding future strategies, or, at least, signposts to the future for more effective sustainable uses of soils. Iain Young and John Crawford, while conceding that soil is "the most complex biomaterial on the planet," see hope in interdisciplinary approaches to understanding this material that particularly focus on the underground architecture of roots within heterogeneous soil, and an emphasis on soil function rather than soil composition. They believe that seeing—and modeling—soils as systems should enable predictions of their "emergent behavior" aimed at helping design soils with desirable biophysical characteristics that come about by manipulating appropriate microbial processes. These authors' emphasis on the biological nature of soils is expanded upon by Karl Ritz, for whom soil resembles a motion picture set with a myriad of players who are extraordinarily different from one another in form, habitat, and function. One of them is featured in Figure I.1. Others include some fungi which are "the largest organisms in both terms of mass and extent on the planet by far," much larger even than elephants. Together, however, the soil biota is responsible for the many services, both regulating and cultural, which soils provide for life above ground, including that of humans. "Life on Earth thus clearly relies on life in earth." The author identifies most farmers as "essentially applied soil ecologists, and increasingly sophisticated biotechnologists." Their challenge—and ours—to provide more effective, but sustainable production systems, requires us to manage soils with an improved understanding of "how the soil biological engine functions." Quoting ancient Roman literature, Garrison Sposito refers to "the genius of soils." His wide-ranging analysis of the prospects for the considerable increases in food production that a global population of 9–10 billion people will demand concludes that neither increasing areas of land under cropping, nor increased use of water for irrigation (with the so-called "blue water") can contribute much at all. Nonetheless, prospects for the more effective accumulation and use of the so-called "green water," or the water retained within soil from rainfall, give cause for optimism, and data from field trials in semi-arid sub-Saharan Africa support this contention. In accord with both Young and Crawford's and Ritz's analyses, Sposito posits that the challenge consequent upon this analysis involves the creation and maintenance of soils as healthy ecosystems. To this end, they need to be provided with an adequate "diet," as inputs of organic matter, to enable their more effective storage and transmission of green water. Thus, "science-based interventions directed at improving the productive flow of green water increase crop productivity by restoring 'the genius of soil'."

"Above all, it is the effort, enthusiasm, and expertise of its authors that have given the book its character and vitality. As its editors, we are grateful to each of them for their efforts and cooperation. Particular thanks go to Jae Yang, current president of the International Union of Soil Sciences (IUSS) for arranging the several chapters from authors in Northeast Asia, an appropriate emphasis considering the location of the 20th World Soils Congress in Korea in 2014. This book has been produced under the auspices of IUSS Commission 4.5 (*History, Philosophy and Sociology of Soil Science*) whose work over the past three decades has been to document the history and heritage of soil science, and to help place both soil science and the soil resource within the context of the broader sphere of human activities. We are the grateful successors of a tradition of editors representing this Commission: Yaalon and Berkowicz (1997), Warkentin (2006), and Landa and Feller (2010). We are also very grateful to Jay Stratton Noller for generously providing a piece of his soil art for the cover of the book, and to Peter Kareiva for his preface. Finally, we are indebted to Randy Brehm and her colleagues at CRC Press for their production of the book and helpful advice to that end.

We hope the aggregate of all of the authors' efforts displayed here will be greater than the sum of its parts—that it will entertain, enlighten, and open doors to new questions, and will help provide new answers to our shared challenges.

REFERENCES

Cribb, J. 2010. *The Coming Famine. The Global Food Crisis and What We Can Do to Avoid It.* Melbourne: CSIRO Publishing.

FAO 2012. *The State of Food Insecurity in the World. Economic Growth is Necessary But Not Sufficient to Accelerate Reduction of Hunger and Malnutrition.* Rome: Food and Agriculture Organization of the United Nations.

Kerrigan, J. and J.D. Rogers. 2013. Biology, ecology and ultrastructure of *Ascobotryozyma* and *Botryozyma*, unique commensal nematode-associated yeasts. *Mycologia* 105: 34–51.

Lal, R. and B.A. Stewart. 2010. *Food Security and Soil Quality.* Boca Raton, FL: CRC Press.

Landa, E.R. 2010. In a supporting role: Soil and the cinema. In *Soil and Culture*, eds. E.R. Landa and C. Feller, 83–105. Dordrecht: Springer.

Landa, E.R. and C. Feller (eds.). 2010. *Soil and Culture.* Dordrecht: Springer.

Lenné, J.M. 2011. Food security and agrobiodiversity management. In *Agrobiodiversity Management for Food Security,* eds. J.M. Lenné and D. Wood, Wallingford: CABI Publishing.

Lenné, J.M. and D. Wood (eds.). 2011. *Agrobiodiversity Management for Food Security.* Wallingford: CABI Publishing.

National Public Radio 2013 http://www.npr.org/2013/03/05/173559220/extinction-looms-for-forest-elephants-due-to-poaching?sc=emaf)

Rosin, C., P. Stock, and H. Campbell et al. 2012. *Food Systems Failure. The Global Food Crisis and the Future of Agriculture.* London: Earthscan.

Stuart, K. 1984. My friend the soil. A conversation with Hans Jenny. *Journal of Soil and Water Conservation* 39: 158–159 and 180–181.

Ugolini, F.C. 2005. Dynamic pedology. In *Encyclopedia of Soils in the Environment*, ed. D. Hillel, 156–165. Amsterdam: Elsevier.

Warkentin, B.P. (ed.). 2006. *Footprints in the Soil—People and Ideas in Soil History.* Amsterdam: Elsevier.

Weis, T. 2007. *The Global Food Economy: The Battle for the Future of Farming.* London: Zed Books, and Halifax: Fernwood Publishing.

Yaalon, D.H. and S. Berkowicz (eds.) 1997. *History of Soil Science—International Perspectives.* Advances in Geoecology, 29. Reiskirchen: Catena Verlag.

Cover Story

Sea Stack I

Jay Stratton Noller

STORY

What's your story? This is my favorite question to students. It is one that I encourage them to ask because their scientific study of a soil profile in the field ultimately has to relate back to the story of this place, the story they need to tell others. The cover art for this book—Sea Stack I—is one of my stories of a place known as Seal Rock along the Oregon coast. The editors of this book were kind enough to pick it out from my gallery of paintings of soilscapes; it tells a great story about this soil at Seal Rock.

If a soil site hasn't an accessible storyline, I am less inclined to consider it for an artwork study. My field experience provides an immediate read on the science questions—what is and is not exposed here? What is the (litho-, pedo-, bio-, chrono-)stratigraphy? What are the meanings of the intersecting lines (surfaces, unconformities, horizon boundaries, etc.)? Are all five state factors (of the soil paradigm) evident here—climate, organisms, relief, parent material, and time? In the end, though, it comes down to a single question: Is this a good learning moment? And by moment, I mean the serendipity of the instant I saw this view of the site. And in the glow of that moment, this view (this artwork) brings to my mind a story to engage the learner, the audience.

IMAGINATION

If this view on the soilscape is a good learning moment, then I want to capture it with sketch, digital image, and or video. In pedology, a subfield of soil science, so many of our process terms consist of a root word being the phenomenon we are describing, such as *podzol*, appended by "i_ation" to mean the process by which the phenomenon formed. In the case of Sea Stack I, *podzolization*—one of the longest studied and enduring process terms in soil science. This process term exists in our usage until we actually figure out how the phenomenon arises in the soil, which may or may not result in a new process term and abandonment of the old. For this pedologist, the process of making the *image* is thought of as *imagination*. For now, this use of "imagination" will suffice for the process(es) involved between the moment of recognizing the story and when the created artwork is complete; at least until I can figure out how it really happens.

DYNAMISM

I look for dynamism in a soilscape. Usually, flat, parallel soil horizon lines don't do this for me. Rather, the features of the soil profile, for example, illuvial accumulations of red clay over yellowish brown peds or the achromatic eluvial (E) horizon erased across a soil that is obviously busy accumulating history. Soil features provide triads and tetrads in the artist's color space that are unique and authentic—one cannot make this stuff up. A couple of years ago, when I was looking back through my portfolio of works I noticed that I tend to paint subjects that naturally have a swirling pattern to them. The swirl is a natural symbol for dynamism. My swirls include tectonic deformation, freeze-thaw, tree throw and other pedoturbation features, as well as the atmosphere and landscape. The subject is dynamic in another sense. Seacliff erosion along this coast is measured in meters per decade. This means that any look into the seacliff—its freshly exposed soil and substratum—is a fleeting glimpse of Earth as it "calves" into the sea, much as glaciers do farther up the Pacific shore.

ART SCIENCE

My August, 2010, field work on soils exposed in seacliffs between Wakonda and Yachats, Oregon, has yielded the most subject matter I have painted to date. I studied 106 stations, each typically involving a soilscape (distant perspective) or soil profile (close up). Of these stations, I shot 364 photographic images worth keeping, which led to at least 19 artwork studies, then 16 canvasses being prepared. Twelve paintings have been completed to date and all have been juried (peer reviewed by professional artists) into art shows one to three times each. One of the first prepared and completed is Sea Stack I, the main subject of this writing. Each painting captures a unique field station, leading to 11.3% representation of one field season's science work represented in artworks. Three paintings remain in various stages of preparation. One painting was started and became unwieldy as I painted; so, it was abandoned, painted over with several coats of fresh gesso, and its canvas used for another.

As artist, I used two media in the production of this artwork. First, the seacliff was photographed under high-overcast sunlight using a Nikkor AF-S DX 18–105 mm, f/3.5–5.6 lens on a Nikon D300s set at 27 mm, f/3.5. From these images I selected Number 365, which I felt provided the closest perspective and aesthetic to my mind's vision of the soilscape at this location. Second, this digital image was used as the reference for my painting, the application of commercial acrylic paints to canvas I previously glued to a hardwood panel. The panelled canvas provides tooth for the brushwork and rigidity for blending colors on the spot. The cloudy sky beyond the soil profile has more to do with my recollection of the day than the moment it was shot with my camera. The resulting painting captures my perceptions of this isolate, soil ecosystem. It is a small seastack at the backedge of the beach, exposed nearly all way round and it will soon disappear—likely catastrophically eroded into a stormy Pacific.

In the course of painting, I almost always encounter something new, something unanticipated, which leads to a characteristic self-inflicted critique "why didn't I see that before?" That a second, deeper look should surprise me is not exceptional. There are always details, even the big picture sometimes, that can be missed during the field work. The scientific research continues today and has yet to mature to the manuscript stage. Oh, how painting has shifted my perception of the field site!

Editors, Foreword Writer, and Illustrator

G. Jock Churchman obtained degrees in chemistry from Otago University in his native New Zealand. He studied the physical chemistry of a clay mineral, halloysite, for his PhD, under a fellowship from the New Zealand pottery and ceramics industry, and carried out research for this industry for a short time before beginning a two-year postdoctoral fellowship in soil science at the University of Wisconsin. He was employed in the New Zealand Soil Bureau, DSIR for 16 years and in CSIRO Division of Soils (later Land and Water) for 14 years and has held visiting fellowships at Reading University (1 year) and the University of Western Australia (6 months). Currently he is adjunct senior lecturer in soils at the University of Adelaide and part-time associate professor in the Centre for Environment Risk Assessment and Remediation at the University of South Australia. His research has been centered on clays. In 2005, he completed a BA (Hons) in philosophy from Flinders University of South Australia with a thesis on the philosophical status of soil science and was awarded the Johan Bouma Prize for best multidisciplinary paper integrating "hard" and "soft" science at the Australia and New Zealand soils conference in 2008. He was elected the 2009 Fellow of the Association Internationale pour l'Etude des Argiles. He was chair of the Commission for Soil Mineralogy of the International Union of Soil Sciences (IUSS) (2006–2010) and is currently (2010–2014) chair of the IUSS Commission on the History, Philosophy and Sociology of Soil Science. His other interests range from walking, especially in the bush and mountains, to the English language. This has led him into editing, including of the journal *Applied Clay Science* for 5 years, writing scripts for satirical workplace plays, volunteering for practice conversations with foreign students, and reading.

Edward R. Landa holds an MS and PhD in soil science from the University of Minnesota and is an adjunct professor in the Department of Environmental Science and Technology at the University of Maryland. His work at the U.S. Geological Survey from 1978 to 2013 focused on the fate and transport of radionuclides and metals in soils and aquatic systems and has spanned a breadth of contaminants introduced into the surficial environment by human activities; these have included tire wear particles and radioisotopes used in medicine. He participated in the International Atomic Energy Agency's International Chernobyl Project and in studies on radioactive contamination in the Arctic regions. Throughout his career, Ed has had an active interest in the history of science and technology and has published on the radium extraction industry, description of color in science and art, and depictions of soils in films. Besides these continuing activities, he has enjoyed brief ventures into new terrains, including writing a country-western song about the plight of a truck driver in New York City and a screenplay *Artists*, dealing with the radium dial painters, that won the Governor's Screenwriting Competition at the 1991 Virginia Festival of Film. He coedited *Soil and Culture* (Springer, 2010) and has served as the chair (2006–2010) and vice chair (2010–2014) of the Commission on the History, Philosophy, and Sociology of Soil Science of the International Union of Soil Sciences.

Peter Kareiva became passionate about scientific research and ecology as a result of a term at Duke Marine Lab as an undergraduate. Since then he has moved from marine ecology, to insect ecology, to mathematical biology, and then on to conservation. He has worked as a private consultant, as a university professor, as a director of conservation biology at NOAA's Fishery Center in Seattle, and now as chief scientist for The Nature Conservancy. Peter thinks that the scientific method is one of humankind's greatest inventions, because it relentlessly leads to progress by challenging existing ideas in favor of better explanations of the world and better understanding of what works and does not work as solutions to society's problems. He enjoys mentoring young scientists because he feels he owes any success he has had to a series of great mentors beginning with Janis Antonovics and John Sutherland at Duke, to Simon Levin and Richard Root at Cornell, and then on to Bob Paine at University of Washington and Gretchen Daily at Stanford University. With Michelle Marvier, he is now revising their conservation science textbook, which takes the viewpoint that conservation depends on factoring human wellbeing into its strategies for biodiversity protection.

Jay Stratton Noller lives and works in the worlds of science and art. He is a professor of landscape pedology at Oregon State University, where he teaches courses in soil morphology and classification, digital mapping of soilscapes, soil geomorphology, and machine learning of shape and form. He is also an accomplished painter, http://soilscapestudio.com/index.html:

> I have a passion to dig beneath the surface, revealing flow, energy and color from that which is all dark. I create large artworks that weave scientific research and mixed-media 2D art to guide, to inspire and to foster dialog between nature and culture. Interests are in the relationship of humans to soils, and seek to explore, through art-science means, the connections of shape, form, process and function—what I call morphologistics.
>
> What I do would be categorized as Neo-Land Art, whereby the Earth Artist authentically represents ecosystems. Subject matter and art media converge on soils as nexus for evocative imagery and scientific exploration. Using scientifically based techniques of pigment creation for the paintings, the site is permanently and resiliently embodied in the artwork. Site history is conveyed in artwork design. Artwork accompanies scientific research at lab bench and field study site, with synthetic publications for outreach and discourse. Current projects involve soilscapes on several continents.

Jay has had numerous shows and commissioned works. He has been a frequent contributor at the Burningman Art Festival in Black Rock City, Nevada.

Contributor Biographies

Jean-Paul Aeschlimann has been in charge, as a principal research scientist with the Australian CSIRO (Canberra), of a whole series of research programs on all the continents aimed at biologically controlling various weeds, plant diseases, and noxious animals. He was educated at the Swiss Federal Institute of Technology (Zürich) and concluded in 1969 with a PhD in environmental protection. Author of several books in the fields of agroecology and the history of agriculture, he has also assumed the role of coeditor of important scientific journals over the last 15 years, *Agriculture, Ecosystems and Environment* in particular. Since 2003, he has been acting as a resource person for Agropolis Museum (Montpellier), and for the newly created Musée des Civilisations (MuCEM, Marseilles), two leading institutions in France in terms of disseminating scientific knowledge, a discipline in which he has developed a keen interest.

Erhan Akça, born in 1969 in Adana, Turkey, is an associate professor at Adiyaman University Technical Programs, Adiyaman, Turkey. His profession is soil mineralogy, archaeometry, and natural resource management. He has written and contributed to more than 100 scientific publications (papers, chapters, books) on soil science, archaeometry, land use, and desertification. He has participated in projects supported by NATO, EU, JSPS, RIHN, TUBITAK, Mitsui Environmental Fund, FAO, and GEF. He has also contributed as an editor and reviewer to journals on archaeology, archaeometry, mineralogy and micromorphology.

Bruno Basso is an associate professor of agricultural modeling at Michigan State University. He began his professional scientific career at the University of Basilicata, Italy. He is the codeveloper of the SALUS model (System Approach for Land Use Sustainability). Among other applications, this has led to important insights for the likely effects of climate change on carbon and water footprints of future cropping systems. During his career, Dr. Basso has participated as PI and co-PI in several international projects. He is the author of more than 150 publications and several invited keynote lectures. He is the recipient of the 2010 "Pierre Robert" Precision Agriculture Award; 2008 "L. Frederick Lloyd Soil Teaching Award" and the 2007 Soil Science Society of America "L.R. Ahuja Agricultural System Modeling Award."

Winfried E.H. Blum is emeritus Professor at BOKU University, Vienna, Austria. He has research and teaching experience in countries in all continents and publications in 12 languages. In his early academic career, he played concert violin in different orchestras and conducted a chamber orchestra himself. He still maintains personal contacts with artists in many countries, collects paintings on soil, nature, and human relationships, and has published about soil and world cultures.

Vanessa Brazier (Kirkbride) is a geomorphologist working for Scottish Natural Heritage, seeking practical ways to conserve dynamic environments and promote the interpretation of the curious and engaging stories revealed in the rocks and landforms of Scotland. She has been involved in research in New Zealand, Scotland, and Norway on the readjustment of the landscape to climate changes, including glacier and ice sheet responses, mountain slopes, and landslides. The Cairngorm Mountains and the cold azure sea off the Assynt coast are among her favorite places.

Patricia M.C. Bruneau is a soil science advisor at Scottish Natural Heritage. She has worked on a range of environmental and land-use issues promoting better understanding of the multiple functions, conservation, and sustainable use of soils both in Scotland and at the UK level. She is involved in national and international soil policy development and recently coedited the *State of Scotland's Soil* report (2011).

Zueng-Sang Chen is the distinguished professor of pedology and soil environmental quality (2007 to now), Department of Agricultural Chemistry (DAC) of National Taiwan University (NTU). He was the associate dean of College of Bioresources and Agriculture of NTU in 2007–2011 and department head of DAC/NTU in 2004–2007. He was awarded the Distinguished Agricultural Expert Award of Council of Agriculture of Taiwan in 2012, the ESAFS (East and Southeastern Federation of Soil Science Societies) Distinguished Award in 2009, NTU Distinguished Social Service Award in 2009, and the Kiwanis International Distinguished Agricultural Expert Award in 2007. He has primarily studied soil genesis, soil environmental quality, the behavior and bioavailability of heavy metals in the soil–crop system, and phytoremediation of metals-contaminated sites. He co-organized and hosted the *2nd ICOBTE* in 1993, and the *6th International Conference on the ESAFS* in 2003 as well as the *14th International Conference on Heavy Metals in the Environment (ICHMET)* in 2008. He was further awarded the Distinguished Teaching Professor Award of NTU in 2004 and the Distinguished Society Award of Chinese Society of Soil and Fertilizer Sciences (CSSFS) in 2008.

David E. Clay is the director of the South Dakota Drought Tolerance Center, professor of plant sciences, and Fellow of the American Society of Agronomy. Currently he is serving as technical editor for the *Agronomy Journal* and associate editor for the *Journal of Plant Nutrition*, *International Journal of Agriculture*, and *Precision Agriculture*. He was the editor for a number of books that include: *South Dakota Best Management Manual* (SDSU), *Site-Specific Farming Guidelines Manual* (PPI), *GIS Applications in Agriculture* (CRC Press), *Soil Science: A Step-by-Step Field Analysis* (SSAJ), and *GIS Applications in Agronomy: Nutrient Management for Improved Energy Efficiency*. He is also an author of the new book *Mathematics and Calculations for Agronomists and Soil Scientists* that is being published by the International Plant Nutrition Institute. He has published over 170 refereed papers.

Brent Clothier is a soil scientist. Actually, he started out as a mathematician, but became a soil scientist. He also has interests in environmental systems, maintaining our natural capital stocks and improving our ecological infrastructures. In his spare time he can either be found watching rugby, out on his in-line skates, or cooking in his wood-fired pizza oven.

John W. Crawford holds the Judith and David Coffey chair in Sustainable Agriculture at the University of Sydney. With a background in physics, his interest is in the origins of organization in cells and communities and has led him to uncover properties in complex living systems that sustain their function in the face of disruption. His current projects range from modeling of soil, to working with a whole range of community and business representatives on projects that focus on the social and cultural systems of food production, food distribution, and food consumption. His collaborators are as diverse as Tongan rugby players, immigrant survivors of torture, major food production companies and indigenous communities, as well as academics and researchers from a diverse range of fields. In the United Kingdom, he chaired the main funding committee responsible for supporting research in sustainable agriculture, diet, and health.

Carmelo Dazzi is a full professor of pedology at the University of Palermo, Italy and honorary professor of the Universidad Nacional de San Agustin de Arequipa, Peru. In recent years, his field of interest has considered mainly soil degradation and desertification, anthropogenic soils, pedodiversity, and pedoethnology. He is vice president of the Italian Society of Soil Science (SISS) and president of the European Society for Soil Conservation (ESSC).

Richard B. Doyle is a deputy head of the School of Agricultural Science at UTAS (University of Tasmania) in Tasmania, Australia. He has over 25 years teaching and research experience in universities and government agencies in Australia, New Zealand, and Namibia. Richard currently teaches three soil science units and over-sees the undergraduate teaching program in agricultural science at UTAS.

He has been awarded multiple teaching and speaking awards including a Rotary Teaching Fellowship to the University of Namibia. Richard is a certified professional soil scientist—Level 3 and is currently serving as president of Soil Science Australia. He has undertaken research in soil carbon, pesticide fate in soils, soil salinity, mapping of erosion, soil mapping, rural tree decline, and wastewater reuse.

Christian Feller is an emeritus soil scientist and the former director of research at the "Institut de Recherche pour le Développement" (IRD) in Montpellier, France. He earned his MS (1969) and PhD in organic chemistry (1972) from the Sorbonne University (Faculty of Sciences) in Paris, and his Doctorate of Science (1994) in soil science from the Louis Pasteur University in Strasbourg. His research focuses on soil organic matter studies applied to soil fertility and environmental services—in particular, the impact of agroecological practices on soil–plant carbon sequestration in tropical and subtropical areas; he has worked extensively in Senegal, French West Indies (Martinique), Brazil, and most recently, Madagascar. Christian is a member of the French Academy of Agriculture, and was the first recipient of the Soil Science Society of America's Nyle C. Brady Frontiers of Soil Science Lectureship in 2006. He served as vice chair (2006–2010) of the Commission on the History, Philosophy, and Sociology of Soil Science of the International Union of Soil Sciences (IUSS). He is also a collector of antiquarian soil science books, and these books help him to publish extensively on the history of soil science and agronomy. He now serves as chair of IUSS Division 4 that deals with the role of soils in sustaining society and the environment.

Deborah Koons Garcia is the director, writer, and producer of the film *Symphony of the Soil*, http://www.symphonyofthesoil.com/ that had its world premiere at the Washington DC Environmental Film Festival in March 2012. Deborah has called Northern California home for over 30 years. Her film *The Future of Food* (2004) premiered theatrically at Film Forum in New York, examines the alarming issues surrounding the rapidly increasing corporate domination of our food supply. It is the first major film to cover the history and technology of genetic engineering and the complex implications of releasing such crops into the food environment and food supply. It has been shown all over the world in theaters, food, farming, and film festivals and by citizens seeking to inform and inspire fellow citizens to take action. *Symphony of the Soil* has been screened worldwide at events as varied as the Soil Science Society of America annual meeting and the National Heirloom Seed Festival. Its New York premiere was at the IFC Center in July 2013.

Richard Gibbs career began in 1986 after completing his PhD in soil science at the University of Canterbury in New Zealand, having won a Commonwealth Scholarship from the United Kingdom in 1982. He then returned to the United Kingdom to work as a turfgrass research scientist and lecturer before emigrating to New Zealand in 1991 to work as a full-time sports surface consultant. In 2012, Richard accepted an offer to come full circle and return to the United Kingdom where he now runs the sports surface design services of the Sports Turf Research Institute in Bingley, West Yorkshire.

John E. Gordon has worked in geoheritage conservation for over 30 years. His research interests also include glacial and periglacial geomorphology, glacier fluctuations and climate change, and the Quaternary history of Scotland. He has a longstanding interest in mountains and glaciers, which he has visited and studied in many parts of the world. Currently he is an honorary professor in the School of Geography & Geosciences at the University of St. Andrews in Scotland and is exploring how the cultural connections between geodiversity, people and landscape can enable the rediscovery of a sense of wonder about the natural world.

Norman Habel is a professorial fellow at Flinders University. He identifies himself as an Earth being with special interest in eco-justice and ecological hermeneutics. His Earth Bible project involves reading the biblical tradition from the perspective of Earth and Earth beings.

Peter K. Haff is a professor of geology and civil engineering at Duke University. He is interested in the dynamics of natural landforms and studies the emergence of humans and technology as the next phase in Earth's geologic evolution. He holds the belief, about which he is philosophic, that humans have little control over the technological world they have created, which instead unfolds according to its own rules.

Sharon J. Hall is an ecosystem ecologist in the School of Life Sciences at Arizona State University who focuses her energy on the complex relationship between humans and the environment. For her PhD, she was trained in soil science under the legacy of Hans Jenny at the University of California, Berkeley. Sharon was raised in the inner city of Oakland, California, but came to love the outdoors through diving in the coastal kelp forests and hiking in the Rocky Mountains of Colorado. She is mom to a cat and a dog of above-average intelligence.

Garth Harmsworth is a senior environmental scientist with Landcare Research NZ Ltd, Palmerston North, New Zealand, and affiliates to the central North Island Māori tribes: Te Arawa, Ngāti Tūwharetoa, and Ngāti Raukawa. He has a broad background in earth sciences, land resource assessment, environmental databases and geographic information systems (GIS), ecological and cultural monitoring and restoration, and applies Māori concepts and knowledge in all aspects of his work. He works from national government to community level in many interdisciplinary inter-agency science projects bridging Western science, indigenous knowledge, participatory research, planning, and policy.

Zeng-Yei Hseu is a professor of soil and environmental sciences in the Department of Environmental Science and Engineering, National Pingtung University of Science and Technology, Taiwan. He received his PhD in pedology from National Taiwan University, Taiwan in 1997. Professor Hseu has been working in the field of soil science since 1990s. Being an assistant professor, he became an independent researcher in 2000. His major topics of interest are heavy metal dynamics and mineralogy of serpentine soil, morphology and genesis of wetland soil, soil chronosequences on river and marine terraces, and soil heavy metal contamination and remediation. He is a member of the Soil Science Society of America and International Union of Soil Science, and The Clay Minerals Society. Professor Hseu is the author or coauthor of numerous internationally scientific papers and book chapters. He had been employed as a guest professor in Kyoto University in 2010 and in Meiji University in 2011, Japan. He was also elected chair of the Department of Environmental Science and Engineering (August 2013 to July 2016), National Pingtung University of Science and Technology.

Bruce R. James is a professor of soil chemistry at the University of Maryland, College Park, Maryland. In addition to endeavors in this field, he pursues transdisciplinary teaching and research on the relationships between soils and ancient civilizations. He takes great pleasure in travelling across soil and water landscapes while hiking and climbing, bicycling, kayaking, and skiing cross country.

Selim Kapur was born in 1946 in Ankara, Turkey. Chairman of the Department of Archaeometry at the University of Çukurova, Adana, Turkey, he is the scientific committee member of the European Soil Bureau Network of the EU-JRC in Ispra, Milan, Italy. He was the earlier secretary of the International Working Group of Land Degradation and Desertification. He has acted as the Science and Technology correspondent of Turkey for the UNCCD (United Nations Convention to Combat Desertification), led projects supported by NATO (The North Atlantic Treaty Organization), EU (European Union), JSPS (Japan Society for the Promotion of Science), TUBITAK (The Scientific and Technological Research Council of Turkey), Mitsui Environmental Fund, FAO (The Food and Agriculture Organization of the United Nations) and GEF (Global Environment Fund) concerning soil protection and environmental issues. He has authored numerous papers in cited and authored/edited books on land degradation and desertification, archaeometry, and micromorphology.

E. Jin Kim is a senior student of Cheongshim International Academy, Gapyeong, South Korea. Her interests are in environmental protection and organic farming, and she plans to major in agricultural economics and management. Her research involves the Korean Demilitarized Zone (DMZ) with the help of the DMZ Ecology Research Institution, and this increases people's awareness of this region as the 10th Eco-generation Regional Ambassador.

Rog-Young Kim is a soil scientist working for the Kangwon National University in Chuncheon, South Korea. She received her BA from the Konkuk University in Seoul, her Diploma (German equivalent to BA and MA combined) and PhD were from the University of Bonn, Germany. Her main research interests are soil micronutrients and heavy metals. She is also interested in soil genesis and classification.

Su-Jung Kim is a professor of the Department of Biological and Environmental Science at the Dongguk University in Seoul, South Korea. Her research interests are soil microbiology, microbial interaction, and bioremediation of metal- and hydrocarbon-contaminated soils.

Mary Beth Kirkham is a professor in the Department of Agronomy at Kansas State University. Dr. Kirkham has written two textbooks dealing with soil–plant–water relations. In addition to research and teaching, Professor Kirkham travels to national and international professional meetings and is a member of the Executive Committee and Council of the International Union of Soil Sciences.

Nikolaus J. Kuhn received his first degree in physical geography in his native country Germany from the University of Trier. He moved to the University of Toronto to complete a PhD in geography. This was followed by postdoctoral and lecturing appointments at the Hebrew University Jerusalem, Clark University in the United States, and the University of Exeter as lecturer in the United Kingdom. In 2007, he was appointed as professor for geography at the University of Basel in Switzerland. His interest focuses on untangling questions of Earth System Science ranging from soil erosion and climate to the fate of water on Mars.

Edward R. Landa is a soil scientist with the Department of Environmental Science and Technology at the University of Maryland. He was a Fulbright senior specialist in Kyrgyzstan in 2007, and spent more than 35 years with the National Research Program/Water Mission Area of the U.S. Geological Survey. Among the spinoffs from Ed's historical work on the early days of soil science research at the U.S. Department of Agriculture has been a look at botanical studies by Lyman Briggs (1874–1963), F.H. King's fellow soil physicist at the Bureau of Soils. Briggs' work from the 1950s was on the pollen-flinging mechanism of Mountain Laurel flowers. This recent reconstruction and analysis of Briggs' partially completed, unpublished research is being done in collaboration with Mary Beth Kirkham (author, *this volume*) and colleagues from the United States and Argentina.

Rebecca Lines-Kelly is an agricultural environment specialist with New South Wales Department of Primary Industries. She is fascinated by the human connection to soils, and always looking for ways to make soil science so enthralling it will change people's lives. She received the inaugural Johan Bouma Prize for best multi-disciplinary paper integrating "hard" and "soft" science presented at *SuperSoil 2004* and the inaugural Australian Society of Soil Science Incorporated LJH Teakle award for soils communication in 2010.

Francisco Mamani-Pati is a soil scientist and professor of soil sciences at Bolivian Catholic University "San Pablo" UAC-Carmen Pampa Campus and Public University of El Alto. He has worked as a visiting research scientist at South Dakota State University. His research has generated a book, chapters in books, scientific articles, and abstracts.

Masanori Okazaki is an environmental soil scientist at Tokyo University of Agriculture and Technology till 2013 and Ishikawa Prefectural University since 2013. He is interested in the ion behavior in soil and rehabilitation of soil polluted with heavy metals, and has a particular interest in rammed earth fences or walls, such as in the samurai residence areas of Nagamachi, Kanazawa City in Japan.

Xinhua Peng is a soil scientist in Institute of Soil Science, Chinese Academy of Sciences (CAS), Nanjing, P. R. China. He obtained bachelor's in geography at Hunan Normal University in 1995, master's in soil science at Hunan Agricultural University in 2000, and PhD at Institute of Soil Science, CAS, in 2003. His main research interests are (i) soil structure: formation, stabilization, and its functions in ecosytems; (ii) the fate and transport of water and nutrients in agricultural watersheds.

Pichu Rengasamy is a senior research fellow in the University of Adelaide. He is a soil scientist with research interests in salt-affected soils, particularly dealing with abiotic stress in saline and sodic soils, soil structural stability as influenced by different cations, and carbon sequestration in high pH soils.

Karl Ritz is a soil ecologist, convinced that soil is the most remarkable, complex and fascinating material on the planet, as well as absolutely fundamental to our past and future civilizations. He holds a chair in soil biology at the National Soil Resources Institute at Cranfield University. Previously he led research programs, groups, and departments at the Macaulay Institute for Soil Research and the Scottish Crop Research Institute in Scotland. He is a chief editor of the journal *Soil Biology & Biochemistry*, and has edited several books, including *Criminal and Environmental Soil Forensics*, and more recently *Architecture and Biology of Soils*. A keen photographer, cyclist, and dog walker (of necessity), in common with several other contributors to this volume, he also particularly enjoys cartoons and comics, which perhaps is a portent for his next editorial idea....

Nick Roskruge is a senior lecturer in the Institute of Agriculture and Environment, Massey University, Palmerston North, New Zealand. His tribal affiliations are Te Atiawa and Ngāti Tama-Ariki. He is a leading authority and expert on Māori traditional crops and works with many Māori communities and organizations, such as Tahuri Whenua the national Māori Vegetable Growers collective, to preserve and expand this knowledge. He is recognized nationally and internationally for his expertise and has published several books and papers on horticulture, traditional crops, and soils.

Mary E. Savina is a professor of geology at Carleton College, in Northfield, Minnesota, where she earned undergraduate degrees in history and geology. Geoarchaeology and landscape history combine these interests nicely. Her doctorate is from the University of California, Berkeley. In addition to her teaching and geology research, she works on curriculum and program assessment, and facilitates workshops about environmental sustainability in college curricula. She loves to travel and usually manages to find opera performances wherever she goes, in addition to fascinating landscapes and people. Richard Doyle, first author, was Mary's graduate student when she taught for three years at Victoria University in Wellington, New Zealand. She studies voice and sings in a choir in her spare time. Music makes the world go round.

Hugh Smeltekop is the director general of the Unidad Académica Campesina-Carmen Pampa, a small college in the cloud forest region of northwest Bolivia, founded to educate the children of impoverished rural farmers. He believes that equality in education will lead to sustainable food systems, better health for people and the environment, and ultimately greater dignity for all people.

Alvin J. M. Smucker is a professor of soil biophysics at Michigan State University. He has taken sabbaticals at the University of Wisconsin-Madison, CSIRO in Perth, Australia, Argonne National Laboratory, Illinois, and CAU in Kiel, Germany. His research, invited seminars, and lecture activities have taken him to 33 countries, where he has identified soil water as one of the most limiting natural resources to agricultural production. His research has brought many honors from governments and universities in Argentina, Australia, Austria, Finland, Germany, Scotland, and Taiwan. He is the recipient of four fellows awards from major scientific societies in the United States. During the past several years, Dr. Smucker has been developing new technologies and associated plant prescription management practices to maximize plant production for farmers with a focus on smallholder farmers experiencing subsistence levels of nutritious food production.

Garrison Sposito holds the Betty & Isaac Barshad Chair in Soil Science at the University of California at Berkeley. Professor Sposito, whose academic degrees are in agriculture, is the author of more than 600 publications and has been elected a fellow of six international scientific societies. He is the recipient of several awards for research in soil science and hydrology, including the Horton Medal of the American Geophysical Union for "outstanding contributions to the geophysical aspects of hydrology." In 2008, he was designated a Legend in Environmental Chemistry by the American Chemical Society.

Richard Stirzaker is a soil and water scientist whose research interests were forged in his childhood vegetable garden. His research includes irrigation, salinity, water use of farming systems, and developing monitoring tools for farmers to aid in learning and adaptive management. He currently holds the position of principal research scientist at the CSIRO Division of Land and Water based in Canberra, Australia and extra-ordinary professor at the University of Pretoria, South Africa.

Kevin R. Tate is currently an honorary research associate at Landcare Research, having retired in 2005 following a number of years leading several research programs on greenhouse gas exchange with the terrestrial biosphere. Like his coauthor (Benny Theng), he began his career in soil science at the New Zealand Soil Bureau, DSIR. During this early period, he took his family to the United Kingdom, first for a year in Aberdeen to work at the Macaulay Institute (now The James Hutton Institute), and later to Rothamsted in England, where he was privileged to work with the late Professor David Jenkinson. In 1995, he was elected a Fellow of the Royal Society of New Zealand, and in 2005, he won the New Zealand Marsden Medal for his contribution to climate change research. After retirement, he is applying his research to the development of mitigation technologies for greenhouse gas emissions from agriculture, and when asked gives talks to community groups on global change. For relaxation, he enjoys cycling, walking, and listening to music, especially jazz. He and his wife Heather also enjoy gardening, entertaining visitors, and spending time with their children and grandchildren.

Benny K. G. Theng is an honorary research associate with Landcare Research, New Zealand. For many years before that, he was a scientist at the New Zealand Soil Bureau, DSIR. Born in Indonesia, he won a Colombo Plan scholarship to study at the University of Adelaide, Australia, where he received his PhD degree. He has worked as a research fellow and visiting scientist in Australia, Belgium, Chile, China, France, Germany, and Japan. His research has focused on the behavior of organic molecules and polymers at clay mineral surfaces. His book *Formation and Properties of Clay–Polymer Complexes, 2nd Edition* has recently been published by Elsevier. He enjoys reading historical books, playing bridge, and listening to chamber music.

Alexandra R. Toland grew up in Boston, Massachusetts, and received a bachelor of arts in 1997 from the University of Wisconsin–Madison and a master of fine arts in 2001 from the Dutch Art Institute. In 2009, she received a Diplom-Ingenieur in environmental planning from the Technische

Universität Berlin and began a doctoral program as a DFG research fellow in the Graduate Research Group "Perspectives in Urban Ecology" at the TU Berlin. She was a recipient of an Andrea von Braun scholarship in 2013, and has lectured at the Technische Universität Berlin, Leuphana University, and the University of Arts, Berlin. She has cochaired the German Soil Science Society's (DBG) Commission on Soils in Education and Society since 2011 with Gerd Wessolek and has been an active member of the Einstein Foundation, the International Environmental Communication Association, the Eco Arts Collective, and the Forest Management Association of Chorin/Senftenberg. Her research interests include environmental art, ecocinema, soil protection, and soil awareness. She regularly exhibits and collaborates with artists in Europe and the United States and is an enthusiastic red-worm composter and wild-food forager.

Gerd Wessolek studied soil science, philosophy, and art in Göttingen and Kassel, Germany. Since 1997, he has been working as a full professor for soil protection at the University of Technology Berlin. His "traditional research fields" are soil physics and the vadose zone of urban soils. For more than 10 years, he has been trying to introduce soil and art into the soil science community by paintings, movies, conceptual projects, and exhibitions. In the 1990s, Gerd guided the commission "Soil Technology," and since 2011, he has been the chair of the commission "Soil in Society and Education" of the German Soil Science Society.

Antoinette M.G.A. WinklerPrins is a geographer who is the academic program director for environmental studies at Johns Hopkins University. Originally from The Netherlands, she considers all landscapes to be infused with human activities, albeit with varying degrees of anthropogenesis. Once upon a time, she and her husband bicycled across the United States, living in and observing landscapes. This confirmed her goal to be a geographer, and as such she has focused on the intersection of people and the environment and has conducted research on smallholder livelihoods, urban agriculture, and local environmental knowledge in Brazil, Kenya, and Mexico.

Jae E. Yang is a professor of soil environmental chemistry in the Department of Biological Environment, Kangwon National University, South Korea. His research interests are on soil pollution assessment and remediation using organic and inorganic ameliorants. Currently, he serves as the president of the International Union of Soil Sciences (IUSS).

Koyo Yonebayashi is a soil scientist whose main research interests focus on the nature of soil organic matter are soil humic substances. His research has employed a variety of advanced analytical techniques to elucidate the functions and chemical structure of humic acids. He is an art lover and a copper engraving artist.

Iain M. Young is a professor and the head of School of Environmental and Rural Science at the University of New England. Graduating from Aberdeen University with a degree and a PhD in soil science, Iain has held several senior positions in the United Kingdom, including coordinating terrestrial carbon research for Scottish Universities, director of Sensation Science Centre, and a committee member of the main Agriculture Research Funding Councils. He is currently the president of the Australian Council of Deans of Agriculture, and serves on Sir Michael Jefferies National Soil Advisory Committee. He has a keen interest in how biology operates in soils.

Contributors

Jean-Paul Aeschlimann
Agropolis Museum
Montpellier, France

Erhan Akça
Adıyaman University
Adıyaman, Turkey

Bruno Basso
Department of Geological Sciences
Michigan State University
East Lansing, Michigan

Winfried E.H. Blum
Institute of Soil Research
Department of Forest and Soil Sciences
University of Natural Resources and Life Sciences (BOKU)
Vienna, Austria

Vanessa Brazier
Scottish Natural Heritage
Perth, Scotland, United Kingdom

Patricia M.C. Bruneau
Scottish Natural Heritage
Edinburgh, Scotland, United Kingdom

Zueng-Sang Chen
Department of Agricultural Chemistry
National Taiwan University
Taipei, Taiwan

David E. Clay
South Dakota State University
Brookings, South Dakota

Brent Clothier
Systems Modelling
The New Zealand Institute for Plant & Food Research Limited
Palmerston North, New Zealand

John W. Crawford
Charles Perkins Centre & Faculty of Agriculture and the Environment
The University of Sydney
New South Wales, Australia

Carmelo Dazzi
Department of Agricultural and Forest Sciences
University of Palermo
Palermo, Italy

Richard B. Doyle
Tasmanian Institute of Agriculture
School of Land and Food
University of Tasmania
Hobart, Tasmania, Australia

Christian Feller
UMR Eco&Sols IRD
SupAgro Cirad-Inra-IRD
Montpellier, France

Deborah Koons Garcia
Lily Films
Mill Valley, California

Richard Gibbs
The Sports Turf Research Institute
Bingley, West Yorkshire, United Kingdom

John E. Gordon
School of Geography and Geosciences
University of St. Andrews
St. Andrews, Scotland, United Kingdom

Norman Habel
Flinders University
Adelaide, Australia

Peter K. Haff
Nicholas School of the Environment
Duke University
Durham, North Carolina

Sharon J. Hall
School of Life Sciences
Arizona State University
Tempe, Arizona

Garth Harmsworth
Landcare Research NZ Ltd – Manaaki Whenua
Palmerston North, New Zealand

Zeng-Yei Hseu
Department of Environmental Science and Engineering
National Pingtung University of Science and Technology
Pingtung, Taiwan

Bruce R. James
Department of Environmental Science and Technology
University of Maryland
College Park, Maryland

Selim Kapur
Departments of Soil Science and Archaeometry
University of Çukurova
Adana, Turkey

E. Jin Kim
Cheongshim International Academy
Gapyeong, Republic of Korea

Rog-Young Kim
Department of Biological Environment
Kangwon National University
Chuncheon, Republic of Korea

Su-Jung Kim
Department of Biological and Environmental Science
Dongguk University
Seoul, Republic of Korea

Mary Beth Kirkham
Department of Agronomy
Kansas State University
Manhattan, Kansas

Nikolaus J. Kuhn
Physical Geography and Environmental Change Research Group
Department of Environmental Sciences
University of Basel
Basel, Switzerland

Edward R. Landa
Department of Environmental Science and Technology
University of Maryland
College Park, Maryland

Rebecca Lines-Kelly
Wollongbar Primary Industries Institute
NSW Department of Primary Industries
Wollongbar, New South Wales, Australia

Francisco Mamani-Pati
Catholic University of Bolivia "San Pablo"
La Paz, Bolivia

Masanori Okazaki
Ishikawa Prefectural University
Ishikawa, Japan

Xinhua Peng
State Key Laboratory of Soil and Sustainable Agriculture
Institute of Soil Science
Nanjing, People's Republic of China

Pichu Rengasamy
School of Agriculture, Food and Wine
The University of Adelaide
Adelaide, South Australia

Karl Ritz
National Soil Resources Institute
School of Applied Sciences
Cranfield University
Cranfield, United Kingdom

Nick Roskruge
Institute of Natural Resources
Massey University
Palmerston North, New Zealand

Mary E. Savina
Department of Geology
Carleton College
Northfield, Minnesota

Hugh Smeltekop
Catholic University of Bolivia "San Pablo"
La Paz, Bolivia

Alvin J.M. Smucker
Department of Plant, Soil and Microbial Sciences
Michigan State University
East Lansing, Michigan

Garrison Sposito
Department of Environmental Science, Policy and Management
University of California at Berkeley
Berkeley, California

Richard Stirzaker
CSIRO Land and Water
Canberra, Australia

Kevin R. Tate
Landcare Research NZ Ltd – Manaaki Whenua
Palmerston North, New Zealand

Benny K.G. Theng
Landcare Research NZ Ltd – Manaaki Whenua
Palmerston North, New Zealand

Alexandra R. Toland
Department of Soil Protection
Institute of Ecology
Berlin University of Technology
Berlin, Germany

Gerd Wessolek
Department of Soil Protection
Institute of Ecology
Berlin University of Technology
Berlin, Germany

Antoinette M.G.A. WinklerPrins
Department of Earth and Planetary Sciences
Johns Hopkins University
Baltimore, Maryland

Jae E. Yang
Department of Biological Environment
Kangwon National University
Chuncheon, Republic of Korea

Koyo Yonebayashi
Ishikawa Prefectural University
Ishikawa, Japan

Iain M. Young
School of Environmental & Rural Science
University of New England
Armidale, New South Wales, Australia

Section I

Future Challenges

"Now that I am 82 years old, I find the remaining challenge in the exploration of the issues that are so important: How to achieve and promote soil and water use efficiency, especially in the face of impending climate change." Daniel Hillel (2012 World Food Prize Winner), *CSA News (Soil Science Society of America)*, 57 (11):4-8 (2012).

1 Climate Change
An Underfoot Perspective?

Kevin R. Tate[*] *and Benny K. G. Theng*

CONTENTS

There can be no life without soil and no soil without life; they have evolved together.

Charles E. Kellogg
USDA Yearbook of Agriculture, 1938

1.1 THE CHALLENGES OF THE FUTURE

1.1.1 Toward an Uncertain Future

The term "Anthropocene" is now increasingly used to denote the current era in the earth's history, where human activities have become the major driver of environmental change. This term was coined by the ecologist Eugene F. Stoermer and introduced into the literature by Paul J. Crutzen (2002), who won the 1995 Nobel Prize for his research on atmospheric chemistry and stratospheric ozone depletion (the so-called "ozone hole"). The Holocene era, which fostered the birth of civilizations over the past 10,000 years, has, in this century, given way to a warming earth that is also overcrowded and severely deforested, with many of its productive soils being lost at an unprecedented rate. The immediate future poses many challenges, but also presents many opportunities for innovation and change. In this chapter, we will elaborate on climate change—the biggest challenge facing humankind—as we move from the Holocene to the Anthropocene. It is in this context that we consider how our use of soils can be both a cause of, and a potential answer to meeting, this challenge.

Climate change has been portrayed by James E. Hansen (2009) as the biggest moral challenge of the twenty-first century, in that the impact of our economic system and Western lifestyles on the atmosphere today will be most severely felt later this century by our offspring. So, if it isn't already too late, our challenge is to change from the path we are currently on, to one where we live within

[*] Landcare Research NZ Ltd - Manaaki Whenua, Private Bag 11052, Palmerston North 4442, New Zealand; Email: tatek@landcareresearch.co.nz

the limits of the earth's finite resources, in order to secure a sustainable planet for future generations. Attempts have been made to define these limits and whether or not they have already been exceeded (Rockström et al. 2009). For example, rapidly rising atmospheric carbon dioxide (CO_2) concentrations, ascribed to fossil fuel burning and land-use change and management, have already exceeded 350 parts per million by volume (ppmv), the value that could have limited the rise in global mean temperature to less than 2°C by 2100.

The impacts from a rapidly changing climate are already being experienced around the world in the form of extreme weather events, such as flooding, droughts, heat waves, and the accelerating melting of glaciers and ice caps (Figure 1.1). There is, therefore, a great deal of urgency to adapt to these changes, and to replace our use of fossil fuels with renewable forms of energy. In addition, we need to halt deforestation, and the loss of productive soils, as well as improve the resilience of our land management practices, including our soils. Accelerating sea level rise caused by the melting of Arctic and Antarctic ice, together with the retreat of glaciers, also poses threats to coastal regions and river deltas where many of our large cities, such as Buenos Aires, Calcutta, Dhaka, Jakarta, Rotterdam, and Shanghai, are located, and where much of our food is grown. As our soils are under increasing pressure to grow food for a hungry world, as well as to regulate climate, water supplies and biodiversity, they need to be protected against degradation, inundation, erosion, and desertification. At the same time, we need to mitigate emissions of greenhouse gases, particularly CO_2, methane (CH_4), and nitrous oxide (N_2O) (Figure 1.1).

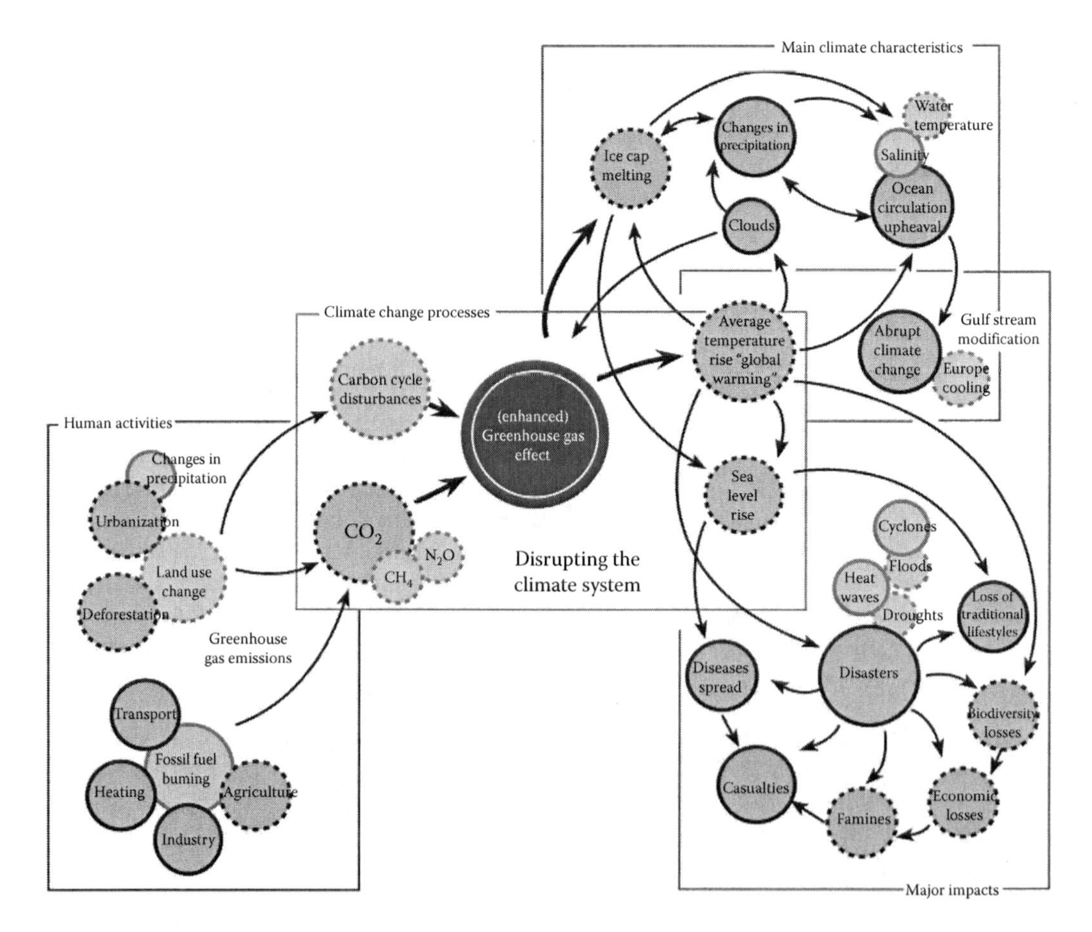

FIGURE 1.1 Diagram showing how human activities affect Earth's climate and vice versa. Aspects where soil plays a prominent role in the human–climate interaction are shown by broken circles. (Adapted from UNEP 2005. Vital Climate Change Graphics. UNEP/GRID-Arendal, Norway.)

Currently, about one-third of the CO_2 that we emit from burning fossil fuels is removed from the atmosphere by terrestrial ecosystems, while another third is taken up by the oceans. The residual CO_2 remains in the atmosphere, where its concentration continues to rise by nearly 2 ppmv per year, partly because of the reduced capacity of land and ocean to absorb it (Canadell et al. 2007). It is the increased combustion of fossil fuels, however, that is largely responsible for the 29% rise in CO_2 emissions between 2000 and 2008 (Le Quéré et al. 2009). After being relatively stable for some years, the atmospheric concentration of CH_4 has also risen steeply in recent years. Although the reasons for this recent increase are as yet unclear, the release of CH_4 from the arctic tundra and from the sea bed as the arctic sea ice retreats may be a contributing factor. We might also add that vast amounts of organic carbon (C) (*ca.* 1700 billion metric tonnes, or Gt) are stored in tundra soils. Its partial release as both CO_2 and CH_4 becomes more likely as microbial activity increases with thawing of the permafrost (Schuur et al. 2011). With respect to N_2O, its concentration in the atmosphere is also rising, mainly because of the excessive and inefficient use of nitrogen fertilizers. Indeed, emissions of this potent greenhouse gas are a clear indication that the concentration of reactive nitrogen (N) in the global environment is greatly in excess of global limits (Rockström et al. 2009).

Soils are contributing to the increased concentrations of all three greenhouse gases in the atmosphere. At the same time, soils are by far the largest reservoir of (organic) C, containing at least 2200 Gt in the top 1 m (Batjes 1996), as compared with 750 Gt C in the atmosphere, and 610 Gt C in the above-ground vegetation worldwide.

It is generally accepted that photosynthesis is the principal mechanism for removing CO_2 from the atmosphere, while soils are the main conduit for storing C, and for returning terrestrial C (as CO_2) to the atmosphere through organic matter decomposition and root respiration. Even small changes in soil C stocks may, therefore, have a large effect on the burden of CO_2 in the atmosphere. The way we manage soils can also strongly influence emissions of "agricultural" greenhouse gases, notably N_2O and CH_4. According to Smith et al. (2008), agriculture accounts for the emission of 52% and 84% of global anthropogenic N_2O and CH_4, respectively. These gases are more potent than CO_2 in that they warm the atmosphere more strongly over a 100-year time period by factors of 320 (N_2O) and at least 25 (CH_4), with unity being for CO_2. As we shall see later, accounting for the relative emissions of these gases is both necessary and complex when we assess the relative benefits of land-use and management systems.

We begin with a brief discussion of how soils function, focussing on their role as a source and sink for greenhouse gases.

1.1.2 From a Black Box to a Living Ecosystem

It is only about 30 years ago that soil scientists began to embrace the concept that soil is a complex living ecosystem, containing a myriad of macro and microorganisms rather than a mere collection of mineral particles, biopolymers, and mineral–organic associations. This concept largely derives from the pioneering work by David S. Jenkinson who proposed that the soil microbial biomass, a labile pool of living microorganisms, played a central role in the cycling and provision of C and plant nutrients (Jenkinson and Powlson 1976). The microbial biomass may thus be likened to the "eye of the needle" through which all C and nutrients must pass before becoming available to plants.

Although much of this C (and nutrients) is released immediately by respiration as CO_2 (and also as CH_4 and N_2O in very wet soils), a proportion is stabilized in soil organo-mineral associations, often for long periods before being released as CO_2, N_2O, and CH_4. Some of the nutrients are taken up by plants, enabling more C to be fixed by photosynthesis. This cycling-storage process in soils is central to our understanding of how soils act as a sink for atmospheric CO_2. The question arises whether the function of this sink, and its capacity, will change as the climate warms.

Adoption of this new paradigm by which soil is regarded as a living ecosystem has blurred the boundaries that used to exist between soil biology, chemistry, mineralogy, and the plant sciences. Its significance is attested by the proliferation over the past 20 years or so of soil-related journals with a focus on biological processes, and the exponential growth in peer-reviewed papers on soil biological and ecological processes. Besides revealing the inherent resilience of soils, and the diversity of the life they support, this new understanding helps us define limits beyond which soils would degrade unless they are carefully managed.

Some years ago, we stumbled on the astonishing discovery that most of the microbial population in soil can survive very long periods of food scarcity. Measurement of the adenylate energy charge (indicative of the collective metabolic activity) of this population (Brookes et al. 1983) suggested that soil microorganisms did not need to rely on resting spore formation, that is, the ability to lie dormant for a long period, when food was in short supply. Rather, they remain in a metabolic state more typical of cells undergoing exponential growth. Subsequent investigations (e.g., De Nobili et al. 2001) have identified small molecules that can "trigger" the microbial population into activity. Further, the ability of soil microorganisms to remain viable in a generally substrate-poor environment would give them a distinct evolutionary advantage over other soil organisms.

This unexpected finding illustrates not only how well-adapted soil microorganisms are to the often harsh conditions in their immediate environment, but also how they can contribute to the resilience of soils in different environments. In other words, soil microorganisms are able to withstand adverse conditions (e.g., drought, flooding), albeit for short periods, and resume activity when these conditions subside. Survival, however, becomes problematic when such conditions persist for long periods, as they do in desert soils. The question arises whether soils differ in their ability to withstand the increased frequency and intensity of extreme conditions that arise from global warming.

The concept that soil organic matter consists of pools (e.g., microbial biomass, organo-mineral associations) with variable stability has improved our understanding of soil processes, and helped us appreciate the vital role organic matter plays in sustaining agricultural systems and food production. The development of process-based mathematical models such as ROTHC and CENTURY, and many others (e.g., see: http://yacorba.res.bbsrc.ac.uk/egi-bin/somnet; http://eco.wiz.uni-kassel.de/ecobas.html), including some for non-CO_2 greenhouse gases, notably the DNDC model (Li et al. 1992), has enabled us to assess the effects of land-use change on soil organic matter stocks as well as to simulate greenhouse gas emissions and uptake by soils. More generalized models capable of running large datasets over expanded spatial and temporal scales (e.g., Potter et al. 1996) have also been developed in response to questions about global processes and changes, and future impacts of global warming on terrestrial ecosystems. Nevertheless, there are large uncertainties in modeling changes in soil C resulting from the combined effects of CO_2-enhanced photosynthesis and climate change. Although there is general agreement among current coupled climate–carbon cycle models that the extra CO_2 from the net loss of soil C will accelerate the rate of warming, the magnitude of this additional warming is far from clear (Luke and Cox 2011). In this respect, one major area of uncertainty is the extent to which elevated atmospheric CO_2 levels may increase soil C storage through enhanced photosynthetic activity to offset the expected loss of soil C as a result of global warming. According to a meta-analysis of the results from many different studies (Sillen and Dieleman 2012), any increase in soil C may be limited, at least in managed grasslands. Indeed, soil C could decrease as the CO_2 concentration in the atmosphere increases because arbuscular mycorrhizal fungi that form symbiotic relationships with most plants can cause the soil to become a net source of CO_2 by stimulating additional decomposition (Cheng et al. 2012; Kowalchuk 2012). Clearly, we need to improve our understanding of how environmental and management practices affect decomposition processes, if we are to maintain, if not increase, current soil C stocks.

Underlying these modeling approaches have been advances in understanding ecological processes that often span periods of thousands to millions of years (e.g., Peltzer et al. 2010). Besides raising our awareness of how long-term pedogenic changes drive short-term biological processes, these advances provide insights into how soils and ecosystems might respond to future disturbances arising from extreme weather events. We need to understand how complex ecosystems work over time and space if we are to predict the impacts of future climate changes on terrestrial ecosystems. The science behind maintaining ecosystem health and function is highly complex. Likewise, the application of complexity theory to financial institutions would indicate that financial systems can become more resilient by diversifying their activities, rather than relying on strong global networking (Markose et al. 2012).

Because of its inherent complexity, the soil underfoot is arguably one of the least understood ecosystems on earth. Recent developments in molecular biology have revealed the enormous diversity of soil organisms, most of which have not yet been described, let alone characterized. We are still only at the early stages of unraveling this diversity and the role of soil organisms in regulating key soil processes including net greenhouse gas emissions. One example is the recent finding that the soil microorganisms involved in denitrification (a key mechanism for releasing N_2O from soils) and N-fixing (a process that captures atmospheric nitrogen for use by plants) are related *in situ* (Singh et al. 2011). Land-use changes such as afforestation of grassland may, therefore, have a significant impact on soil N transformations and other ecosystem processes (Singh et al. 2011). Later, we will give further examples to show how these new findings may be used to develop technologies for reducing greenhouse gas emissions.

In a nutshell, these advances in our understanding of how soils function over varying spatial and temporal scales have provided an insight into managing soils and ecosystems in order to sustain their essential functions in providing food, fiber, and water, and in reducing greenhouse gas emissions over the coming decades.

1.1.3 "Business-as-Usual" Will Fail to Provide a Path to a Sustainable Future

Soils have sustained life on earth since its emergence about 4.5 billion years ago, while clays might have played a key role in catalyzing the formation of pre-biotic RNA (Brack 2006; Ferris 2006). The possibility that soils on Mars and other planets in our solar system could, by analogy, serve as a nursery for life has stimulated current exploration of planetary bodies. Here on Earth, ice core analysis has provided evidence to show that large natural changes in climate have occurred over successive glacial and interglacial periods for at least the past 1 million years. Soil processes have apparently influenced these changes in that the concentration of atmospheric CO_2 increased by *ca.* 100 ppmv when Earth emerged from the last glacial maximum into the pre-industrial Holocene. This increase has been attributed to the disappearance of an inert C pool, probably induced by the decomposition of organic matter in the frozen soil of northern regions as the ice retreated (Ciais et al. 2012). The increasing exposure of C-rich soils at high latitudes, due to permafrost melting, indicates that such an event may again be occurring.

For most of the past 10,000 years, however, the earth's climate has been in an interglacial phase (often referred to as the "Goldilocks zone"). The prevailing warm stable climate, in conjunction with the availability of fertile soils, was conducive to plant growth, enabling net primary productivity (NPP) of terrestrial ecosystems to double that achieved during the Last Glacial Maximum. The move from hunting/gathering to farming also allowed communities to form, and civilizations to flourish, often until their immediate land, soil, and water resources had seriously declined, or were largely depleted.

Where close attention was given to agriculture, such as in Egypt, the use of primitive Neolithic tools on unmodified natural soils gave way to that of the plough, while irrigation and crop rotations became common. These practices undoubtedly contributed to the endurance of Egyptian civilization. As in Egypt, other civilizations in Mesopotamia and the Indus Valley were based on an equitable climate and recent alluvial soils of good structure and high fertility. Despite continued improvement in soil management, food production could not keep pace with the needs of an expanding population, causing these civilizations to decline and ultimately collapse. More often than not, however, it was soil degradation that led to land abandonment, as exemplified by Easter Island, and medieval Greenland (Diamond 2005). The development of the dust bowl during the 1930s in the Great Plains of America is a twentieth century example (Worster 2004; Montgomery 2007).

The concentrations of atmospheric greenhouse gases during the Holocene were relatively stable at about 280 ppmv for CO_2, 700 parts per billion by volume (ppbv) for CH_4, and about 270 ppbv for N_2O. Nonetheless, recent studies by Eriksson et al. (2012) (well summarized in *New Scientist*,

September 22, 2012, p. 12) have shown that climate change was a key regulator of successive human migrations and global expansion over the past 120,000 years. Likewise, the large-scale human crises, including economic downturns, during the period of 1500–1800 AD were primarily caused by natural climate changes (Zhang et al. 2011). Interestingly, Sapart et al. (2012) have found large centennial variations in the C isotope content of CH_4 between 100 BC and 1600 AD that can be ascribed to changes in biogenic and pyrogenic sources. These source changes correlate not only with natural climate variability but also with alterations in human population (e.g., decline of the Roman Empire and the Han dynasty), and land use (e.g., expansion of rice agriculture).

With the start of the industrial revolution in the mid-eighteenth century, rural people in Europe migrated to urban centers to work in the factories. Since these "dark satanic mills" were powered by burning wood and coal, their expansion and proliferation necessitated the "old world" to seek external resources and markets, many of which were provided by territories overseas. In order to feed the expanding human and animal populations, forests were cleared for agriculture, exposing the soils beneath to erosion by water and wind. This change in land use has continued, particularly in tropical South America and Asia where during the 1990s deforestation accounted for 26% and 13%, respectively, of total CO_2 emissions (0.5–3.0 Gt C per year) (Houghton and Goodale 2004). From the 1920s onward, the extraction and use of oil rapidly expanded, causing atmospheric CO_2 concentrations to rise even faster. The latest comprehensive record of such changes to the world's atmospheric composition can be found in the Fourth Assessment Report of the Intergovernmental Panel on Climate Change (IPCC 2007); the Fifth Assessment Report is currently being prepared. By 2012, atmospheric concentrations of the three main greenhouse gases have risen from 280 (in 1750) to 393 ppmv (for CO_2), from 700 to *ca.* 1800 ppbv (for CH_4) and from 270 to 320 ppbv (for N_2O). The most rapid increases in atmospheric CO_2 concentration have occurred since about 2002, owing largely to fossil fuel-powered growth of the giant emerging economies of China and India. We should also add that the large increase in CO_2 emissions from these two countries is partly associated with the production of cheap consumer goods for export to developed countries. As already mentioned, some of this increase may be ascribed to a decline in the capacity of the oceans and land to take up CO_2 (e.g., Canadell et al. 2007; Le Quéré et al. 2009), although large uncertainties are associated with estimating the size and dynamics of oceanic and terrestrial C sinks. Indeed, recent monitoring of these sinks has indicated that the terrestrial component may be increasing (Raupach 2011; Beaulieu et al. 2012). Expansion of the land sink, if confirmed, would have a major effect on the rate of climate change. There is clearly a need to understand why and how the land and ocean sinks are changing. For example, the reason for an apparent abrupt increase in the mean net land uptake of C in 1988 (Beaulieu et al. 2012) is unclear, but this finding is unlikely to be associated with variations in soil organic C stocks.

Although a change in land management can cause soil organic matter to lose C abruptly, the accumulation of C in soils is always a slow process, becoming progressively slower as a new equilibrium is established. In the example from Rothamsted shown in Figure 1.2, the ROTHC model simulation over 144 years indicated that the annual rate of soil C increase over the first 20 years of Farm Yard Manure (FYM) application was 1.0 t $ha^{-1}y^{-1}$. This rate slowed down to 0.1 t $ha^{-1}y^{-1}$ over the last 20 years, illustrating how soil C moved to a new equilibrium level with changes in management. Further, only a small fraction of the added organic C was stored in the soil with most of it being microbially decomposed and respired as CO_2 (Powlson et al. 2011). The slow build-up of soil C stocks also underlines the importance of managing soils to maintain or enhance soil organic matter levels because soils are naturally limited in their capacity to store C (e.g., Kool et al. 2007).

1.1.4 Land-Use Change and Management: Causes, Consequences, Effects, and Solutions

Atmospheric carbon dioxide measurements in the Northern Hemisphere (Mauna Loa, Hawaii) were initiated in 1950 by Charles D. Keeling (Harris 2010). Similar data were later obtained by Southern Hemisphere monitoring stations, notably at Baring Head, New Zealand (Brailsford et al. 2012).

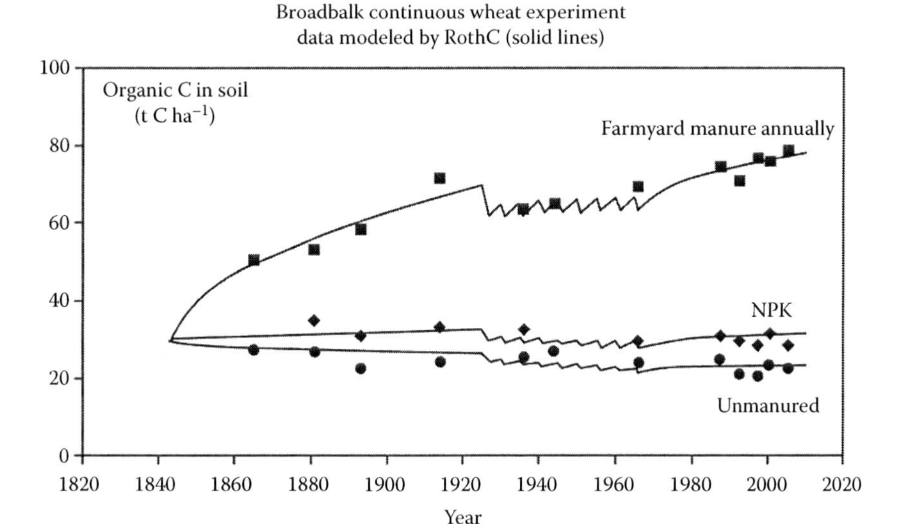

FIGURE 1.2 Changes in soil C in the plough layer (0–23 cm) in three treatments of the Broadbalk Wheat Experiment at Rothamsted Research, UK. Annual treatments are: •, no fertilizer or manure applied since 1844; ♦, PKMg plus 144 kg N ha^{-1} since 1852; ▪, farmyard manure at 35 t ha^{-1} fresh weight applied since 1885 plus 96 kg N ha^{-1} since 1968. (Powlson, D.S., A.P. Whitmore, and K.W.T. Goulding. Soil carbon sequestration to mitigate climate change: A critical re-examination to identify the true and the false. *European Journal of Soil Science.* 2011. 62: 42–55. Copyright Wiley-VCH Verlag GmbH & Co. KGaA. Reproduced with permission.)

Although the two datasets show similar upward trends over time, the curve from the Northern Hemisphere (where most land is located) has a distinct annual saw tooth pattern caused by the seasonality of photosynthesis and soil respiration, while this pattern is only weakly expressed in the data from the Southern Hemisphere record. These atmospheric measurements have been extended and supplemented using surrogates such as tree rings and ice cores.

Land-use changes since *ca.* 1750 have initially made large contributions to atmospheric CO_2 concentrations. Houghton and Goodale (2004), for example, estimated that 156 Gt C was released to the atmosphere from 1850 to 1990, about half of which came from fossil fuel combustion. Soils accounted for about 25% of the total CO_2-C released during this period. Recent evidence suggests emissions from land-use change have decreased during the 1980–2008 period (Raupach 2011), although CO_2 emissions from biomass clearing/burning and soil organic matter decomposition still account for a large global CO_2 flux of *ca.* 0.4–2 Gt C y^{-1} (Denman et al. 2007).

Much of the CO_2 released from soils results from conventional agricultural practices. For example, cultivation of previously untilled soils typically leads to C losses of about 30% over the first few years (Davidson and Ackerman 1993). Besides being affected by soil type and management practice, the response of soil to cultivation is strongly influenced by climate. The effect of climate, at a regional scale, clearly shows up when we compare the responses of soil organic C and N to cultivation in India with those in the Great Plains of the United States (Miller et al. 2004). Whereas temperature was primarily responsible for soil C losses in India, precipitation was the dominant climatic factor controlling soil C losses in the Great Plains. Apart from climatic effects, the differential response of soil C and N to cultivation in each region is also influenced by differences in management practices, crop cycles, and fertilizer use. Currently, between 10% and 15% of total anthropogenic CO_2 emissions are attributable to changes in land-use (Friedlingstein et al. 2010). The demands of a growing human population for food, fiber, and shelter together with climate change and decreased water and nutrient availability would clearly affect CO_2 emissions from soils. At the same time, the ability of soils to provide the services we expect, including food production and water storage, would diminish. Since the costs, arising from these reduced services, are not directly factored into conventional economic analyses, they effectively represent a loss of subsidy from the environment.

At global as well as regional scales, land-use change remains a critical issue in relation to how much more land should be used for cropping (Rockström et al. 2009), using soil degradation or loss as a useful index of terrestrial health (Bass 2009). For example, soil C losses from one-quarter of the global land area over the past 2–3 decades have caused productivity to decline (Bal et al. 2008). The way we manage land, however, is at least as important as how much land should be reserved for food production. In this context, we should also mention that soil type, and its associated clay mineralogy, can strongly influence the storage and retention of C with cultivation. Parfitt et al. (1997), for example, found that an Andisol, rich in allophane (and ferrihydrite) under old pasture, could store twice as much C as an Inceptisol of similar texture but where mica was the dominant clay mineral. They further noted that the Andisol (derived from andesitic tephra) lost only half as much C as did the Inceptisol (formed in quartz/feldspar alluvium), after cultivation for 20 years. These observations are consistent with the ability of allophane, a short-range order aluminosilicate, to retain and stabilize organic matter (Theng and Yuan 2008; Calabi-Floody et al. 2011).

Although proven management practices are available for the sustainable management of different land uses including cropland (e.g., Ostie et al. 2009), their rate of adoption is slow. The issue is complex, as multiple benefits need to be considered, including erosion protection, fertility maintenance, and the need to build resilience into farming systems. Organic agriculture generally provides these and other benefits (FAO 2002). Intensified use of land for crop production can also contribute to emissions reduction in crop production (Burney et al. 2010), and solutions tailored to specific land systems and soil types are available.

For example, Powlson et al. (2012) have pointed out that soil C stocks in England and Wales could be increased through reduced tillage or organic material additions to soils. Any gains in soil C, however, had to be weighed against potential increases in N_2O emissions in some years. In reviewing the topic of soil C sequestration as a means of mitigating climate change, Powlson et al. (2011) have concluded that (i) the capacity of soil to store C is finite, (ii) the process is reversible in that soil can store or lose C, and (iii) any increase in soil C storage may be accompanied by changes in fluxes of N_2O and CH_4. Thus, converting cropland into forest, grassland, or perennial crops does genuinely sequester atmospheric CO_2, but these changes in land use may also lead to the emission of the more potent greenhouse gases.

An emerging technology that may achieve multiple benefits, including increased soil C storage and reduced N_2O emissions, is the use of biochar, a charcoal-like material, formed by heating plants at ~500°C in an oxygen-limited atmosphere. In making biochar, a proportion of the plant C is retained for long-term storage in soil, while the energy produced can reduce, or even replace, fossil fuel use. The technology is based on the ancient practice of adding charcoal to soil and, so far, has mainly been used in C-depleted arable soils (WinklerPrins 2014). When added to soil, biochar can improve the yield and productivity of some crops, sequester C into stable soil pools, and markedly reduce emissions of N_2O and NH_3 (Taghizadeh-Toosi et al. 2011). Ammonia retained by the biochar is available to plants, and contributes to the N nutrition of crops (Taghizadeh-Toosi et al. 2012).

Using New Zealand as a case study, we have been able to assess the atmospheric impacts of land-use changes from forest and grassland and vice versa by accounting for the associated changes in N_2O and CH_4 emissions together with the albedo effects (Kirschbaum et al. 2012). A number of atmospheric impacts were identified, but only their direction could be confidently predicted since their severity varied with site conditions (e.g., soil type and management, microclimate, post-harvest use of biomass). Land-use changes also induce large below-ground changes at the microbial level, as indicated by shifts between fungal and bacterially dominated populations which, in turn, influence both C sequestration and nutrient cycling processes (Macdonald et al. 2009).

On the whole, however, afforestation in temperate and tropical regions is clearly beneficial in terms of increasing soil C stocks, particularly when compared with using land as pastures for grazing animals. While grassland soils can store large amounts of C, mainly through prolific root turnover, the introduction of grazing ruminants produces large amounts of CH_4 (from enteric fermentation and waste) and N_2O (from dung and urine deposition on soil). As a result, these grazed pastoral systems can adversely affect the atmosphere, and water quality. Globally, the use of land for grazing by ruminants is increasing, with substantially larger increases being recorded for grazing by monogastric animals (e.g., pigs, poultry). The latter trend can dramatically reduce environmental impacts associated with livestock production (Steinfeld and Gerber 2010).

1.1.5 Reducing Agricultural Greenhouse Gas Emissions Using Existing Technologies and Management Practices

The global expansion and intensification of modern agriculture to meet the food demands of over 7 billion people lie behind the large-scale use of N (and phosphorus) fertilizers (e.g., Gruben and Galloway 2008). As a result, the release of reactive N, including N_2O and NH_3, has increased beyond the planet's capacity for their assimilation (Rockström et al. 2009). Fertilizer use and the associated release of reactive N can be reduced by changing management practices such as minimizing the time animals spent grazing in the field (Ledgard and Luo 2008), while incorporation of nitrification and/or urease inhibitors into fertilizers (Watson et al. 2009) would promote the retention of N as ammonium (NH_4^+) and/or reduce the hydrolysis of urea-N (Figure 1.3). Thus, the application of dicyandiamide, a nitrification inhibitor, to dairy pastures across New Zealand could decrease emissions of N_2O from urine patches by 35–45% (de Klein et al. 2011).

The practicality of implementing these and other strategies will depend on many factors, including cost and the environmental regulation of N use and emissions. For example, Steinfeld and Gerber (2010) have argued that responsible intensification of livestock production is achievable without resorting to deforestation by using emerging technologies and aligning agricultural policies to the need for reversing current global environmental changes. These technologies may be added to the already long list of more conventional management tools that dairy and other livestock farmers (e.g., Luo et al. 2010) can adopt to reduce N fertilizer use and associated emissions of these potent greenhouse gases. In the absence of regulatory guidelines, however, many of these management tools would not be taken up.

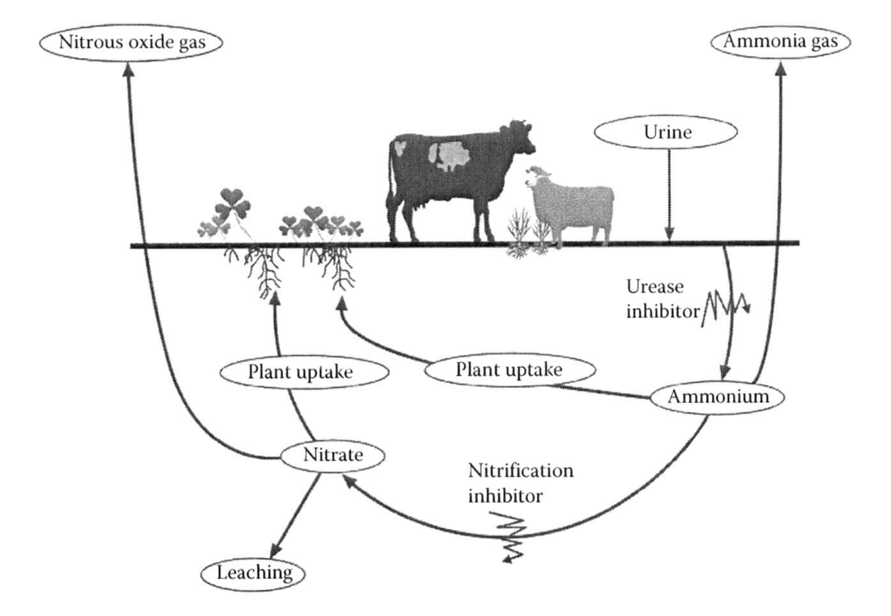

FIGURE 1.3 Diagram showing the main forms and flows of N in a dairy-grazed pasture, and where the application of nitrification and urease inhibitors can limit N losses and, in some cases, increase pasture production.

Recent advances in nano- and biotechnologies, inspired by nanoscale processes in nature, including soils (Theng and Yuan 2008), have made it possible to progressively displace many of the fossil-fuel-based technologies of the last century by "green" methods of production, some examples of which are shown in the next section.

1.2 SOIL-INSPIRED SOLUTIONS FOR A BETTER FUTURE

Besides being a medium for growing crops for food and fiber, soils have been used for millennia to provide shelter (mud bricks, trees) and fuel (peat, wood). Soils have also been recognized as a rich source of antibiotics and have long been eaten (to suppress hunger) during times of famine. Although our knowledge of soil processes is incomplete, some of these may be applied for mitigating greenhouse gases, and for adapting to the future impacts of climate change. Although clays *per se* have rarely been used to mitigate greenhouse gas emissions, their large surface area and peculiar charge characteristics are conducive not only to complexing and stabilizing soil organic matter (Theng 2012), but also to reducing greenhouse gas emissions and adapting to climate change (Yuan 2010; Yuan and Wada 2012). We have recently screened a large number of clays for their potential in mitigating enteric CH_4 emissions from ruminant animals by incubating them with cow rumen inoculum *in vitro* under anaerobic conditions. We found that some clays could substantially reduce CH_4 emissions, although the mechanisms involved are not yet understood (Figure 1.4).

Similarly, soil microorganisms are potentially useful in developing technologies to mitigate greenhouse gas emissions as well as for adapting to climate change. Two further examples from agriculture include methanogens (bacteria that produce CH_4) that can be managed to produce maximum amounts of CH_4 for energy production (e.g., Prapaspongsa et al. 2010), and methanotrophs (bacteria that consume CH_4) for removing CH_4 where source emissions are sufficiently below economic levels to warrant capture for energy production. A promising example of the latter is biofiltration, a well-proven and cost-effective technology for removing atmospheric pollutants. We have found that CH_4 emissions from dairy farm animal effluent stored in ponds under anaerobic conditions can be substantially reduced using biofilters containing a very active population of soil methanotrophs (Pratt et al. 2012; Pratt et al. 2013) (Figure 1.5).

Other examples involving specific microorganisms include the production of biofuel products from CO in fugitive flue gases emitted by industrial plants (Holmgren 2012), and the harnessing of electrical energy from anaerobic soil processes in rice paddies (Takanezawa et al. 2010).

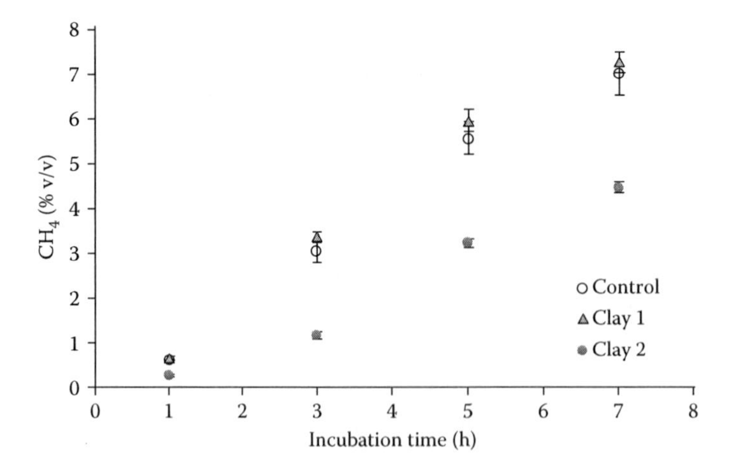

FIGURE 1.4 Diagram showing the ability of some clays to reduce CH_4 emissions from cow rumens, on incubation with rumen liquor inoculum *in vitro*. Data represent the means of five replicates, and error bars indicate standard deviation (K.R. Tate, unpublished results).

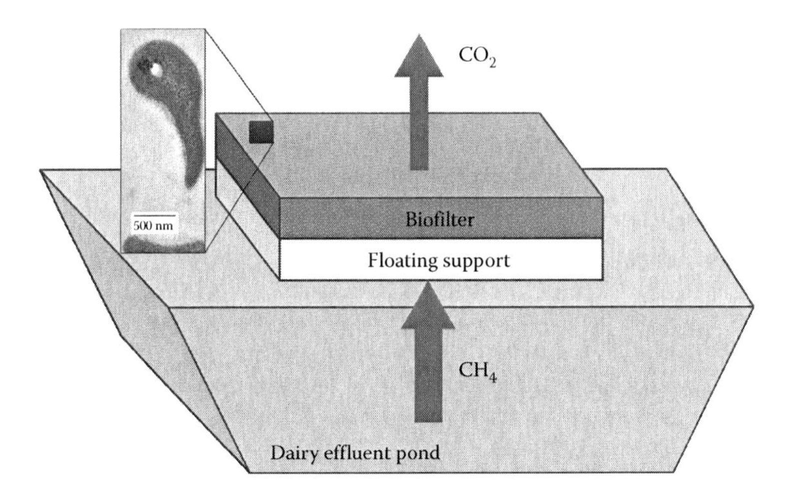

FIGURE 1.5 **(See color insert.)** Diagram showing how an active soil methanotroph population can be used in a cover-biofiliter to remove CH_4 emissions from a dairy effluent pond. (Adapted from Pratt, C. et al. 2013. *Environmental Science & Technology* 47:526–532.). An electron micrograph of one of the methanotrophs (*Methylosinus trichosporium*) is shown on the left. (Courtesy of Dr. Réal Roy and Brent Gowan, Victoria University, BC, Canada.)

We have already noted that understanding soil biological processes in an ecosystem context has raised our awareness of how soils and ecosystems might respond to future extreme weather events. In this context, we should add that some soil microorganisms have the ability to enhance the tolerance of plants (e.g., rice) to environmental stress through symbiosis with fungal endophytes (Redman et al. 2011). This exciting observation suggests the potential of symbiotic technology for adapting plants to the impacts of climate change, facilitating the expansion of agriculture into marginal lands.

These and other examples demonstrate that despite our inadequate understanding of soil processes, there are means available for managing soils more sustainably so that future demands for food and water could be met, and greenhouse gas emissions reduced. We have also indicated that the net loss of soil C from global warming will not only challenge our ability to manage soils in future but also increase the rate of warming.

In the 1970s, the Club of Rome warned us that we were treading a dangerous path unless we adopted a more sustainable way of managing the earth's finite resources. Continued procrastination by many governments to take urgent collective action over climate change, and the failure of the recent Rio + 20 conference to address this issue indicate that short-term economics and political expediency are preventing the adoption of sensible, evidence-based policies to avert an approaching climate crisis. We have reached a point now where the question is not whether the climate is changing due to increasing greenhouse gas emissions but rather, what can we do about it, and whether we should bear the cost now or later. It is inevitable that future generations will have to deal with global warming and its consequences, as global mean temperatures now seem likely to be 4–5°C above present by the end of this century. It is imperative that we learn to manage our planet prudently so as to sustain life, in all its diversity, into the near future. The following saying in Sanskrit, recorded about 4000 years ago (Johnston et al. 2009, p. 2), provides a stark reminder of the importance of managing our soils wisely, and the likely consequences of not doing so:

> Upon this handful of soil our survival depends. Husband it and it will grow our food, our fuel and our shelter and surround us with beauty. Abuse it and the soil will collapse and die, taking man with it.

We have many opportunities to adopt a new paradigm that ceases to rely on fossil fuels and subsidies from the environment, including the soil. In 2007, the IPCC stated that we already have many options for making development paths more sustainable and for achieving reductions of greenhouse gas emissions (IPCC 2007). With careful management and innovation, our soils can continue to provide many of the services we need. It only requires us to make the appropriate changes to our "underfoot" perspective.

ACKNOWLEDGMENTS

We are grateful to Nicolette Faville for help with graphical support, and to Anne Austin for final editing of the manuscript.

REFERENCES

Bal, Z., D.L. Dent, J. Olsson, and M.E. Schaepman. 2008. Proxy global assessment of land degradation. *Soil Use and Management* 24: 223–234.

Bass, S. 2009. Keep off the grass. *Nature Reports Climate Change* 3: 113–114.

Batjes, N.H. 1996. Total carbon and nitrogen in the soils of the world. *European Journal of Soil Science* 47: 151–163.

Beaulieu, C., J.L. Sarmiento, S.E. Mikaloff Fletcher, J. Chen, and D. Medvigy. 2012. Identification and characterization of abrupt changes in the land uptake of carbon. *Global Biogeochemical Cycles* 26: GB1007, doi: 10.1029/2010GB004024, 2012.

Brack, A. 2006. Clay minerals and the origin of life. In *Handbook of Clay Science,* eds. F. Bergaya, B.K.G., Theng, and G. Lagaly, 371–391. Amsterdam: Elsevier.

Brailsford, G.W., B.B. Stephens, A.J. Gomez et al. 2012. Long-term continuous atmospheric CO_2 measurements at Baring Head, New Zealand. *Atmospheric Measurement Techniques Discussions* 5: 5889–5912.

Brookes, P.C., K.R. Tate, and D.S. Jenkinson. 1983. The adenylate energy charge of the soil microbial biomass. *Soil Biology & Biochemistry* 15: 9–16.

Burney, J.A., S.J. Davis, and D.B. Lobell. 2010. Greenhouse gas mitigation by agricultural intensification. *Proceedings of the National Academy of Sciences* 107: 12052–12057.

Calabi-Floody, M., J.S. Bendall, A.A. Jara et al. 2011. Nanoclays from an Andisol: Extraction, properties and carbon stabilization. *Geoderma* 161: 159–167.

Canadell, J.G., D.E. Pataki, R. Gifford et al. 2007. Saturation of the terrestrial carbon sink. In *Terrestrial Ecosystems in a Changing World, The IGBP Series*, eds. J.G. Canadell, D. Pataki, and L. Pitelka, 59–78. Berlin: Springer Verlag.

Cheng, L., F.L. Booker, C. Tu et al. 2012. Arbuscular mycorrhizal fungi increase organic carbon decomposition under elevated CO_2. *Science* 337: 1084–1087.

Ciais, P., A. Tagliabue, M. Cuntz et al. 2012. Large inert carbon pool in the terrestrial biosphere during the Last Glacial Maximum. *Nature Geoscience* 5: 74–79.

Crutzen, P.J. 2002. Geology of mankind. *Nature* 415: 23.

Davidson, E.A. and I.L. Ackerman. 1993. Changes in soil carbon inventories following cultivation of previously untilled soils. *Biogeochemistry* 20: 161–193.

de Klein, C.A.M., K.C. Cameron, H.J. Di et al. 2011. Repeated annual use of the nitrification inhibitor, dicyandiamide (DCD) does not alter its effectiveness in reducing N_2O emissions from cow urine. *Animal Feed Science & Technology* 166–167: 480–491.

De Nobili, M., M. Contin, C. Mondini, and P.C. Brookes. 2001. Soil microbial biomass is triggered into activity by trace amounts of substrate. *Soil Biology & Biochemistry* 33: 1163–1170.

Denman, K.L., A. Brasseur, A. Chidthaisong et al. 2007. Couplings between changes in the climate system and biogeochemistry. In *Climate Change 2007: The Physical Science Basis,* eds. S. Solomon, D. Oin, M. Manning et al. Contribution of Working Group I to the Fourth Assessment Report of the Intergovernmental Panel on Climate Change. Cambridge: Cambridge University Press.

Diamond, J. 2005. *Collapse: How Societies Choose to Fail or Succeed.* New York: Viking Press, Penguin Group.

Eriksson, A., L. Betti, A.D. Friend et al. 2012. Late Pleistocene climate change and the global expansion of anatomically modern humans. *Proceedings of the National Academy of Sciences* 109: 16089–16094.

FAO 2002. *Organic Agriculture, Environment and Security.* Brazil: Environment and Natural Management Series, Version 4.

Ferris, J.P. 2006. Montmorillonite-catalysed formation of RNA oligomers: The possible role of catalysis in the origins of life. *Philosophical Transactions of the Royal Society B* 361: 1777–1786.

Friedlingstein, P., R.A. Houghton, G., Marland et al. 2010. Update on CO_2 emissions. *Nature Geoscience* 3: 811–812.

Gruben, N. and J.N. Galloway. 2008. An earth-system perspective of the global nitrogen cycle. *Nature* 451: 293–296.

Hansen, J. 2009. *Storms of My Grandchildren: The Truth about the Coming Climate Catastrophe and Our Last Chance to Save Humanity*. London: Bloomsbury Publishing Plc.

Harris, D.C. 2010. Charles David Keeling and the story of atmospheric CO_2 measurements. *Analytical Chemistry* 82: 7865–7870.

Holmgren, J. 2012. Waste flue gas CO to innovative biofuel production. In *Nanotechnology 2012. Biosensors, Instruments, Medical, Environment and Energy, Volume 3, Carbon Capture, Biomaterials and Biofuels (Chapter 7)*, 479–480. Austin: The NanoScience and Technology Institute (NSTI).

Houghton, R.A. and C.L. Goodale. 2004. Effects of land-use change on the carbon balance of terrestrial ecosystems. *Ecosystems and Land-Use Change Series* 153: 85–97.

IPCC. 2007. Summary for policy makers. In *Climate Change 2007: Mitigation*, eds. B. Metz, O.R. Davidson, P.R. Bosch, R. Dave, and L.A. Meyer, Contribution of Working Group III to the Fourth Assessment Report of the Intergovernmental Panel on Climate Change. Cambridge: Cambridge University Press.

Jenkinson, D.S. and D.S. Powlson. 1976. The effects of biocidal treatments on metabolism in soil. V. A method for measuring soil biomass. *Soil Biology & Biochemistry* 8: 209–213.

Johnston, A.E., P.R. Poulton, and K. Coleman. 2009. Soil organic matter: Its importance in sustainable agriculture and carbon dioxide fluxes. *Advances in Agronomy* 101: 1–57.

Kirschbaum, M.U.F., S. Saggar, K.R. Tate et al. 2012. Comprehensive evaluation of the climate-change implications of shifting land use between forest and grassland: New Zealand case study. *Agriculture, Ecosystems & Environment* 150: 123–138.

Kool, D.M., H. Chung, K.R. Tate, D.J. Ross, P.C.D. Newton, and J. Six. 2007. Hierarchical saturation of soil carbon pools near a natural CO_2 spring. *Global Change Biology* 13: 1282–1293.

Kowalchuk, G.A. 2012. Bad news for soil carbon sequestration. *Science* 337:1049–1050.

Le Quéré, C., M.R. Raupach, J.G., Canadell et al. 2009. Trends in the sources and sinks of carbon dioxide. *Nature Geoscience* 2: 831–836.

Ledgard, S.F. and J. Luo. 2008. Nitrogen cycling in intensively grazed pastures and practices to reduce whole-farm nitrogen losses. In *Multifunctioning Grasslands in a Changing World*, ed. Organising Committee of 2008, IGC/IRC Conference, 292–297. Guangdong: People's Publishing House.

Li, C., S. Frolking, and T.A. Frolking. 1992. A model of nitrous oxide evolution from soil driven by rainfall events. I. Model structure and sensitivity. *Journal of Geophysical Research* 97: 9759–9776.

Luke, C.M. and P.M. Cox. 2011. Soil carbon and climate change: From the Jenkinson effect to the compost-bomb instability. *European Journal of Soil Science* 62: 5–12.

Luo, J., C.A.M. de Klein, S.F. Ledgard, and S. Saggar. 2010. Management options to reduce nitrous oxide emissions from intensively grazed pastures. *Agriculture, Ecosystems & Environment* 136: 282–291.

Macdonald, C.A., N. Thomas, L. Robinson et al. 2009. Physiological, biochemical and molecular responses of the soil microbial community after afforestation of pastures with *Pinus radiata*. *Soil Biology & Biochemistry* 41: 1642–1651.

Markose, S., S. Giansante, and A.R. Shaghaghi. 2012. "Too interconnected to fail" financial network of US CDS market: Topological fragility and systemic risk. *Journal of Economic Behavior & Organization,* 83: 627–646.

Miller, A.J., R. Amundson, I.C. Burke, and C. Yonker. 2004. The effect of climate and cultivation on soil organic C and N. *Biogeochemistry* 67: 57–72.

Montgomery, D.R. 2007. *Dirt, The Erosion of Civilizations*. Berkeley: University of California Press.

Ostie, N.J., P.E. Levy, C.D. Evans, and P. Smith. 2009. UK land use and soil carbon sequestration. *Land Use Policy* 265: S274–S283.

Parfitt, R.L., B.K.G. Theng, J.S. Whitton, and T.G. Shepherd. 1997. Effects of clay minerals and land use on organic matter pools. *Geoderma* 75: 1–12.

Peltzer, D.A., D.A. Wardle, V.J. Allison et al. 2010. Understanding ecosystem retrogression. *Ecological Monographs* 80(4): 509–529.

Potter, C.S., P.A. Matson, P.M. Vitousek, and E.A. Davidson. 1996. Process modeling of controls on nitrogen trace gas emissions from soils worldwide. *Journal of Geophysical Research* 101: 1361–1377.

Powlson, D.S., A. Bhogal, B.J. Chambers et al. 2012. The potential to increase soil carbon stocks through reduced tillage or organic material additions in England and Wales. A case study. *Agriculture, Ecosystems & Environment* 146: 23–33.

Powlson, D.S., A.P. Whitmore, and K.W.T. Goulding. 2011. Soil carbon sequestration to mitigate climate change: A critical re-examination to identify the true and the false. *European Journal of Soil Science* 62: 42–55.

Prapaspongsa, T., T.G. Poulsen, J.A. Hansen, and P. Christensen. 2010. Energy production, nutrient recovery and greenhouse gas emission potentials from integrated pig manure management systems. *Waste Management & Research* 28: 411–422.

Pratt, C., A.S. Walcroft, K.R. Tate et al. 2012. Biofiltration of methane emissions from a dairy farm effluent pond. *Agriculture, Ecosystems & Environment* 153: 33–39.

Pratt, C., J. Deslippe, and K.R. Tate. 2013. Testing a biofiliter cover design to mitigate dairy farm effluent pond methane emissions. *Environmental Science & Technology* 47: 526–532.

Raupach, M.R. 2011. Pinning down the land carbon sink. *Nature Climate Change* 1: 148–149.

Redman, R.S., Y.O. Kim, C.J.D.A. Woodward et al. 2011. Increased fitness of rice plants to abiotic stress via habitat adapted symbiosis: A strategy for mitigating impacts of climate change. *PLoS One* 6: 1–10.

Rockström, J., W.M. Steffen, K. Noone et al. 2009. A safe operating space for humanity. *Nature* 461: 472–475.

Sapart, C.J., G. Montell, M. Prokopiou et al. 2012. Natural and anthropogenic variations in methane sources during the past two millennia. *Nature* 490: 85–88.

Schuur, E.A.G., B.W. Abbott, W.B. Bowden et al. 2011. High risk of permafrost thaw. *Nature* 480: 32–33.

Sillen, W.M.A. and W.I.J. Dieleman. 2012. Effects of elevated CO_2 and N fertilization on plant and soil carbon pools of managed grasslands: A meta-analysis. *Biogeosciences* 9: 2247–2258.

Singh, B.K., K. Tate, N. Thomas, D. Ross, and J. Singh. 2011. Differential effect of afforestation on nitrogen-fixing and denitrifying communities and potential applications for nitrogen cycling. *Soil Biology & Biochemistry* 43: 1426–1433.

Smith, P., D. Martino, Z. Cai et al. 2008. Greenhouse gas mitigation in agriculture. *Philosophical Transactions of the Royal Society B* 363: 789–813.

Steinfeld, H. and P. Gerber. 2010. Livestock production and the global environment: Consume less or produce more? *Proceedings of the National Academy of Sciences* 107: 18237–18238.

Taghizadeh-Toosi, A., T.J. Clough, L.M. Condron, R.R. et al. 2011. Biochar incorporation into pasture soil suppresses *in situ* nitrous oxide emissions from ruminant urine patches. *Journal of Environmental Quality* 40: 468–476.

Taghizadeh-Toosi, A., T.J. Clough, R.R. Sherlock, and L.M. Condron. 2012. Biochar adsorbed ammonia is bioavailable. *Plant and Soil* 350: 57–69.

Takanezawa, K., K. Nishio, S. Kato, K. Hashimoto, and K. Watanabe. 2010. Factors affecting electric output from rice-paddy microbial fuel cells. *Bioscience, Biotechnology and Biochemistry* 74: 1271–1273.

Theng, B.K.G. 2012. *Formation and Properties of Clay-Polymer Complexes*. 2nd edition. Amsterdam: Elsevier.

Theng, B.K.G. and G. Yuan. 2008. Nanoparticles in the soil environment. *Elements* 4: 395–399.

UNEP 2005. Vital Climate Change Graphics. UNEP/GRID-Arendal, Norway.

Watson, C.J., R.J. Laughlin, and K.L. McGeough. 2009. Modification of nitrogen fertilisers using inhibitors: Opportunities and potentials for improving nitrogen efficiency. *Proceedings International Fertiliser Society* No. 658.

WinklerPrins, Antoinette M.G.A. 2014. Terra Preta: The mysterious soils of the Amazon. In *The Soil Underfoot*, eds. G.J. Churchman and E.R. Landa, 235–245. Boca Raton: CRC Press.

Worster, D. 2004. *Dust Bowl: The Southern Plains in the 1930s*. New York: Oxford University Press.

Yuan, G. and S.-I. Wada. 2012. Allophane and imogolite nanoparticles in soil and their environmental applications. In *Nature's Nanostructures*, eds. A.S. Barnard and H. Guo, 494–516. Singapore: Pan Stanford.

Yuan, G. 2010. Natural and modified nanomaterials for environmental applications. In: *Biomaterials for the Life Sciences, Vol. 7. Biomimetic and Bioinspired Nanomaterials*, ed. C. Kumar, 429–457. Weinheim: Wiley-VCH.

Zhang, D.D., H.F. Lee, C. Wang et al. 2011. The causality analysis of climate change and large-scale human crisis. *Proceedings of the National Academy of Sciences* 108: 17296–17301.

2 Soils and the Future of Food
Challenges and Opportunities for Feeding Nine Billion People

*Sharon J. Hall**

CONTENTS

In 2011, our planet welcomed the seven billionth human, born into a global society that is more long lived, technologically connected, and urban than any time in Earth's history. Depending on where she is raised, this child will experience a future of almost limitless possibilities, joined effortlessly to a stream of new choices and ideas—or she may strive simply for access to food, the most basic of human needs. But regardless of her geographic origin or socioeconomic status, it is very likely that she will live in a bustling city as she grows and will depend on a nearby market for sustenance, physically and mentally disconnected from the soil and natural resources that nourish her. The start of the twenty-first century buzzed with optimism, as scientific advances allowed humanity to consider a once-unthinkable future: urban skyscrapers that harbor vertical hydroponic farms, livestock fed on algae biofuel by-products, and meat cultured in vitro from living stem cells. However, despite the real promise and possibilities that these technologies provide, soil remains the single most important medium in which we grow food, now and for the foreseeable future. By 2045, when the seven billionth child reaches age 34, she will experience a world that is substantially more crowded, warmer, less biodiverse, and less forested than when she arrived, as she shares the soil's limited bounty with two billion new people who also dream of a long and happy life. Her charge then, as ours, will be to ensure that expansion and intensification of agricultural production will meet the needs of the future without compromising the very life support systems on which her children and grandchildren depend.

* School of Life Sciences, Arizona State University, Tempe, AZ, USA; Email: sharonjhall@asu.edu

> Caring for soil may seem like an old-fashioned concept to technologically minded city dwellers, but even in the information-rich 21st century, soil is still the most important resource on our planet for growing food.

Since the early part of the twentieth century, food production has kept pace with our exponentially growing population, due to significant advances in agricultural technology. These "green revolution" innovations included the development of synthetic fertilizers and pesticides, irrigation systems, and mechanized farm processes, along with monoculture planting of high-yielding, hybrid crop varieties that flourish with intensive nutrient and water supplements. Green revolution advances have miraculously allowed us to more than double grain yield with only a ~9% increase in land area over the last 50 years (Godfray et al. 2010). This accomplishment, one of civilization's most significant, was recognized by the international community with a Nobel Peace Prize in 1970 to Norman Borlaug, an American agronomist from Iowa. Yet, decades later, the real-world outcomes and unforeseen trade-offs of this important transition in world history have become abundantly clear. Despite current global production of enough food to feed every human ~2700 kcal per day, one in six children still remains hungry due to complex and interrelated social forces that create recalcitrant barriers to food security, one of humanity's most "wicked problems" (WHO 2012, Rittel and Webber 1973). Agriculture has transformed over one-third of Earth's ice-free land area, triggering losses of species in forests, grasslands, and wetlands (Foley et al. 2011)—which is by far the most significant and irreversible human signature on Earth. Our food system contributes over one-third annually of the human-produced greenhouse gas emissions that are changing our climate, double the amount generated by all transportation sources combined (in CO_2-equivalents; ITF 2010, Vermeulen et al. 2012). Copious fertilizer use—one of the very reasons the green revolution was a success in terms of food production—is now known to cause widespread pollution of rivers, lakes, and coastal ecosystems that support some of our most important fisheries (Smith 2003). And more than a thousand different types of agricultural pesticides are in circulation worldwide, some of which threaten the health of humans and other animals that share the planet with us, as these compounds help to combat insects and disease organisms that take advantage of hectares of identical plants in monoculture (WHO 2010).

To successfully produce food in a mere 30 years for nine billion people—30% more than we have today—global crop production will need to *double* from current rates (Figure 2.1, Matson 2011, Tilman et al. 2011). This disproportionate increase in per capita demand occurs in part because people eat more meat as they rise out of poverty, which we expect will occur, and because livestock production uses plant material inefficiently. However, most of the best arable land worldwide is already in production, and the rest supports important ecosystem services for humanity, such as clean air and water, biodiversity, and climate regulation (Clothier and Kirkham, 2014). To reach our production goal sustainably against the backdrop of an increasingly uncertain global climate, without accepting trade-offs between yield and human–environmental health, we will need to find the political will to adopt flexible, multiscale, and multidisciplinary solutions.

Several recent publications have discussed the numerous challenges we face to increase food security for a growing population in our current globalized economy (see Further Readings). Central to this discussion, yet less explored, is the importance of soil and ecological knowledge that underlies our path toward sustainable food production. In this chapter, we will explore key features of the plant–soil relationship that must shape any discussion of future agricultural potential. Then, we will explore the costly problems—particularly those emerging in the twenty-first century—that are created by conventional agricultural practices due to mismanagement and misunderstanding of this fundamental ecological relationship. Finally, we will discuss a suite of creative solutions that have been proposed to help achieve sustainable agricultural production under our new global reality of social and ecological uncertainty.

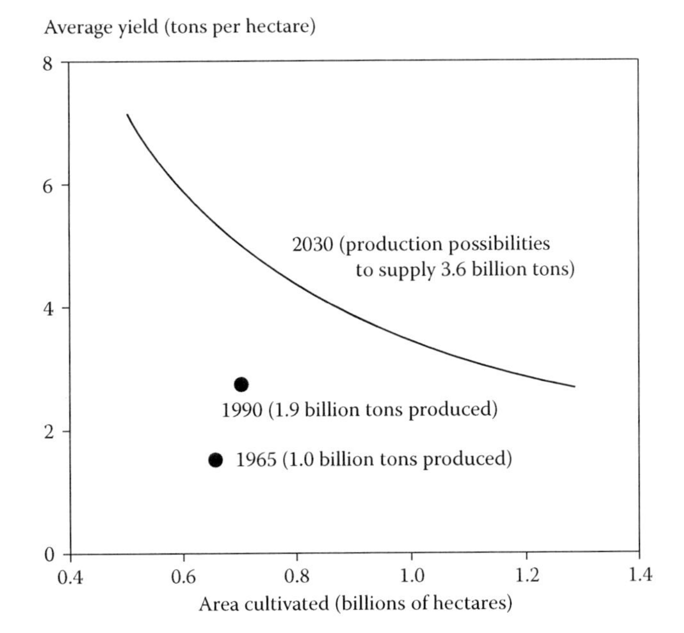

FIGURE 2.1 World production of cereal crops in the past and future. To supply enough food for a growing population, agriculture can either expand in area—threatening natural ecosystems and the life-supporting services they provide—or intensify on already cultivated land, leading to indirect problems caused by soil degradation and freshwater, fertilizer, and pesticide use. (From Matson, P.A. 2011. *Seeds of Sustainability: Lessons from the Birthplace of the Green Revolution in Agriculture.* Washington D.C.: Island Press.)

2.1 WHY ARE SOILS SO IMPORTANT FOR CROP GROWTH?

Less than 1% of human-consumed calories are harvested from the ocean or aquatic ecosystems, with the remaining 99% produced in soils on land (FAO 2009). As primary producers, plants require water and nutrients from the soil, along with sunlight and atmospheric carbon dioxide (CO_2) to make biomass that is consumed by all the animals on Earth, including humans and livestock raised for food. Because crops generally have plentiful CO_2 and sunlight, production is often limited by the health of their root systems and the resources that are taken up by roots through soil. These soil resources include freshwater and elements that are central to plant metabolism, including some nutrients that are required in relatively small quantities (calcium, magnesium, sulfur, and seven others), and some that are in very high demand due to their use in photosynthesis and other essential plant processes (nitrogen [N], phosphorus [P], and potassium [K]). In natural systems, a diversity of root structures from many different plant species inhabit any given patch of soil, competing for and collectively using scarce water and nutrients at a range of different depths and abilities with incredible efficiency. Conventional agricultural systems are not as efficient with resources, and crop plants depend on regular water and nutrient inputs—such as N, P, and K fertilizers—from farmers. However, even in the most sophisticated agricultural operations, whether in fields of corn in Mexico, soybeans in the Amazon, or rice in the Riverina of Australia, the retention and sustained availability of these limiting (and expensive) resources to plants is controlled by soil characteristics.

A handful of properties most often limit a soil's ability to provide belowground resources for plants, including the high-yielding, green revolution crop varieties that are most commonly in use today. The natural "parent material" of soil—the rocks or sediments from which the soil evolves—along with topography, the original vegetation, and climate over time (together, called the "soil-forming factors") generates very large differences in soil fertility at both local and landscape scales, primarily through their effects on the composition of materials that supply nutrients and retain them in a form that plants can use (Jenny 1941). These soil-forming factors also modify the acidity or alkalinity of soil water, which can affect beneficial soil microbes and lead to chemical reactions that render some nutrients unavailable for plant use. For example, some of the richest agricultural land in North America and central Asia is derived from geologically young, carbon-rich grassland soils that developed on fine, crushed minerals from former glaciers or dust (called "loess"). In contrast, highly weathered soils that underlie some forests in tropical areas such as Brazil or central Africa no longer contain large pools of mineral-derived nutrients from their parent materials, and the acidic clays that remain are either unable to hold many nutrients or are "sticky," binding the essential nutrient P into forms that plants cannot access. This natural heterogeneity of soils foreshadows the potential successes or problems of future crop growth—and indeed, has driven the rise and decline of civilizations and cultures (Montgomery 2007). Variation in soil properties leads to significant differences in agricultural production capacity that require quite different management strategies to produce high yields, whether across a farmer's field, a farming landscape, or beyond.

In addition to natural processes, farming activities themselves significantly affect a soil's ability to provide resources to crop plants. Approximately a third of all land area globally is considered degraded to some degree due to mismanagement, with ~80% of agricultural land suffering from significant soil erosion (Pimentel 2000, Bindraban et al. 2012, Kuhn 2014). This "yield penalty" from loss of soil productivity can be substantial, leading to tropical deforestation (from pressure to find more productive land), and it further sets us back from our mid-twenty-first century food production goals. For example, the yield penalty from soil degradation in China averaged 9% in 2005 and is projected to rise to 40% by 2050 with intensified "business-as-usual" practices (Ye and Van Ranst 2009). Some soils are considered degraded because they have become too salty ("saline") from intensive irrigation in dry climates, like in Western Australia or Iraq. Paradoxically, soil salinization from overirrigation *reduces* the ability of soil to provide water for plants because salt binds water—making it hard for roots to extract—and can clog soil pores that are required for water percolation. In fact, the decline of entire cities in ancient Mesopotamia from 2400–1700 BC is linked to diminishing yields due to soil salinity from long-term irrigation along the Tigris and Euphrates river floodplains (Jacobsen and Adams 1958).

Most often, however, soil degradation is caused by wind and water erosion or accelerated biological decomposition processes that lead to loss of soil organic matter (SOM), a critical property that has comprehensive benefits for plant growth. Composed of the decayed remains of roots and other living organisms, SOM binds minerals into granules that reduce erosion, promote air circulation that allows roots to breathe, and absorb and retain water at just the right tension for plant uptake. Carbon compounds in SOM create energy-rich havens for a diversity of soil animals that can keep belowground pests in check, and they prevent nutrients from being sequestered by soil minerals. Most importantly, carbon from SOM and roots together feed talented microbial communities that transform locked-up essential nutrients into plant-available form—similar to a "slow-release" fertilizer—initiating a close, mutually beneficial relationship between plants, soil animals, microorganisms, and the soil matrix in which they live. Common conventional practices such as continuous cultivation (e.g., no fallow), high-density livestock grazing, and tillage accelerate rates of net SOM removal through erosion and microbial organic matter decomposition, degrading the land's resource base. For example, on average about half the organic carbon in soils is lost after ~25 years of conventional cultivation in temperate climates such as North America and Europe, while the same amount can be lost in only 5 years in warm tropical regions such as in Southeast Asia (Matson et al. 1997). Restoration of degraded soils is difficult, as they become trapped in a self-reinforcing cycle of "desertification," in which poor water retention in degraded soils leads to drought, which accelerates soil erosion by wind and water and further decreases organic matter content. As an alternative, the principles of conservation agriculture promote management activities that accumulate SOM in order to take advantage of its extensive benefits on yield (FAO 2012b). For example, no-till farming reduces SOM decomposition, cover cropping prevents topsoil erosion, and rotation of multiple crop species with different root structures increases SOM supply. Soil restoration leads to comprehensive win–win outcomes, with benefits for yield and environmental services. For example, improvements in SOM content stimulate crop growth by supporting the plant–soil relationship, reduce airborne dust from soil erosion, and increase soil carbon sequestration to help offset agriculture's vast contribution to global climate change (Lal 2010).

Soils have supported the growth of plants for consumers for millions of years, developing a highly efficient, cyclical system composed of minerals, water, atmospheric compounds, soil animals, and microorganisms that feed (and are fed by) plant roots. Driven by the increasing demand for sustenance, humans cleverly found ways to speed this cycle by decoupling the tight plant–soil relationship through the development of a small group of extraordinarily productive crop varieties that are able to grow tremendous amounts of food with external pesticide, water, and nutrient inputs. However, our successful efforts to augment yields over the twentieth century are now known to cause major, unanticipated problems that are sacrificing future production for the present, limiting our ability to reach even higher output in the coming decades.

2.2 TOO MUCH, YET TOO LITTLE

Green revolution grain crops were developed to maximize aboveground yield, so these cultivars have relatively shallow root systems that perform poorly compared to their wild relatives when water or nutrients are in short supply (Chapin 1980). Moreover, high nutrient demand by fast-growing varieties depletes soils of essential resources, and intensive cultivation practices accelerate their losses. Consequently, to increase yield, most agricultural operations supplement their soils with water and substantial inputs of N, P, or K in an inorganic, "quick-release" form that is easy for plants to use. Some technologically sophisticated farms practice "precision" agricultural techniques to improve yield and reduce costly nutrient excess by carefully matching irrigation and fertilization to the exact time, place, and amount of nutrients that each plant needs. However, these advanced, high-tech practices have large up-front costs and require considerable expertise, and so may be profitable primarily with economies of scope and scale (Koch and Khosla 2003, Fernandez-Cornejo 2007). Instead, ~80% of farmland worldwide is not even irrigated (i.e., it is "rain-fed," Rockström et al. 2003), and if farmers have access to fertilizers, they apply them liberally as "insurance" to maximize the probability that they will get a good harvest (Good and Beatty 2011). Industrial agricultural practices often focus on production efficiency at the expense of nutrient-use efficiency within the plant–soil system, in part because fertilizers are commonly subsidized and thus represent a good investment for farmers—especially without the true costs of environmental degradation included. For example, maize crops in the United States receive on average ~200 kg of N/ha each year (and up to three times this rate in some areas of the world), although only about *half* that amount is used by the plants (Davidson et al. 2012). Similarly, low amounts of P fertilizers make it into crop harvests, averaging 50% of the P applied, and as little as 10% of the P added to fields ends up in food in some regions of China where inputs are among the highest in the world (Cordell et al. 2009, Ma et al. 2011). Fertilization practices that are not optimized to crop uptake and soil characteristics virtually ensure that nutrients will occur in excess of plant demand at some point across a farmer's field. This excess is particularly common when soils are planted in monoculture with genetically identical crop varieties that lack the efficient roots systems of wild, diverse plant communities.

What happens to all of these leftover nutrients? In many regions of the world, excess N and P is washed into streams, lakes, groundwater, or to the ocean, causing harmful algal growth and low oxygen concentrations (a process called "eutrophication"), which disrupts the health of fisheries and other aquatic ecosystem services. These losses are extremely costly, both environmentally and economically. For example, eutrophication in the U.S. freshwaters alone is estimated to cause losses of $2.2 billion per year (Dodds et al. 2009). Unused N in soil is also easily transformed into gases that affect local air quality and global climate. Ammonia (NH_3) and nitrogen oxide (NO_x) reactive gases are commonly emitted from livestock yards and agricultural fields as well as from combustion, contributing to acid rain and "nitrogen deposition," a process that unintentionally fertilizes downwind ecosystems from the atmosphere. For example, upward of 30 kg N/ha are deposited annually from the atmosphere to forests in some areas of northern Europe compared to a "preindustrial" rate of ~2 kg N/ha per year, causing soil acidification and other problems as ecosystems become "nitrogen saturated" (Dentener et al. 2006). Nitrous oxide (N_2O) is also emitted from soil and functions as a potent greenhouse gas with a global warming potential 300 times that of the usual suspect, CO_2. Recent research shows that the exponential increase in atmospheric N_2O concentration over the twentieth century is largely due to agricultural practices (Davidson 2009, Park et al. 2012). Together, leftover nutrients from agriculture—resulting primarily from the mismanagement of the naturally efficient plant–soil relationship—ultimately represent a major cost that society is left to bear. For example, the European Union estimates that the consequences of overabundant agricultural N cost €20–150 billion per year (~€40–350 per person annually), more than the net economic returns on yield that are gained by farmers from increasing fertilizer use (Sutton et al. 2011). Improvements in fertilizer management to more closely match crop needs will go a long way to reducing these costs while maintaining high yields for the future.

While nutrient excess in agriculture leads to serious environmental and economic problems, at the same time, farmers and the people they feed in many rapidly developing countries are suffering from just the opposite: *not enough*. This troubling mismatch exemplifies the real-world complexity that challenges our journey toward sustainable agricultural production for a rapidly increasing population. Worldwide, agricultural land is estimated to produce roughly 50% of its potential due in part to inadequate fertilizer access and use (Lobell et al. 2009). Disturbingly, this "crop yield gap" dips as low as 20% of potential production capacity in some of the world's most vulnerable communities, such as in tropical lowland Africa, where per capita production today is the same as it was in 1961 (Godfray et al. 2010). Vexing social barriers sustain the yield gap, such as lack of technical knowledge and adequate credit to purchase fertilizer, poor infrastructure, and numerous biophysical factors such as climate uncertainty, seed choice, pests, and soil degradation. In low-productivity locations such as sub-Saharan Africa, yields could increase immensely if fertilizers and seed varieties were simply accessible to most farmers. But in other cases, yields can be maintained and even increased while using much *less* fertilizer through changes in management to incorporate ecological knowledge of the plant–soil system (Matson et al. 1998, Chen et al. 2011).

2.3 EMERGING PROBLEMS THAT CHALLENGE THE LIMITS OF OUR CURRENT FOOD SYSTEM

While food production has soared since 1960, the consequences of our narrow, short-term focus on yield have increasingly degraded environmental quality and future agricultural potential. Yet, as demand rises, our ability to keep up is again challenged by a suite of new, complex problems that stem from a disjointed approach to our global agricultural system. The severity of these emerging challenges—including P scarcity, nonfood uses of agricultural land, and climate change—will ultimately force a "radical redesign" of our food system to prevent unnecessary human and environmental harm (Foresight 2011). Meeting this challenge will require a fundamental reconsideration of the merits and trade-offs of green revolution agricultural production practices in use today, in addition to development of realistic, systems-based alternatives.

2.3.1 Peak Phosphorus and the Geopolitics of Fertilizer Security

Unlike N fertilizers, which are most often synthesized from an unlimited source of atmospheric nitrogen gas (N_2, with a substantial energy cost), P fertilizer is currently mined from finite, naturally occurring phosphate rock deposits that occur in only a handful of countries. For example, ~80% of global deposits of phosphate rock are located in Morocco and Western Sahara (Cooper et al. 2011). At the same time, demand for P fertilizer is increasing at 2% per year due to population pressure, higher per capita consumption of meat (which requires proportionally more P to produce than plants), nutrient-inefficient crops, and an accelerating expansion of agriculture in tropical regions that have naturally P-poor soils. If we follow business-as-usual trends, this growth rate translates into only about 35 years until the consumption of phosphate fertilizer doubles worldwide, leading to projections of "peak phosphorus" and subsequent scarcity in the near future when our population needs it most (Cordell et al. 2009). Because there are no substitutes for P in living organisms, it is likely that P fertilizer scarcity (and thus higher prices) will encourage the development of conservation or recycling technologies. Yet, in advance, the geographic bottleneck of current P supply has raised international concern about the looming geopolitics of "fertilizer security" (Elser and White 2010). For example, Cooper and colleagues (2011) estimate that the phosphate rock reserves the United States and China—currently the top two producers in the world—will be used up in 50–60 years at current rates of extraction, leaving just 7 countries with recoverable phosphate rock resources by 2100. Indeed, global reliance on a few, limited sources of a key crop element unnecessarily complicates political and trade negotiations—and it is unsurprisingly reminiscent of current international and energy security issues created by humanity's oil addiction. If P scarcity drives up fertilizer and food prices, ultimately these effects will be felt most by the world's vulnerable populations, including smallholder farmers and the poor.

To meet the challenge of increasing agricultural production given the naturally finite P supply, we must fix the enormous leaks in the human food system. Only a fraction of P fertilizer makes it into the food that we eat, while P losses from agricultural soil erosion and livestock waste (~15 million tonnes per year globally) are roughly equal to the amount of P fertilizer added to arable land worldwide (~14 million tonnes per year; Cordell et al. 2009). Many countries with advanced wastewater systems are already removing P from their human waste streams, but to capture all that is lost will require a comprehensive system-level reform to expand our range of plausible options. These reforms include adoption of conservative nutrient management practices, concerted investment in P-capture biotechnology, or a change in dietary preferences (Elser 2012). For example, development of efficient P-scavenging crops could increase nutrient use efficiency and potentially reduce soil P losses (see Section 2.4.2). Additionally, new chemical and biological technologies may make it easier and more profitable to recover P from human and livestock organic waste streams for reuse. These methods include, among others, adding hydrated lime (calcium hydroxide) or forming struvite (magnesium ammonium phosphate) crystals in wastewater, and cultivating nutrient-scavenging algae that can be used as both fertilizer and biofuel (Jaffer et al. 2002, Vanotti and Szogi 2009, Cai et al. 2013).

2.3.2 Competing Uses of Prime Agricultural Land

Given our global food production pressures, humanity is now faced with the daunting choice of bringing more land into production or intensifying on current land to meet the needs of the future. Yet, in some ways, we are slipping backward, as some of our most productive land is lost not just to soil degradation, but also to competing nonfood uses, including urban development and, most recently, biofuels. The steady pull of cities has long provoked concern about rural land cover and land use change, but the resource demands of nine billion have stimulated this controversy anew. Now home to more than half of the world's people, cities cover a small fraction—less than 3%—of Earth's terrestrial surface. But the urban land footprint is expanding fast, at ~5% per year (equivalent to a 14-year doubling time), even outpacing the rapid rate of urban population growth (Seto et al. 2011, UNPD 2012). This expansion is disproportionately affecting our best farmland, as human settlements are generally situated near the most fertile soils, as they have been for millennia (Mumford 1956). For example, the rich valley soils of California supply over half the fruits and vegetables consumed in the United States, yet nearly three-quarters of the urban growth in California since 1990 has occurred on prime agricultural soil (Thompson 2007, USDA 2012).

Compounding this progressive loss to urban development, a new threat has developed over the last decade stemming from legitimate worries about national energy security and climate change. Biofuels are "renewable" products derived from plants that are grown for combustion rather than consumption. The commercially viable biofuels in production today are made from sugar, starch, and oil that come from food crops such as corn, sugarcane, and oil palm (called "first-generation biofuels"). Advanced second-generation fuels are derived from nonfood sources such as the woody or structural material of plants (e.g., trees, crop residues, and various grasses, such as switchgrass and *Miscanthus*), and third-generation biofuels come from algae, but most of these products are still in development. World annual production of biofuel from food crops has exploded over the last 10 years, increasing sixfold between 2000 and 2011 primarily due to government mandates in the United States, European Union, and Brazil that were intended to promote energy security, support rural economies, and reduce greenhouse gas emissions (BP 2012). As a result, biofuel production now uses ~17.5 million ha of farmland in the United States and Canada, ~8 million ha in Europe, and ~6 million ha in Latin America (UNEP 2009). Even by the most conservative projections, the total area used for biofuel will need to *double* to reach government production mandates by 2030, which is an additional area of agricultural land roughly equivalent in size to Germany. In total, the social and environmental effects of biofuel production have been extraordinary, creating far-reaching ripples even beyond the food system. For example, ~30% of the U.S. corn crop went to ethanol production in 2012 instead of food (after accounting for coproduction of livestock feed; USDA 2013), and recent analyses show that fuel production from some crops actually *increases* greenhouse gas emissions relative to fossil fuels when considering the new land that is cleared (Searchinger et al. 2008). Moreover, production of first-generation biofuels has led to real human and environmental consequences. Current biofuel policies have contributed to international food price spikes and volatility, multimillion dollar losses in pest biocontrol benefits due to landscape simplification, and intensified deforestation in some of the world's most biodiverse and ecologically sensitive tropical areas (Laurance 2007, Landis et al. 2008, Wilcove and Koh 2010, Naylor 2012).

2.3.3 Climate Change and Agricultural Production

Among the numerous emerging threats to agriculture, climate change is the most serious due to its comprehensive effects on the entirety of the food system at local to global scales. Increased variability in patterns of rainfall and temperature is one of the primary expectations for our future climate due to the continued rise in greenhouse gas concentrations in the atmosphere. This change is expected to affect not just food quantity but also quality, storage, delivery, and food safety (IPCC 2007, Vermeulen et al. 2012). Agriculture developed as a part of human civilization during the last 10,000 years of relative climate stability because, at its most basic level, long-term crop success depends on management practices that carefully match the timing of planting, harvest, and other activities with water availability and the length of the frost-free growing season (Bettinger et al. 2009). Yet, even as technologically sophisticated as modern agriculture is today, with irrigation infrastructure and greenhouses in many parts of the world, climate variability will transform the production landscape because food and feed markets are deeply interconnected—hydrologically, ecologically, and economically—across geography and time (Naylor 2012). Some land area will inevitably become better for crop production as temperatures warm and CO_2 concentrations rise, including temperate regions such as high-latitude and high-altitude regions of Europe (Olesen et al. 2007). Other areas in the tropics will become too hot or dry for crop production, leading to a projected 10–50% yield reduction in major grain crops in places that are already vulnerable to food insecurity, such as sub-Saharan Africa and South Asia (Nelson et al. 2009). Overall, without the development of major adaptation strategies, climate change is expected to significantly affect worldwide food security due to the direct effects of temperature and water stress combined with less predictable, indirect effects of agricultural pests, rising sea level, more frequent extreme events (e.g., storms and drought), and exacerbation of yet unsolved problems such as soil degradation and nutrient pollution (Rosenzweig et al. 2001, Vermeulen et al. 2012).

As countries become more aware of the severity of current and forthcoming climate change problems, the push toward adaptation is no longer an *if*, but rather a *how*. The World Bank estimates that $70 billion to $100 billion per year will be required to buffer agricultural systems against a 2°C change in global mean temperature by 2050, although the probability of a much larger change by 2100, ~3–4°C, is becoming increasingly likely (World Bank 2010, 2012).

2.4 DIVERSE SOLUTIONS FOR GROWING POPULATION AND RAPIDLY CHANGING PLANET

Feeding nine billion people by 2045 is, alone, an enormous challenge for humankind, given that Earth's soil and other finite resources also provide critical ecosystem services—clean water and air, climate regulation, waste detoxification, and biodiversity—that underpin the sustainability of human life. This challenge of the next 30 years becomes much more difficult with the addition of numerous environmental problems that are hindering our journey forward. Some of these problems have been truly unforeseen, particularly by the nations and communities whose attentions have been focused on providing basic human services. But most of these problems result from billions of independent decisions aggregated by civic institutions that fail to grasp the ecological and economic connections that bind our species together. While the challenges are great, the consequences of doing nothing amount to a global experiment that many people are simply not willing to carry out. We can and must do better.

The sustainable path forward will require a foundational shift from our current, narrow approach to problem solving, in which we develop single-solution answers to individual problems as they arise in separate places. Instead, we must work across disciplines to develop systems-based solutions based on knowledge of the integrated ecological, political, and economic connections and feedbacks within our food system. Numerous opportunities exist to meet our growing agricultural challenge, both technological and organic, that draw upon human ingenuity and the best ideas from the natural ecological systems of which we are a part. These possibilities include ways to increase agricultural efficiency across two fronts—efficiency of production on current land, and efficiency of nutrient capture by crop plants. Further gains can be made through changes in consumer demand by using incentives that target the drivers of human decisions. Additionally, all parts of the supply chain—from production activities to distribution, infrastructure, markets, and waste—must adopt principles of resilience to buffer the threats of climate variability using successful examples from traditional cultures and natural ecosystems. Finally—although beyond the scope of this chapter—to realistically transform promise into practice, all solutions must be designed in tandem with programs to address the intractable yet pervasive economic, political, and social barriers that can prevent knowledge transfer and change. Over the long term, the success of these proposed solutions will ultimately require an advancement in our understanding of agriculture as a complex system—an agro-ecosystem—composed of interconnected biophysical and social components that interact with one another across scales. Indeed, the thriving field of "agroecology" uses ecological concepts and a whole-systems approach to advance a triple goal of sustainable production, healthy environments, and viable food and farming communities (Gliessman 2013).

2.4.1 Increase the Efficiency of Production on Current Land

Despite the incredible gains in average per hectare production over the last 50 years, global analyses of crop yield gaps and soil degradation suggest that we have not yet reached the limits of how much food can be grown on the land area that is already in cultivation. Through systematic efforts to restore soil productivity and improve agricultural management, it is estimated that we can produce an additional 30–60% of food on the land used today even with current technology, without clearing additional, mostly forested, land that provides indispensable environmental services (Foley et al.

2011). Approximately 13 million ha of biodiverse, carbon-rich tropical forests are currently lost to agricultural expansion every year, but these lands produce on average half of the crop yields of farmland in temperate regions and lead to twice the carbon emissions when cleared (West et al. 2010, FAO 2012a). With appropriate economic and political tools and a whole system-based approach, this productivity could be replaced with restoration of the roughly 10 million ha of agricultural soils that are degraded and abandoned annually, with additional economic benefits from ameliorating environmental quality (Pimentel 2006). A systems perspective may help decision makers to understand the value of protecting the most productive agricultural soils within our nations for food, and allocating lands that are naturally marginal for agriculture instead to important nonfood uses such as urban development, waste disposal, or ecological reserves.

Improvements in land-use efficiency can also lead to win–win outcomes for yield and environmental quality at the field level, profiting individual farmers. Some of these strategies lead to long-term benefits but may require a substantial revision of farm operations, including well-tested conservation agriculture (or agroecological) practices such as crop diversification. For example, instead of planting fields in monoculture, different type of crops can be grown together in polyculture as in natural systems, such as herbaceous crops with fruit trees. Diverse communities of plants create a vast network of roots in the soil that captures water and nutrients at different depths and different times of the growing season, increasing the efficiency of resource use across fields and landscapes, and augmenting soil fertility. Some of these strategies have been used to maintain soil health for thousands of years, such as crop rotation (planting different species at different times in one field) or cover cropping with nitrogen-fixing legumes, because they maximize long-term yield by limiting soil erosion, nutrient losses, and pest problems (Davis et al. 2012). Furthermore, any management practice that increases SOM (e.g., reduced tillage, manure and crop residue inputs) will broadly benefit plant growth by enhancing soil water-holding capacity, aeration, and nutrient supply.

Using a different approach, a suite of other proposed solutions can yield major benefits by simply modifying practices that are currently used by farmers, such as changing the timing and use of fertilizers to coincide more closely with plant demands, improving husbandry practices within mixed crop-livestock farms, and—importantly—creating incentives within the local socioeconomic system that allow farmers to profit directly from ecologically sound choices (Matson et al. 1998, Herrero et al. 2010). For example, field trials in China showed that maize yields can be *doubled* with no increase in N fertilizer use through relatively straightforward agronomic modifications, such as increasing planting density and splitting fertilizer applications across the growing season (Chen et al. 2011). Other new programs across India are training community members to help small farmers improve livestock nutrition, leading to higher production of better-quality milk per kg of feed, improved animal reproductive efficiency, higher immunity to disease, 10–15% lower emissions of enteric methane (a potent greenhouse gas), and a ~10% increase in daily monetary profits (Garg 2012). With removal of economic and sociopolitical barriers to implementation, clear recommendations such as these can boost yield and income while improving environmental quality, and are applicable to both large agribusinesses as well as smallholder famers that supply the bulk of food in developing countries.

2.4.2 Increase the Efficiency of Resource Use by Plants

Although the role of technology in food production has become increasingly polarized in the public sphere, both agroecologists and biotechnologists alike agree that the majority of plants that are currently used for food and livestock feed are not nearly as efficient with soil nutrients and water as they could or should be, especially under the threat of climate variability. As a result, one of the most active areas of agronomic research is focused on improving the ability of crops to withstand drought, pests, and low soil fertility using a suite of methods, including traditional breeding and transgenic technology. Humans have bred, or "domesticated," plants since the beginning of agriculture by selecting and propagating individuals with desirable traits, such as bigger fruits or more tasty and nutritious edible material. Indeed, most of the crops we eat today—even those certified as "organic"—began their journey to our plates after artificial selection by humans long ago (Cox 2009). Following the discoveries by Gregor Mendel in the late nineteenth century, scientists began interbreeding distant plant relatives into new "hybrid" varieties with characteristics that were better for cultivation, eventually resulting in the high-yielding wheat, maize, and rice cultivars that helped bring the human population to its size today. Techniques in modern plant breeding have progressed over the last several decades, and now include the use of genetic markers to rapidly screen populations of plants for beneficial characteristics (called "marker-assisted selection"), and adding or removing DNA (commonly called transgenic technology or genetic engineering) to develop favorable traits, such as herbicide- and pest-resistance, or improved root growth.

Globally, one of the highest priority goals for crop science programs is to increase belowground resource-use efficiency of livestock-feed plants and species such as wheat, maize, and rice, because grains supply most of the calories consumed by people worldwide. For example, new genetically engineered varieties of rice are in development that produce nearly twice the root mass compared to conventional varieties under low-phosphorus conditions, due to an inserted gene that enhances root proliferation and activity (Figure 2.2a; Yang et al. 2007). This technology, and others like it, could increase the productivity of crops in dry conditions or on marginal lands. Other avenues of research are using traditional (nontransgenic) breeding methods to develop perennial grain and grass crops as alternatives to the annuals that make up our current crop palette (Glover et al. 2010). For example, researchers at The Land Institute in Kansas (USA) are developing a perennial wheatgrass

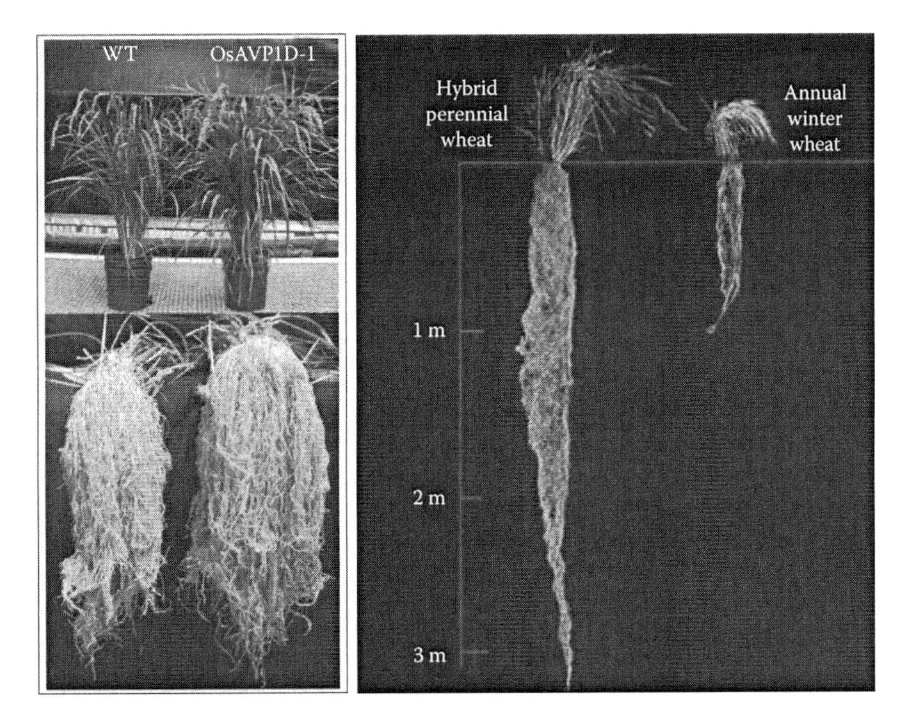

FIGURE 2.2 (See color insert.) Root systems of conventional and nutrient-efficient crop plants. (a) (Left) Transgenic rice plants (right-most plant in panel, labeled OsAVP1D-1) have larger root systems that use phosphorus more efficiently than conventional rice cultivars (left-most plant in panel, labeled WT). (Adapted from Yang, H. et al. 2007. *Plant Biotechnology Journal* 5:735–745.) (b) (Right) Root systems of perennial wheat (intermediate wheatgrass) (left-most plant in panel) have much longer and more efficient root systems than conventional annual wheat (right-most plant in panel). (Adapted from The Land Institute. 2013. http://www.landinstitute.org/vnews/display.v/SEC/Press%20Room (accessed February 2013).)

(*Thinopyrum intermedium*) to produce higher seed yield (Figure 2.2b). Perennial plants produce food over multiple years using tremendous root systems. Owing to their large investment in belowground biomass, plant species with a perennial life strategy confer numerous benefits to agricultural soils—particularly degraded or low-fertility soils—by increasing SOM content, reducing erosion and nutrient losses, and stimulating the growth of helpful microorganisms. Although research efforts are progressing quickly, much work is still needed to develop high-yielding perennial and annual crops with more effective root systems than current varieties, as well as to evaluate the short- and long-term environmental costs and benefits of these varieties in the field (Pimentel et al. 2012).

The phrase "genetically modified" food (i.e., GM food, or GMO for genetically modified organism) refers to transgenic technology specifically but has resulted in controversy and confusion. In fact, people have modified all domesticated crops genetically from their original progenitors, because humans intentionally selected and bred plants that contained genes for particular traits. Furthermore, genetic engineering underlies many medical advances that have broad public acceptance, such as transgenic bacteria that produce human insulin, or gene therapy (Lewis 1998). Beneath this unfortunate misunderstanding of the science, however, is significant public concern and mistrust about the commercialization and domination of food biotechnology by the private sector in rich, industrialized countries as well as the simultaneous decline in funding for public and nonprofit plant breeding programs (Conway 2000, Morris et al. 2006). These political and ethical disagreements have been exacerbated by some early, poor choices made by the biotech industry to protect intellectual property by engineering "terminator genes" into patented crops that prevent farmers from using seed for the next generation of planting as they have done for millennia. Since then, mistrust on both sides has deepened as large corporations face off against nonprofit environmental organizations over GM food labeling legislation, and members of both groups attempt to discredit legitimate scientific research on plant molecular biology and environmental consequences of transgenic crops (Sarewitz 2004). Indeed, complex ethical issues surround intellectual property and biotechnology (i.e., patents on "living" material), and these become especially contentious around public health issues such as environmental quality or seed provisioning for the world's poorest farmers. To reach beyond this impasse to advance food security will require continued efforts among ecologists and plant biologists to increase public literacy about genetics and the environment, and among government agencies to minimize risks through development of strong, trustworthy biosafety programs. But it will also require increased cooperation and joint research and training between the private sector, which can encourage and speed innovation, and public/nonprofit plant breeding programs, with firm commitments from all parties to develop affordable and ecologically sustainable options to farmers and their communities over the coming few decades (Brummer et al. 2011).

2.4.3 Incentivize Change in Consumer Demand

To produce more food more efficiently, it is clear that we need to close the crop yield gap—the difference between current and potential yield on a piece of land—which is particularly large in developing countries where population is expected to increase most rapidly. However, global analyses have also revealed a significant "diet gap" that exists primarily in developed countries, defined as the difference between intrinsic calories produced if all crops were eaten by people, and delivered food calories that are actually available for consumption (accounting for the material used for animal feed and other uses; Figure 2.3). If the diet gap was closed—for example, if all people became vegetarian—an additional 50% of food calories would be available for consumption every year (Foley 2011). Even under a more realistic scenario, such as a smaller but widespread shift in meat preference, food availability and environmental quality could increase substantially because different animals convert feed into protein with different efficiencies depending on what and where they are fed. For example, production of 1 kg of grain-fed beef requires on average twice as much feed as 1 kg of pork and 5–6 times as much as 1 kg of chicken (Pimentel and Pimentel 2003). Furthermore, beef production leads to ~4 times the global warming potential as these other meat options (De

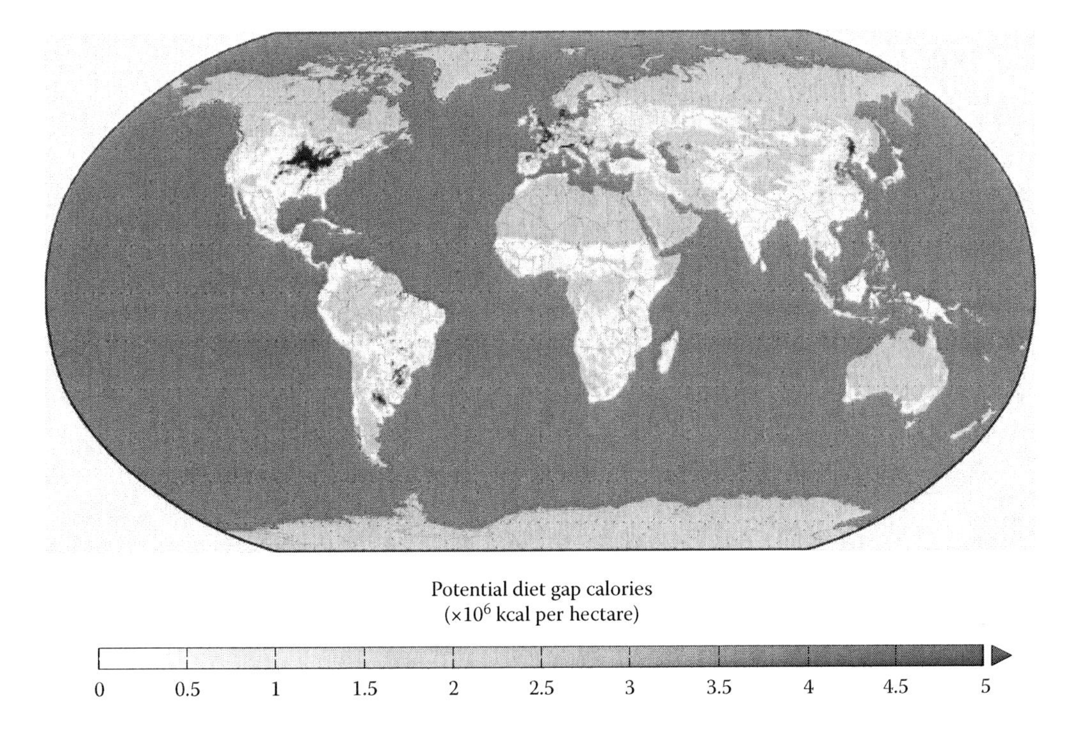

FIGURE 2.3 **(See color insert.)** The global "diet gap" for 16 of the most important staple crops. This analysis shows how much more food energy could be directly available to people if crop production was not allocated first to livestock or biofuels. [Calculated as the difference between the total available calories from crops (if all calories were consumed by people) and the actual, delivered food calories to people (based on allocation of crop energy to food, animal feed, and other uses)]. (Light gray color is area that is not used in this analysis, i.e., land that is not used for the 16 staple crops studied by Foley et al. 2011). (The scale for potential diet gap is applicable to the color version only). (Adapted from Foley, J.A. 2011. *Nature* 478:337–342.)

Vries and De Boer 2010). At a finer scale, the type of livestock system matters substantially to these statistics, as mixed crop-livestock farms, and grazing on low-productivity land leads to a net gain in available food calories, whereas use of high-productivity cropland for animal feed leads to a net calorie loss (Foley et al. 2011). These analyses highlight the importance of food choices and the difficult trade-off between consumption of animal-based products, which increases proportionally as people become wealthier, and long-term human–environment sustainability.

How do we change consumer demand? Centuries of research by both traditional farmers and scientists have given us the tools to produce food for seven billion people, and the awareness of the extensive environmental consequences. Despite these scientific advances, agricultural and food system sustainability remain elusive due to the complexity of forces that drive human behavior. For example, even though most of us know that regular exercise and a balanced diet are good for our health, on average half of the citizens in the industrialized countries of the OECD are now considered overweight (OECD 2010). If there is one thing we have learned from this obesity crisis, it is that simple knowledge is not enough. We may know that our food choices are unhealthy or drive environmentally costly agricultural practices, but education or even concern about these far-off or far-away outcomes does not necessarily change our consumption patterns because our decisions are also influenced by powerful economic, cultural, psychological, and institutional forces (Moran 2006, Frisk and Larson 2011). As a result, changing consumer demand will require a variety of different incentives. Market-based options that can incentivize sustainable food choices include government subsidies to producers for adopting forward-thinking, whole-systems activities (e.g., conservation agricultural practices, or conservation of prime agricultural land), and permit trading systems combined with monitoring to efficiently internalize problematic externalities (e.g., cap-and-trade of agricultural nutrient pollutants). Other market-based solutions include comprehensive labeling and third-party certification programs that, along with other incentives, help people to make informed yet ultimately private choices (e.g., USDA Organic certification or Forest Stewardship Council certification; NRC 2010). Consumer demand can also be influenced by formal measures, such as laws or ordinances within cities or districts, and informal guidance, such as by social norms that play surprisingly large roles in shaping people's preferences and choices (Ostrom 2000). Finally, strategies to change consumer demand may work most effectively when they are community-based and participatory, in which technical experts, farmers, and their families are treated as equals in decisions that lead to tangible, private benefits.

2.4.4 Increase System Resilience to Prepare for an Uncertain Future

Although climate models are becoming more sophisticated each year, considerable uncertainty about future climate remains due to the complexity of interconnections and feedbacks between the biosphere, atmosphere, and social systems, particularly at the small spatial and temporal scales relevant to agriculture. Furthermore, changes in global mean temperature or rainfall patterns may lead to numerous, indirect consequences for food production that are difficult to predict. As a result, climate change adaptation plans are in development, or at least being discussed to varying degrees, within agricultural agencies across the world, including those in the European Union and United States, India, China, South Africa, the United Nations, and CGIAR (the Consultative Group on International Agricultural Research), among many others. Significant biophysical, economic, and sociocultural differences across the globe make it clear that a one-size-fits-all solution will not be effective. Nevertheless, all food systems—including those within villages, cities, nations, and across international markets—must become more resilient to economic and environmental shocks to adequately prepare for climate uncertainty.

The concept of resilience for both people and nature is defined as the capacity of a system to reorganize and rebound to its original state after a disturbance (Walker et al. 2004). Natural ecosystems that bounce back from extreme weather events such as droughts or hurricanes contain many different species and landscape types with connected functions (Tilman et al. 2001, Gunderson 2010). Similarly, some traditional societies have long understood how to maintain food production amid climate uncertainty, such as traditional Zuni farmers in the semiarid Southwestern United States who continue to grow crops on rain-fed, runoff fields with unpredictable rainfall as they have for a thousand years (Sandor et al. 2007). Central to traditional agricultural strategies are diversification and some level of duplication among food types, crop genetics, and planting location, and inclusion of responsive management practices that mitigate environmental variability (Altieri 2004). Across a range of social and natural settings, these examples and others have shown us that the most resilient systems contain diverse, connected components; their important functions are insured with some redundancy; and they are flexible and able to respond to rapid feedbacks (Biggs et al. 2012).

What do these characteristics mean for resilience in modern agriculture? One of our most important priorities should be to encourage and build diversity into the way we produce food at all scales, from the level of the genome to the plant–soil system, and across fields, landscapes, livelihood strategies, and governance structures (Lin 2011). Of the 7000 different food plants that are in cultivation, just five of these (wheat, rice, maize, potatoes, and cassava) provide nearly 50% of the total calories consumed by our population globally, and 90% of cropland is planted with just ~100 species (FAO 2009, Tomich et al. 2011). Furthermore, many of these important crops are planted in monoculture, resulting in landscape simplification, rather than diversity, across large areas (Meehan et al. 2011). Agroecological strategies that promote crop, field, and landscape diversity to conserve nutrients and increase land use efficiency also create a redundancy of ecological functions within food production systems, providing insurance for farmers against risk. Moreover, restoration of SOM using conservation agriculture techniques promotes feedbacks within the plant–soil relationship that encourages root health, and buffers crops against drought and erosion (Lal et al. 2011). We could also diversify our food system by investing in the development of other nonsoil agricultural options, although currently these sources produce a much smaller fraction of our global food supply. For example, hydroponic technology (growing plants in a nutrient solution) may be well adapted for specialty crops in urban areas, and aquaculture (i.e., fish and shellfish farming), if sustainably managed, can provide another, redundant source of protein (in addition to livestock) for people (Naylor et al. 2000). Diversification in modes of production can only enhance resilience, however, if important changes are also made in the social and governance structures that underlie food systems. Among the most important change globally will be to increase the efficiency by which knowledge is transferred to rural farmers and communities through locally optimized, participatory research and training programs (i.e., building connectivity). These connections can be facilitated further through expansion of trade networks and development of flexible institutions and infrastructure to buffer climatically vulnerable regions.

2.5 THE WAY FORWARD

In the 1960s, Norman Borlaug and others recognized that one of humanity's greatest challenges was to feed our rapidly growing population to avert worldwide famine. The green revolution resulted from this challenge, and it quickly increased global agricultural output due to research and dissemination by other leading scientists and agronomists from around the world. In the early part of our new century, however, the central challenge has shifted. Our goal is no longer simply to feed a growing population, but to do so in sustainable ways that meet current and near-term needs without sacrificing future productivity and the essential ecosystem services on which humanity depends.

To feed nine billion people with Earth's finite resources, we must draw upon the infinite capacity of humans to learn and adapt. We now have the technology to connect citizens in all regions of Earth to images of our collective activities from space, and we can connect to each other with virtual conversations in real time, regardless of our location. We must use this new talent to transition to a whole systems-based perspective of agricultural production—an "evergreen revolution"—as advocated by M.S. Swaminathan, a plant geneticist and colleague of Borlaug who brought the green revolution to Asia in the late 1960s (Swaminathan 2006). As diversity in agricultural systems can promote resilience and sustainability of food production, a diversity of approaches forms the core of this vision, including traditional knowledge, participatory research with farming communities, and modern science. Such an agroecological approach appears to recall a previous era when farmers understood the opportunities and limitations of the plant–soil relationship or risked failure. Yet the foundations of an evergreen revolution will require the best knowledge from the past, present, and indeed the future, to develop solutions that are resilient to environmental change and locally optimized to work across geography and generations.

> It is important to harness all the tools that traditional wisdom and contemporary science can offer in order to usher in an era of bio-happiness. The first requirement for bio-happiness is nutrition and water security for all and forever.
>
> **M.S. Swaminathan**
> *In Search of Biohappiness: Biodiversity and Food, Health and Livelihood Security (2011)*

ACKNOWLEDGMENTS

SJH is grateful to P.A. Matson and the Blaustein Fellowship Program at Stanford University for their support while writing this chapter. Thanks to Alison Flanagan and John Watson for comments on an earlier draft of this work, and to Roz Naylor and Peter Vitousek for stimulating discussions and ideas.

FURTHER READING

1. *Science Magazine* special online collection on Food Security (2010). These papers and multimedia content are freely available at http://www.sciencemag.org/site/special/foodsecurity/.
2. A special feature of the international journal, *Ecology and Society*, devoted to sustainable agriculture, titled: A social-ecological analysis of diversified farming systems: Benefits, costs, obstacles, and enabling policy frameworks, edited by C. Kremen, A. Iles, and C. Bacon. Freely available online at http://www.ecologyandsociety.org.
3. Three excellent papers that summarize the current challenges and opportunities for agricultural sustainability: (a) Jonathan Foley and others, 2011. Solutions for a cultivated planet. *Nature* 478:337–342. (b) David Tilman and others, 2002. Agricultural sustainability and intensive production practices. *Nature* 418, no 6898:671–677. (c) Rattan Lal, 2010. Managing soils and ecosystems for mitigating anthropogenic carbon emissions and advancing global food security. *Bioscience* 60, no 9:708–721.
4. Two books on agroecology, the study of agricultural ecosystems: *Agroecology: Ecological Processes in Sustainable Agriculture* (2000), by Stephen R. Gliessman, Eric Engles, and Robin Krieger; and *Agroecology: The Science of Sustainable Agriculture* (1995), by Miguel A. Altieri (both books also available in Spanish).
5. Two lively articles on different sides of food politics and our agricultural future: Paul Collier, 2008. The politics of hunger how illusion and greed fan the food crisis. *Foreign Affairs* 87, no 6:67–79; and Vandana Shiva, 2004. The future of food: Countering globalisation and recolonisation of Indian agriculture. *Futures* 36, no 6–7:715–732.

REFERENCES

Altieri, M.A. 2004. Linking ecologists and traditional farmers in the search for sustainable agriculture. *Frontiers in Ecology and the Environment* 2:35–42.

Bettinger, R., P. Richerson, and R. Boyd. 2009. Constraints on the development of agriculture. *Current Anthropology* 50:627–631.

Biggs, R., M. Schlüter, D. Biggs et al. 2012. Toward principles for enhancing the resilience of ecosystem services. *Annual Review of Environment and Resources* 37:421–448.

Bindraban, P.S., M. van der Velde, L. Ye et al. 2012. Assessing the impact of soil degradation on food production. *Current Opinion in Environmental Sustainability* 4:478–488.

BP. 2012. *BP Statistical Review of World Energy*. London: BP.

Brummer, E.C., W.T. Barber, S.M. Collier et al. 2011. Plant breeding for harmony between agriculture and the environment. *Frontiers in Ecology and the Environment*, doi:10.1890/100225.

Cai, T., S.Y. Park, and Y.B. Li. 2013. Nutrient recovery from wastewater streams by microalgae: Status and prospects. *Renewable & Sustainable Energy Reviews* 19:360–369.

Chapin, F.S. 1980. Mineral nutrition of wild plants. *Annual Review of Ecology and Systematics* 11:293–296.

Chen, X.-P., Z.-L. Cui, P.M. Vitousek et al. 2011. Integrated soil-crop system management for food security. *Proceedings of the National Academy of Sciences of the United States of America* 108:6399–6404.

Clothier, B. and M.B. Kirkham. 2014. Soil: Natural capital supplying valuable ecosystem services. In *The Soil Underfoot*, eds. G.J. Churchman and E.R. Landa. 135–149, Boca Raton: CRC Press.

Conway, G. 2000. Genetically modified crops: Risks and promise. *Conservation Ecology* 4:2 [online].

Cooper, J., R. Lombardi, D. Boardman, and C. Carliell-Marquet. 2011. The future distribution and production of global phosphate rock reserves. *Resources, Conservation and Recycling* 57:78–86.

Cordell, D., J.O. Drangert, and S. White. 2009. The story of phosphorus: Global food security and food for thought. *Global Environmental Change—Human and Policy Dimensions* 19:292–305.

Cox, T.S. 2009. Crop domestication and the first plant breeders. In *Plant Breeding and Farmer Participation,* eds. S. Ceccarelli, E.P. Guimar, and E. Weltizien. 1026. Rome: U.N. Food and Agriculture Organization.

Davidson, E.A. 2009. The contribution of manure and fertilizer nitrogen to atmospheric nitrous oxide since 1860. *Nature Geoscience* 2:659–662.

Davidson, E.A., M.B. David, J.N. Galloway et al. 2012. Excess nitrogen in the U.S. environment: Trends, risks, and solutions. *Issues in Ecology*, Volume 15, Ecological Society of America.

Davis, A.S., J.D. Hill, C.A. Chase et al. 2012. Increasing cropping system diversity balances productivity, profitability and environmental health. *PLoS ONE* 7: e47149.

De Vries, M. and I.J.M. De Boer. 2010. Comparing environmental impacts for livestock products: A review of life cycle assessments. *Livestock Science* 128:1–11.

Dentener, F., J. Drevet, J.F. Lamarque et al. 2006. Nitrogen and sulfur deposition on regional and global scales: A multimodel evaluation. *Global Biogeochemical Cycles* 20:21.

Dodds, W.K., W.W. Bouska, J.L. Eitzmann et al. 2009. Eutrophication of U.S. freshwaters: Analysis of potential economic damages. *Environmental Science and Technology* 43:12–19.

Elser, J. and S. White. 2010. Peak phosphorus. *Foreign Policy Magazine* Online, April 22 2010.

Elser, J.J. 2012. Phosphorus: A limiting nutrient for humanity? *Current Opinion in Biotechnology* 23:833–838.

FAO. 2009. FAOSTAT, Food Balance Sheets. Food and Agricultural Organization, United Nations. http://faostat.fao.org (accessed January 2013).

FAO. 2012a. *State of the World's Forests 2012.* Food and Agricultural Organization of the United Nations, Rome, Italy.

FAO. 2012b. *Conservation Agriculture.* Food and Agricultural Organization of the United Nations, http://www.fao.org/ag/ca/1a.html (accessed January 2013).

Fernandez-Cornejo, J. 2007. *Off-Farm Income, Technology Adoption, and Farm Economic Performance.* USDA Economic Research Report Number 36.

Foley, J.A. 2011. Can we feed the world and sustain the planet? *Scientific American* 305:60–65.

Foley, J.A., N. Ramankutty, K.A. Brauman et al. 2011. Solutions for a cultivated planet. *Nature* 478:337–342.

Foresight. 2011. *The Future of Food and Farming: Challenges and Choices for Global Sustainability.* The Government Office for Science, London, UK.

Frisk, E. and K.L. Larson. 2011. Educating for sustainability: Competencies & practices for transformative action. *Journal of Sustainability Education* 2, March 2011. Available at http://www.jsedimensions.org/wordpress/wp-content/uploads/2011/03/FriskLarson2011.pdf

Garg, M.R. 2012. *Balanced Feeding for Improving Livestock Productivity: Increase in Milk Production and Nutrient Use Efficiency and Decrease in Methane Emission.* U.N. Food and Agricultural Organization, Rome.

Gliessman, S.R. 2013. *Agroecology.* URL: http://www.agroecology.org.

Glover, J.D., J.P. Reganold, L.W. Bell et al. 2010. Increased food and ecosystem security via perennial grains. *Science* 328:1638–1639.

Godfray, H.C.J., J.R. Beddington, I.R. Crute et al. 2010. Food security: The challenge of feeding 9 billion people. *Science* 327:812–818.

Good, A.G. and P.H. Beatty. 2011. Fertilizing nature: A tragedy of excess in the commons. *PLoS Biology* 9: e1001124, doi:1001110.1001371/journal.pbio.1001124.

Gunderson, L. 2010. Ecological and human community resilience in response to natural disasters. *Ecology and Society* 15:18 [online].

Herrero, M., P.K. Thornton, A.M. Notenbaert et al. 2010. Smart investments in sustainable food production: Revisiting mixed crop-livestock systems. *Science* 327:822–825.

IPCC. 2007. *Climate Change 2007, Synthesis Report.* Intergovermental Panel on Climate Change, Geneva.

ITF. 2010. *Transport Greenhouse Gas Emissions: Country Data 2010.* International Transport Forum: Organisation for Economic Co-Operation and Development (OECD), Paris.

Jacobsen, T. and R.M. Adams. 1958. Salt and silt in ancient Mesopotamian agriculture. *Science* 128:1251–1258.

Jaffer, Y., T.A. Clark, P. Pearce, and S.A. Parsons. 2002. Potential phosphorus recovery by struvite formation. *Water Research* 36, no 7:1834–1842.

Jenny, H. 1941. *Factors of Soil Formation: A System of Quantitative Pedology.* New York: McGraw-Hill.

Koch, B. and R. Khosla. 2003. The role of precision agriculture in cropping systems. *Journal of Crop Production* 9:361–381.

Kuhn, N.J. 2014. Soil loss. In *The Soil Underfoot*, eds. G.J Churchman and E.B. Landa. 37–48, Boca Raton: CRC Press.

Lal, R. 2010. Managing soils and ecosystems for mitigating anthropogenic carbon emissions and advancing global food security. *Bioscience* 60:708–721.

Lal, R., J.A. Delgado, P.M. Groffman et al. 2011. Management to mitigate and adapt to climate change. *Journal of Soil and Water Conservation* 66:276–285.

Landis, D.A., M.M. Gardiner, W. van der Werf, and S.M. Swinton. 2008. Increasing corn for biofuel production reduces biocontrol services in agricultural landscapes. *Proceedings of the National Academy of Sciences of the United States of America* 105:20552–20557.

Laurance, W.F. 2007. Switch to corn promotes Amazon deforestation. *Science* 318:1721–1721.

Lewis, R. 1998. Public expectations, fears reflect biotech's diversity. *The Scientist* 12:184.

Lin, B.B. 2011. Resilience in agriculture through crop diversification: Adaptive management for environmental change. *Bioscience* 61:183–193.

Lobell, D.B., K.G. Cassman, and C.B. Field. 2009. Crop yield gaps: Their importance, magnitudes, and causes. *Annual Review of Environment and Resources* 34:179–204.

Ma, W.Q., L. Ma, J.H. Li et al. 2011. Phosphorus flows and use efficiencies in production and consumption of wheat, rice, and maize in China. *Chemosphere* 84:814–821.

Matson, P.A. 2011. *Seeds of Sustainability: Lessons from the Birthplace of the Green Revolution in Agriculture.* Washington D.C.: Island Press.

Matson, P.A., R. Naylor, and I. Ortiz-Monasterio. 1998. Integration of environmental, agronomic, and economic aspects of fertilizer management. *Science* 280:112.

Matson, P.A., W.J. Parton, A.G. Power, and M.J. Swift. 1997. Agricultural intensification and ecosystem properties. *Science* 277:504–509.

Meehan, T.D., B.P. Werling, D.A. Landis, and C. Grattona. 2011. Agricultural landscape simplification and insecticide use in the Midwestern United States. *Proceedings of the National Academy of Sciences of the United States of America* 108:11500–11505.

Montgomery, D.R. 2007. *Dirt: The Erosion of Civilizations.* Berkeley: University of California Press.

Moran, E.F. 2006. *People and Nature: An Introduction to Human Ecological Relations.* Malden, MA: Blackwell.

Morris, M., G. Edmeades, and E. Pehu. 2006. The global need for plant breeding capacity: What roles for the public and private sectors? *Hortscience* 41:30–39.

Mumford, L. 1956. The natural history of urbanization. In *Man's Role in Changing the Face of the Earth*, ed. W.L. Thomas, 382–398. Chicago: University of Chicago Press.

Naylor, R. 2012. *Biofuels, Rural Development, and the Changing Nature of Agricultural Demand.* Stanford Symposium Series on Global Food Policy and Food Security in the 21st Century. Center on Food Security and the Environment, Stanford, CA.

Naylor, R.L., R.J. Goldburg, J.H. Primavera et al. 2000. Effect of aquaculture on world fish supplies. *Nature* 405:1017–1024.

Nelson, G.C., M.W. Rosegrant, J. Koo et al. 2009. *Climate Change: Impact on Agriculture and Costs of Adaptation.* Washington D.C: International Food Policy Research Institute.

NRC. 2010. *Certifiably Sustainable?: The Role of Third-Party Certification Systems.* Report of a Workshop by the Committee on Certification of Sustainable Products and Services, National Research Council. The National Academies Press, Washington D.C.

OECD. 2010. *Obesity and the Economics of Prevention: Fit Not Fat.* Paris: The Organisation for Economic Co-operation and Development (OECD).

Olesen, J.E., T.R. Carter, C.H. Diaz-Ambrona et al. 2007. Uncertainties in projected impacts of climate change on European agriculture and terrestrial ecosystems based on scenarios from regional climate models. *Climatic Change* 81:123–143.

Ostrom, E. 2000. Collective action and the evolution of social norms. *The Journal of Economic Perspectives* 14:137–158.

Park, S., P. Croteau, K.A. Boering et al. 2012. Trends and seasonal cycles in the isotopic composition of nitrous oxide since 1940. *Nature Geoscience* 5:261–265.

Pimentel, D. 2000. Soil erosion and the threat to food security and the environment. *Ecosystem Health* 6:221–226.

Pimentel, D. 2006. Soil erosion: A food and environmental threat. *Environment, Development and Sustainability* 8:119–137.

Pimentel, D. and M. Pimentel. 2003. Sustainability of meat-based and plant-based diets and the environment. *American Journal of Clinical Nutrition* 78:660S–663S.

Pimentel, D., D. Cerasale, R.C. Stanley et al. 2012. Annual vs. perennial grain production. *Agriculture Ecosystems & Environment* 161:1–9.

Rittel, H. and M. Webber. 1973. Dilemmas in a general theory of planning. *Policy Sciences* 4:155–169.

Rockström, J., J. Barron, and P. Fox. 2003. Water Productivity in rain-fed agriculture: Challenges and opportunities for smallholder farmers in drought-prone tropical agroecosystems. In *Water Productivity in Agriculture: Limits and Opportunities for Improvement*, eds. J.W. Kijne, R. Barker, and D. Molden, 145–162. Wallingford: CAB International.

Rosenzweig, C., A. Iglesius, X.B. Yang et al. 2001. *Climate Change and Extreme Weather Events—Implications for Food Production, Plant Diseases, and Pests.* NASA Publications, Paper 24. National Aeronautics and Space Administration, New York, NY.

Sandor, J.A., J.B. Norton, J.A. Homburg et al. 2007. Biogeochemical studies of a native American runoff agroecosystem. *Geoarchaeology* 22:359–386.

Sarewitz, D. 2004. How science makes environmental controversies worse. *Environmental Science and Policy* 7:385–403.

Searchinger, T., R. Heimlich, R.A. Houghton et al. 2008. Use of U.S. croplands for biofuels increases greenhouse gases through emissions from land-use change. *Science* 319:1238–1240.

Seto, K., M. Fragkias, B. Güneralp, and M. Reilly. 2011. A meta-analysis of global urban land expansion. *PLoS ONE* 6: e23777.

Smith, V.H. 2003. Eutrophication of freshwater and coastal marine ecosystems—A global problem. *Environmental Science and Pollution Research* 10:126–139.

Sutton, M.A., H. Van Grinsven, C.M. Howard et al., editors. 2011. *The European Nitrogen Assessment, Sources, Effects and Policy Perspectives.* Cambridge: Cambridge University Press.

Swaminathan, M.S. 2006. An evergreen revolution. *Crop Science* 46:2293–2303.

The Land Institute. 2013. http://www.landinstitute.org/vnews/display.v/SEC/Press%20Room (accessed February 2013)

Thompson, E. Jr. 2007. *Paving Paradise: A New Perspective on California Farmland Conversion.* Davis, CA: American Farmland Trust.

Tilman, D., C. Balzer, J. Hill, and B.L. Befort. 2011. Global food demand and the sustainable intensification of agriculture. *Proceedings of the National Academy of Sciences of the United States of America* 108:20260–20264.

Tilman, D., P.B. Reich, J. Knops et al. 2001. Diversity and productivity in a long-term grassland experiment. *Science* 294:843–845.

Tomich, T.P., S. Brodt, H. Ferris et al. 2011. Agroecology: A review from a global-change perspective. *Annual Review of Environment and Resources* 36:193–222.

UNEP. 2009. *Towards Sustainable Production and Use of Resources: Assessing Biofuels.* United Nations Environment Programme, Nairobi, Kenya.

UNPD. 2012. *World Urbanization Prospects: The 2011 Revision.* United Nations Population Division, Department of Economic and Social Affairs, http://esa.un.org/unup/index.html (accessed February 2013).

USDA. 2012. *California Agricultural Statistics, Crop Year 2011.* National Agricultural Statistics Service, California Field Office, Sacramento, CA.

USDA. 2013. Feed Grains Database. Economic Research Service, United States Department of Agriculture, http://www.ers.usda.gov/data-products/feed-grains-database.aspx (accessed Feburary 2013).

Vanotti, M. and A. Szogi. 2009. Technology for recovery of phosphorus from animal wastewater through calcium phosphate precipitation. In *International Conference on Nutrient Recovery from Wastewater Streams*, ed. K. Ashley, 459–468. London: IWA Publishing.

Vermeulen, S.J., B.M. Campbell, and J.S.I. Ingram. 2012. Climate change and food systems. In *Annual Review of Environment and Resources,* eds. A. Gadgil and D.M. Liverman, 195–222. Palo Alto: Annual Reviews.

Walker, B., C. S. Holling, S.R. Carpenter, and A. Kinzig. 2004. Resilience, adaptability and transformability in social–ecological systems. *Ecology and Society* 9:5 [online]

West, P.C., H.K. Gibbs, C. Monfreda et al. 2010. Trading carbon for food: Global comparison of carbon stocks vs. crop yields on agricultural land. *Proceedings of the National Academy of Sciences of the United States of America* 107:19645–19648.

WHO. 2010. *The WHO Recommended Classification of Pesticides by Hazard and Guidelines to Classification: 2009.* Geneva, Switzerland: World Health Organization.

WHO. 2012. *World Health Statistics.* World Health Organization, http://www.who.int/gho/publications/world_health_statistics/2012/en/index.html (accessed January 2013).

Wilcove, D.S. and L.P. Koh. 2010. Addressing the threats to biodiversity from oil-palm agriculture. *Biodiversity and Conservation* 19:999–1007.

World Bank. 2010. *The Costs to Developing Countries of Adapting to Climate Change: New Methods and Estimates.* The Global Report of the Economics of Adaptation to Climate Change Study. The World Bank Group, Washington D.C.

World Bank. 2012. *Turn Down the Heat: Why a 4°C Warmer World Must Be Avoided.* Washington D.C: The World Bank.

Yang, H., J. Knapp, P. Koirala et al. 2007. Enhanced phosphorus nutrition in monocots and dicots over-expressing a phosphorus-responsive type I H+-pyrophosphatase. *Plant Biotechnology Journal* 5:735–745.

Ye, L. and E. Van Ranst. 2009. Production scenarios and the effect of soil degradation on long-term food security in China. *Global Environmental Change* 19:464–481.

3 Soil Loss

*Nikolaus J. Kuhn**

CONTENTS

3.1 SOIL: THE THREATENED SPHERE

The 2012 yearbook of the United Nations Environmental Programme (UNEP 2012, v) identifies the depletion of soil and the growing number of end-of-life nuclear power reactors as some of the most pressing environmental issues to be solved by humankind in the coming decades (Figure 3.1). While intuitively unrelated, the two issues are closely connected through the future role of biomass for energy production, contributing the so-called energy–food–water nexus for land (UNCCD 2012). Efforts to protect the atmosphere and mitigate global warming do already draw heavily on soil resources, for example, in Brazil or Southeast Asia, where sugarcane and palm oil production have been increased significantly during the past 20 years. While the production of ethanol from sugarcane in Brazil aims at limiting soil loss, the pressure to clear forest to replace the land "lost" to biofuel production may harm the soils there (Lapola et al. 2010). In southeast Asia, clearing of tropical rain forests has more immediate effects not only on vegetation, but also on soils (Fargione et al. 2008). Making matters worse, the positive effect on greenhouse gas emissions achieved by growing these biofuels is questionable because of emissions generated, among others, from the loss of living biomass after land clearance and the subsequent depletion of the soil organic carbon (OC)

* Physical Geography and Environmental Change Research Group, University of Basel, Switzerland; Email: Nikolaus.kuhn@unibas.ch

FIGURE 3.1 **(See color insert.)** Soil erosion by water on cropland in the Eifel region of Germany. The risk of soil loss by water erosion is increasing in many European countries as a consequence of the use of heavy machinery, larger fields, and increasing cultivation of maize to produce biomass. (Image by Nikolaus K. Kuhn.)

stocks. While the ecological amortization with regard to emissions of these biofuels is, in at least some cases, questionable, the damage to the soil and its ecosystem services is obvious: the removal of the natural grass- and tree cover leads to a reduction in organic matter, which is mostly released as CO_2 into the atmosphere. As a consequence, the soil loses structural stability, that is, the crumbs fall apart into mineral particles, resulting in an increased risk of runoff and erosion. Over time, the nutrient pool in the soil is depleted and productivity can only be maintained by adding fertilizers, which, in many cases, have been produced causing considerable emissions themselves or consume nonrenewable resources, such as phosphorus. The conversion of land cover also carries the risk of biodiversity loss and environmental pollution. Even in regions where no land cover conversion is required, the production of biomass to generate energy can have negative consequences for the soil. The prime crop to generate the so-called biogas, methane, in large slurry-filled reactors such as those subsidized heavily in Germany, is maize. The production of maize carries a strong risk of soil erosion (Weidanz and Mosimann 2008). One must therefore wonder whether the protection of the atmosphere at the cost of another sphere of the planet Earth, the soil, is a sustainable way forward at a time when secure food supplies are already at risk (Rosegrant and Cline 2003).

3.2 SOIL DEGRADATION AND SOIL LOSS

3.2.1 Soil Degradation

The negative effects of land use on soil can be summarized as soil degradation. Soil degradation describes the deterioration of the productivity and further environmental services of a soil as a result of human activity. Soil degradation includes the destruction of the topsoil by wind and water erosion, but an "invisible" loss of soil through lack of nutrients, structural damage, or introduction of pollutants is also included. Figure 3.2 shows one of the most recent assessments of soil degradation on a global scale (Oldeman et al. 1991); illustrating that, between 30% and 50% of the arable land of the planet has experienced some degree of degradation. While the processes that lead to this soil loss are manifold, the basic and most immediate consequence is the same—a loss of productivity (Morgan 2005). A key concern with such a loss is the limited ability to restore the productivity

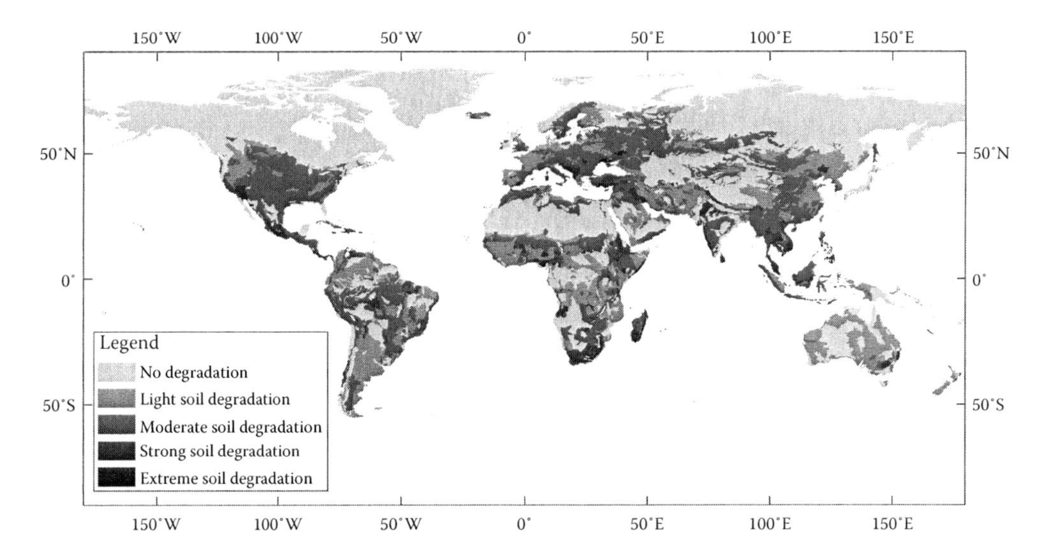

FIGURE 3.2 The effect of soil loss on productivity based on data from GLASOD. *Light*: The terrain has somewhat reduced agricultural suitability, but is suitable for use in local farming systems. Restoration to full productivity is possible by modifications of the management system. Original biotic functions are still largely intact. *Moderate*: The terrain has greatly reduced agricultural productivity but is still suitable for use in local farming systems. Major improvements are required to restore productivity. Original biotic functions are partially destroyed. *Strong*: The terrain is nonreclaimable at farm level. Major engineering works are required for terrain restoration. Original biotic functions are largely destroyed. *Extreme*: The terrain is irreclaimable and beyond restoration. Original biotic functions are fully destroyed.

and further ecosystem services of the soil quickly or at all. In regions where the topsoil that acts as the substrate for plant roots is shallow, a physical removal of this layer (properly referred to as horizon) by wind or water erosion leads to a quasi-permanent loss of the fertile layer because the redevelopment is very slow, if at all. Where the depth of loose soil material is greater, the loss of the topsoil still leads to a disturbance of the productivity that has to be compensated by fertilization and/or irrigation. Apart from these costs, off-site effects of soil loss also cause environmental damage, such as the sedimentation of reservoirs, increased runoff and flooding, nutrient loss and eutrophication, loss of soil organic C by erosion, and greenhouse gas emissions.

3.2.2 Visible Processes of Soil Loss

3.2.2.1 Splash and Rainwash

The destruction of soil is commonly associated with running water, cutting rills, and gullies into the topsoil layer. However, there is a range of processes that act on the topsoil leading to a loss of productivity either by removing the soil or destroying its function as a substrate for plant growth. In most cases, human activity introduces or accelerates these processes. The removal of vegetation to prepare a piece of land for crop farming carries the risk of splash, crust formation, and rainwash. Rainsplash describes the effects of raindrops impacting the soil surface. On impact, the kinetic energy of the falling drop is expended on the soil surface. As a consequence, soil crumbs at the soil surface are destroyed and compacted, and the ability of the soil to absorb water is reduced (Figure 3.3). In addition, small amounts of soil material can be dislodged and "jump" over short distances. As soon as the ability to absorb water is overcome by rainfall intensity, patches of runoff develop on the soil surface. Such runoff is too weak to cause much erosion, but facilitates the transport of material dislodged by raindrop impact and is referred to as rainwash. Overall, the immediate loss of soil caused by splash and rainwash is limited and rarely causes permanent on-site damage. However, the material that

FIGURE 3.3 A dense soil surface with a crust formed by raindrop impact during an erosion simulation experiment at the University of Basel (left), which inhibits infiltration, increases the risk of erosion and drought. On the right, a less degraded soil with more stable crumbs resists destruction and crust formation. The difference in erosion is also visible in the beakers collecting sediment from runoff (outside) and splash (inside). The opaque suspension in the beaker on the left is indicative of the greater concentrations of eroded sediment compared to the beaker on the right. (Image by Yaxian Hu, University of Basel.)

is eroded is often enriched in nutrients and organic matter, causing pollution of rivers and lakes. Furthermore, the reduced infiltration of water into the soil can lead to drought stress later on.

3.2.2.2 Rill and Gully Erosion

On large fields or slopes with concavities, excess runoff water can concentrate. Such concentration of runoff increases the energy of the flow significantly, leading to rill and gully erosion. Rills and gullies are linear, downslope "cuts" into the topsoil layer caused by concentrated surface runoff. The main distinction between rills and gullies is their size: rills can be obliterated again by tillage, while gullies are too large to be filled again easily and therefore, often develop into a permanent feature in the landscape. Both cause a considerable damage to the soil. Gullies destroy arable land immediately, while the soil loss caused by rill erosion removes the fertile topsoil layer around the initial channel of flow, regularly forming rills over time.

On a global scale, soil erosion by running water amounts to 28 petagrams per year (Quinton et al. 2010). A petagram is a one followed by 15 zeros, or one thousand million metric tons. The mass of 28 petagrams corresponds to a pile of dry sand of approximately 17.5 cubic kilometers or a cube with a side length of 2.6 km. For soil, the volume is about 25% greater because of its lower density. Spread over the approximately 48 million square kilometers of agricultural land on Earth, this layer is about 0.5 mm thick, that is, on average, half a millimeter of fertile soil is eroded by water annually. While this amount may seem small, over decades and centuries, such soil loss can lead to the destruction of the fertile topsoil. Furthermore, the so-called off-site effects cause further damage, including the clogging of rivers with sediment, overloading of watercourses with nutrients, and perturbations of water and carbon cycles.

3.2.2.3 Wind Erosion and Dust Pollution

Another prominent process of soil loss is wind erosion, infamous since the days of the dust bowl in the American Midwest in the 1930s. Wind erosion acts similar to erosion by water mostly on

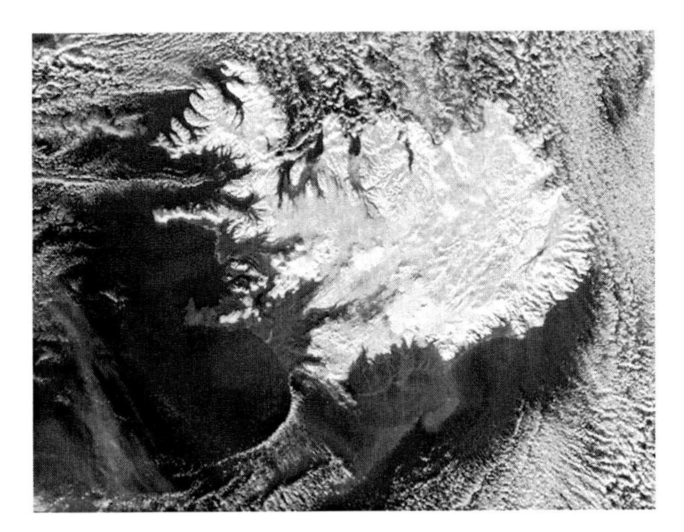

FIGURE 3.4 Dust storm over Iceland in 2002 with plume extending over the south coast and then out over the North Atlantic. During such storms, dust originates from glacial outwash plains, fine-grained glacial deposits exposed after retreat of glaciers, and fresh ash deposits. The destruction of vegetation that stabilized such surfaces in the interior also contributes to the dust production. (Image from NASA Earth Observatory.)

bare and dry soil, removing particularly the fine and light particles, such as clay or small debris of litter, which are vital for the fertility of the soil. Globally, wind erosion removes approximately 2 petagrams of soil annually, which is <10% of the amount mobilized by water erosion. However, wind erosion is largely concentrated in drylands and can thus cause considerable damage to soil. In addition, the off-site effects of wind erosion are considerable, leading to dust concentrations that are unhealthy for humans.

Prominent cases for such off-site effects are the cities of Beijing in China and Reykjavik in Iceland (Figure 3.4). In Beijing, industrial air pollution is worsened by dust blown in from overgrazed regions in the northern drylands of China. In Iceland, centuries of overgrazing of the interior by the Viking settlers has contributed to the destruction of the tree and grass cover present at settlement some 1100 years ago (Diamond 2005, p. 178). The exposed light and loose volcanic debris can be moved easily by wind, causing soil destruction, on- and off-site, and air pollution. The destruction of the soil by grazing and wind was also a factor in the near collapse of the Icelandic society in the seventeenth century. Dust can be generated by wind erosion from deposits of volcanic ash and especially from areas with material left exposed after the retreat of glaciers or as sediments from glacial outburst floods even if overgrazing has not occurred, although it is accelerated by this cause. Wind erosion may also influence climate because dust leads to a dimming of the atmosphere, which has a cooling effect. However, most atmospheric dust originates from areas that are naturally prone to wind erosion, such as dry lakebeds in deserts, and is therefore not associated with land use-induced soil loss.

3.2.2.4 Mass Wasting and Erosion by Tillage

A final, easily visible process of soil loss is mass wasting. The movement of loose surface material is common and natural in mountainous regions with slopes steeper than the natural angle of repose. Land use, such as grazing in the European Alps, is therefore likely to affect the stabilization caused by the anchoring of soil to the lower-lying solid bedrock by roots. However, since the anchored material on steep slopes is prone to move at some point anyway, the effect of land use in mountains is not so much a change in the average long-term erosion rate, but a shift in the frequency of mass-movement events. After a change of vegetation cover, that is, from trees to grass, the reduced anchoring effect leads to an initial rapid increase of mass wasting, followed by a phase of reduced

FIGURE 3.5 Soil erosion by tillage: plowing moves soil downslope, exposing less fertile sandy material, central Jutland, Denmark. (Photo by Philip Greenwood.)

frequency after much of the material on steep slopes has been moved following the loss of its root anchoring (Guthrie and Brown 2008). While the inherent instability of soil on steep slopes limits soil development in mountains, the rapid destruction of the soil after clearing can nonetheless have significant consequences. First of all, the soil resources are lost for the farmers who expected to use them. Second, the off-site effects are substantial because the rivers cannot remove the wave of eroded soil moved into the valleys, leading to a clogging of the rivers with sediment. As a consequence, the course of riverbeds can move and flood risks increase due to the sediment filling up the channel. Reservoirs are also filled more quickly than anticipated when assuming natural erosion rates, leading to a loss of storage capacity and potential for energy production. Finally, the river ecosystem can suffer from too much sediment; for example, the loss of salmon-spawning river reaches observed after intense logging in British Columbia.

On croplands outside mountain regions, mass wasting can also lead to soil loss, amounting to 5 petagrams per year. Land suitable for machine-based, large-scale agriculture is mostly flat because steep slopes limit the use of machinery or require the construction of terraces. Therefore, natural mass-wasting processes hardly occur on such croplands. However, the tractors pulling tillage tools can deliver the energy required to induce soil movement, as in plowing. When soil is moved, gravity can act on the loose soil material when it settles. This movement is balanced to either side of the plow on a flat slope, but is more pronounced in a downslope direction on a hillside. Over time, even on relatively flat land suitable for crop production, soil is moved downhill (Figure 3.5). At the foot of slopes, soil builds up, while at the top, the fertile topsoil layer is destroyed.

3.2.3 Invisible Soil Loss: Salinization, Waterlogging, Nutrient Depletion, and the Loss of Soil Structure

3.2.3.1 The Threat of Irrigation

Apart from erosion by water, wind, and gravity, other processes of soil loss exist that do not destroy the topsoil layer physically, but reduce the capacity of the soil to act as a substrate for plant growth and agricultural use. These processes are also a consequence of land use. First and foremost, irrigation has often led to the destruction of fertile soil. The large-scale application of water to cropland to ensure sufficient production of food is one of the triggers for the development of centrally organized government and urban civilization. However, it has also led to the downfall of many of these societies. Irrigation water contains dissolved substances. When the water evaporates from the irrigated field, these substances precipitate and salts accumulate in the soil. Over time, some of these substances, for example, sodium chloride, accumulate in the topsoil and can reach toxic levels for the crops (Johnson and Lewis 2006, p. 248).

In theory, the accumulation of salts can be avoided by over-irrigation, inducing a downward movement of water, washing the salts out of the soil. Over-irrigation is not intuitive in drylands where great efforts have to be made to move water onto agricultural fields. Besides, it carries the risk of washing nutrients out of the soil, reducing yields. However, even where the risks associated with irrigation are known, the means to reduce salinization can be limited. Traditional flood irrigation requires flat fields, which lack a natural slope, but an artificial gradient to enable water to move under gravity can be introduced by digging drainage ditches between the irrigated fields. Nonetheless, leaching is limited by the ease of movement of water through soil, and its inhibition may lead to waterlogging. Limited drainage may also cause an enrichment of salts in the topsoil. However, the lack of air in the topsoil limits plant growth and thus yields, even if salinization has not reached toxic levels. As well, when the salt *per se* is washed out, some sodium ions may displace other cations by cation exchange due to their enhanced concentration in solution. When the rain comes, or further irrigation is carried out with relatively fresh water, the sodium-exchanged (clay) particles in the soil swell and/or disperse, block pores, and also lead to waterlogging, possibly followed by hardsetting and crusting when the soil dries, and these effects can also induce erosion. This phenomenon is known as "sodicity" (Gupta and Abrol 1990) and is common in areas with high, salty water tables such as southern and western Australia and also in irrigated areas such as parts of California. More than 25% of the total land area in Australia is affected by sodicity (Northcote and Skene 1972).

Over-irrigation can also have an off-site impact. Downslope of the irrigated land, the drained soil water accumulates in valleys and depressions, leading to rising groundwater, waterlogging, salinization, and eutrophication.

Salinization and waterlogging have long been thought to be responsible for the decline of some of Mesopotamia's urban centers in the second and third millennium BC (Jacobsen and Adams 1958). Most of them were set up in the flat alluvial plains of the Euphrates and Tigris rivers. While ideally suited for flood irrigation due to their naturally flat topography, they also suffered from limited drainage and thus salinization and waterlogging. In contrast, traditional flood irrigation systems on the terraces of the Nile river or rice paddies in mountainous regions of eastern Asia appear not to have suffered from problems arising from salinization and waterlogging.

Modern irrigation systems depend less on natural topography than their ancient predecessors because the water can be pumped using other energy sources than gravity, humans, or animal power. Sprinklers, inducing artificial rain, can also be installed on sloping cropland. However, the risks of salinization and waterlogging still threaten the sustainability of irrigation farming today. The negative effects of the rapid development of irrigation have caused serious environmental and economic damage in the region of the Murray-Darling in eastern–central Australia (Clarke et al. 2002).

The risks for soils associated with irrigation are considered critical for many government and NGO (nongovernmental organization) strategies that rely on expanding irrigation to meet food demands in drylands (Figure 3.6).

3.2.3.2 Loss of Soil Fertility

The invisible loss of soil does not necessarily require drastic alterations to the ecology of a landscape such as irrigation. Sustainable agriculture aims at ensuring a sufficient supply of water and nutrients to crops and maintaining the soil as a substrate for these plants to grow in. Fertilizers can replace nutrients consumed by crops and removed from the fields. If these are not available or insufficient, a fallow period can be used to ensure a "refilling" of the soil with nutrients. Such regeneration of the soil fertility requires a certain duration of the fallow period as well as a minimum nutrient stock at

FIGURE 3.6 Modern flood irrigation plots in Inner Mongolia. (Photo by Nikolaus J. Kuhn.)

the beginning of the fallow period. If the duration of the crop production is too long or the fallow is too short, nutrient stocks may be depleted to a level where the ability of the soil to recover is permanently damaged (Figure 3.7). In such a case, the addition of natural or artificial fertilizer or a careful management of green manures to recover soil fertility is required (Slaymaker and Spencer 1998).

The risk of overuse of soil is not limited to nutrient depletion. Reduced productivity and removal of crops also leads to a decline of soil organic matter. A key effect of soil organic matter is the facilitation of the development of soil crumbs, called aggregates. The presence of stable aggregates

FIGURE 3.7 **(See color insert.)** Nutrient-depleted soils in northern Namibia; as a consequence of continued sorghum production, the soils have a strongly reduced productivity of <100 kg per hectare. (Photo by Nikolaus J. Kuhn.)

larger than 2–5 mm improves the ability of soil to store water and function as the habitat for soil flora and fauna. The latter improves the turnover of litter into nutrients, the former ensures water supply, and limits the risk erosion. The overuse of a soil can seriously harm the soil structure and thus the ability of a soil to recover from use. If combined with erosion and the physical destruction of the topsoil, the soil is lost for the foreseeable future. A classic case of overuse combined with a natural stress is the drought and famine in the Sahel between 1967 and 1973 (Nicholson et al. 1998). After several decades of fairly high and regular rainfall, intensification of agriculture had led to an overstocking of the land. Drought then caused a decline in productivity, which combined with overgrazing led to a destruction of the vegetation cover. The lack of vegetation made the soil vulnerable to wind and water erosion. The depletion of nutrients and lack of litter input resulted in a further decline of productivity and a deterioration of the soil structure. In the course of 5 years, large tracts of land changed from a fertile patchwork or grazing land and crop fields into what was perceived at the time as expanding deserts. In many cases, such desertification can be avoided with careful soil management and can be reversed when rainfall reaches average levels again, but at the risk of at least a temporary loss of soil (Slaymaker and Spencer 1998).

3.2.3.3 Soil Loss by Urbanization and Industrialization

The growing world population, the increasing demand on a limited space for living as well as for working, and the resulting urban sprawl contribute to the destruction of soil. Complete data are difficult to find, but some case studies indicate that a significant degree of fertile land is lost. The loss is even more pronounced since many urban centers have developed in regions with fertile soils. The Greater Toronto area (GTA) on the fertile—for Canadian climatic conditions—northern shore of Lake Ontario has experienced an unprecedented growth during the past decades. This urban sprawl has largely consumed valuable and limited cropland while a green belt around the city was conserved. Within the 7124 km^2 area covered by the GTA, the extent of the farmland dropped by 4.2% between the 2001 and 2006 census, corresponding to a loss of 11,523 ha of farmland. But, even in regions with much older urban settlement than Canada, cropland is lost to the growing cities. In Germany, urbanization in the 1990s reduced the wheat productivity by 1.4 million tons per year (EEA 2012). This corresponds to 6.5% of the 2000 wheat production and 97% of the increase in wheat production achieved between 2000 and 2010 based on the average harvest during the first decade of the twenty-first century. In other words, an amount of wheat slightly less than the increase in production was lost due to destruction of cropland by urbanization.

A final invisible cause of soil loss is contamination with pollutants causing a deterioration or loss of one or more soil functions. Both Europe and the United States have experienced such contamination as a consequence of the lack of waste management during their industrialization, triggering large cleanup efforts, most notably, the multi-billion dollar Superfund in the United States. A current example of soil loss caused by contamination is occurring in China. The immense economic growth of the People's Republic of China during the past 40 years has caused dramatic soil pollution. The State Environmental Protection Administration estimates that 100,000 square kilometers of cultivated land have been polluted. Furthermore, contaminated water has been or is used to irrigate approximately 22,000 square kilometers, while 1300 square kilometers are covered or destroyed by solid waste. The area lost for cultivation by soil contamination adds up to about 10% of China's cultivatable land.

3.3 CONSEQUENCES OF SOIL LOSS

3.3.1 Soil Erosion versus Soil Formation

Soil formation, that is, the (re-)development of new and fertile topsoil, occurs on the order of one-tenth of a millimeter per year. Even the average erosion rate estimated above is therefore not matched by soil formation. The difference leads to complete and permanent destruction of shallow soils within decades. Even on soils with a deep profile, the loss of the natural surface horizons leads to a decline in productivity and requires increased inputs, such as water, energy, and fertilizer, to maintain productivity. These inputs themselves (or their production) may be harmful to the environment or consume limited resources, for example, phosphorus. And the downslope deposition of the eroded fertile soil material also does not improve the picture. A great proportion of the material ends up in rivers and lakes, lost for agricultural production, causing riverbed clogging and pollution. The rest is deposited in lower-lying parts of the landscape and shallower slope sections. However, a deeper soil profile does not necessarily increase productivity above the levels of a simply healthy soil at a site of deposition.

The extent of the global land area and the limits imposed by climate on agriculture render soil a limited, eroding, resource. On a regional scale, further limitations to using soil are set by the economics of food production and affordable market prices. Soil loss can therefore be seen as the consumption of a nonrenewable resource. While available, achieving access to new soil can compensate the destruction or overexploitation. However, a decline lurks when no more new soil is available. Two classic examples of such a development that are described in detail by Montgomery (2007) and Wrench (1946) and summarized below are the Roman Empire and the Old South in the United States.

3.3.2 The End of an Empire

The Mediterranean landscape we know today has experienced severe degradation, most notably, soil erosion, in the past. The Roman agricultural system developed in a region called Latium, roughly situated at the knee of the Italian boot. Soil degradation spread from Latium with the increasing size of the empire. Varro recognized fields abandoned after degradation in Latium during the third century before the Common Era. At about 60 AD Columella, known for his writings on agriculture, recognized that the people of Latium heavily relied on imported grain because soil degradation had severely limited their ability to feed themselves. Compensating for the loss of soil, the Roman armies conquered land outside, giving control to the Roman farmers and their nonsustainable practices. The soils of Latium, Campania, Sardinia, Sicily, Spain, and Northern Africa, each a breadbasket for Romans at a time, were destroyed. The end of the Western Roman Empire is attributed to the limitations placed on governing by the sheer extent the empire had reached by the fourth century. The threat of the westward migration of German tribes at the end of the relatively warm and humid Roman climate optimum could simply not be managed from Rome. In his book *Dirt*, Montgomery (2007) illustrates the effect of erosion on the longevity of an economic system that relies on exploiting soils such as the Romans did. Without erosion, fertile topsoil layers in the Mediterranean are on the order of 1 m deep. At an erosion rate of 1 mm per year on cropland, these soils are eroded within 1000 years, which strikingly correspond to the length of time between the beginning of the initial expansion of the Roman republic in the sixth century BC and its end in 476 AD.

3.3.3 Soil and Slavery in the Old South of the United States

In the early nineteenth century, the invention of the cotton gin greatly increased the efficiency of cotton harvesting. This focused the economic development of the southeastern United States on cotton production, which also relied heavily on inhumane slave labor. As has become well known to soil scientists and agronomists, the fact that there was a monoculture on the farms and plantations enhanced soil degradation. Two processes contributed to the degradation: the monoculture, without refreshment of the soil by green manuring, let alone any crop rotation, led to a depletion of nutrients and loss of soil structure, inducing runoff and erosion. The slavery on the plantations also offered no motivation to preserve the soil as a resource to act against the degradation. With large tracts of land in the western part of the continent perceived as empty and available for settlement, moving the cotton fields to productive land compensated for the destruction of the soil. The westward expansion of the cotton industry also spread slavery from its established base into new territories and states during the first half of the nineteenth century. The expansion of slavery met resistance within the union and eventually led to secession of the confederate states after the election of Abraham Lincoln. The Old South saw this as a severe threat to its economic interests by the limitations to consuming more land and soil for cotton production. The following civil war ended slavery in the United States and the spread of soil degradation caused by the southern cotton industry. However, the devastation of the land caused by the unsustainable production system was a major obstacle for the recovery of the South. As late as the 1930s, as part of the New Deal introduced by Franklin D. Roosevelt, land restoration efforts were initiated, most notably, the Tennessee Valley Authority led by Robert E. Horton, whose work was of fundamental importance and whose influence continues to this day.

3.4 PEAK SOIL: THE ULTIMATE LIMIT

3.4.1 Soil: Losing a Resource

The destruction of the soil resource in the Old South mirrors the need of the Romans to find new land to maintain their wealth and feed the masses. Eventually, both systems broke down, not exclusively as a result of soil loss, but certainly as one of the consequences of the depletion of the soil resource. The examples of the Roman Empire and the land degradation in the Old South highlight the relevance of losing soil as a resource. Since the introduction of agriculture during the Neolithic some 10,000 years ago, about 20 million square kilometers of arable land have suffered from degradation (UNCCD 2012). The loss of soil is not limited to these examples from the past, but continues to deplete a nonrenewable resource. During the past 40 years, approximately 30% of the arable land has become unproductive due to erosion, mainly in Asia and Africa. The loss of soil is reaching a critical dimension (UNCCD 2012): every minute, 25 ha of tropical rain forest are converted into agricultural land, 10 ha of soil suffer from significant degradation, 23 ha suffer from the effects of desertification, and 5.5 ha are built over. While an increase in the productivity or use of so far unused land is still possible, the sustainable use of the existing cropland balances such gains. Furthermore, the greenhouse gas emissions associated with land cover change, soil degradation, fertilizer production, as well as the use of nonrenewable fertilizers such as phosphorus place a limit on the sustainable increases in productivity on agricultural land (Kuhn 2010). As a consequence, the number of people that can be fed is limited. Some studies suggest that about 10 billion humans could be sustained on Earth, while others, more skeptical ones, identify 7 billion as a peak and 5 billion as a sustainable limit (Sverdrup and Ragnarsdottir 2011). If these are right, the number of humans has already grown beyond the carrying capacity of the planet. Food production also experiences increasing competition for soil from biomass production, perceived as a green energy and thus both are sustainable and climate friendly. Finally, some potentially arable land has other ecosystem functions that are equally important to our well-being; for example, tropical forests and wetland acting as sinks for carbon dioxide and hot spots of biodiversity. The threat illustrated by these trends and calculations is underlined by the land grabbing observed especially in Africa, aimed at securing soil to produce food, fiber, and energy for those who can afford it (Hall 2011).

3.4.2 Stopping Soil Loss

The current state of the world's soils, the historic examples of the consequences of soil loss, and the increasing demand for use of the soil resource indicate that soil is indeed a limited and nonrenewable resource that is already used at or beyond its limits of productivity. To ensure food production and other ecosystem services of soil, the United Nations Convention to Combat Desertification has therefore adopted a strategy aimed at zero net soil loss (UNCCD 2012). The goal of this strategy focuses on "achieving a sustainable land use in agriculture, forestry, energy, and urbanization for all and by all." Targets encompass zero net land and forest degradation by 2030 and the establishment of an intergovernmental panel on land and soil, taking on a role similar to the Intergovernmental Panel on Climate Change in identifying risks to the soil resource and suggesting pathways for protecting this vital resource. While recognizing that the soil underfoot should not remain a forgotten sphere, soil loss continues at a rate of 550 km^2 (square kilometers) per day. Compared to the threat of climate change or the pollution of water, soil loss must be considered an equally or even more serious threat, because air and water can be cleaned, while soil may require thousands of years until it has redeveloped. In human terms, soil loss is forever.

REFERENCES

Clarke, C.J., R.J. George, R.W. Bell, and T.J. Hatton. 2002. Dryland salinity in south-western Australia: Its origins, remedies, and future research directions. *Australian Journal of Soil Research* 40: 93–113.

Diamond, J. 2005. *Collapse: How Societies Choose to Fail or Succeed.* New York: Penguin.

EEA (European Enviroment Agency). 2012. Annual impact of soil losses due to urbanisation, http://www.eea.europa.eu/data-and-maps/figures/annual-impact-of-soil-losses (accessed January 23, 2013).

Fargione, J., J. Hill, D. Tilman, S. Polasky, and P. Hawthorne. 2008. Land clearing and the biofuel carbon debt. *Science* 319 (5867): 1235–1238.

Gupta, R.K. and I.P. Abrol. 1990. Salt-affected soils: Their reclamation and management for crop production. *Advances in Soil Science* 11: 224–288.

Guthrie, R.H. and K.J. Brown. 2008. Denudation and landslides in coastal mountain watersheds: 10,000 years of erosion. *Geographica Helvetica* 63(1): 26–35.

Hall, R. 2011. Land grabbing in Southern Africa: The many faces of the investor rush. *Review of African Political Economy* 38 (128): 193–214.

Jacobsen, T. and R.M. Adams. 1958. Salt and silt in ancient Mesopotamian agriculture. *Science* 128 (3334): 1251–1258.

Johnson, D. and L. Lewis. 2006. *Land Degradation: Creation and Destruction*, 318pp. Lanham: Rowman and Littlefield.

Kuhn, N.J. 2010. Erosion and climate. *Nature Geoscience* 3: 738.

Lapola, D.M., R. Schaldach, J. Alcamo, A. Bondeau, J. Koch, C. Koelking, and J.A. Priess. 2010. Indirect land-use changes can overcome carbon savings from biofuels in Brazil. *Proceedings of the National Academy of Sciences of the United States of America* 107(8): 3388–3393.

Montgomery, D.R. 2007. *Dirt: The Erosion of Civilizations.* Berkeley and Los Angeles: University of California Press.

Morgan, R.P.C. 2005. *Soil Erosion and Conservation.* Oxford: Blackwell Publishing.

Nicholson, E., C.J. Tucker, and M.B. Ba. 1998. Desertification, drought, and surface vegetation: An example from the West African Sahel. *Bulletin of the American Meteorological Society* 79(5): 815–829.

Northcote, K.H. and J.K.M. Skene. 1972. *Soils with Saline and Sodic Properties.* Soil Publication 27. Melbourne: CSIRO Australia.

Oldeman, L.R., R.T.A. Hakkeling, and W.G. Sombroek. 1991. *World Map of the Status of Human-Induced Soil Degradation.* Wageningen: Global Assessment of Soil Degradation (GLASOD).

Quinton, J.N., G. Govers, K. Van Oost, and R.D. Bardgett. 2010. The impact of agricultural soil erosion on biogeochemical cycling. *Nature Geoscience* 3: 311–314.

Rosegrant, M.W. and S.A. Cline. 2003. Global food security and policies. *Science* 2003: 1917–1919.

Slaymaker, O. and T. Spencer. 1998. *Physical Geography and Global Environmental Change.* Essex, UK: Pearson.

Sverdrup, H.U. and K.V. Ragnarsdottir. 2011. Challenging the planetary boundaries II: Assessing the sustainable global population and phosphate supply, using a systems dynamics assessment model. *Applied Geochemistry* 26 (2011): S307–S310.

UNCCD. 2012. *Zero Net Land Degradation—A Sustainable Development Goal for Rio + 20.* Bonn, Germany: United Nations Convention to Combat Desertification.

UNEP. 2012. *United Nations Environmental Yearbook 2012: Emerging Issues in Our Global Environment.* Nairobi, Kenya: United Nations Environmental Programme.

Weidanz, J. and T. Mosimann. 2008. Auswirkungvon Maisanbauzur Produktionvon Biogasaufdie Bodenerosion. *Wasser und Abfall* 7–8: 16–20.

Wrench, G.T. 1946. *Reconstruction by Way of the Soil.* London: Faber and Faber Ltd.

4 The Finite Soil Resource for Sustainable Development
The Case of Taiwan

Zeng-Yei Hseu and Zueng-Sang Chen*

CONTENTS

4.1 INTRODUCTION

4.1.1 Taiwan and World Food Requirements

Various challenges and problems will be encountered while trying to greatly increase the world's food production to meet the nutrition requirements of about 9 billion people. To anticipate the kinds of challenges and problems that we may face, it is pertinent to examine the case of Taiwan, which is a relatively small island nation in area, but with a high density of population and which is rather

* Department of Environmental Science and Engineering, National Pingtung University of Science and Technology, Pingtung 91201, TAIWAN; Email: zyhseu@mail.npust.edu.tw

highly developed, both agriculturally and industrially. As a result, the soils and those who manage them have encountered various challenges and problems. This case study of the way that these are being met by soil scientists and society in Taiwan can provide lessons for us all for the future.

4.1.2 Geographical Background of Taiwan

Taiwan is a humid tropical and highly industrialized country. It is an island located about 150 km off the southeast coast of mainland China. It covers 36,000 km^2 in area with more than 23 million residents. The Central Mountain Ridge is the most prominent feature in the landscape. There are more than 200 peaks over 3000 m in altitude. The high mountains have resulted from an arc-continent collision between the Eurasian plate and the Philippine Sea plate since the Pliocene and Pleistocene, that is, since about 2.6 million years ago (Ho 1988). The landscape decreases in altitude westward and eastward from the north–south trending ridge. In addition to the mountain area, the hills and table lands account for 40%, and alluvial plains for the remaining 30% of the total area in Taiwan (Figure 4.1). The environment of the whole country supports a rich fauna, flora, and biodiversity. There are over 4000 vascular plant species and a spectrum of six forest ecosystem types through the country.

The Tropic of Cancer (23.5°N) running across Taiwan's middle section divides the island into two climatic zones, tropical in the south and subtropical in the north. The island's average annual temperature is about 24°C in the south and about 22°C in the north. Summer and winter monsoons also bring intermittent rainfall to Taiwan's hills and central mountains. In general, there is more than 2500 mm of rainfall every year. The various land forms and vegetation types yield a diversity of soil types in Taiwan (Chen 1992a, 1993).

4.1.3 Soil Diversity of Taiwan

Human-induced processes that markedly change soil properties and result in diagnostic horizons or properties are termed "anthropogenic processes." In the case of soils affected by human activity in large-scale farming, the anthropogenic processes have operated in Taiwan for approximately 400

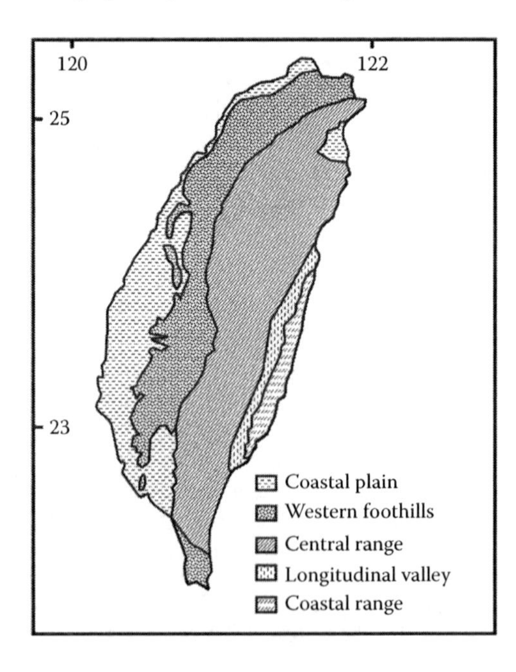

FIGURE 4.1 Landscape domains of Taiwan.

years. They have particularly involved the numerous agricultural activities in soils, such as puddling of surface soils (mechanically stirring and mixing surface soil with water and making it into a muddy paste) for paddy rice production by human beings. Four hundred years ago, the migrants from China began to reclaim the land of Taiwan. Before this period, a few aborigines formed the majority group of people on this island, and hunting and fishing were their major occupation, and hence the soils had negligible impact from human beings. The migrants brought farming techniques into the island, so that the forests became crop lands, particularly in the western part of Taiwan including the alluvial plains and mountains with low elevation. Because paddy rice is the major staple food for Chinese, the crop lands were dominated by paddy fields. Therefore, frequent submergence and plowing for rice production have been the major factors in soil development by human activity in Taiwan since the eighteenth century. With increasing population and the discovery of mechanical power, the area of paddy soils has expanded, particularly since the mid-twentieth century. In addition to paddy rice, the farming of sugar cane and orchards has become more common since the nineteenth century.

Paddy soils (rice-growing soils) are developed from various other soils. Land leveling and terracing produce changes in the soil moisture regime for paddy rice production. The recognition of paddy soils is mainly based on the obvious impacts of anthropogenic activities on the soils. Artificial and seasonal water saturation often leads to the formation of special layers characterized by Fe and Mn distribution. Puddling is often a necessary practice for easy transplanting of seedlings, and it results in a poor soil structure in the plow layer and a compacted plow pan. This latter is good for saving irrigation water as long as it does not adversely affect the growth of roots. Therefore, a principal distinction of paddy soils from upland soils is the variation in their oxidation and reduction status.

Overall, the soils of Taiwan consist of 11 of the 12 Soil Orders (Hseu et al. 2011) based on USDA *Soil Taxonomy* (Soil Survey Staff 2010). Only Gelisols, which are soils found in cold climates that include permafrost, are not present. Four of the Soil Orders dominate the rural soils (Table 4.1).

TABLE 4.1
Distribution of Soil Orders[a] in Cultivated Soils of Taiwan

Soil Orders	Approximate Area (km^2)	% of Land Use (%)
Inceptisols	8590	51.0
Alfisols	3668	21.8
Ultisols	1624	9.6
Entisols	1142	6.8
Andisols	195	1.2
Mollisols	191	1.1
Oxisols	50	0.3
Histosols	6	0.04
Vertisols	1	<0.01
Aridisols	1	<0.01
Miscellaneous lands	1363	8.1
Subtotal	16,830	100.0

Source: Soil Survey Staff. 2010. *Keys to Soil Taxonomy*. 11th edition. U.S. Dep. Agric., Soil Conservation Service, Washington, DC.

[a] According to the *Soil Taxonomy* classification system, the major soils reported in Taiwan are Inceptisols: soils with weakly developed horizons (soil layers); Entisols: soils showing little or no differentiation with soil depth; Alfisols: moderately leached soils with clay-rich subsoils; and Ultisols: strongly leached soils.

4.2 SOIL DEGRADATION AND CONTAMINATION IN TAIWAN

Agricultural soils have been intensively cultivated for crop production to fulfill the needs of the large population in Taiwan, especially in the last four decades. Soil degradation has occurred as a result of both natural hazards and human activities. The total area of cultivated soils in Taiwan is 0.85 million ha (24% of the total area in 2012), while more than 30% of the cultivated soils are potentially degraded. The major types of degradation of soils in Taiwan are soil erosion, soil acidification, soil compaction, high salinity, and heavy metal contamination, especially in the last three decades (Guo et al. 2005; Hseu et al. 2010).

4.2.1 Soil Erosion

Soil erosion in Taiwan is mainly driven by the high slopes and the occurrence of heavy rains in the unstable geological environment during the typhoon season and, to a lesser extent, of wind erosion (Guo et al. 2005). Regardless of human activities, coastal sand dunes, young river terraces, mudstone bad lands, and slope lands are strongly eroded in different landscapes. However, if these lands are illegally cultivated for vegetables or orchards, the extent of soil erosion is accelerated due to the reduction of land cover. The soil erosion of normal land cover is always less than 5 tons/ha/year, but serious soil erosion always releases more than 100 tons/ha/year, especially since the typhoon comes 2–3 times per year and also carries a heavy load of rain of 300–500 mm over 3 days each time.

4.2.2 Soil Acidification

Acid rain and long-term chemical fertilization are major factors causing soil acidification in Taiwan (Guo et al. 2005). Approximately 70% of the rainfall events monitored by 20 metrological stations through Taiwan have a pH lower than 5.6. The chemical composition of rainwater in Taiwan is mainly affected by external materials from other regions, especially from mainland China. Moreover, serious acid rain in Taiwan with mean pH values of 4.5–5.5 often occurs in the highly urbanized and industrialized areas.

The Taiwan Environmental Protection Administration (EPA) has established many monitoring stations for acid deposition in Taiwan since 1990. According to their database, the main constituents of rainwater were sulfate, chloride, sodium, and ammonium ions, followed by nitrate and calcium ions. The concentration ratio of sulfate to nitrate in acid deposition in Taiwan is about 5. The mean annual total deposition of sulfate from rain water in northern Taiwan is higher than 100 kilograms (kg)/ha/year, which is more than twice that of eastern U.S.A. (Chen et al. 1998). Monitoring results also indicate that about 70 kg/ha/year of sulfate is deposited in southern Taiwan. The mean annual total deposition of nitrate from rainwater in northern Taiwan and western Taiwan is 40–60 and <30 kg/ha/year in eastern Taiwan and southern Taiwan. The mean annual total deposition of ammonium from rainwater is highest in southwestern Taiwan, ranging from 20 to 30 kg/ha/year and <10 kg/ha/year in eastern Taiwan and southern Taiwan (Chen et al. 1998).

Over-application of chemical fertilizers of nitrogen (N), phosphorus (P), and potassium (K) mostly result from the activities of farmers on agricultural soils in Taiwan, and this causes a clear increase of acidity in soils under high cropping intensities (Guo et al. 2005). The general application rate of fertilizers for rice production is always higher than 150 kg N/ha, 35 kg P/ha, and 75 kg K/ha per crop, and the general application rate of fertilizers for corn production is always higher than 180 kg N/ha, 22 kg P/ha, and 48 kg K/ha per crop in Taiwan (Guo et al. 2005).

4.2.3 Soil Compaction

After long-term and continuous cultivation on the cropped lands, the porosity of subsurface soil becomes very low to give a firm and platy soil structure, poor aeration, and drainage, with difficulty for root penetration, particularly in clayey soils (Guo et al. 2005). In some clayey soils from aged lacustrine (i.e., lake-derived) deposits on the plains of southwestern Taiwan, soil compaction from long-term cultivation for lowland rice production largely results from tillage to a depth of only 15–18 cm from the surface. In general, for rice-growing soils, these compact layers are found at a depth of 20–40 cm. Soil compaction in rice-growing fields is also one of the reasons for poor drainage. Rice growing has occupied approximately 50% (0.45 million ha) of the agricultural lands in Taiwan over the past decade. Lowland rice has been cultivated for a period between 50 and 350 years, particularly on the numerous alluviums and terraces, and its growth has been aided by complete irrigation systems since the early and middle stages of the twentieth century (Guo et al. 2005). Because Taiwan is a member of the World Trade Organization and has signed the international trade liberalization of rice, the government of Taiwan changed the policy of land use by decreasing rice production, accepting rice imported from other countries, and also increasing the production of upland crops such as corn and soybean in original rice-growing soils. However, changing the land use from rice-growing soils into nonwaterlogged cropping systems produced some problems. These problems included (1) root growth was impeded by the subsurface compact layers of the soil profile, (2) a reduction in the level of the groundwater table, and (3) a decrease of biota habitat and microclimate change (Guo et al. 2005). Additionally, the remaining irrigation water has sometimes become mixed with industrial wastewater discharges, producing contamination by heavy metals of the surface 20–30 cm depth and affecting the safety of food chains. An example was the cadmium(Cd)-rice contaminated by chemical plant wastewater in northern Taiwan in the 1980s (Chen 1991)—see Section 4.2.5.

4.2.4 The Process of Salinization

Land subsidence by over pumping of groundwater for freshwater aquaculture has occurred locally along the coastal regions of western Taiwan, and thus the coastal soils are usually flooded by seawater particularly in storms during the summer season (Guo et al. 2005). Therefore, soils of these regions have become salt-affected. Overfertilization by livestock compost is another cause of saline accumulation in the surface soils of seashore regions of southwestern Taiwan. The total area of salt-affected soils (>4 deci-Siemens (dS)/m electric conductivity of a saturated soil paste extract) is about 25,000 ha (Guo et al. 2005).

4.2.5 Contamination of Soils with Heavy Metals

The major anthropogenic sources of heavy metals in the rice-growing soils of Taiwan are illegal wastewaters discharged from industrial plants (such as for electroplating or pigment production) of western Taiwan. For instance, a famous cadmium-contaminated rice event occurred in 1987 in Taoyuan county, northwestern Taiwan. Over 100 ha of rice-growing soils had been contaminated by cadmium (Cd) and lead (Pb) by illegal discharge of wastewater from a plastic-stabilizer producing plant nearby the rice fields in 1987 (Chen 1991). After the measurement of soils and brown rice, Cd concentration generally ranged from 10 to 30 milligrams (mg)/kg in soils and 0.5–2.5 mg/kg of brown rice in the contaminated sites, respectively (Chen 1991; Hseu et al. 2002). However, the standards for the maximum desirable levels for human health in Taiwan for Cd of soils and brown rice are 5.0 and 0.4 mg/kg, respectively. This event clearly impacted the modification of Cd regulations in soils and in brown rice and drew attention toward food safety in Taiwan, especially for rice. More than 30 ha of seriously Cd-contaminated soils still remaines within the site until now; we require better communication with the community farmers to find the best remediation and management strategies to solve this problem. As a result, surveys of contaminated soils in Taiwan have been performed in four stages by the Taiwan Environmental Protection Administration (Taiwan EPA) since 1984 (Taiwan EPA 2013).

On the basis of the results of a continuing detailed rural soil survey begun in 2002, the Taiwan EPA reported that the areas of metals-contaminated rural soil, for which the concentration of heavy metals is higher than the regulation values (or control standards) for heavy metals in soils, were 159 ha of nickel (Ni)-contaminated soils, 148 ha of copper (Cu)-contaminated soils, 127 ha of chromium (Cr)-contaminated soils, 113 ha of zinc (Zn)-contaminated soils, 17 ha of Cd-contaminated soils, 4 ha of Pb-contaminated soils, and 0.3 ha of mercury (Hg)-contaminated soils. Moreover, the total area of contaminated rural soils was about 251 ha (Taiwan EPA 2013). We also know that several metal-contaminated soils have still not been found in western Taiwan where the irrigation water has been partially mixed with wastewater discharged from the industrial parks of Taiwan.

4.3 SOIL PROTECTION IN TAIWAN

4.3.1 The Soil and Groundwater Pollution Remediation Act

The Soil and Groundwater Pollution Remediation Act of the Legislature (SGWPR Act 2000) fully addressed soil contamination and established a management system where soil pollution sites were divided into two categories: control sites and remediation sites (Taiwan EPA 2013). When levels of soil contaminants exceed the regulation of pollutants at a site, this site will be listed as a "control site," and the competent authority will be charged with taking steps to avoid further deterioration by contamination. If the "control site" was assessed to show clear risks for environmental quality and human health, then this site will be further announced as a "remediation site" (Taiwan EPA 2013). Most of the remediation sites for organic pollutants are gasoline stations and petroleum storage tanks. However, approximately 90% of the total number of control sites occurred on rural soils contaminated with heavy metals.

4.3.2 Social Awareness of the Importance for Soil Protection in Taiwan and Globally

Hseu et al. (2010) provide a schematic summary of the identification, transport, and different remediation techniques for heavy metals in relation to their agroenvironmental impact on rice-growing soils in Taiwan. According to the regulation of soil pollutants in the SGWPR Act in Taiwan, the source, fate, and remediation of heavy metals for food safety is a dynamic outcome of scientific information integrated over time. However, site-specific and health risk-based assessments are deemed to provide the most reliable and practical approaches for resolving the problem of heavy metal-contaminated soils in Taiwan, especially for Cd-contaminated soils. To understand the bioavailability of heavy metals in various soil types after long periods of time and to provide reliable soil parameters for health risk-based assessments, more scientific evidence is needed from different research cases, especially from long-term field trials involving all kinds of key conditions and factors (Su and Chen 2009, 2010). Eventually, the compilation of a reliable database on heavy metal concentrations in soils and different crops must be given the utmost importance. From the database, a variety of useful information and methodology can be developed toward achieving the ultimate goal of providing health and high-quality food for human beings in the future (Hseu et al. 2010).

Rapid development makes soil contamination an inevitable problem and a big challenge for scientists and environmental policy makers. To sustain soil health for future generations around the whole globe, the soil resource worldwide should be protected against slow and insidious poisoning by contaminants released from industrial development and intensive agricultural activities. While abundant literature on soil pollution is available, the guidelines established by individual countries worldwide to control the pollution of agricultural soils are not consistent and standardized. This reflects the complexity of the behavior of contaminants in agroenvironmental systems in various climatic, geological, and hydrological conditions, and also the political and nonscientific factors that affect the establishment of regulations in different countries.

4.4 STRATEGIES FOR SUSTAINABLE MANAGEMENT OF THE SOIL RESOURCES

4.4.1 Supporting High Soil Quality for Human Health

Hseu et al. (1998) selected key indicators for the evaluation of soil quality in Taiwan. The key indicators are physical, chemical, and biological soil characteristics (Karlen et al. 1997). The physical indicators included surface soil depth, soil texture classes, or contents (%) of clay, silt, or sand particles, bulk density, available water content (%), and the stability of soil aggregates at a depth of 30 cm. It is easy to understand that measuring the bulk density, soil texture, and penetration resistance (or infiltration) rate can provide useful indicators of the status of soil compactness, the movement of water and air, and the spread of roots. Measurements of infiltration rate and hydraulic conductivity (rate of travel of water through soil) are also very useful soil data, but are seldom available, due to the difficulty and expense of making enough measurements to obtain a reliable mean value of these soil characteristics. Measuring aggregate stability can give us valuable data about soil structural degradation, which is often affected by soil pollution (e.g., sodium) and loss of soil organic matter. This showed that visual assessment of the soil profile, such as soil color and its feel, is a very easy way of assessing the physical condition of the soil, and whether there is a need for soil reclamation or remediation to maintain soil functions (Arshad et al. 1996).

The soil chemical indicators include soil pH, electric conductivity (EC), organic carbon%, extractable available N, P, and K, and extractable available trace elements (Cu, Zn, Cd, and Pb) (Karlen et al. 1997; Lee et al. 2006; Jien et al. 2012). The standard soil fertility attributes—soil pH, organic carbon, available N, P, and K—are the most important factors in terms of plant growth, crop production and microbial diversity, and soil function (Hseu et al. 1998; Tsai et al. 2009). It is well known that these chemical parameters are generally sensitive to soil management. For those of polluted or degraded soils, these soil fertility indicators are regarded as part of a minimum data set of soil chemical indicators.

The soil biological indicators include potential nitrogen mineralization (extent of conversion by microbes of organic forms of nitrogen into forms that plants can uptake), microbial biomass of carbon, nitrogen, and phosphorus (the amount, by weight, of each of these elements that is present in living microorganisms), soil respiration (the amount of carbon dioxide produced by soil organisms), the number of earthworms, and crop yield (Karlen et al. 1997; Lee et al. 2006; Jien et al. 2012). Soil biological parameters are potentially early, sensitive indicators of soil degradation and contamination.

4.4.2 National Network of Soil Surveys

4.4.2.1 Soil Survey Techniques Developed for Boundaries of Contaminated Soils

For saving costs and understanding the detailed boundaries of contaminated soils, many soil survey techniques have been developed over the last decade. Since 2000, some new and innovative techniques have been developed to clean up contaminated soils in Taiwan (Taiwan EPA 2013) (http://ww2.epa.gov.tw/SoilGW/en/index.htm). Many workshops have been organized over the last 10 years to demonstrate these techniques to the relevant scientists, consultants, and community people (TASGEP 2013) (http://www.tasgep.org.tw). The objective of the soil survey project of contaminated sites is to understand the total volumes of the contaminated soils in the site (Hseu et al. 2010).

4.4.2.2 National Soil Information System for More Soil Interpretation in Taiwan

A national soil information system has been established in the computer center of the Taiwan Agricultural Research Institute (TARI) (http://www.tari.gov.tw/index_in.htm) for combining the soil databases of rural soils, hill lands, and forest soils of Taiwan (Guo and Chen 2009). The information is almost completely digitized to provide more easily available services to clients. These databases and digital soil maps can now be retrieved through the Internet. Many soil survey reports and soil maps have been printed for technology transfer for different applications from this information system for sustainable agriculture, the planning and management of lands, and also for environmental impact assessments in Taiwan (Guo et al. 2009).

4.4.2.3 Soil Museum for Education on Soil Functions

A national museum was established at an international level in 2005 at TARI for service to students, farmers, teachers, production communities, land planners, and government officials (http://www.tari.gov.tw/index_in.htm). More than 200 soil monoliths collected from different landscapes and land uses in Taiwan are demonstrated in this soil museum. Soil museums were also established at a smaller, university-level scale, and the first of these was opened in 2005 at National Taiwan University (http://Lab.ac.ntu.edu.tw/soilsc/) (Figures 4.2 through 4.5). It provided more than 100 soil monoliths to show the typical soil characteristics of different major soil groups along with valuable posters to demonstrate soil properties and how to use these characteristics and properties for

FIGURE 4.2 Soil museum and exhibition at National Taiwan University: Typical alluvial soils of Taiwan (20 × 120 cm, 20 kg).

FIGURE 4.3 Soil museum and exhibition at National Taiwan University: Black color soils (20 × 120 cm, 20 kg) and large soil monoliths (200 kg).

FIGURE 4.4 Soil museum and exhibition at National Taiwan University: Red soils, alluvial soils, and old soil monoliths (20 × 120 cm).

FIGURE 4.5 **(See color insert.)** Soil museum and exhibition at National Taiwan University: Soil monolith exhibition for students and farmers.

soil interpretation for soil management in order to maintain soil quality, crop quality, and environmental quality in the future (Chen 1992b; Lee et al. 2006).

4.4.3 Maintaining Soil, Crop, and Environmental Quality

Because the SGPR Act was officially enacted in 2000, the regulations of pollutants in soil and groundwater are going to be revised soon in Taiwan. Recently expanded soil databases covering the concentrations of the various pollutants and their bioavailability of soils, and the development of new models linking metals concentrations in soils and crops with health risk-based assessments, should help to revise the new regulations of pollutants in soils, crops, and groundwater (Hseu et al. 2010; Lai et al. 2010; Jien et al. 2012).

4.4.4 We Are Standing at the Crossroads

There were always about 20 soil scientists, including graduate students involved in national soil survey projects from 1974 until recent years. But currently, there are only 10 soil scientists working at the Taiwan Agricultural Research Institute and only 6 pedologists—scientists who carry out soil surveys to obtain information on soil morphology, soil formation, and soil classification, and make soil interpretations for land use—have studied at the four universities in Taiwan in the past decade. More than 50 soil surveyors were trained for national forest soil survey projects from 1993 to 2002. Currently, only few graduate students are available to take part in pedology studies or soil survey projects for different purposes. Therefore, only about 10 active pedologists remain in Taiwan. This means that the number of people with the basic knowledge of soils in Taiwan is certain to decline in the near future, especially for updates of digital soil mapping and for soil interpretation for land use of soil resources in Taiwan.

We are standing at the crossroads for the development of pedology in Taiwan. Basic pedology courses, including basic soil science, soil morphology, genesis and classification, and soil survey techniques, are still offered by the four national universities of Taiwan. However, during the last decade in Taiwan, most graduate students have shifted their interests to other fields in order to find jobs, especially on soil survey, and on the prevention and remediation of potentially highly contaminated sites for environmental consultant companies in Taiwan.

REFERENCES

Arshad, M.A., B. Lowery, and B. Grossman. 1996. Methods for assessing soil quality. In *Physical Tests for Monitoring Soil Quality*, eds. J.W. Doran and A.J. Jones, 123–141. Madison: SSSA Spec. Publ. 49.

Chen, Z.S. 1991. Cadmium and lead contamination of soils near plastic stabilizing materials producing plants in northern Taiwan. *Water Air and Soil Pollution* 57–58: 745–754.

Chen, Z.S. 1992a. Morphological characteristics, pedogenic processes, and classification of wet soils in Taiwan. In *Proceedings of the 8th International Soil Classification Workshop*, ed. J. M. Kimble, 53–59. Louisiana and Texas, USA, October 6–21, 1990.

Chen, Z.S. 1992b. Metal contamination of flooded soils, rice plants, and surface waters in Asia. In *Biogeochemistry of Trace Metals.* ed. D. C. Adriano, 85–107. Orlando, FL: Lewis Publishers Inc.

Chen, Z.S. 1993. The soil–topographic relationships of forest soils in some regions of Taiwan. In *Proceedings of the 8th International Management Workshop: Utilization of Soil Survey Information for Sustainable Land Use*, ed. J. M. Kimble, 87–100. Oregon, California, and Nevada of United States of America, July 11–24, 1992.

Chen, Z.S., J. C. Liu, and J. Y. Cheng. 1998. Acid deposition influences on the dynamics of heavy metals in soils and biological accumulation of heavy metals in the crops and vegetables in Taiwan. In *Acid Deposition and Ecosystem Sensitivity in East Asia*, eds. V. N. Bashkin and S.-U. Park, 189–225. NOVA Science Publishers, Inc., Commack, New York, USA (ISBN 1-56072-611-3).

Guo, H.Y., C.H. Liu, C.L.Chu, C.H. Chiang, and M.C. Yei. 2005. Taiwan soil information system. Taiwan Agricultural Research Institute, Council of Agriculture, Taiwan. (http://www.tari.gov.tw/index_in.htm)

Guo, H.Y., Z.Y. Hseu, Z.S. Chen, C.C. Tsai, and S.H. Jien. 2009. Integrating soil information system with agro-environmental application in Taiwan. In *Proceedings of the 9th International Conference of East and Southeast Asia Federation of Soil Science Societies (ESAFS).* Ed. J.S. Suh, K.H. Kim, J.H. Yoo, J.K. Kim, M.K. Paik, and W.I. Kim. Seoul, Korea, Oct. 27–30, 2009. Korean Society of Soil Science and Fertilizer, 117–122. ISBN: 978-89-480-03901.

Guo, H.Y. and Z.S. Chen. 2009. Soil information system and its application in Taiwan. In *Proceedings of the 9th International Conference of East and Southeast Asia Federation of Soil Science Societies (ESAFS)*, eds. J.S. Suh, K.H. Kim, J.H. Yoo, J.K. Kim, M. K. Paik, and W. I. Kim. Seoul, Korea, Oct 27–30, 2009. Published by Korean Society of Soil Science and Fertilizer, 403–405. ISBN: 978-89-480-03901.

Ho, C.S. 1988. *An Introduction to the Geology of Taiwan and Explanation Text of the Geological Map of Taiwan.* 2nd edition. Taipei: The Ministry of Economics Affairs, Taiwan.

Hseu, Z.Y., Z.S. Chen, and C.C. Tsai. 1998. Selected indicators and conceptual framework for assessment methods of soil quality in arable soils of Taiwan. *Soil and Environment* 1: 12–20.

Hseu, Z.Y., Z.S. Chen, C.C. Tsai, C.C. Tsui, S.F. Cheng, C.L. Liu, and H.T. Lin. 2002. Comparison of different digestion methods on the total contents of heavy metals in sediments and soils. *Water, Air, and Soil Pollution* 141: 189–205.

Hseu, Z.Y., S.W. Su, H.Y. Lai, H.Y. Guo, T.C. Chen, and Z.S. Chen. 2010. Remediation techniques and heavy metals uptake by different rice varieties in metals-contaminated soils of Taiwan: New aspects for food safety regulation and sustainable agriculture. *Soil Science and Plant Nutrition* 56:31–52.

Hseu, Z.Y., C.C. Tsai, H. Tsai, Z.S. Chen, and H. Eswaran. 2011. Asian Anthroscapes: China and Taiwan. In *Sustainable Land Management: Learning from the Past for the Future,* eds. S. Kapurke, H. Eswaran, and W.E.H. Blum, 205–241. Elsevier: Amsterdam.

Jien S.H., Z.Y. Hseu, C.C. Tsai, H.Y. Guo, and Z.S. Chen. 2012. Soil carbon stocks in soils of rural soils and forest soils in Taiwan. In *Soil Organic Matter: Ecology, Environmental Impact and Management*, eds. P. A. Bjorklund, and F. V. Mello, 63–88, Hauppauge, New York: Nova Science.

Karlen, D. L., M.J. Mausbach, J.W. Doran, R.G. Cline, R.F. Harris, and G. E. Schuman. 1997. Soil quality: A concept, definition, and framework for evaluation. *Soil Science Society of America Journal* 61: 4–10.

Lai, H.Y., Z.Y. Hseu, T.C. Chen, B.C, Chen, H.Y. Guo, and Z.S. Chen. 2010. Health risk-based assessment and management of heavy metals-contaminated soil sites in Taiwan. *International Journal of Environmental Research and Public Health* 7: 3595–3614.

Lee, C.H., M.Y. Wu, V.B. Asio, and Z.S. Chen. 2006. Using soil quality index to assess the effects of applying manure compost on the soil quality under a crop rotation system in Taiwan. *Soil Science* 171: 210–222.

Soil Survey Staff. 2010. *Keys to Soil Taxonomy*. 11th edition. Washington, DC: U.S. Dep. Agric., Soil Conservation Service.

Su, S.W. and Z.S. Chen. 2009. Cadmium exposure from plant-based food consumption in global population especially in Asia and techniques to reduce the risks to human health. In *Proceedings of the 9th International Conference of East and Southeast Asia Federation of Soil Science Societies (ESAFS)*. Seoul, Korea, Oct 27–30, 2009, eds. J.S. Suh, K.H. Kim, J.H. Yoo, J.K. Kim, M.K. Paik, and W.I. Kim, 54–60. Seoul: Korean Society of Soil Science and Fertilizer.

Su, S.W. and Z.S. Chen. 2010. Liming and P addition slightly increase As availability in As-contaminated soils. In *Arsenic in Geosphere and Human Diseases As 2010: The Third International Congress on Arsenic in the Environment*, ed. J. S. Jean, 172–173. National Cheng Kung University, Tainan. Taiwan. May 17–21, 2010.

Taiwan EPA (Environmental Protection Administration). 2013. Soil remedial sites of Taiwan. (http://sgw.epa.gov.tw/public/En/index.htm) (Verified on April 5, 2013).

TASGEP (Taiwan Association of Soil and Groundwater Environmental Protection). 2013. The website of TASGEP. http://www.tasgep.org.tw (Verified on April 5, 2013).

Tsai, C.C., T.E. Hu, K.C. Lin, and Z.S. Chen. 2009. Estimation of soil organic carbon stocks in plantation forest soils of northern Taiwan. *Taiwan Journal of Forest Science* 24: 103–115.

5 The Far Future of Soil

*Peter K. Haff**

CONTENTS

5.1 INTRODUCTION

The word "Far" in the title of this chapter indicates a focus on the state of the Earth's soils far enough in the future that present trends and recent experience with patterns of soil change and function begin to lose their predictive value. The time span of interest is decades to a century or more. Despite the difficulties of long-term prediction, it is worthwhile pondering the possible states of soil over a time period that will be experienced by our grandchildren, or even our children. On geologic time scales, measured in millions or billions of years, a century is very short. However, the apparently inexorable rise of technology as a new force in the world means that we should not be surprised to see transformative change in soils over this period of time. In the next section, we discuss how technology is likely to set the stage for future evolution of soil.

Thinking about the far future of soil one must move beyond a veil of uncertainty that clouds all attempts to predict the distant future. It is impossible to know how the background environment that controls or defines the conditions for soil evolution might change. One obvious background control variable is climate, whose own future is highly uncertain. This defect cannot be cured by running climate models because these are subject to the same limitation as prediction of soil futures—we cannot be sure of background assumptions (no none knows if China's economic boom and production of atmospheric carbon will continue over decades or not). The upshot is that, over long enough periods of time, the assumption of a known background is always going to fail, and so will predictions based on such an assumption. So at the outset of an endeavor to discuss the far future of soils, we seem stymied by an inherent barrier to prediction.

Then what can we say? One approach is to develop scenarios—that is, to tell stories. Of all the ways in which the future might unfold, we pick one and sketch out how it might plausibly unfold, given all that we know about the system. Scenarios are sometimes used in business and government for building awareness of possible futures that cannot be predicted with any certainty (van der Heijden 1996). A scenario is not a prediction. However, being alert to what might happen, although less satisfying than reliable prediction of what will happen, is better than not thinking about future

* Nicholas School of the Environment, Duke University, Durham, North Carolina 27708 USA; Email: haff@duke.edu

possibilities, and far better than believing the distant future can be predicted. A scenario can suggest measures to avoid problems that might occur, and can prepare us to act when new opportunities arise.

Below we outline several scenarios for the future of soil. We focus on how technology is defining a new set of driving forces that may have large consequence for soil conditions in the future. What technologies will emerge over the coming decades is an even more impossible question than the future of soils, so we look for a general way to approach the question of the interaction of soil and technology that appeals to general principles that are independent of the precise path over which the future unfolds. We begin by situating the emergence of technology and technological change within a geologic framework of Earth evolution in order to give us a clearer picture of the relation between technology, soils, and people.

5.2 THE ROLE OF TECHNOLOGY: A NEW EARTH PARADIGM

The starting point of each of our scenarios is the assumption that technology will be an important determinant of the properties of the future soils of the world. This assumption is of course part of the scenario. The main point is not just the increasing impact on soils of familiar technological processes like farming and urbanization, which have led to large-scale erosion, scrambling of soil horizons, and alteration of natural mineral compositions through addition of fertilizers and other chemicals. These are important issues, especially for the near- and intermediate-future prospects for soil productivity; but in a technological world, there is a larger dynamic in play that may have an even more profound effect on the future of soils.

The grandest earthly processes, those of the largest geographical scope, and those that consume the most energy and require the greatest quantity of Earth resources to support their function, belong to the class of geological phenomena that can be called "spheres" or "paradigms." The classical spheres include the atmosphere, the hydrosphere, the lithosphere, and the biosphere. To this we add the realm of soil that by virtue of its global extent merits the appellation of "pedosphere." In a broad-brush perspective, the dynamics of these geological paradigms has to a large extent defined the natural world for billions of years. Today, however, a new sphere has arisen, defined by the combined actions of humans and technology. This can be called the *anthroposphere*, or, the term used here to emphasize the role of technology, the *technosphere* (Haff 2012).

To put the potential transformative power of technology on soils in perspective, it is instructive to compare the properties of the different spheres, including the pedosphere, with those of the technosphere. First, all the spheres are global in extent. Processing of water by the hydrologic cycle, the slow movement of solid earth materials in the rock cycle (plate tectonics, subduction, and continent building), the flow of winds in the great atmospheric circulation cells, the production of soil by weathering from the underlying bedrock and soil loss by erosion, and the biologic cycle of growth and decay all operate at continental scale or larger. Similarly, the influence of the technosphere extends broadly across the surface of the Earth in the form of communication, transportation, and power networks, all connected to a profusion of localized technologic hotspots from individual dwellings to cities which in turn are decorated with myriad constructed artifacts and devices. The fact that the technosphere has spread across the surface of the Earth in a few centuries, and continues to grow, should alert us to the possibility that this new phenomenon may have disruptive effects on the other spheres, including the pedosphere.

In addition to being expansive, the spheres are also massive. In order to compare the masses—the amount of material—in each sphere we imagine the following. First, the matter that comprises each sphere is assumed to be distributed over the surface of the Earth as a perfect spherical shell. Second, to make the comparison easier, each shell—representing the atmosphere, hydrosphere, biosphere, and technosphere, respectively—is taken to be squashed down so that its density is the same as the bulk density of an average soil. A nominal value of soil density is taken here as 2 metric tons per cubic meter. This makes it easy to visualize the relative amounts of materials, like air and soil, that have different densities, and otherwise are hard to compare. For example, if we tossed a shoveful of soil containing 1 kilogram of mass into an empty cubical box measuring 1 cubic meter in volume, it might not be obvious that the mass of air in a box of the same size and shape had about the same mass as the soil. But if we squashed down the air to the bulk density of soil, and smoothed the soil out over the bottom of its box, then we would see that soil and (compressed) air formed layers of similar thickness (on the order of 1 mm), that is, they had equivalent mass. In squashing down the various spheres, the surface of the Earth plays the role of the "box" in the previous example. The following numerical estimates on properties of the spheres are mostly order of magnitude and intended for purposes of illustration only. In these terms the relative thickness of the spheres is: the lithosphere, 100 km; the hydrosphere, including the water in the ocean, 1000 m; the fresh water part of the hydrosphere, 30 m; the atmosphere, 5 m; the pedosphere, including saprolite or weathered bedrock, 1 m; the most biologically active part of the pedosphere, 25 cm; the biosphere, taken here to represent the mass of living organisms, 1 mm; the technosphere, whose mass represents that of all built things, including buildings, highways, and manufactured goods, perhaps 0.1 mm.

The thickness of the technosphere seems paltry compared to that of the hydrosphere and atmosphere, and even compared to the modest 1 m of the pedosphere. How could such a small mass have any significant effect on a much larger pedosphere, or on the biosphere for that matter? One reason lies in the geometry of the spheres—that they are layers. We know from many examples that a small amount of material spread as a thin layer spread across a large surface area can have effects that are far out of proportion to its mass. A plastic sheet 0.025 mm thick effectively preserves the freshness of a sandwich that is 25 mm, or 1000 times, as thick. The technosphere of course is not literally an impervious sheet, but like the biosphere, its coverage is still sufficiently dense to have a profound impact on the much thicker pedosphere that it overlays.

Another reason for the importance of the technosphere is its profligate energy consumption. Power-wise, the technosphere dominates the pedosphere and presently is beginning to challenge the power level of the biosphere and of the freshwater part of the hydrosphere. The approximate energy budget (Hermann 2006) of the spheres as they affect the land surface looks like this: the atmosphere, via wind friction with the land surface, 150 TW (terawatts, 10^{12} W); the biosphere, via production of chemical energy stored in plant matter, 90 TW; the technosphere, 80% of whose power use is from fossil fuels, 17 TW; the hydrosphere, via river flow, 7 TW; the hydrosphere, via raindrop impact on land, 0.01 TW.

The pedosphere is low man on the metabolic totem pole. The energy available to the soil to do work includes its gravitational potential energy, which is consumed through downslope movement at a rate of about 0.01 TW. Much more energy is available to the technosphere than to either the pedosphere or the fluvial part of the hydrosphere to do work on the environment. In view of this ranking, we should not be surprised that technology can be an effective agent of landscape change, especially as it affects rivers and soils. Moreover, the imbalance is becoming more extreme. While the pedosphere and other natural spheres are on average relatively stable in their level of energy consumption, energy use by the technosphere is currently increasing by more than 2% per year (BP 2012).

To get a better sense of the direct impact of technology on soils, it is useful to consider the amount of soil and rock moved by technological agency per year, compared to the amount displaced by natural causes. Hooke [1994; see also Haff (2010)] estimated, in order of magnitude, the mass of material displaced by highway construction, house building, mining, and plowing, as well as the mass displaced by a variety of natural processes. According to these calculations, the total soil displacement for technology-enabled activities, except plowing, was estimated to be on the order of 30 Gty^{-1} (gigatons per year; one gigaton is 10^9 tons; for scale, a gigaton of water occupies a volume of one cubic kilometer). More soil is moved by plowing than by all other technological processes combined, about 1500 Gty^{-1}, quantifying the dominance of agriculture as the premier technological soil-moving phenomenon in terms of mass displaced per year.

Soil and rock displacement by natural processes such as transport down a river channel, lateral displacement by river meandering, downslope creep on hillslopes, and movement of soil by the wind was estimated to be about 50 Gty^{-1}. The rate of technology-enabled disturbance of the Earth's surface in terms of mass moved per year is thus more than an order of magnitude greater than the rate of displacement by major natural processes. (For convenience the word "natural" is used here with its traditional implication of origin independent of human or technological activities. However, technology and the technosphere are also natural in being products of Earth evolution; see final paragraph of this section).

Technological impact on the pedosphere tends to be long lasting. This is, soil has a memory. If the wake of a boat mars the surface of a river, the effects of the disturbance are transitory. The wake disperses and the river rapidly returns to its original state. But when a plow tills a row of virgin soil, disrupting no larger a volume of material than does the boat cutting through the water, the soil configuration does not quickly, if ever, return to its original state. Soil horizons are overturned or scrambled, and the broken soil, assuming it does not succumb to erosion, retains an enduring memory of its encounter with the steel technology of the plow. The soil records its own history. Without memory, soil would return to its initial state and the passage of the plow would soon be erased, as the ruffled water surface quickly calms and forgets the passage of the boat.

Memory is erased from the river in minutes, but from soil only over millennia or longer. There is little chance for recovery of native soil once technology has disturbed it, except over very long time spans. Of course, agricultural soils are repeatedly plowed; each plowing, if we neglect erosion and loss of fertility, leaves the soil in the same desirable state, receptive to the seed. The recurring plowed state is an example of what in nonlinear systems theory is called an "attractor"; of all the possible states that a system (the soil) may occupy under given conditions, there is one toward which it tends to return when gently disturbed. Luckily for humans, this attractor state is one that will support an abundant growth of food crops. However, it is not a natural state of the soil. In fact, one could argue that plowed soil is really a construct that belongs to the technosphere as one of its parts, in the same way that buildings constructed from wood are no longer considered natural systems or parts of the biosphere.

Up to the present time, the main impact of the technosphere on soils has been the proliferation of the plowed profile (and of follow-on effects such as erosion, nutrient depletion or augmentation, and revegetation of abandoned soils). However, the memory capacity of soil is not eradicated by plowing or any other process that does not physically remove the soil. Under the influence of a high-energy, rapidly changing technosphere, new kinds of information much richer than the incidental recording of the passage of a plow could potentially be downloaded into the memory of the pedosphere. In a long-time perspective, we should not be surprised to see the emergence of highly altered soils with properties that are novel and surprising. This possibility is the basis for the discussion of the smart soil scenario below.

A final and critical fact about the technosphere that bears heavily on the future of soil is that the technosphere is not under human control. Normally we think of technology as a set of human "inventions" that we design, manufacture, deploy, and maintain in an attempt to make our environment conform more to our desires. A furnace may warm our house; an automobile transport us long distances; a farm provide food for the table, relieving us of the need to hunt and forage. This is true as far as it goes. However, technology-in-the-large grew organically with no overall plan or management. As the different pieces of the technosphere were assembled and networked, the size and complexity of the resulting system became too large for understanding by any one person or group of humans. The only way for such a system to function coherently over extended periods of time is if its dynamics—that is, its rules of behaviors and patterns of organization—become independent of direct human control. That is, the technosphere had to become autonomous in order to function in all its complexity. Despite its human origins and dependence on its human components, the technosphere can now be seen as a newly emergent geologic system—that is, a product of Earth evolution—that functions according to its own dynamical laws and that now helps define the environment within which humans must live (Haff 2012). The future of soil will be affected by the development of technology, but the fact of technological autonomy means that we have no assurance that the impact of technology on the pedosphere will bear any necessary or direct relation to human design or planning. Therefore, we expect that the future of soil will be determined as much by the autonomous dynamics of the technosphere as by human intention. Under these conditions, what can one say about soil's future?

5.3 EXISTENTIAL SOIL AND NATURAL CAPITAL

At the outset, let us narrow our discussion of the future of soil by focusing on a set of possible outcomes that are not so extreme that human existence becomes impossible. Fifty years ago such a qualification would have seemed ludicrous, but today, global warming is leading us into *terra incognita* whose possibilities, however unlikely, include end states like a run-away greenhouse that might transform Earth into the next, uninhabitable, Venus. If we restrict consideration to a future in which our grandchildren might be able to live, what can be said about possible "existential soils" of such a world? That is, what does soil provide that is essential to human existence, and human well-being?

The essential product that soil provides for us is natural capital. Natural capital (Daily 1997; Clothier and Kirkham 2014) is the source of all the goods and services arising in the nontechnological world that are essential for our existence and well-being but which fall outside the standard accounting of economics. Natural capital inherent in soil includes its fertility; its water absorbing, filtering, and transporting capacity; the support of food production through the mechanical foundation, nutrients, and moisture that soil provides to plants; its fine granularity that provides easy penetrability to roots, and the sheltered microenvironment it offers to essential microbial organisms. Soil also offers us less concrete goods and services that are nonetheless required for our well-being, if not for our existence. These include the delighting layers of sand in which the child plays, the forgiving surface of the athletic field (Gibbs 2014), the expressive media of clay for the use of artist and potter, the carpet that sustains the forest stands to whose bowers we turn for renewal from the complexities of everyday life, and the soils of tribal, hereditary, or religious tradition from which solace and inspiration well up in those attached to them.

Below we develop four scenarios for possible soil futures with an eye to the role of soil as a source of natural capital.

5.3.1 Logistic Soils

In this scenario, we consider the soils likely to develop in the future if the technosphere continues its current pattern of resource use. This is the business-as-usual scenario. The word logistic is applied here in the same sense that the term is used in biology, to describe the relation of the rate of growth of an animal population to the finite resources available to support that growth (e.g., Nowak 2006). If the rate of increase is not too fast, then population growth slows down and eventually ceases at the logistic limit or "carrying capacity" of the system. Fertile soils and arable land represent one type of logistic resource that defines a carrying capacity for the human population. It is useful to consider the technosphere, rather than its human components, as the primary consumer of natural capital, in somewhat the same sense that a human individual rather than her cells (where the actual chemical release of energy is performed) is the agent responsible for consuming food calories flowing from farmed land. The logistic limit then applies to the growth of the technosphere as well as to human population.

In the logistic picture, when a population increases in size rapidly enough, the population overshoots its carrying capacity. The result is usually one of two conditions. In the first, the population fluctuates between a size that exceeds its carrying capacity and a size below carrying capacity, from which the population then recovers to begin a new expansion, repeating the cycle. In the second, the dive is so fast and the resulting population so small, that remaining organisms die out before recovery is possible. The human population probably reached its carrying capacity with respect to biosphere resources in the 1980s, and by about 2000 was ~20% in excess of the carrying capacity (Wackernagel et al. 2001).

With markets, at least in developed countries, always seemingly overladen with food, one might think otherwise—that the world's soils are well able to supply whatever we require. After all, even if technology uses up natural capital, it also enhances it; for example, in the fertilizers that make each hectare of soil more productive than it would be otherwise. It is tempting to ask, can't we take the idea of fertilization to its extreme and just bootstrap ourselves out of harm's way by applying technology to continuously renew and even regenerate soils, and other elements of threatened natural capital? We return to a further consideration of this question in the last scenario, but the fact is that today the answer is a flat out "no." A glance at the map of the nighttime lights of the world (Elvidge et al. 2001), which provide a proxy for the geographic distribution of the main energy-consuming elements of technology, shows that the physical realization of the technosphere—and of the human population which is tethered to it by short apron strings—is confined primarily to two mid-latitude bands. This is where a golden conjunction of ample rainfall, equable temperatures, and fertile soil provides abundant natural capital without which technology and humans cannot survive. It is not just that the technosphere depends intimately on good soil; it is that for the most part, that relation must be local. Aside from outposts like Phoenix, Arizona, in the United States, or Dubai in the United Arab Emirates, which divert to their own use from distant sources requisite products of natural capital, the technosphere has not been able on the whole to extend its base (as opposed to its extractive arms) away from regions of productive soils. Technology is robust and powerful as long as it remains rooted close to the soil that sustains it.

What then would soils look like if the time came that large negative logistic oscillations led to dramatic shrinkage of the technosphere? If the technosphere collapses, remaining soils will at first be quickly degraded or abandoned. In the absence of a robust technosphere, large-scale and intensive chemical fertilization, irrigation, pesticide usage, sowing and harvesting, mechanized plowing and other soil manipulation protocols, and long-distance, fast, and cheap connections to distant markets (should such markets against all expectation continue to exist) would become impossible. Soils will quickly be degraded through overuse as their fertility is lost, fallow and cover cropping is abandoned, water and wind erosion of exhausted, bare soil strips remaining top soil, and gullies emerge to put the final seal of destruction on once fertile arable ground. A preview of this scenario at smaller scale has played out before in many regions. A good example (Trimble 1974) is the piedmont zone of the southeastern United States. Here the edges of an expanding technosphere marked by increased population, proliferation of axe and plow, and growth of trade routes to Europe—the then epicenter of technology—pushed inward across virgin American soils in the late eighteenth and the nineteenth centuries. After a burst of productivity, a parcel of cleared and then farmed land would quickly become exhausted, and then, left unprotected from erosion, become scored by deep gullies. A layer of topsoil on the order of 10 cm thick was stripped from the fields of the southeastern U.S. piedmont and deposited at the base of local slopes and in streams.

The loss of good soils from the piedmont was a direct consequence of the technosphere exploiting the appetite of humans for cotton, tobacco, and other agricultural products, and of the application of slave labor to working the land—black human beings being even more enslaved to the requirements of the technosphere than their white owners, who were happily ignorant of the fact of their own indenture to the same technological master as their slaves. Failing soils were temporarily rescued by technological means when fertilizer became available for widespread application, but, as indicated above, technology cannot produce more natural capital than it devours, and land productivity finally declined to the point of widespread abandonment. The abandoned soils of the southern piedmont are now slowly recovering in a regime of mixed natural and industrial reforestation. What has saved the land from a logistic fate of complete ruin was the fact that new soils came online, for example, in the American Midwest, and, better managed and more extensive, relieved the pressure of production on piedmont soils. But, on the global scale, there is little reserve of additional soil to bring into production, and if a business-as-usual logistic scenario plays out we may find future soils to be those of exhaustion and abandonment.

5.3.2 Picture Postcard Soils

The logistic scenario assumes that humans will not react quickly or decisively enough to avoid catastrophic damage to the world's soils. That humans are deeply dependent on the technosphere, and cannot control it directly, however, does not mean that they have no influence on how the future unfolds. Despite their subordination to its dynamics, humans are collectively both the cognitive and reproductive organs of the technosphere. Many people, and certainly most of the readers of this volume, understand that preserving soil natural capital to secure the goods and services that flow from it is essential to our future well-being. Clearly the well-being of the technosphere—that is, its viability—also depends on this same flow because, without humans as consumers and maintainers, the technosphere would cease to function. The technosphere is just as dependent on maintenance of good soils as are its human components.

One goal of responsible environmentalism is to move from our current logistic trajectory toward a path that leads to a future where both humans and the technosphere they belong to can function "sustainably," that is, without net destruction of essential natural capital. Humans have developed policy ideas—which are vastly easier to implement than policy actions because they have no physical effects or costs—to halt or reverse the ongoing technology-driven changes that are responsible for increasing pressure on soils and other components of natural capital. In a nutshell, these policies would attempt to rein in the technosphere, to alter its metabolism through more conservative energy and resource use in order to arrive at a state in which soils and other forms of natural capital can be regenerated as fast as they are depleted. The goal would be to return to a state where large tracts of soils and ecosystems and wetlands were protected from further damage and restored to function according to the ancient natural laws they had always followed. In this scenario, the surface, of the Earth would become more like it "used to be"—like the scene in the picture postcard sent back to the city from a national park vacation or a trip to the countryside. The soils that would populate this future would be the deep and fertile soils of the Holocene (the last classical epoch in the geologic pantheon of time prior to the emergence of the man- and machine-dominated Anthropocene of the present day).

This future may be a difficult one to reach despite apparently desirable characteristics for both people and the technosphere. The most fundamental problem is energy use. In principle, one way to decrease the chance of a logistic fate is to use less energy. Tamping down, or even moderating, the metabolic rate of the technosphere is, however, fundamentally difficult because it goes against the grain of technosphere dynamics. The technosphere is a high-metabolic-rate phenomenon that continues to increase its rate of energy consumption. The classic natural spheres have already spun up to maximum energy use, but the technosphere is still on the rising limb of the logistic curve. This is because it was only recently in geologic history (human intelligence was required) that strong positive feedback loops emerged, powered by hitherto stranded chemical energy resources—the world's pools of oil, stores of gas, and seams of coal. The positive feedback that drives the technosphere is built on incentivization of its human parts; the technosphere gives us not only what we desire, but what we require for our well-being. It provides food, security, better health and other desiderata, and we do what is necessary to get what it offers us. Positive feedback loops are phenomena that outcompete other mechanisms for use of resources from their environment. They are robust. Human attempts to dial back energy use directly oppose this basic dynamic of technospheric function. Given the technosphere's intrinsically aggressive appetite for power and resistance to restraint, it is not at all clear that efforts to moderate global technological energy use will be successful in time (or at all) to avert catastrophic damage to soils and other critical elements of natural capital. The route to picture post card soils in the future is a rocky one and progress is likely to be slow.

5.3.3 Geoengineered Soils

Geoengineering is the term usually applied to plans for the willful human manipulation of the world climate system—specifically, to methods to dial-back the mean atmospheric temperature toward historic (Holocene) norms (Keith 2000). An attempt to directly manage climate in order to counteract the warming effect of carbon dioxide presumably would be a response to a perceived existential threat to a technologically capable country—for example, repeated grain failures attributed to the upward trend of atmospheric temperature. Geoengineering would amount to a desperate attempt to preserve natural capital by cancelling out the most serious consequence (atmospheric warming) of the failure of technology to recycle its own carbon waste stream. Geoengineering is unlikely only if an existential threat to society is unlikely. Some scientists think the threat is real (e.g., Ehrlich and Ehrlich, 2013).

Many ideas have been floated to accomplish an engineered cooling of the planet, and a few exploratory field experiments have been attempted (e.g., stimulation of phytoplankton blooms by iron fertilization of the ocean) (e.g., Coale et al. 1996). However, the geoengineering method most likely to be deployed first at large-scale is alteration of planetary albedo by increasing the concentration of aerosols in the upper atmosphere (Crutzen 2006). We will refer to the soils resulting from a geoengineered climate as geoengineered soils for simplicity, even though soil is not the direct target of the geoengineering efforts.

That climate has a determinative effect on soil properties was formally encoded by the great soil scientist Hans Jenny in his soil state function (Jenny 1941), which indicates the dependence of soil properties on climate together with other critical determinants like organismal activity, relief (slope), parent material, and time. If the consequence of geoengineering were simply to return climate to its state of say a century ago, then we would be back on a logistic trajectory, with continuing drawdown of natural resources including soils (because the metabolism of the technosphere would not have been decreased), but perhaps at a slower pace lacking the accelerant of rapidly increasing air temperature. However, the effects of a geoengineered climate would not necessarily be uniformly distributed across the globe. Some research suggests that the distribution of precipitation would likely be different from the historic Holocene record. For example, increasing concentration of CO_2 in the absence of climate warming could lead to closing of leaf stomata, causing a significant reduction in evapotranspiration and thus large decreases in precipitation over land, especially in the tropics, with implications for local soil moisture and surface runoff conditions (Matthews and Caldeira 2007). Sudden changes in climate, either wetter or dryer, would lead to soils that were out of equilibrium with their formative processes as described by Jenny's soil state function—for example, by waterlogging of desert aridisols, or increased erosion of previously stable soil horizons.

Geoengineering may seem like a science fiction adventure that, if it should occur, would do so in the distant future. However, cooling the climate with aerosols is likely to be cheap (perhaps US$50 billion per year), doable with essentially off the shelf technology (such as a fleet of balloons), and is reversible (particulates would fall out of the atmosphere in a few years) (Crutzen 2006). Moreover, the barriers that resist the transition to a picture postcard planet would be lower in the case of geoengineering. Geoengineering is a technosphere-compliant action that does not infringe on energy consumption. Aerosol loading of the atmosphere may even lead to increases in consumption, as people and governments feel they are let off the hook regarding serious action to cut back on carbon dioxide emissions. This is one of the principal environmental arguments against going the geoengineering route. However, the technosphere-compliant nature of geoengineering increases the probability of its occurrence. Geoengineered soils might be closer than we think.

5.3.4 Smart Soils

In the above scenarios on existential soil futures, it was assumed that soil would remain an important source of natural capital. However, in the far future one could imagine a technosphere of sufficient complexity and power that technological systems could substitute for most of the natural systems that currently serve as sources of natural capital. This is a future where technology finally relegates natural capital to an anachronism, and natural soil, rivers, and even much of biology become unnecessary as suppliers of hitherto indispensable natural goods and services. We have shown that technology today is not remotely powerful enough to replace natural capital provided by soil and other natural systems, but in view of the rapidity of technological change we are led to consider the fate of soil on a future technology-dominated planet. What would be the properties of soil in such a world?

One observation from geologic history is that great earth changes—for example, the emergence of vascular plants onto the land surface in the Silurian period—tended to modify, but not to eliminate, the basic elements of the preexisting world. For example, soils were thinner in the pre-vegetated world because of the role of plants in the physical and chemical breakdown of bedrock, because plant roots help hold soil in place, and because plant canopies absorb some of the momentum of the wind that might otherwise lead to soil erosion. Also, the appearance of bank vegetation tended to confine rivers to a single well-defined channel, discouraging their pre-Silurian tendency to split and recombine in a braided pattern. Despite these changes, both soils and rivers survived, in modified form, after the continents had been colonized by plants. The same conservatism may prevail in a world in which the land surface has become densely colonized by technology, with a soil blanket still in place but with properties modified or repurposed by the new dynamic regime. What a technologized soil would actually look like, what its properties would be, no one knows. But as an idea of how far the soils of Earth might be expected to stray from their current functionality should the technosphere finally cement its status as a geologic paradigm, we offer the following thought experiment.

Currently, something like 10^{21} transistors have been manufactured and dispersed into the environment, most of these in recent years (extrapolating from data in Moore 2006). These tiny artifacts appeared first in large discrete devices like radios. With increasing miniaturization of circuits, faster speeds, lower costs, and the development of microsensors (of light, temperature, vibration, and so on), small clusters of transistors (chips) migrated to watches, cell phones, toys, medical devices, coffee pots, toasters, refrigerators, and more generally into the wider environment as parts of sensors for environmental monitoring and surveillance and for security networks. Meanwhile, a project to barcode all the world's species is underway (iBOL 2013), and an artificial leaf that mimics photosynthesis (Nocera 2012) points the way toward techno-transformation of the most basic process—plant growth—by which soil natural capital benefits humans. The growing conjunction of artifacts with computer chips and wireless communication is the opening wedge of the so-called "Internet of Things" (Atzori et al. 2010), where many elements in our environment become computerized. The natural endpoint of the proliferation of miniaturization and granulated computation across the surface of the Earth may end with the computerization of soils.

The land area of the Earth is in round numbers 10^{14} m^2. If computing power were apportioned uniformly over this area, there would be about 10^7 transistors assignable today to each square meter of soil. A modern laptop contains on the order of 10^8 transistors, so in the not-so distant future more computing power than is now available in a powerful laptop computer would be available per square meter of land surface. Two decades in the future, if Moore's law—that computing power approximately doubles every year (Moore 2006)—continues to hold, computing power per square meter would be measured in units of millions of today's laptop computers. We imagine that some fraction of this computation base is actually deployed on the land surface. Because of the inexorable shrinkage in size of electronic circuits many sand-grain-size or smaller computers might populate each square meter. So we imagine the following picture. Soil-grain-sized computers are liberally scattered over each square meter of the Earth's land surface. Power is provided from ambient energy sources. For example, average incident solar power measures on the order of 100 W/m^2 at the Earth's surface. Wind, seismic vibrations, and other low-power sources are also available. Individual soil computers are wirelessly networked to their near neighbors, so that information can be shuttled in and out of each local area, or are hard wired together with filamentary cables—a communication strategy already pioneered by biology in the form of fungal hyphae. Cables would also provide mechanical stability to soil computing elements which otherwise would be susceptible to erosion by wind or water. Because of ambient power limitations, large-scale data crunching would be shifted offsite to centralized computing centers dedicated to processing continuous data streams coming in from the field.

What would these computers be computing? In a world with a dense and relatively uniformly distributed set of miniaturized computers, sensors, and actuators (systems that can exert physical force on their surroundings), local computation would be used to collect and relay environmental information about the state of the soil—temperature, moisture, relative motion of near-surface particles—and to coordinate with instructions relayed back from central computing for activation and control of local actuators. The actuators would be able to change the configuration of nearby soil particles to create a modified surface whose parameters would then be fed back again to central computing for further updating. For example, actuators might be directed to move themselves and nearby grains incrementally upslope to counteract the effects of local erosion. To move a square-meter-size monolayer of sand-grain-sized particles upslope a distance of 1 cm on a slope of 4° would require about 0.05 J of energy to overcome the force of gravity, and perhaps 1 J to overcome friction with underlying grains. One joule of energy is delivered on average by sunlight to a square meter of the Earth's surface in about one-hundredth of a second. In other words, with sufficient computing power and development of suitable microsensors and actuators, the dynamics of the Earth's soil can be divorced from the classic forces of gravity, rain, and other environmental forces, and in turn become subject to information-based forcing by the technosphere.

In 1959, the soil scientist Nikiforoff (1959) sketched a picture of soil as the "excited skin" of the Earth—a natural dynamical system consuming energy and interacting with its environment. This metaphor is continued in the present section to include the effects of technology—the computerized soil of the Earth becoming in a sense aware and responsive, regenerative and reactive like our own skin, acknowledging stimuli and the analog of injury and pain, able to do much of its own self-repair, but requiring intervention of greater computing power and application of force from an outside source to deal with larger scale stresses and threats to integrity, such as flooding, that arise from beyond the local region. This is similar to the intervention of our brain and other parts of our central nervous system in activating muscles when, for example, we burn ourselves, changing the entire environment of the damaged skin by mechanisms (like moving our hand away from the fire) that have nothing to do directly with skin repair and maintenance processes. Generating a hierarchy of computation and control, and a back and forth flow of information between the whole and its parts, is one of the signature features of the evolving technosphere, and we should not be surprised at some point in time to see the soils of the world become integrated into such a hierarchy.

This scenario may seem farfetched, but in reality many of its elements are already in place, only in a cruder and less distributed way than in the example above. From the eyes of the farmer, hunter, and surveyor, to the eyes of Google Earth and military satellites, the surface of the Earth is already under high-resolution scrutiny like never before in Earth history. The resulting information is processed to determine what forces or actions should be applied to the land—to plow, to log, to plant, to harvest, to clear, to grade, to build, to invade, to occupy, and to protect from plowing, logging, building, invasion, and other impacts. Local computation is still mostly sparse, but data collection is widespread. With the dropping price and increasing power of computer chips, sensors, and actuators, it can be expected that much of what one is tempted to call management of soil, land, water, and vegetation (but which, more fundamentally, is cooption of people and of the land, soils, and other elements of the landscape into the sphere of technology) will continue to increase to the point where soil and other surficial Earth systems are recognized more as technological artifacts than natural systems.

What would a smart soil world look like? One expects that the form and function of soil captured by the technosphere will evolve to support the metabolism of the technosphere, as this is the most fundamental need of technology. Already a large fraction of the Earth's soils have been transformed to this end through agricultural technology, that is, to underwrite the production of food, a necessary power source for essential components of the technosphere. However, today's soil is not yet smart. As the computational power of soil emerges, soil might move beyond agriculture to host other globally distributed functions, such as distributed recycling of technology-generated wastes (the most important today being carbon dioxide) as the technosphere works to replace natural capital services that soil historically provided. Global forests of sunlight-powered artificial trees designed to optimize chemical capture of carbon might cover the landscape, with transistorized, reactive soil programmed to chemically sequester the captured carbon as carbonate or other semi-stable compound within the soil matrix. Perhaps the emergence of smart soil is also the most likely path for a return to something resembling a picture postcard landscape. As long as humans remain essential to the function of the technosphere, they must be incentivized to cooperate with it. The biophilia hypothesis of E. O. Wilson (1984) suggests that humans are deeply inclined to explore and affiliate with life, hence our attraction to the biology-rich picture postcard landscape. If future soil turns out to be a programmable medium, then its program might include the generation and support of faux vegetation—fake, but in large stands attractive and soothing, or even inspiring, to humans at the same time it acts as a powerful sequestering agent of carbon or performs other, once natural, services.

5.4 CONCLUSION

The far future of soil cannot be predicted, even approximately. The best we can do is play out scenarios that are consistent with general principles that can be inferred from what appear to be long-term controlling factors of Earth evolution. The trajectory of Earth evolution over the past few centuries appears to point toward the continuing growth and increasing dominance of large-scale technological networks. These together with all their smaller parts, including their human components, constitute the technosphere, which is argued to be an emerging geologic paradigm with features in common with the biosphere, hydrosphere, and other great Earth paradigms. Technology acts as a kind of universal acid that dissolves and recrystallizes into its own form, everything it comes into contact with, including nature. Neither the future of soil nor of any other natural Earth system should be considered immune to the technologizing process. In the narrative above, four possible futures of the Earth's soil mantle are discussed, each of which is consistent with the reconfiguration of the Earth by technology, and which illustrate the need to preserve (or, eventually, to substitute for) essential natural capital embodied in the Earth's soil resources. To be ready to deal with the future in a world that is being transformed faster and more radically than ever before in human history, it is necessary to recognize the fact that both humans and soils are now subject to the actions of an overarching system, the technosphere, which is not under direct human control, and whose future is not predictable. Mathematical models and other methods of traditional science cannot tell us how these forces may play out. However, we can tell stories. Stories alert and help prepare, even if they cannot predict.

ACKNOWLEDGMENTS

I thank Stacey Worman for her insightful comments on soil intelligence and her tenacious argumentation about the nature of technology. I thank David Jon Furbish for a critical reading of the manuscript.

REFERENCES

Atzori, L., A. Iera, and G. Morabito. 2010. The Internet of things: A survey. *Computer Networks* 54: 2787–805, doi:10.1016/j.comnet.2010.05.010.

BP. 2012. BP Statistical review of world energy, June 2012. bp.com/statisticalreview (accessed January 19, 2013).

Clothier, B. and M.B. Kirkham. 2014. Soil: Natural capital supplying valuable ecosystem services. In *The Soil Underfoot*, eds. G.J. Churchman and E.R. Landa. 135–149, Boca Raton, FL: CRC Press.

Coale, K.H., K.S. Johnson, S.E. Fitzwater et al. 1996. A massive phytoplankton bloom induced by an ecosystem-scale iron fertilization experiment in the equatorial Pacific Ocean. *Nature* 383: 495–501, doi:10.1038/383495a0.

Crutzen, P.J. 2006. Albedo enhancement by stratospheric sulfur injections: A contribution to resolve a policy dilemma? *Climatic Change* 77: 211–19, doi: 10.1007/s10584-006-9101-y.

Daily, G.C. (ed.) 1997. *Nature's Services: Societal Dependence on Natural Ecosystems*. Covelo, CA: Island Press.

Ehrlich, P.R. and A.H. Ehrlich. 2013. Can a collapse of global civilization be avoided? *Proceedings of the Royal Society* B 280: 20122845. http://rspb.royalsocietypublishing.org/content/280/1754/20122845.full.pdf (accessed March 12, 2013).

Elvidge, C.D., M.L. Imhoff, K.E. Baugh et al. 2001. Night-time lights of the world: 1994–1995. *ISPRS Journal of Photogrammetry and Remote Sensing* 56: 81–99.

Gibbs, R. 2014. Sports surface design: The purposeful manipulation of soils. In *The Soil Underfoot*, eds. G.J. Churchman and E.R. Landa. 351–368, Boca Raton, FL: CRC Press.

Haff, P.K., 2010. Hillslopes, rivers, plows, and trucks: Mass transport on Earth's surface by natural and technological processes. *Earth Surface Processes and Landforms* 35: 1157–66, doi: 10.1002/esp.1902.

Haff, P.K., 2012. Technology and human purpose: The problem of solids transport on the Earth's surface. *Earth System Dynamics* 3: 417–31.

Hermann, W.A., 2006. Quantifying global exergy resources. *Energy* 31: 1685–702.

Hooke, R. LeB., 1994. On the efficacy of humans as geologic agents. *GSA Today* 4: 217, 224–25.

iBOL. 2013. International Barcode of Life Project. http://ibol.org/ (accessed January 18, 2013).

Jenny, H., 1941. *Factors of Soil Formation: A System of Quantitative Pedology*. New York: McGraw-Hill.

Keith, D.W. 2000. Geoengineering the climate: History and prospect. *Annual Review of Energy and Environment* 25: 245–84.

Matthews, H.D. and K. Caldeira 2007. Transient climate: Carbon simulations of planetary geoengineering. *Proceedings of the National Academy of Sciences USA* 104: 9949–54.

Moore, G.E. 2006. Moore's law at 40. In *Understanding Moore's Law: Four Decades of Innovation*, ed. D.C. Brock, 67–84, Philadelphia, Pennsylvania: Chemical Heritage Foundation.

Nikiforoff, C.C. 1959. Reappraisal of the soil. *Science* 129: 186–96.

Nocera, D.G. 2012. The artificial leaf. *Accounts of Chemical Research* 45: 767–76.

Nowak, M.A. 2006. *Evolutionary Dynamics: Exploring the Equations of Life*. Cambridge, MA: Harvard University Press.

Trimble, S.W. 1974. *Man-Induced Soil Erosion on the Southern Piedmont: 1700–1970*. Ankeny, IA: Soil Conservation Society of America.

Van der Heijden, K. 1996. *Scenarios: The Art of Strategic Conversation*. New York: Wiley.

Wackernagel, M., N.B. Schulz, D. Deumling et al. 2001. Tracking the ecological overshoot of the human economy. *Proceedings of the National Academy of Sciences USA* 99: 9266–71, doi: 10.1073/pnas.142033699.

Wilson, E.O. 1984. *Biophilia*. Cambridge, MA: Harvard University Press.

Section II

Valuing Soils

"... for only rarely have we stood back and celebrated our soils as something beautiful, and perhaps even mysterious. For what other natural body, worldwide in its distribution, has so many interesting secrets to reveal to the patient observer" Les Molloy (From Preface, p.10) in L. Molloy. *Soils in the New Zealand Landscape: the Living Mantle*. Wellington: Mallinson-Rendel (1988).

6 Seeing Soil

Deborah Koons Garcia[*]

CONTENTS

Most people are soil blind. They walk on soil, they gaze at it on the horizon, they gain pleasure and sustenance from its bounty, but soil itself goes unseen, unappreciated. Modern life conspires to remove us from any connection to or awareness of soil. We spend a lot of time looking down, not at the soil, but at the various devices that connect us to our techno-fied world. Those of us who understand the importance of soil can bemoan the sad state of affairs as evidenced by the story of the schoolchild who visited a farm for the first time and, when presented with a carrot freshly pulled from the earth, proclaimed, "I'm not going to eat that—it's been in the ground!"

A few years ago, I decided the world was ready for a good film on soil, and I was the filmmaker to make it. I have to admit my bias was toward seeing soil as an agricultural medium. As I went through the 4-year process of researching, shooting, editing, and completing the film, I found that my attitude toward soil completely changed. At a certain stage, I did not want to have any agriculture in the film. I empathized with the attitude of some Native American tribes—why would I cut into my mother, the earth? The plow, I had learned, had caused more damage to this planet than the sword. I became protective of the soil and wanted to move away from the assumption that soil is a thing and we humans' primary concern should be what can we get out of it—what is in it for us. I wanted to support and encourage a healthier relationship with what I came to see as a miraculous substance and bring that awareness to a wide audience.

In preparing to make the film, I bought a bunch of alarmingly thick soil textbooks, some non-academic books about soil, and befriended soil scientists. I love the research phase of filmmaking. At that point, anything and everything is possible. I absorbed massive amounts of information and my head filled up with many more interesting ideas than I could possibly fit into a 90-minute film. Watching a good film is primarily an emotional experience—we want movies to move us. Even if a movie is information heavy, we want that information to engage us, to affect our hearts and minds. I realized that distilling all I was learning into an audience-friendly piece would be a real challenge. How could I take soil, a medium that is dark and seemingly inert, and marry it to film, a medium that is all about movement and light? When I started shooting, I had my cameraman set up a shot focused on a patch of soil. I called out "Action!" He turned to me and said "There's nothing happening." "Oh," I replied, "There is, there is! There's so much happening! How will we fit it all in!"

My task as a filmmaker was to allow, invite, encourage, fascinate, and seduce my audience into understanding and feeling a connection with soil.

Since film is essentially visual, I finally had to figure out exactly what was going to be up there on the screen. Ultimately, my work was to figure out how to see the soil. My vision expanded. I would

[*] Lily Films, Mill Valley, California, USA; Email: deborah@lilyfilms.com

see soil in many different ways, and by presenting this variety of points of view, the nature of soil in all its glory would be revealed. I realized that for me to hold a piece of soil in my hand and say this is soil would be like holding a drop of seawater in my hand and saying this is the ocean.

How to see soil: As an entity, as an organism, as an ecosystem, as a community, as a collection of cycling nutrients, as a place where billions of microorganisms thrive, as a plant-growing medium, as a living system. It's alive! We can see soil in time and in space. We see soil in close-up, medium, and long shot. Ultimately, I had let people to see soil in all these ways and more. Most importantly, I decided to present soil as a protagonist of our planetary story, and the protagonist of my film.

Since I am an experienced filmmaker with a classical bent, I decided to write the story of soil in three acts. Act 1: Soil itself, its multidimensional nature. Act 2: Soil in relationship, primarily in relationship with humans. Inevitably, that relationship must focus on agriculture. Act 3: Soil and big picture ideas such as soil and global warming, soil and water, soil and feeding the world, soil and metaphysics.

Having digested so much material and having debated with myself and my colleagues about what should be in the film, I came up with a wonderful title: *Symphony of the Soil*. The piece would be complex, made up of many different instruments and parts, creating a satisfying whole. Soil is a symphony of elements and processes, and my film would evoke that. This also allowed me to bring music into the mix. My editor Vivien Hillgrove and my composer Todd Boekelheide are so talented, and working with them is such a fulfilling part of making a film. I was confident that embracing music as a strong, primary element would sweeten and deepen the film-going experience. One of my very favorite parts of *Symphony* is the treatment of the 12 different soil orders (see Editors' Note). The melody, musical phrasing, and instrumentation for each of the orders echoes the "personality" of the order itself: Ultisols' music has an old man tuba, Oxisols' has tropical maracas, Histosols' sounds, well, evocatively wet, and on through the orders. It is brilliant, if I do say so myself, and it is exactly what I wanted. Every time I see this section, I smile. Soil scientists and soil freaks who know the orders see tell me excitedly that they love how we dealt with the taxonomy. They get it! It is fun and fresh. As I told a nonscientist friend: yes, you will sit through all 12 soil orders and you will like it. Which she did.

To further my goal of promoting soil consciousness, I decided to mix science, art, and activism, to create a kind of hybrid that is all too rare in film these days. Music, of course, is an essential part of this artistic treatment. In addition to this, I found a young artist who painted hundreds of watercolors to animate various processes that I wanted to explain, such as photosynthesis and nitrogen cycle. This tactic was highly successful and adds variety and texture to the film. In all the decisions I made, I committed to a process in which science is embraced and revealed through artistic means, which then moves people to perhaps change their behavior in some way.

Unfortunately, the absurd antiscience bias that exists in American culture today has made its way into the film world as well. Filmmakers are often wary of too much science. I actually believe people enjoy science. The word science comes from the Latin word *sciere*, to know—science is about what we know. And soil science is a field where much is known and much remains to be discovered. Perhaps, more than in other disciplines, soil scientists appreciate what they do not know, yet. Audiences want to know how things work, especially what happens in the natural world, but the science has to be intriguing. Some filmmakers dumb things down in the hope that will please people. My experience is that smarten up is a better strategy. People are hungry for information that helps them make sense of the world. Giving people a new way of seeing allows them a deeper understanding of the world around them, and challenges them intellectually. They learn and then they feel good about that; they gain confidence, and want to learn more. This is actually the way our brains work—our positive reinforcing loop. I seek to respect and nourish the intelligence of people watching my films. Fortunately, the topic of soil provides fertile ground for this kind of exploration and active engagement with audiences. To fully appreciate soil, you have to understand it.

Soil is a transformational substance. It is the place on earth that transforms life into death and death back into life. In the film, scientists dig soil pits, and then stand in them. During the first few screenings for lay audiences, I was surprised by the laughter that occurred when the soil pit shots came up on the screen. Then I realized that there is a subconscious connection between a soil pit and a grave. The laughter was a reaction to that. Examining a soil profile revealed by a soil pit in a curious way makes us face our own mortality. Coming face to face with the soil profile reminds us that we arise from the soil and we return to it.

Because of technological advancements such as electron microscopes and satellites, we are able to see soil from more vantage points than ever before in history. Soil in extreme close-up: We can see soil microscopically, and start to learn about billions of microorganisms that are at home in a healthy soil. We are just beginning to understand the relationships between these microorganisms and their dynamics within the soil. One of the things I learned about in my soil studies that struck me as totally cool was cation exchange. The idea that on the most microscopic level, ions were constantly moving back and forth between the soil solids and water held in pores just blew my mind. One day, as I was in the middle of making the film, I went for a walk near my home, looked out on the hills, and felt fully conscious of how much movement, how much activity was happening in my own soil community. From mountain lions, deer, foxes, voles, worms, dragonflies, and dung beetles all the way down to electrically charged atoms bouncing around: it was all about movement. Somehow realizing how much activity was going on that was invisible to me, which was happening on the most elemental level, expanded my vision exponentially. When we finally showed the film to test audiences, it turned out that the cation-exchange part was just too much information; we had to leave it out, and I felt it was almost tragic. I told one of the scientists I was working with: "I have bad news—we have to leave out cation-exchange because pretty much everyone agrees it's just overwhelms! No matter how we cut it, we lose them on that so out it goes!" Anything that pulls the audience out of the film has to go—that is basic. So, my editor and I waved a sad goodbye to cation exchange. This is something viewers get to discover on their own.

How to see soil: We can see soil from the human point of view, walking on it, looking around to observe the kind of plants and creatures that live on, and in our own soil community.

We can see soil in very long shots—from satellites, from space—and track the larger changes that climate, the environment, and our human practices make on soil. We can also see soil in time using technology to understand what happened in the past and how that affects what's happening now.

6.1 SOIL AS A STORY: THE STRUCTURE OF *SYMPHONY OF THE SOIL*

6.1.1 Act 1: Soil as a Protagonist of Our Planetary Story

Soil is alive. Soil is busy being born and busy dying. Soil has parents just as we have parents, as John Reganold of Washington State University states in the film. Soil forms from parent material. I decided that to blow people's minds about what the film about soil would be, I would have to steer clear of agriculture early in the film. So, I start on a glacier in Norway and Ignacio Chapela of the University of California (UC) at Berkeley telling us about the nutrients being pulled from the glacial deposits and made available to plants whose growth and death are then key to the development of the soil. In a nod to the current Hollywood conventional wisdom about what attracts young audiences, I do have an explosion within the first 10 minutes of the film—a volcano! We had a great time tromping around the active volcano and forests in Hawaii with Peter Vitousek of Stanford University and Oliver Chadwick of UC Santa Barbara. Seeing 50-year-old soil one day, and then 5-million-year-old soil the next day, drove home the idea that soil has a life span and different characteristics based on its age. Any viewer who thought *Symphony* was going to be just about plowing fields would leave that idea behind. Soil has parents, a life span, and a life cycle. It has individual characteristics. It contains billions of organisms of all sizes within it. It contains mycorrhiza and fungi, among them, the largest organisms on earth. It is so incredibly full of life. If it dies, if the organisms and processes cease to function, it is no longer soil—it is the four letter D word, *dirt*.

6.1.2 Act 2: Soil and Relationship

Patrick Holden, farmer and for many years director of the Soil Association in the United Kingdom, tells us agriculture is a dance with nature. I am a strong proponent of organic/sustainable/agroecological farming. I have had an organic bias for decades. That bias is why I make the films I make. While organic agriculture tends to be based on a conscious relationship with a living organism, chemical industrial agriculture tends to treat soil as an inert medium that chemicals are poured into and products are pulled out of. I believe we can and must transform agribusiness into agriculture, into a less toxic, more soil-friendly activity. The thousands of toxic chemicals we put on the soil, the overuse of nitrogen fertilizer, problems with soil erosion, and overuse of water—none of this is healthy. I do believe we can and must move away from practices that deform or kill the soil. The more people understand how soil works, the more likely it is that we will begin to move away from chemical agriculture. I showed *Symphony of the Soil* at the annual meeting of the American Society of Agronomy/Crop Science Society of America/Soil Science Society of America. I was honored to be invited to screen there, and after I arrived and saw that Monsanto was the chief sponsor, I was even more amazed and grateful for the organizers' invitation. During the question-and-answer session after the screening, we had a rousing interchange between people who believed we can feed the world organically/ecologically and those people who are absolutely convinced that we cannot. It is a great honor to have Daniel Hillel, winner of the World Food Prize in 2012, in my film. Although he is best known for inventing drip irrigation, I wanted him in the film because I had read his fascinating book *A Natural History of the Bible*. I was intrigued by his philosophy about soil that has evolved during his 80 years of life. I was also honored that he appeared on the panel with me after the screening at this scientific meeting. He really liked the film, thank goodness. He also said "I don't agree with everything in the film but we need to reevaluate our assumptions about agriculture." I absolutely agree. If my film does nothing other than encourage people to reevaluate their assumptions about agriculture, I will feel that my years of work have been valuable.

6.1.3 Act 3: Soil and Big Ideas

The choices we make about how we use or abuse our soil will have impact not only on our human health but the health of our planet. We can sequester carbon, use less water, and grow healthier food if we treat soil right. Why we in the United States use our fertile Mollisols (see Editors' Note) in the mid-west prairie states, our best soils, to grow food for cars and cows instead of people shows how skewed our priorities are. Healthy soil holds onto and filters water better than sick soil. Soil full of nutrients grows food full of nutrients. Fully understanding how soil works, or does not work, encourages people to participate themselves by planting gardens, buying local food, eschewing toxic chemicals, and treating the earth a little more kindly. Some people who have seen *Symphony of the Soil* told me they decided to plant cover crops in their gardens to replenish their soil. At the least, people can compost. Backyard and industrial-scale recycling and composting are a great way to get the nutrients and life back in the soil. I have filmed composting on four continents and right next door. The city of San Francisco near where I live has a program in which green waste and restaurant waste are taken to a huge facility in Vacaville and composted. That compost is then sold to farmers who use it to grow crops that they sell to the restaurants, which recycles that waste back into compost. We need these programs everywhere. Composting is a way of getting the organic matter back in the soil, something our soils desperately need. Just as schoolchildren embarrassed their parents into recycling in the 1970s, schoolchildren these days are learning about composting in school and encouraging their parents to participate in this kind of recycling.

There are other ways to look at soil that have to do with who is doing the looking. A farmer sees soil in a very different way than a soil scientist would, but both of them understand that soil is the fount of life, that we are dependent on it, and we ignore that at our peril. To make this film, I sought the wisdom of farmers, soil scientists, and activists. Soil scientists tend to study a few topics deeply and here in California and the West, they are helping farmers figure out how to protect our precious soil. John Reganold, a soil scientist from Washington State University who was both a participant and advisor on the film, told me that he has worked with many farmers through the years. When he talks to conventional farmers, they tend to talk about their yields, the chemicals they use, and their machinery. When he talks to organic farmers, they just want to talk about their soil. I sought out farmers, primarily organic farmers, and asked them to share with me how they treat their soil. What techniques do these farmers use? Composting, cover crops, crop rotations, and mulch—practices that actually improve the soil year after year. These farmers follow what Sir Albert Howard called the Law of Return. They give back to the soil, return to the soil the organic matter and nutrients they took out of it in harvested crops. They understand we have to feed soil to keep it healthy. I kept going over and over this idea in the hopes that viewers would understand it and change the way they garden, change the way they eat.

One of my favorite stories in the film is told by Warren Weber who has been farming organically in Bolinas, California for over 40 years. He has the first certified organic farm in California and I can tell you from personal experience that the lettuce and vegetables he grows are absolutely delicious. When he wanted information from the local agricultural authorities back in the 1970s about how to grow organically, he was told that there could never be commercial organic farms in California and that Californian soil could not support an organic regime. Now, of course, there are more organic farms and vineyards in California than any place in the United States. These farmers and our temperate climate that allows for year-round growing are why we in the San Francisco Bay Area enjoy the best food in the world. What a joy and relief that Warren Weber did not take this bad science as gospel and kept on to forge a healthy industry and a healthy soil community.

Symphony of the Soil premiered at the Smithsonian Museum in Washington DC in 2012. There continues to be lots of interesting screenings all over the world, some of which I have been able to attend. I have also been asked to give talks about soil that allows me to develop some of the ideas that I could not fit in the film. My talks naturally focus on seeing soil and becoming soil conscious. "Soil as Matrix, Myth, and Metaphor" has become my soil stump speech. We can see soil as a matrix, a word that comes from the Latin word *Mater*, mother, and it originally meant womb. Then it meant a breeding female, as in livestock. A matrix is the substance, the situation, or the environment in which something has its origin—a place that allows something to take form. Soil as a matrix allows life on earth to take form. Soil is a protagonist of our planetary story, and soil is also the mother of our planetary story. A soil scientist friend of mine who is a single mother pointed out to me that you only need one parent material such as a rock; then with the action of microorganisms, plants, weather, and time, a soil develops. So, soil is a single mother.

I also see soil as myth, a traditional story that a culture comes up with to make sense of the world. Creation myths often have soil as a starting point. Man is made by God from clay, and God breathes life into it. Or, a God makes human images from different colors of soil, and that makes different colors of humans. Soil has a role in other aspects of myth. The temples of the Greek and Roman gods were built on soils and terrains based on the realm that God ruled over. The temples of Demeter and Dionysus were placed on land that could grow things. Artemis and Apollo had temples on hunting grounds; Hera and Hermes had temples on cattle-grazing soils. At the most profound level, soil and myth are connected.

Today, there is another meaning of myth that has grown in popularity, and that is myth as a misconception, something that is not true. There are lots of these kinds of myths about soil. One is that rain-forest soils are rich. They are not—all the fertility is in the forest. If you cut the forest down and farm that land, you will run through that soil's limited nutrient supply in a very few years. Another dangerous myth is that soil will detoxify anything, that any toxic chemical can be eaten up, and transformed by soil microorganisms. Soil can detoxify some things but not all. There is still DDT (dichlorodiphenyltrichloroethane) in soil decades after it was banned. There are lots of myths being peddled by corporate ad and chemical companies. The idea that we need genetically modified crops to feed the world is the biggest one of these. It is hard to believe that if corporations control all the seeds on this planet and the plants that grow from those patented, engineered seeds, are designed to be sprayed with toxic chemicals, we are then going to be living in harmony with nature. As we now know, we are using more toxic chemicals in the genetic engineering regime than we were before it started 20 years ago. Weeds are becoming ever more resistant to pesticides; so, the corporations advise using even more toxic chemicals to control them. This toxic regime is dangerous and unnecessary. I was pleased to have Hans Herren, a World Food Prize winner, in the film talking about his hands-on experience in Africa and around the world. Organic and sustainable agriculture can feed the world, as the International Assessment of Agricultural Knowledge, Science, and Technology for Development (IAASTD) 2009 report he cochaired proved (http://www.unep.org/dewa/Assessments/Ecosystems/IAASTD/tabid/105853/Default.aspx). After 3 years of work, 900 participants and 110 countries agreed—the healthy future for our planet is in sustainable, not chemical/industrial, agriculture.

Another helpful way to see soil is soil as a metaphor. Soil shapes culture. Culture shapes soil. As an American who grew up in the Midwest in the 1950s, I am very aware of the mythic quality inherent in the American consciousness. What I have learned in the last few years is that the United States is especially blessed with really good soil. Over 40% of our soils are the two most productive orders Mollisols and Alfisols (see Editors' Note). That our soils are so rich and plentiful has shaped our national character. We can use up our resources and move on. We do not like limits. And, because our resources seemed unlimited, we did not have to recognize limits in the first 200 years of our history.

I was visiting China a few years ago to show my work, and had dinner with a senior scientist who advises the Chinese government. He told me that China does not have a lot of good soil, maybe 11% or so of the soil is good. China also does not have an abundance of water. The Chinese people have had to work together as a collective culture to make good use of the resources that they have and that has shaped their character.

Americans are now at the point in our history when we need to recognize limits. But in some circles, it is seen as positively un-American to do so. Within this view, imagining that we do not have limits will somehow change the harsh reality that if we keep treating our soil the way we do, we will be out of topsoil in 30 years. I can only hope that this blind arrogance can be superseded by another aspect of our national character—we like to solve problems. If someone tells us what we want to do is going to be really difficult, and we might not be able to accomplish it, we want to try harder. Moving toward healthier, more resilient ways of treating our soil is essential. Our future depends on it.

There is another metaphor I think about when I think about soil, and that has to do with the use of synthetic nitrogen, one of the most problematic aspects of conventional agriculture today. Ignacio Chapela, when teaching his students at UC Berkeley about nitrogen, compares it to a happy marriage. Nitrogen in its natural state is very stable. It takes a lot of energy to break it apart; but once it is broken apart, it runs around combining with the wrong things, causing a lot of trouble. A couple of years before *Symphony of the Soil* was completed, I was asked to show it as the Hans Jenny Memorial Lecture at UC Berkeley. I was honored by this invitation, and I wanted to take advantage of this opportunity; so, we made a one hour version of the film that looked finished but was not. In this particular version, I came down pretty hard on synthetic nitrogen. My attitude was that it causes a lot of trouble, and that we need to find other ways, healthier ways, of getting nitrogen into the soil. The film screening at UC Berkeley went well, and I was invited to show that version in a few other places, including Yale University and Fresno State University as part of their 100th anniversary celebration. Fresno is in the center of the San Joaquin Valley of California, and more food is grown in the San Joaquin Valley than any place in the world. About 600 people showed up for the screening in Fresno on a stormy evening. Afterward, there was a panel with me and five longtime ranchers and farmers from the San Joaquin Valley. They were not happy with the way that I treated synthetic nitrogen, and they were quite vocal about that. In addition, a few of them wrote long critiques of the film and sent them to me afterward. I was pleased that they had taken so much time to help me make a better film. For one thing, I did not want my film to be dismissed by the majority of farmers who live in this country. I worked hard to reconfigure the sections on nitrogen. With the help of Ignacio Chapela, I presented the idea that for the 10,000 years of agriculture, nitrogen had been a limiting factor. With the development of synthetic nitrogen for munitions and fertilizer production in the early twentieth century, nitrogen was no longer a limiting factor for crop growth, but there are consequences from the manufacture and use of synthetic nitrogen. Increased energy usage, greenhouse gases, and dead zones that are the results of using this regime are a few of these. The Green Revolution was made possible because of the Haber-Bosch discovery that allows for the creation of synthetic N. I see the Green Revolution itself as a metaphor: putting synthetic nitrogen on regular crops makes these crops grow too much: the stalks get too tall and weak, the plants fall over (they "lodge"), and are ruined. So, the trick of the Green Revolution was to use dwarf plants and overfertilize them. The Green Revolution is about putting dwarf plants "on steroids," so that they look normal. This seems to be a curious and true metaphor for the most unhealthy kind of agriculture. This unnatural regime has to be propped up with pesticides, and uses more water than non-Green Revolution regimes. I was talking to a senior scientist who studies nitrogen use about how to present N in the film, as I was aware of how damaging synthetic nitrogen is to the soil, water, and air. She told me that if we had not developed synthetic nitrogen, we would have 2 billion less people on earth. I replied that is not a bad idea. Of course, she could not say that, but I am an artist; so, I can and did in a private conversation, not in my film. In the final film, I treated nitrogen in a much more thoughtful way than I originally had. I did not disparage or criticize people for using synthetic N, but simply pointed out the negative consequences of overuse. I also point out that using cover crops can cut down on nitrogen leakage up to 70%. One of the ranchers from the San Joaquin Valley who was most critical of the Hans Jenny Memorial Lecture version came to see *Symphony* when it was finished. He was very moved by the final film, almost to tears. He told me how much better the film was, and I felt that I had accomplished a kind of victory. I did not promote nitrogen fertilizer, but I explained it. I laid out what happens if we continue to overuse it. I hope many farmers all over the world see this film and use less synthetic nitrogen fertilizer. Maybe they will even go organic. We have had several farmers order screening copies of *Symphony* so that they can show the film in their communities, in England, Iowa, or India, to convince other farmers to move away from the chemical regime. One of the most satisfying aspects of presenting *Symphony of the Soil* to the public is that people are moved by it and they feel hopeful. If we changed our practices and treated soil well, we could heal our soil in just a few years. While it is true that I had to include in the film examples of destructive practices and a montage of misery of what things could be like in the future, ultimately, soil is resilient. Our hopes for a healthy future are based on healthy soil. *Symphony of the Soil* is ultimately hopeful.

I decided to make a film on soil because I like a good challenge. I had made a film called *The Future of Food* that was very well received when it came out in 2004 and continues to be popular all over the world. That film primarily dealt with the corporatization of agriculture and wholesome alternatives to that development. When I chose to make a film on soil, I assumed I would make a piece that primarily dealt with soil and agriculture. As I have learned and developed my understanding, I have moved far away from that position. I realize that soil is its own vast and mysterious realm, and agriculture is a part of that realm. I learned about soil from cutting-edge science and from ancient practices, such as those we filmed in Egypt and India. I have made a serious film that is easy to watch and user friendly. During the time I was making the film and since, I have attended several conferences in which the speaker, usually a soil scientist, despairs about the soil blindness of the general public: if people really understood how soil works, everything would change. They say that soil needs an advocate who can present its virtues to people—soil needs a good PR (public relations) agent. Every time I heard this, I would say to the colleague I was sitting with "I'm trying! We're trying!".

So now, the *Symphony of the Soil* is done and is making its way in the world. There are two sets of comments I have heard a few times from my audience members. One is that the film is surprising.

I love hearing that. Surprising is good! The other is that the film reminds farmers and people who work with the soil, why they do what they do. These people want to show it to their relatives who cannot understand why anyone would want to farm or study soil; so they can appreciate why they do what they do. *Symphony of the Soil* heartens and fortifies. As one viewer put it, the film brings people together in a community of discovery.

It was a wonderful challenge for me and my team of collaborators to use a highly technological medium to connect people to nature. This connection transcends our relationship to agriculture. Agriculture has been practiced for only 10,000 years. Our relationship with soil goes back to hundreds of thousands of years. All generations of humans have walked on soil, observed it, learned from it, and felt part of it. Soil provides food, clay for creating art, bricks for building homes, antibiotics for healing, and gardens for beauty. Soil creates the very ground we live on. We humans are true creatures of the soil in every way. *Symphony of the Soil* is a celebration of soil. We made soil sexy and we made soil seen. Soil is a gift and giver. Soil is a hero and humble worker. Soil is life. Soil is sacred. Soil is home, http://www.symphonyofthesoil.com/.

6.2 EDITORS' NOTE

The *Soil Taxonomy* is a system for classifying and naming types of soils that was developed in the United States (by the U.S. Department of Agriculture) around 50 years ago and that has been refined by trials worldwide to become one of the main international systems for soil classification. In this scheme, every soil is placed into one of the 12 soil orders. Those mentioned in this chapter are Ultisols: strongly leached soils; Oxisols: intensively weathered tropical and subtropical soils; Histosols: organic matter-rich soils, for example, peats; Mollisols: grassland soils with a high nutrient content; and Alfisols: moderately leached soils with clay-rich subsoils.

7 Picturing Soil

Aesthetic Approaches to Raising Soil Awareness in Contemporary Art

Alexandra R. Toland and Gerd Wessolek*

CONTENTS

7.1 AESTHETIC ENTRY

Covered by layers of rotting leaves on the forest floor, the wavy stubble of fields and meadows, or the weed-and-litter-fringed concrete slabs of city sidewalks, soil is hidden by default. Pictures of the soil are necessary, however, to help to ensure its protection. The most common images we have of the soil come from two main sources—the scientific community, with its numeric graphs and analytical descriptions, and popular media. Both are powerful and informative, but insufficient in providing imagery to truly inspire stewardship or stop soil degradation.

According to soil communications specialist, Rebecca Lines-Kelly (2004), "Science has played a major role in disconnecting society and culture from the soil that sustains it because it has studied soil in isolation ... concentrated primarily on aspects of soil that make money ... and has used technical, complex language that makes soils inaccessible to non-scientists. It has attempted to remove emotion and spirit from the understanding of the substance that underpins our existence and gives us life, by reducing it to part of a scientific equation ..." (see also the approach of Lines-Kelly 2014).

Popular media, on the other hand, typically rely on scare tactics and rural clichés in what Secretary General of the German Council for Sustainable Development, Günther Bachmann (2001, p. 6), calls the down-cycling of soil imagery: "The all too frequent use of archived pictures and recycled imagery by environmental media features is what environmental policy calls down-cycling—a

* Berlin University of Technology, Institute of Ecology, Department of Soil Protection Ernst Reuter Platz 1, 10587 Berlin, Germany; Email: a.r.toland@mailbox.tu-berlin.de

reuse of materials at the expense of their value ... As a consequence, environmental programs are often not innovative, without a unique artistic signature, without their own aesthetic" (translation from German by the authors).

In the practice of image production, contemporary artists are expert craftspeople who use innovative techniques to make the invisible visible. Bachmann (2001, ibid, pp. 8–9) emphasizes the ability of artists to bring to the surface more than mere facsimiles of the soil profile: "The aesthetic portrayal of soil is more than just representation of what is found in nature, rather it is an image charged with metaphors and meanings. ... To bring the sound of the earth to vibration ... to make the color of time visible, to learn to read the soil again—the arts could show us the way ..."

This chapter presents a range of contemporary artworks with the aim of expanding the scientific portrait of soil as a pedogenetic body. Together, the works deconstruct conventions of genetic classification by inventing new methodologies of understanding defined by sensory experience, cultural contextualization, and an intuitive collaboration with the soil itself as material guide. More importantly, these artworks restore "a sense of emotion and spirit to the understanding of the substance that underpins our existence" (Lines-Kelly 2004, p. 10), introducing pictures of the "soil underfoot" beyond scientific and economic rationalism or down-cycled popular imagery.

While the following works allude to environmental protection issues and diverse underlying cultural contexts, what they mainly have in common is an *aesthetic entry* to the hidden realms of the pedosphere. With *aesthetic entry* a formalist aesthetic reading is implied, that is, the appreciation of that which may be physically referenced not only in the formal compositions of artworks, but also in the soil environments: color, texture, form, contrast, rhythm, balance, topography, and so on (see Toland and Wessolek 2010; Carlson 2000; and Zangwill 2001 for discussions on aesthetic formalism and the environment). All of these works are produced in highly aesthetic styles to expose the beauty, diversity, and uniqueness of the soil and the landscapes from which soils develop. Other strategies focused on soil protection include, for example, political activism, social engagement, spiritual or ritual practice, creative gardening and farming, and ecological restoration. Such strategies are indispensable for a comprehensive argument on the artistic contribution to soil protection, but are impossible to cover within the scope of this chapter.

The chapter is loosely organized by genre, including: painting, pigmenting, printing, and artists' archives. Beyond these categories, aesthetic awareness is also created with installation, performance, and different types of participatory engagement. By blending a range of contemporary artistic positions with scientific concepts, an attempt is made to reimagine and reevaluate the soil.

Table 7.1 provides an overview of artworks described herein with links to artist and project websites. Due to publishing limitations, not all images of artworks can be reproduced in print, but we are grateful to all artists who generously shared images and personal insights for this publication and encourage their further engagement in the interdisciplinary task of raising soil awareness.

7.2 GENRES AND APPROACHES

7.2.1 Painting with Soil

Soil enjoys a special place in landscape painting (Zika 2001, Feller et al. 2010) and its representation throughout art history has been analyzed (Busch 2002, Hartemink 2009, van Breemen 2010) as well as taken on as an expressive medium within the soil science community (e.g., paintings by Gerd Wessolek, Jay Stratton Noller, and Ken van Rees). Given its primary position in art history, painting can provide relatively easy access to the complex world of contemporary art. In conversations with soil scientists, painting often comes up as a relatively well-known and accepted art form with which to explore the aesthetic properties of the soil. When the Soil Alliance of Lower Austria launched its educational outreach project *Painting with the Colors of the Earth* with artist Irena Racek in 2008, resonance within the soil science community was immense. Soil painting workshops with children and youth groups popped up in Europe and the United States. The Austrian government began

TABLE 7.1
Aesthetic Approaches[a]

Painting with Soil	
Ulrike Arnold: Earth Paintings	http://www.ulrikearnold.com/
Mario Reis: Nature Watercolors	http://www.marioreis.de/seiten/nature3.htm
Jesse Graves: Mud Stencils	http://mudstencils.com/
Alison Keogh: Clay Drawings	http://www.alisonkeogh.com/
Irena Racek: Painting with the Colors of the Earth	http://www.irena-racek.at/
Rainer Sieverding: Boden Bilder (Soil Pictures)	http://www.rainer-sieverding.de/
Printing and Other Techniques of Soil Transfer	
Betty Beier: Earth Print Archive	http://www.erdschollenarchiv.de/
Herman deVries: From Earth	http://www.hermandevries.org/
Ekkeland Götze: Terragraphy	http://www.ekkeland.de/
Anneli Ketterer: Decrustate	http://decrustate.net/
Daro Montag: This Earth, Bioglyphs	http://www.microbialart.com/galleries/daro-montag/
Pigmenting and Soil Color Collection	
Ulrike Arnold: Earth Paintings	http://www.ulrikearnold.com/
Margaret Boozer: Red Dirt Studio	http://margaretboozer.com/
Koichi Kurita: Soil Library	http://soillog.exblog.jp/
Albert H. Munsell	http://en.wikipedia.org/wiki/Albert_Henry_Munsell
Peter Ward: Exploration of Earth Pigments in North Devon	http://intim8ecology.wordpress.com/earth/
Elvira Wersche: Sammlung Weltensand	http://www.elvirawersche.com/
Archiving Soil Beauty and Diversity	
Betty Beier: Earth Print Archive	http://www.erdschollenarchiv.de/
Margaret Boozer: Correlation Drawing/Drawing Correlations: A Five Borough Reconnaissance Soil Survey	http://margaretboozer.com/
Herman deVries: Earth Muséum	http://www.hermandevries.org/
Marianne Greve: One Earth Altar	http://www.eine-erde-altar.net/
Ekkeland Götze: Terragraphy	http://www.ekkeland.de/
Sarah Hirneisen: Unfamiliar Ground (and other collections)	http://www.sarahhirneisen.com/
Koichi Kurita: Soil Library	http://soillog.exblog.jp/
Elvira Wersche: Sammlung Weltensand	http://www.elvirawersche.com/
Formal Aesthetic Approaches Beyond the Visual Field	
Lu La Buzz: States and status of clay	http://www.plymouth.ac.uk/pages/view.asp?page=34011
Olle Corneer: Harvest (for Terrafon)	http://www.ollecorneer.com/
Florian Dombois: Auditory Seismology	http://www.auditory-seismology.org/version2004/
Alison Keogh: Clay Drawings	http://www.alisonkeogh.com/
Laura Parker: Taste of Place	http://www.lauraparkerstudio.com/tasteofplace/project.html
Land Art, Earth Art, Art in Nature	
Walter de Maria: Earth Room	http://www.earthroom.org/
Herman deVries: From Earth	http://www.hermandevries.org/
Agnes Denes: Wheatfield; Tree Mountain	http://greenmuseum.org/content/artist_index/artist_id-63.html
Andy Goldsworthy: Rivers and Tides	http://www.artnet.com/artists/andy-goldsworthy/
Michael Heizer: Double Negative	http://www.diaart.org/exhibitions/introduction/83
Richard Long: Mud Circles	www.richardlong.org/
Ana Mendieta: Silueta Series	http://www.moca.org/pc/viewArtWork.php?id=87
Robert Smithson: Spiral Jetty	http://www.robertsmithson.com
Alan Sonfist: Time Landscape	http://www.alansonfist.com

continued

TABLE 7.1 (continued)
Aesthetic Approaches[a]

Socially Engaged Art and Ecological Restoration Art	
Jackie Brookner: Urban Rain; Fargo Project	http://www.jackiebrookner.net/
Mel Chin: Revival Field; Operation Paydirt	http://www.melchin.org/
Future Farmers Artist Collective; Soil Kitchen	http://www.futurefarmers.com/soilkitchen/
Nancy Klehm: Soil Garden; Ground Rules; Humble Pile	http://socialecologies.net/about/projects/
Kultivator: Guerilla Composting; Friend Farm	kultivator.org/
Daniel McCormick: Watershed Sculptures	http://danielmccormick.blogspot.com/
Rebar Studio: Parking Day	http://parkingday.org/

[a] Selected case studies and mentioned artists are grouped by genre and appear in multiple genres where applicable. For more examples visit www.soilarts.org

distributing soil painting kits online (Szlezak 2009). Sand painting and pigment making workshops appeared at soil science and geological conferences and on soil information websites (e.g., artists Rainer Sieverding, Karl Böttcher, Peter Ward, and Irena Racek).

Through a growing interest in painting with soil, a wider conversation on the aesthetic properties of soil has opened within the soil science community. Intrinsic values of beauty, diversity, and uniqueness are gaining traction alongside instrumental values such as those typified by soil functions and ecosystem services. This development falls within the scope of "raising soil awareness"—a strategy of preventative soil protection that is gathering support not only in soil science circles but also in environmental policy and decision-making contexts (e.g., within the European Network of Soil Awareness ENSA or the initiatives of the European Commission's Joint Research Centre on *Raising Soil Awareness*).

Contemporary painter Ulrike Arnold has been raising soil awareness for over 30 years. Traveling to remote landscapes all over the world, Arnold creates abstract paintings with local earth materials on massive canvases or directly on the cave walls and rock ledges where she sometimes takes shelter (Figure 7.1). Arnold is on a life quest to bring about awareness of soil to a greater public. Using the

FIGURE 7.1 **(See color insert.)** Ulrike Arnold, Earth Paintings, Bryce Canyon, 2003. (Courtesy of the artist.)

landscape as her studio in what could be called a dynamic dialog with the earth, Arnold's work is all about eliciting the beauty and diversity of earth materials—something immeasurable and unclassifiable by science. According to Arnold, "I'm a cave woman or a mud woman, a woman of the earth. Soil is the strongest of the archaic elements. I feel directly connected to a place by simply touching the earth. I feel immediate contact and then I feel at home, not homeless or lost, even in a totally foreign place ..." (Ulrike Arnold, in an interview with A. Toland on November 6, 2012).

Painting with the earth, however, must be understood in context, Ulrike Arnold warns. The colors and textures brought together in her *Earth Paintings* come from somewhere distinctive and carry with them the aura of place. To complete the picture, Arnold often exhibits detailed photographs and films of rock crevices, caves, muddied surfaces, surrounding landscape features, and local inhabitants, in addition to her *Earth Paintings*. These materials help viewers contextualize the landscapes, climates, and cultural settings from which her works are extracted. This is a valuable lesson for soil scientists looking to reconnect with society and culture in their research. The inclusion of field photography and descriptions of local land-based problems are sometimes included in scholarly articles to underline the relevance of a given study. Such contextualization of research within local environments and communities could be more widely practiced in academic journals, at conferences and in research projects. Artists could be of great assistance in such contextualization efforts.

Given the dominant status of painting in art history, it is up to contemporary artists to push the borders and expectations of what painting is and what painting can do, as well as what color is and where color comes from. Two more examples show the range of possibilities of painting with soil.

On the street level, artist Jesse Graves has been branding bridges, sidewalks, and buildings with mud graffiti for the past several years. Using black compost, red clay deposits from his native Wisconsin, and sediments from the Milwaukee River, Graves' *Mud Stencil* project combines painting and graffiti, art and activism, natural elements, and urban public space. Graves says: "... mud is a pigment. It sticks, it binds, and it works very similar to the way paint does ... Spray paint is really toxic. I want to create environmental messages, so it wouldn't make sense (to use spray paint) ... Earth is the most basic substance. It's what all life grows out of. It's what things breakdown into when they decompose. So I'm using earth to spread messages that I see can help preserve our Earth ... because it's the material that's most logical for my message" (Jesse Graves, in an interview with A. Toland on September 20, 2012).

Typical for street artists, Graves openly shares his methods and designs under creative commons license on his blog, mudstencils.com, and the mud stencil idea has since spread to cities across the United States and abroad. This street art project brings soil back into the urban environment, where it is usually invisible, and takes painting out of studio and exhibition contexts to champion environmental messages over authorship or originality.

Authorship is also challenged in natural landscapes, for example, in the works of Mario Reis. Reis "coauthors" his work with streams and rivers in a unique approach to painting he calls *Nature Watercolors*. By placing stretched canvases in strategic positions and depths in the river (Figure 7.2), Reis captures the signature swirls of waterways in the sediment load unique to each river system. Reis (2004, p. 107) explains, "The rivers in my paintings are both the object and subject of my work. I am not creating an illusion of rivers, but catching some of the real essence of the rivers in my painting. They leave their imprint on the cotton and show us how they are." While the artist chooses the specific sites and painstakingly installs the stretched canvases, sometimes using rocks as counterweights to control the placement of pigment, the river itself completes the painting by depositing sediment residue on the surface of the canvas.

More than anything, Reis's works are a record of time. Eroded sediments from mountains, forests, and farmlands on an exodus to the sea are captured en route in the steady course of the river's flow. "Whatever happens," Reis says, "it's an expression of Nature in its own voice. Each stream has a specific character. Some paint in a really hard-edge manner, others paint more softly. In this way each painting, influenced by the interaction between myself and the river, is a kind of self-portrait of that specific river" (Reis 2004, p. 106).

FIGURE 7.2 Mario Reis, Nature Watercolors, canvas installed in Godebach, Mürlenbach, 2013. (Courtesy of the artist.)

7.2.2 Pigmenting (and Other Forms of Celebrating Soil Color)

Painting is primarily an exploration of color, or the practice of applying color to a surface. Any art form focusing on the properties and potentials of color is in this sense inevitably a derivation of painting. It is no surprise that a painter, Albert H. Munsell, invented the system used today for describing soil color. Commenting on the introduction of Munsell's color description system into soil survey practice in the mid-twentieth century, Ed Landa (2004, p. 88) explains that "the linkage established at that time between worlds of art and science was utilitarian, stemming from a shared need to describe color." The Munsell Color System provides *A Grammar of Color* (Munsell and Cleland 1921), which codes color (hue), lightness (value), and purity of pigment (chroma) into numeric values that are useful for painters as well as scientists for determining soil color.

The following works are impossible to pin down as paintings, because although they are about color, and in part produced by painters, they redefine the approach to and application of color in such a way that may better be described as performance, installation, storytelling, or participatory research. While "pigmenting" is perhaps an insufficient term to describe the artistic processes mentioned here, the following works represent a unique exploration of soil color and succeed in introducing viewers to pedogenetic diversity by means of color.

Elvira Wersche has an extensive collection of soil pigments from all over the world donated by friends and strangers. The ongoing *Sammlung Weltensand* (Figure 7.3) is organized by color (e.g., purples from Switzerland, reds from Angola, yellows from Tunisia, chalk whites from New Mexico, etc.). These are then sifted into various degrees of coarseness, to create a palette of distinct hues and textures.

Wersche brings her sand repository with her to museums, churches, and public buildings to create large ornamental floor works in a performative process of contemplative placement. Where each color is placed in the complex geometric floor pattern is an intuitive decision made in the making. Later, all colors are merged in a simple final performance as sand from all over the world is swept together and then given to members of the audience. From start to finish, bags and plastic bottles of labeled samples line the perimeter of the work, creating literally a frame of view and frame of reference. A spectacular display of color is contrasted with the realization that the color is sourced from something as common and unspectacular as sand. Walking around the work, it is possible to discern faraway deserts and beaches on the handwritten labels. It is clear where the pigments come from,

FIGURE 7.3 **(See color insert.)** Elvira Wersche, Sammlung Weltensand. (Courtesy of the artist.)

but as the artist herself admits, not where the color itself comes from. "I know a little bit about the mineralogy and geology of my materials . . . but I'm searching for someone who can really explain the colors of the earth to me . . . *how* is it possible that some sands are mainly red and others mainly yellow?" (Artist Elvira Wersche in an interview with A. Toland on December 7, 2012).

Soil scientists Ed Landa (2004) and Fiorenzo Ugolini (2010) provide important insights on the history, interpretation, and meaning of soil color, which is important for artists interested in soil. Based on the unique mineral composition of the parent material as well as chemical and biological weathering processes over time, the origins of soil color are as diverse as the resulting palette: "... black colorations result from the transformations, through biological and chemical processes, of the green biomass into black humus, in addition to charcoal produced by fires. Brown, reddish-brown, red, and yellowish-red are due to the presence of iron oxides ... Manganese oxides are responsible for blue and black colors. Among the clay minerals, kaolinite and muscovite produce a white coloration. Glauconite and celadonite are green. Limestone, in spite of its typical white color, can, in a Mediterranean climate, produce a strongly red colored weathering residue" (Ugolini 2010, p. 79).

The use of soil color in everything from clothing to parchment to interior design to weapons (e.g., as camouflage for vehicles and machinery or the decoration of arrowheads and darts with mineral pigments as described in Eerkens et al. 2012) has a long history from hand-hewn indigenous practices to mechanized industrial manufacturing. In a further example, *eARTh—An Exploration of Earth Pigments in North Devon,* environmental artist Peter Ward focuses on the natural and cultural history of soil color. Ward is not only interested in the geographic distribution and natural causes of soil color, but also in the traditional and industrial uses of earth pigments in terms of their cultural, personal, and socioeconomic histories.

Ward's multidisciplinary and multisensory inquiry focuses on his family's native region of North Devon in the southwest corner of the United Kingdom—from Bideford black clay and china clay deposits to what was unearthed from former large-scale iron and copper mining and small-scale coal mining activities. Ward collects earth pigments as well as the stories hidden behind their collection. He then creates paintings on site, and in mixed media installations in nontraditional art venues such as the Ussher Society's annual geological conference in 2011, the Camborne School of Mines, and the Plymouth University Chemistry Department.

Beyond painting, pigment collection, and historical research, Ward designs public pigment-making workshops and field excursions, encouraging people to produce their own pictures and memories of the soil. Ward says his work with soil color has been revelatory: "I was simply a painter before, using pigments to paint pictures, but ... the whole process of working with earth pigments completely changed my work as an artist, form being an object-based maker to really understanding the ideas of process, from processing pigments and understanding where materials come from ... to recording stories ... people's personal recollections and connections to the things around where they live ..." (Peter Ward in an interview with A. Toland, November 15, 2012). Ward's work is also innovative in its multisensory approach to color: "Visual perception is challenged by actually touching things, by tactile engagement ... You can walk past something a hundred times and then one day say, 'I'm going to touch that' and that's when the exploration into color begins ... For me, its exciting to share that."

Ward's work, similar to the *Painting with the Colors of the Earth* workshops in Austria, remind us how effective art can be in engaging the senses, emotions, and personal connections to place. In such participatory workshops, it is not so much the final product, but rather the process and experience of being in the field and completely focused on the qualities of the material substances at hand.

7.2.3 Printing (and Other Forms of Particle Transfer)

Next to painting, printing, or print-making, enjoys an eminent role in the visual arts. Printing refers to the process of copying text and images onto paper or other flat media—literally the transfer of materials from one surface to another in the reproduction of meaningful information. Regarding soil, printing provides an interesting metaphor for transferring that which is amorphous, hidden, and often misunderstood, to the light of cultural reception, contemplation, and critique.

In soil science, pedotransfer functions refer to the transmission and application of easily available soil data (e.g., texture and bulk density) into predictive estimations for properties that are difficult to determine, such as a soil's sorption capacity for heavy metals, or the capillary rise of water in a soil and its subsequent evapotranspiration for entire regions (Bouma and Van Lanen 1987, Wessolek et al. 2011). The increasing fascination among artists in preparing and presenting soil samples and soil data that are generally inaccessible to the public might be called artistic pedotransfer—a kind of printing process for the transference and transformation of soil *in situ* to other forms of encounter. The following approaches draw on the metaphors of printing and pedotransfer to distill three features of the soil generally *not* addressed by pedotransfer functions or other scientific analyses of soil: the sacredness of place, the animism of the soil biome, and the ethical significance of the surface.

Borrowing from the history of other print-making techniques, Ekkeland Götze is skilled in the art of *terragraphy* (Figure 7.4), a printmaking technique which the artist invented to create hundreds of individual abstract impressions of the soil that together make up a composite image of the earth—*earth artwork*. Götze has traveled the world letting intuition, serendipity, and local inhabitants lead him to special sites to harvest a limited number of prints determined solely by the individual structure, color, and aura that characterizes each place. "The technology is absolutely the same every time. You look at the wide range of different colors and structures and it's the most exciting thing. I have an idea or a conception of what will happen, but I can never really know. It happens in the moment of the printing. Each picture is an absolutely unique work, because the soil on the pressure carrier (paper, canvas, fresco) can be only on this one copy. If another work/copy is produced with soil from the same excavating site, a kind of familial relationship is created, but it not the same. It is unique, like a human being!" (Ekkeland Götze in an interview with A. Toland on November 14, 2012).

Beyond producing mere graphic representations of the soil or the sites he visits, Götze sees a spiritual dimension to his work: "*Earth Art Work* is not an impression of the excavating site. It is really an authentic picture of place and contains the spirit and the energy of the site, because it is produced only with the soil of the site, without any other material influence." The sites he chooses to work in are usually culturally or spiritually "charged," including places of historical, mythical, and religious significance such as the biblical Mount Sinai in Egypt, the sacred Mount Kailash in Tibet,

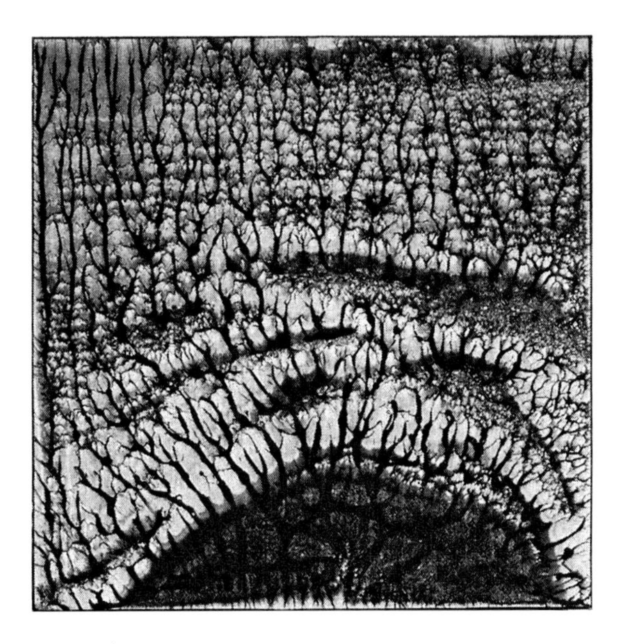

FIGURE 7.4 Ekkeland Götze: Ófærufoss Eldgia Terragrafie-N° 515, 2004. (Courtesy of the artist.)

and the Aboriginal songlines (mythical routes celebrated in traditional song and dance, painting, and story-telling) in Australia. Christopher Balme (2011) describes Götze's role as an artist as that of "medium" in a shamanic sense. The medium (*the artist*) guides the medium (*the soil*) through a process of printing, as if through a door, to a new state of awareness.

Like Ulrike Arnold, Ekkeland Götze also uses photographs and other materials to contextualize his activities within the local surroundings and communities he visits. Connecting with indigenous elders, gathering their stories, and recording their surroundings through photographs are as much a part of Götze's work as collecting and printing the earth materials. In his interactions on site and later with anonymous audiences, the artist sees his work as a means for personal reflection: "This objective picture of the EARTH is a platform for intercultural communication and offers room for association, to be filled by the individual thoughts, memories and hopes of the observer" (Götze 2012).

In the age of Facebook and Instagram, a second printing process to consider in this overview is photographic printing, or the process of producing lifelike images on chemically sensitized paper, usually from a photographic film negative, slide, or digital image. Since the invention of photography in the early nineteenth century, the "truthfulness" of photographic representation has been debated by scholars such as Roland Barthes, Michel Foucault and Susan Sontag, among others. Given the conventional trust in photography, soil scientists have also come to use and take photographs for educational and public awareness building purposes such as the *Soil of the Year* publicity campaign in Germany, the *World Soil Day* Exhibition at the European Union's Joint Research Centre, or the photographic work of Berlin-based soil scientist Mohsen Makki.

Combining knowledge of photography and soil biology, artist Daro Montag collaborates with legions of nematodes, flagellate protozoa, and other soil fauna and flora to create true snapshots of the earth not with a camera, but with the soil itself (Figure 7.5). Rather than photographing the soil, Montag allows soil microorganisms to eat away the gelatin surface of film strips laid directly on the soil profile. The resulting *Bioglyphs* are a record of microbial action in the soil, which varies according to depth, pH value, bulk density, and moisture content.

Montag (2007, p. 45) explains his process: "Five prepared films were encased in wooden boxes with soil cores to create *Bioglyphs*. After four weeks the films were extracted, dried and viewed on a microscope. Photographs of small sections of the film were made with a diameter of approximately

FIGURE 7.5 **(See color insert.)** Daro Montag, 5 Film Strips from *This Earth*, 2007. (Courtesy of the artist.)

2 mm." While the process itself is relatively straightforward, the resulting *Bioglyphs* are stunning portraits of complex communities brought to life by the innovation of the artist and the gaze of the spectator. Montag's work is "composition by decomposition" (Montag 2001, p. 8), an aestheticization of decay, and an introspection on the transformation function of the soil. Like the ruby radiance of stained glass windows in a gothic cathedral, or the illuminated pages of a medieval manuscript, Montag's film-strips glow with a sense of animism, as they light up the interconnected lives of millions of tiny soil organisms.

While Ekkeland Götze uses printmaking techniques to transfer soil particles to paper, and Daro Montag adapts photographic printing techniques to illuminate the soil biome, the works of Betty Beier (Figure 7.6a) and Anneli Ketterer (Figure 7.6b) rather allude to printing as a process of preserving a small selection of the earth's surface as an authentic document of a specific time and place. Both artists are specialists in surface extraction, with Ketterer's works focusing on fleeting natural phenomena such as waves, wind, or animal tracks, and Beier's work concentrating on contemporary cultural phenomena such as traffic, urban development, and climate change. Ketterer's process could be called crusting, or in her words *de-crusting*; Beier's process could be called casting, or in her words *earth-printing*.

In a photorealistic sense, both artists are unwaveringly faithful to the temporal situations they seek to capture. Both create 1:1 scale models of the soil's surface in unique "monoprints" that preserve moments in history with archival precision. Approaches differ in the final "printing" process and site selection. While Ketterer's *decrustation* process uses multiple applications of adhesive to extract the surface as it appears in nature, Beier relies on photographs and detailed field notes to reconstruct the casts she takes on site with meticulous detail-work later in the studio. Ketterer's selected sites are often remote landscapes where the traces of natural forces may be recorded. Waiting days and sometimes weeks for natural elements to sculpt the surface, Ketterer's *Decrustations* are a tribute to the climatic forces above ground before they are sealed as traces of time on the soil surface below.

Beier, meanwhile, follows stories she finds in diverse news sources and seeks out sites "... that are accompanied by controversial discussions" (Katerndahl 2009, p. 11). Typical sites include

(a)

(b)

FIGURE 7.6 Two examples of artistic pedotransfer in action: (a) Betty Beier, Earth Print Archive, the artist installing a plaster frame. (b) Anneli Ketterer, Decrustate, the artist removing a surface peel.

contested locations of planned dams, major construction, and infrastructure projects, and the melting permafrost soils of the frigid zones. Blending journalistic objectivity with a skill for retelling stories through images, the evidence of land lost is documented in Beier's *Earth Print Archive*, like the precious pages of an earthen newspaper. In similar, but at the same time, very different ways, both artists invoke ethical reflection on the soil's surface, documenting the fragile beauty of the natural world as it is, was, and in many cases will cease to be.

7.2.4 Archiving Soil Diversity

Archiving is a fundamental activity of cultural heritage—the act of collecting, processing, and storing information for the benefit of sharing knowledge, artifacts, and understanding with the rest of humanity. From the idea of reproduction in terms of printing, a secondary idea of *collection* emerges. What can be produced or printed once can be reproduced many times. The possibility of reproduction sets in motion a new series of processes—that of collection, archiving, and systemization.

For the soil science community, the idea of archive is embodied in programs such as the FAO's World Reference Base for Soil Resources, the European Commission Joint Research Centre's Soil Portal of the EU, and World Soil Survey Archive and Catalogue (WOSSAC) at Cranfield University in the United Kingdom. Artists' archives complement and expand these extensive collections of soil maps and soil data by focusing on aesthetic experience and tailoring reception to a broader audience.

The works here deviate from those in the previous sections on painting, pigmenting, and printing, in their shift of focus from color and form to a more general preoccupation with capturing and archiving soil diversity. Furthermore, they break out of the two-dimensional picture frame to round out the overview of genres with works of sculpture and installation. The concept of archive may be understood in sculptural terms as a container that may be expanded or collapsed according to scale, size, content, and context. The following works take as points of departure singular sites, material samples, and encounters to create emergent collections in celebration of the complexity, diversity, and collective beauty of soil. As with the printed pages of a book, or the millions of single particles and individual organisms contained in a handful of soil, the following works are distinguished by the whole being greater than the sum of their parts.

One of the first and most renowned artist's soil collections is herman der vries' *erdmuseum* and his related *from earth* works. (Note: Lowercase letters are a signature of de vries' philosophy of objectivity.) Initiated in 1979 in Eschenau, Germany, the *erdmuseum* consists of 7000 soil samples from all over the world, many of which were collected personally by the artist. All samples have been dried, pulverized, packed into bags, and stored in cardboard boxes marked with the date and location of the finding. These samples are further represented as earth rubbings on uniform grids on paper, the *erdkatalog*, which is also organized according to time and place of discovery.

In a minimalist handling of material, a political overtone is imbued in the work. All samples are unique in color and origin. All have been handled equally. None have been emphasized, singled out, or excluded, even samples from history-laden sites like Chernobyl or the concentration camp at Buchenwald. Mel Gooding (2006, p. 97) remarks on de vries' ability to distil the intrinsic qualities of "self-ness" from the materials he collects: "Crushed to a powder, reduced to its basic mineral 'suchness', rubbed down with the tips of the fingers into a simple rectangle of colour, the earth itself is transformed not into the sayable abstraction of the word, but into a material sign of its self-ness."

The *erdmuseum* uses scientific precision in its execution. As de vries (2004) explains in an interview with John Grande, his methodical approach is adapted from his former career as a biologist: "a systematic approach is one possible way artists can work. I learned this discipline from science." But he warns, "science on its own cannot provide us with a complete understanding of the world and our life. art and science can be complementary. by fusing both, the two main creative streams of our culture in relation to our life space can be integrated." de vries thus fuses scientific rigor with the philosophy of nature in his artistic explorations of *random objectivation* (1962–1975), *chance and change* (1972 to present), *natural relations* (1982–1989), and *from earth* (1983–ongoing). By doing so, he also calls for a transdisciplinary approach to soil protection by integrating different ways of knowing and understanding the soil.

Years later, on the other side of the planet, Koichi Kurita's *Soil Library* (2006–ongoing, Figure 7.7) picks up on de vries' minimalist form and meticulous execution with a performance/installation work that celebrates the beauty and diversity of Japan's soils. Kurita has collected over 30,000

FIGURE 7.7 **(See color insert.)** Koichi Kurita, Soil Library, 2012. (Courtesy of the artist.)

samples from around the Japanese archipelago one handful at a time, which he uses to create giant floor installations in a slow and meditative performance. Like de vries, Kurita has collected most of the samples himself and presents them in a uniform manner on square pieces of paper laid out in a grid. What differentiates Kurita's work is his distinct ethno-geographic approach to collection.

In an email to the author on August 27, 2012, Kurita wrote: "In Japan, when we are born and given a name, we visit the Shinto shrine in the village and pay respect to Ubusunagami, 'the god of birth soil.' Everybody has Ubusunagami. We believe we are from the soil and we go back to the soil. So I am collecting the soil from all the villages, towns, and cities, for all of the Japanese people... I've already collected 32593 soil samples in 3213 (out of 3233) villages... Road maps are more important than geological maps for me. I can see the names of villages, information about soil and life from a road map. If people are living there, there is soil there. I am interested in the soil of life, not the soil of geology. The soil becomes the material of research for geologist ... pigment for the painter ... money for architect ... But the soil is life for me."

Kurita provides two insights that are relevant for the context of soil protection. On the one hand, soil (as well as geologic) mapping could be more strongly tied to maps documenting demographic distribution, settlement, sprawl, trade, and transportation routes, but also areas of significant cultural and religious meaning. On the other hand, the fundamental dependence of humans on the soil is inverted to a fundamental dependence of soil on human activity: where there are people, there is soil. It is soil that is ultimately dependent on us for its health and safety. In the age of the Anthropocene, where the effects of damaged soils manifest themselves in worldwide health issues such as cancer, reproductive and respiratory illnesses, healthy soil is an indicator of healthy societies (Handschumacher and Schwartz 2010). Birth soil, *Ubusunagami*, can be seen as an indicator of the heritage built around not only protecting but also worshipping that which gives life and health.

Artistic methods of collection, archival, and documentation of soil diversity vary according to access to samples and sites, geographic focus, physical parameters of storage and presentation, and more than anything else, creative impulse and background intention. Methods vary from individual to group collection in local, regional, and global scales. Collection may occur via personal networks and "snowballed" contacts, to anonymous online networks, and collaboration with local communities and scientific institutions. Those parts of the collections available to the public are determined in various exhibition contexts either by the artist, by museum directors and curators, or by members of the public. Like scientific collections, artists' soil collections usually contain voluminous amounts of material (hundreds to tens of thousands of samples) and can be stored in anything from informal accumulations of bags and boxes to highly organized reference systems.

Where artists such as herman de vries, Koichi Kurita, and Ekkeland Götze collect and process most of their own samples, other artists rely heavily on third parties, as in Elvira Wersche's *Sammlung Weltensand* described above. Another renowned soil collection which draws on the collective efforts and insights of donors worldwide is the *One Earth Altar*, created by Marianne Greve on occasion of the Expo 2000 for the One Earth Church in Schneverdingen. This project merges human and soil diversity by combining up to 7000 soil samples from around the world with the individual impressions of the donors. Soil books, made out of Plexiglas containers with inscribed reference numbers, are contained in a massive bookshelf behind the church's altar. An on-site database links the soil books to corresponding narratives and location data.

In a more regionally focused work, *Correlation Drawing/Drawing Correlations: A Five Borough Reconnaissance Soil Survey* (Figure 7.8), Margaret Boozer collaborated with Dr. Richard Shaw, soil scientist for the Natural Resources Conservation Service (NRCS), to literally draw correlations between the soils, and the people of New York City. Soil samples collected by the New York City Soil Survey were arranged in a Plexiglas grid for the exhibition *Swept Away: Dust, Ashes, and Dirt* at the Museum of Arts and Design and later for the Museum of The City of New York's South Street Seaport location. QR codes (a kind of barcode typically used in advertisements and industry to convey information digitally through a QR reader, often in the form of a smart phone app) near the installation link the location of the soils to the NRCS' National Cooperative Soil Survey, an online

FIGURE 7.8 Margaret Boozer, Correlation Drawing/Drawing Correlations: A Five Borough Reconnaissance Soil Survey, 2012. (Courtesy of the artist.)

database documenting the taxonomic classes, profile descriptions, land use, and other soil information. Visitors are encouraged to find themselves within a complex matrix of color and texture, natural history and urban development, soil information and individual location.

Boozer (2012) explains her motivation behind the work as an act of creating awareness: "with the worldwide population shift toward mega-cities, soil mapping is beginning to focus on urban areas where soils information is useful for restoration and re-vegetation efforts, storm water management, land use decisions and most importantly, studying the effects of human disturbance on the environment. The Survey as a document is meant to increase public awareness and appreciation of a valuable resource." In scientific efforts to understand and evaluate urban soils, artists elsewhere could similarly be integrated into processes of soil archival and documentation.

While collections are generally "objective" in the sense that objects and data are presented in an orderly and unembellished fashion, the politics of the collector—be it a person or institution—are hard to avoid. Collections are influenced by underlying political and philosophical tendencies, as demonstrated in many of the works already mentioned. The concept of archive is also politically charged in soil protection contexts, where the function of soil as natural and cultural archive is protected under law (at least in Germany, under the Federal Soil Protection Act) along with other functions such as habitat, biomass production, and filter functions (Meuser and Greiten 2009). The soil *as archive*, as well as scientific and artistic projects *to archive* the soil are inherently political in their composition and realization.

For artist Sarah Hirneisen, the politics of collections as well as the psychology of collecting are inseparable from the objects collected. Hirneisen has collected the cremated ashes of people's personal belongings (*Lares and Penates,* 2006), dust from vacuum cleaners (*Inventory of Disintegrated Particles from 120 Homes,* 2004, 2007), and contaminated soil from Superfund sites in California (*Contaminated, 20 Active Calsites in Oakland, CA,* 2005). The soil, ashes, and dust in Hirneisen's collections are symbolic variables for cultural compost—residues of the cycles of personal and industrial consumption and waste frozen in time before or upon entering the natural environment.

For the project *Unfamiliar Ground* (Figure 7.9), Hirneisen was interested in archiving the soil of war. In 2007, the artist sent care packages to soldiers stationed in Iraq and Afghanistan via the anysoldier.com website with letters requesting the collection of soil. She then cast the dusty samples in fragile glass pouches engraved with the soldier's name and date and location of collection. Soil is seen here as symbol for personal identity but also for political power structures and the wars that

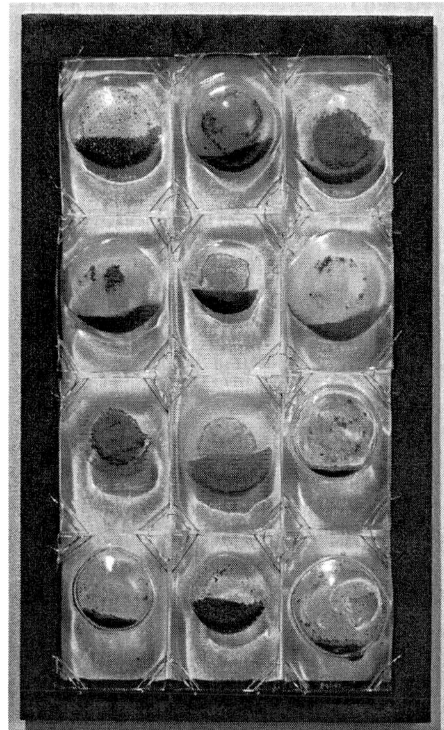

FIGURE 7.9 Sarah Hirneisen, Unfamiliar Ground, 2007. (Courtesy of the artist.)

erupt between nations over soil and land use, property, identity, and rights to resources: "Soil as in war ... I was thinking about mothers having their child away fighting a war. And I was also really interested in the loss of identity that was happening in Iraq ... the soil just became the symbolic reference for all of these things ... being away from home soil, being on foreign soil, stealing soil, trying to claim soil that's not yours" (Sarah Hirneisen in an interview with A. Toland on September 27, 2012).

The soil of war is well documented in movies and literature, ranging from topics of soil moisture effects on soldier mobility and morale in Vietnam (Wood 2006, 2010) to the challenges of digging tunnels out of prisoner of war camps in occupied Poland (Brickhill 1951, Doyle et al. 2010) to the health risks of inhaling microbial-rich dust in Baghdad (Borell 2008, Wood 2010). Apart from the incidental handling of soil in war movies and paperback novels, however, the soil of war has not been well archived in scientific or artistic collections. Sarah Hirneisen's participatory archive of the war-torn soils of Iraq and Afghanistan addresses what is still missing in soil archives, using the vocabulary of sculpture to confront uncomfortable aspects of soil diversity.

7.2.5 Further Genres and Approaches

What is missing from this overview is a handful of works that emphasize aesthetic appreciation *beyond the visual field*, including touch (e.g., Lu La Buzz' performative works with clay and Alison Keogh's gestural paintings with clay), tasting and smelling (e.g., Laura Parker's *Taste of Place* soil tasting workshops and Lynn Peemoeller's *Soil Perfumes*) and hearing (e.g., Florian Dombois' earthquake symphonies and Olle Corneer's field concerts). Such multisensory aesthetic approaches provide access to the soil for the whole body and are especially useful in educational settings and field excursions.

Celebrated examples of Land Art by artists such as Walter de Maria, Michael Heizer, and Robert Smithson, as well as stunning Art in Nature works by Andy Goldsworthy, are also not included but should be recognized as important aesthetic contributions to the field of soil art. Although some of the artists associated with the Earth Art movement (e.g., Agnes Denes, Alan Sonfist, Richard Long) were committed to a larger environmental mission, these genres are primarily a reflection on sculptural practice and the topography of the land rather than an exploration of soil formation or protection issues.

Finally genres such as activist art (Parking Day), socially engaged art (Nancy Klehm), ritualistic art practice (Ana Mendiatta), ecological restoration art (Jackie Brookner, Mel Chin, Daniel McCormick), as well as experimental forms of alternative agriculture and avantgardening (Future Farmers, Kultivator) challenge formal aesthetic approaches in ways that engage not only the senses but also individual and collective sources of identity and understandings of political and social constructs. To this end, a wider survey of approaches would also have to include a more comprehensive discussion on the underlying ethical, social, cultural, and political aspects of soil protection and soil art beyond aesthetic categorization.

7.3 AESTHETIC INTEGRATION

From painting and printing with soil, to multisensory explorations of color, to archiving soil diversity, the artistic positions presented here provide soil protection discourses with new sources of imagery. Regarding soil protection, the integration of aesthetic values and contemporary art may be summarized into three main points.

Aesthetics represents a powerful but largely unused strategy of soil protection planning, policy, and practice. While many soil scientists admit to collecting personal samples for aesthetic reasons (Toland and Wessolek 2010) and properties of the soil such as color and texture are instrumental to scientific fieldwork and taxonomic classification, aesthetics remains at the margins of soil protection paradigms. In a narrower sense, formal aesthetic criteria from environmental protection contexts could be adapted for soil protection purposes. Categories of beauty, diversity, and uniqueness used to protect landscapes on account of their picturesque or recreational value [see, for example, §1 German Federal Nature Conservation Act (known in-country as the "BNatSchG")] could be used to protect soils of particular aesthetic quality, distinctiveness, or rarity, as well as the diversity of soil types across the planet and the biodiversity of the soil biome. In preventative soil protection measures, comparative assessments of soil quality used to estimate soil values for agricultural purposes could also be adapted for purposes of evaluating soil aesthetic properties. In the tradition of Albert H. Munsell, artists could be consulted as important resources in the collection, interpretation, and documentation of soil aesthetic properties.

In a wider sense, cultural heritage, belief systems, and personal identity embedded in the soil constitute a broader set of aesthetic values with which to examine, picture, and protect the soil. Such values are anchored in emotional points of departure, often taboo or trivialized by the scientific community. In soil protection discourses emotions are largely missing but necessary for stewardship to be integrated into society at large. Pleasure and disgust, humor and grief, hope and fear, fun and enjoyment, reconciliation and celebration are as much a part of people's motivations to protect the soil as quantifiable parameters of soil fertility. Emotional currency is generated in the grounding of research and planning at the local level. The history of land use, social structures, cultural norms, governing powers, and environmental factors of change are as integral to soil research as the isolation of discrete variables. Where soil science lacks the tools, time, or incentive to examine and depict both narrower formal aesthetic values as well as to contextualize wider non-formal aesthetic values, artists could provide valuable guidance.

The archive function of the soil offers a unique opportunity for the integration of aesthetic values into soil protection contexts. The understanding of archive as historical narrative, as sculptural container, and repository for countless personal and cultural meanings, imparts the archive function with a special status that is difficult to evaluate with instruments such as are used to measure, for example, buffer and filter functions, or functions of biomass production. This special status of the archive function calls for interdisciplinary collaboration in order to more comprehensively evaluate and protect the soil as archive. Along with the expertise of artists, soil scientists would also benefit from the knowledge of archeologists, historians, cultural theorists, theologians, and indigenous people in their efforts to protect the archive function of the soil. Especially in urban areas where the loss of soil due to sealing and development is one of the biggest threats to soil protection, the archive function takes on new meaning as a fundamental cultural record, requiring investigation by multiple actors in multiple disciplines.

7.4 SOIL PROTECTION AS INTERDISCIPLINARY OPPORTUNITY

Finally, sustainability science argues for the implementation of fundamental research into social contexts (Godemann 2011, Adomßent and Godemann 2011). "Knowledge production is no longer a privilege of a special group of experts. Instead, it takes place in a number of different constellations of actors. In these inter- and transdisciplinary work contexts, not enough attention has been paid to the problem of translating and communicating this knowledge in a way that is adequate to its target groups" (Wardekker et al. 2009, *in* Adomßent and Godemannn 2011, p. 32).

Although soil cultivation is almost as old as human civilization, soil protection is a relatively new field that begins with scientific paradigms and norms, crosses over to policy and planning, and finally, must reach a point of individual stewardship and social acceptance. For this to be achieved, a new culture of knowledge production is necessary. A transdisciplinary approach to soil protection requires what Marcel Hunecke (2006) describes as an integration of perspectives at multiple phases of knowledge production—from the research design to the interpretation, contextualization, communication, and implementation of results. Finally, it calls for time, trust, and a complete overhaul of existing funding structures.

The real challenges of soil degradation, climate change, and food security are pushing soil science and soil protection from the home domains of agro- and geo-science into new fields of sustainability research. Similarly, contemporary art is finding new aims and functions within the field of sustainability communication. Artists provide not only new images and depictions of the soil, but new means of understanding, interpreting, and protecting "the soil underfoot." Artists are experts in tapping into people's aesthetic abilities and emotional states, navigating between different social settings, and targeting different audiences. The arts have always been a driving force in questioning and tearing down disciplinary barriers. As such, they can pave new avenues of raising soil awareness.

REFERENCES

Adomßent, M. and Godemann, J. 2011. Sustainability communication: An integrative approach. In *Sustainability Communication – Interdisciplinary Perspectives and Theoretical Foundations*. eds. J. Godemann and G. Michelsen, 27–37. Dordrecht: Springer.

Bachmann, G. 2001. Terra preciosa: Böden und ihre Wahrnehmung in Kunst und Kultur. http://www.bodenwelten.de/bod_kunst/pdf/preciosa.pdf (accessed November 15, 2012).

Balme, C. 2011. *Ekkeland Götze—Das Bild der Erde. Foreword to Exhibition Catalogue*. Munich: Center for Advanced Studies, Ludwig-Maximilians-Universität München.

Boozer, M. 2012. Correlation Drawing/Drawing Correlations: A Five Borough Reconnaissance Soil Survey. Artist Statement. http://margaretboozer.com/exhibitions.html (accessed November 20, 2012).

Borell, B. 2008. Bagdad Hack. *The Scientist,* 22(12). http://www.the-scientist.com/?articles.view/articleNo/26949/title/Baghdad-hack/ (accessed January 10, 2013).

Bouma, J. and Van Lanen, H. A. J. 1987. Transfer functions and threshold values: From soil characteristics to land qualities. *Proceedings of the International Workshop on Quantified Land Evaluation Procedures*, Washington, DC, 106–110.

Brickhill, P. 1951. *The Great Escape*. London: Faber.

Busch, B. 2002. *Erde*. ed. B. Busch, *Schriftenreihe Forum, Band 11, Elemente des Naturhaushalts III*. Köln: Kunst- und Ausstellungshalle der Bundesrepublik Deutschland GmbH.

Carlson, A. 2000. Formal qualities in the natural environment. In *Aesthetics and the Environment—The Appreciation of Nature, Art and Architecture*, ed. A. Carlson, 72–101. Abingdon: Routledge Taylor & Francis Group. 2007.

de vries, h. 2004. Chance and change. In *Art Nature Dialogues—Interviews with Environmental Artists*, ed. J. Grande, 223–234. Albany: State University of New York Press.

Doyle, P., Babits, L., and Pringle, J. 2010. Yellow sands and penguins: The soil of the great escape. In *Soil and Culture*, eds. C. Feller and E. R. Landa, 417–429. Dordrecht: Springer.

Eerkens, J. W., Gilreath, A. J., and Joy, B. 2012. Chemical composition, mineralogy, and physical structure of pigments on arrow and dart fragments from gypsum cave, Nevada. *Journal of California and Great Basin Anthropology* 32(1): 47–64. http://anthro.dss.ucdavis.edu/people/jweerken/JCGBA2012.pdf (accessed March 23, 2013).

Feller, C., Chapuis-Lardy, L., and Ugolini, F. 2010. The representation of soil in the western art: From genesis to pedogenesis. In *Soil and Culture*, eds. C. Feller and E. R. Landa, 3–21. Dordrecht: Springer.

Godemann, J. 2011. Sustainable communication as an inter- and transdisciplinary discipline. In *Sustainability Communication—Interdisciplinary Perspectives and Theoretical Foundations*. eds. J. Godemann and G. Michelsen, 39–51. Dordrecht: Springer.

Gooding, M. 2006. From earth. In *herman de vries – Chance and Change*, ed. M. Gooding. 96–101. London: Thames and Hudson.

Götze, E. 2012. Terragraphy. http://www.ekkeland.de/2.0.html?&L=1 (accessed November 10, 2012).

Handschumacher, P. and Schwartz, D. 2010. Do pedo-epidemiological systems exist? In *Soil and Culture*, eds. C. Feller and E. R. Landa, 355–369. Dordrecht: Springer.

Hartemink, A. 2009. The depiction of soil profiles since the late 1700s. *Catena* 9: 113–127.

Hunecke, M. 2006. *Eine forschungsmethodologische Heuristik zur Sozialen Ökologie,* Band 3. München: Oekom Verlag.

Katerndahl, J. 2009. From turf to picture. In *Betty Beier—Earth Print Archive, Ministergärten, Karahnjukar, Solheimajökull.* Foreword to Exhibition Catalogue, 11–16. Saarbrücken: Galerie der HBKsaar, K4 forum.

Landa, E. R. 2004. Albert H. Munsell: A sense of color at the interface of art and science. *Soil Science* 169: 83–89.

Lines-Kelly, R. 2004. Soil: Our common ground—A humanities perspective. Keynote Address presented at Super Soil 2004: 3rd Australian New Zealand Soils Conference at the University of Sydney, Australia 5–9. http://www.regional.org.au/au/pdf/asssi/supersoil2004/lineskelly.pdf (accessed December 12, 2012).

Lines-Kelly, R. 2014. Poetry in motions: The soil excreta cycle. In *The Soil Underfoot,* eds. G. J. Churchman, and E. R. Landa. 303–314. Boca Raton, FL: CRC Press.

Meuser, H. and Greiten, U. 2009. Bodenfunktionsbewertung in Osnabrück. In *Umweltberichte* ed. Stadt Osnabrück Fachbereich Umwelt.

Montag, D. 2001. Artist's statement. In *Montag, Bioglyphs, 8. Cross Farm, Diptford.* Totnes, and Devon: Festerman Press (originally published in Sci-Art catalogue, 1998).

Montag, D. 2007. Thinking soil. In *This Earth, 1–55. Cross Farm, Diptford.* Totnes, and Devon: Festerman Press. ISBN 978-0-9544187-4-8.

Munsell, A. H. and Cleland, T. M. 1921. *A Grammar of Color: Arrangements of Strathmore Papers in a Variety of Printed Color Combinations According to The Munsell Color System.* Mittineague, MA: The Strathmore Paper Company.

Reis, M. 2004. Riverwork. In *Art Nature Dialogues—Interviews with Environmental Artists*, ed. J. Grande, 105–116. Albany: State University of New York Press.

Szlezak, E. 2009. Soilart—Painting with the Colours of the Earth. Project description from the Amt der NÖ Landesregierung Abteilung Landentwicklung. http://www.soilart.eu/1-0-Home.htm (accessed December 20, 2011).

Toland, A. and Wessolek, G. 2010. Core samples of the sublime—On the aesthetics of dirt. In *Soil and Culture*, eds. C. Feller and E. R. Landa, 239–260. Dordrecht: Springer.

Ugolini, F. 2010. Soil colors, pigments and clays in paintings. In *Soil and Culture*, eds. C. Feller and E. R. Landa, 67–82. Dordrecht: Springer.

Wardekker, J. A., van der Sluijs, J. P., Janssen, P. H. M., Kloprogge, P., and Petersen, A. C. 2009. Uncertainty communication in environmental assessments: Views from the Dutch science-policy interface. *Environmental Science and Policy* 11, 627–641.

Wessolek, G., Bohne, K., Duijnisveld, W., and Trinks, S. 2011. Development of hydro-pedotransfer functions to predict capillary rise and actual evapotranspiration for grassland sites. *Journal of Hydrology* 400 (3): 429–437.

Wood, C. E. 2006. *Mud: A Military History*. Washington, DC: Potomac Books, Inc.

Wood, C. E. 2010. Soil and warfare. In *Soil and Culture*, eds. C. Feller and E. R. Landa, 401–415. Dordrecht: Springer.

Van Breemen, N. 2010. Transcendental aspects of soil in contemporary arts. In *Soil and Culture*, eds. C. Feller and E. R. Landa, 37–43. Dordrecht: Springer.

Zangwill, N. 2001. Formal natural beauty. *Proceedings of the Aristotelian Society* (Hardback) 101 (1): 209–224.

Zika, A. 2001. ParTerre—Studien und Materialien zur Kulturgeschichte des gestalteten Bodens. PhD Dissertation, Fachbereich 5 der Bergischen Universität, Wuppertal.

8 Principles for Sustaining Sacred Soil

*Norman Habel**

CONTENTS

In this chapter, I explore several principles for sustaining sacred soil, principles that are latent in the biblical tradition. Numerous religious cultures have revered Earth as sacred and many still do. Many Indigenous Australian cultures affirm the spiritual in nature, especially the land and the soil. In the Western world, however, attitudes to the soil as a component of planet Earth have largely been associated with certain interpretations of the biblical tradition that no longer perceive the soil as sacred—interpretations that I would like to challenge. To this end, I will refer to sections of the Bible in a separate manner to other books and papers by the name of the Book, with chapter and verse numbers.

Many of us are familiar with the famous voice addressed to Moses from the burning bush: "Take off your shoes, this is holy ground" (Exodus 3.5). The Hebrew expression for "holy ground" could just as readily be translated "sacred soil." In this tradition, Moses is warned about the sacred nature of the soil on which he stands because of the fiery presence of God in the burning bush. Here, the soil is sacred because of the proximity of the divine presence. My interest in this study, however, is not in locations that become sacred because of special acts of divine intervention, but the very character of soil as an integral part of the world of nature.

My appreciation of the principles I plan to enunciate has grown in recent years by my awareness of ecology as a radical new way of viewing nature, an awareness that has led me to develop an ecological hermeneutic for interpreting the Hebrew and Christian scriptures. These guidelines are outlined in a number of books, including *Exploring Ecological Hermeneutics* published by the Society of Biblical Literature in 2008.

My point of departure for discerning relevant principles in this context is my recent ecological reading of Genesis 1–11 in which I have identified two contrasting primal origin myths that I have designated The *Erets* (Earth) Myth and The *Adamah* (Soil) Myth (Habel 2011). In my analysis in this chapter, I will first focus on key elements of the second myth that are relevant to the topic of "sacred soil." Subsequently, I will explore several other relevant biblical traditions that highlight the living and sacred dimensions of soil.

The two major narratives of Genesis 1–11 are clearly primal myths and not history or science in any contemporary sense. Yet, it needs to be recognized that these myths are not necessarily irrelevant. They have influenced our traditional understanding of the world and while they may not

* Professorial Fellow, Flinders University, Adelaide, Australia; Email: nhabel@bigpond.com

concur with our current understanding of planet Earth as a product of evolution, they incorporate insights about the natural world that are still meaningful and relevant.

8.1 PRINCIPLE ONE: SOIL IS TO BE SERVED NOT SUBDUED

The Soil Myth in the Book of Genesis, Chapter 2 (Genesis 2.4b-6), otherwise known as "Genesis Two," begins with a portrayal of the primordial world as an empty domain where the soil preexists but is devoid of rain, vegetation, or someone/something to "serve" the soil: "...when no plant of the field was yet on Earth, and no herb of the field had yet sprung up—for the Lord God had not caused it to rain on Earth and there was no one to serve/till the soil." The Hebrew verb *abad* here is often translated "till" but its normal meaning is "serve" as when a subject serves a master or a priest serves at a sacred site. The primordial soil in this myth is depicted as lifeless without humans to serve/till it.

In this myth, humans are created to benefit the soil, not vice versa! The soil has primordial priority and intrinsic worth; it was not created for humans!

When the God of this myth creates humans, God takes some of the preexisting soil from the primordial domain and, like a primal potter, molds the soil into the discrete form of a human being. In this myth, humans are the first entities created. The Hebrew word for soil/ground is *adamah*; the related word for human being is *adam*. A human (*adam*) is a living son of the soil (*adamah*)—a soil being not a celestial being.

God then plants a forest where the humans are to live. While this location is usually translated the "garden" of Eden, the myth itself describes a forest with two types of trees—those with fruit for food and those that are characterized by beauty or grandeur. The soil is the primal source out of which the forest of Eden emerges at the instigation of God. This soil is ultimately the sacred source of all life in this myth.

To make the specific role of the human clear, God takes the first human and locates him in the forest of Eden with the specific task of "serving and preserving" (Genesis 2.15): "Then the Lord God took the man (*adam*) and put him in the garden (forest) of Eden to serve (*abad*) and preserve (*shamar*) it." While the Hebrew verb *abad* may sometimes have a nuanced meaning suggesting "till" as well as "serve," the supplementary verb *shamar* clearly means to "preserve" or "keep." The mission of humans is much more than "tilling" as such. Their primal mission is to serve, care for, protect, and sustain that primordial source of life we call soil! They are created to be soil keepers not soil exploiters, soil servants not soil rulers.

The significance of emphasizing this primal mission lies in the fact that Western Christianity has largely ignored this mythic tradition in favor of the myth in the Book of Genesis, Chapter 1, or "Genesis One." On the sixth day of creation according to the alternative Genesis One myth, the God of this myth says, in no uncertain terms, "Let us make humans in our image and let them have dominion..." (Genesis 1.26). After creating humans, God then blesses them and says to them, "Be fruitful and multiply, and fill the Earth and subdue it and have dominion over the fish of the sea and over the birds of the air and over every living thing that moves upon Earth" (Genesis 1.28).

A close comparison of this mission with the mission of Genesis 2.15 reveals two radically divergent perspectives. The verb for "serve" in Genesis Two is the diametric opposite to the verb for "rule" in Genesis One. And the verb for "subdue" in Genesis One is the diametric opposite of the verb for "preserve" in Genesis Two. In these two chapters, we have two opposing traditions about the role of humans—the one is a mission to serve the soil, the other a mission to subdue the Earth. In another context, I designate the Genesis Two text a green text that values nature and the Genesis One text as a grey text that devalues nature (Habel 2009).

Western Christianity has, in recent years, generally supported the perspective of a society bent on subduing nature and taming the forces of Earth. Earth has been viewed as a domain that God has given to humans to exploit for their benefit. Like kings, they believe that have a mandate to dominate.

This mandate to dominate has provided the grounds for the so-called harnessing of nature—the destruction of species, the relentless clearing of forests and the perpetual exploitation of soils.

Emulation of the God who acts violently in history with accompanying deeds of violence to domains of creation, has led to collateral damage in nature.

Reading from an ecological perspective, I would argue that we ought to recognize the mission of Genesis Two as integral to our very nature and purpose. We are soil beings with a responsibility to serve and sustain the soil as the very source of our being. We are kin with all that has evolved from the soil and ought to be concerned not only with our kin but with mother soil. In this tradition, we are called to sustain the soil as sacred.

For those with a specifically Christian orientation, I would point to the principle as enunciated by Jesus in Mark Chapter Ten. In this text, the explicit way of Jesus of Nazareth is to "serve" and not to "rule" like the tyrants of the Gentile world. Jesus' disciples are commission to serve, following the lead of Jesus, the Son of Man, who gives his life in service as a ransom (Mark 10.42–45). The act of God in Christ giving his life as a ransom is the epitome of the Gospel. "Serving" not "ruling" is the implication for all of life—including our life with creation (Habel 2009, p. 119).

If we are to sustain the soil as sacred, those of us who still find meaning in the biblical tradition need to affirm the principle implied in the mission of Genesis Two rather than Genesis One. To sustain the soil as sacred in this context means that we relate to the soil as an essential source for the life, sustenance, and celebration of all soil beings on Earth. Given this orientation, I would elaborate the first principle as follows:

I: Soil is to be served as a priest would serve and sustain a sacred site and not be subdued as a king who dominates his subjects.

This is an *a priori* principle that should guide those interested in sustaining the soil. How this might be applied in practical and ethical terms is clearly the task of sensitive soil scientists and ethical agriculturalists. The development of best practice, however, is ultimately dependent on the underlying values of the scientists involved. Recognizing the soil as sacred, I contend, adds a valuable dimension to our understanding of nature.

8.2 PRINCIPLE TWO: SOIL IS TO BE REJUVENATED WITH REST

In the biblical tradition, rest is associated with Sabbath, a dimension of reality we first meet on the seventh day of creation and frequently linked with worship on the seventh day. On the seventh day God rests, blesses, and sanctifies.

By the act of resting, God, as the primal impulse in the cosmos, relates to the completed cosmos through the act of blessing. That blessing, significantly, relates to the seventh day. Prior to that day, blessing has been dispensed to activate life as such. Now, time is blessed with the inherent capacity to initiate, sustain, and restore life that has emerged from the soil, seas, and air.

Many scholars who explore the themes of Sabbath and rest, however, are anthropocentric in their orientation. Rest is viewed as "time out" from the rat race of contemporary society, the 24/7 appropriation of time for human exploitation of Earth's resources, including soil. Or in the words of Darby Ray, "Sabbath experiences of self-care and self-worth, of deep rest and sustained reflection, support the development of critical consciousness in relation to the dehumanizing dynamics of corporate capitalism and 24/7 temporality" (Ray 2006, p. 171). Valuable as human "time out" may be, the need for "time out" or "deep rest" for the soil is crucial in societies obsessed with the exploitation of nature.

The most significant biblical tradition relating to soil, Sabbath, and rest is articulated in Leviticus 25–26. In this tradition, we meet a portrayal of God as the custodian of the land who is concerned about the sustaining power of the soil. A pivotal injunction of this custodian is to "keep my Sabbaths and reverence my sanctuary" (Leviticus 26.2).

The initial step of many scholars has been to assume that the sanctuary refers to the temple or some central sacred site. Upon closer investigation, however, it becomes clear that the sanctuary here is the land where God resides and wanders like a concerned custodian, just as God wandered through the forests of Eden (Genesis 3.8). In Leviticus 26, the sanctuary is not a building that is sacred but the soil that sustains life. No images are to be erected anywhere in the land presumably because all of the land is sacred and filled with God's presence (Habel 1995, p. 101).

Another biblical tradition that portrays Earth as a sanctuary is found in the call of the biblical prophet Isaiah (Isaiah 6.3). In this account, the seraphim cry, "Holy, holy, holy is the Lord of hosts. The whole Earth is filled with his *kabod*." The *kabod* of God is the vibrant presence of God that appeared on Mt Sinai and later filled the tabernacle, the sanctuary for Israel in the wilderness. In the Isaiah text, that presence fills the planet. If Earth is viewed as a sanctuary, then its various domains, of which soil is a significant part, are sacred.

The complementary injunction in the text of Leviticus 26.2 is to "keep my Sabbaths." At first reading, it may appear that this is a reiteration of the well-known commandment for the Israelites to keep the seventh day holy by desisting from work and celebrating their liberation from slavery in Egypt (Deuteronomy 5.12–15). In the Leviticus tradition, however, the focus is on "a Sabbath of complete rest for the land" not rest for humans (Leviticus 25.2–4).

This rest is specified as a complete absence of any agriculture every seventh year. In the sixth year, the people of the land are expected to gather enough food to meet their needs during the seventh year. The focus of the Sabbath has moved from rest on the seventh day to complete rest in the seventh year. In the seventh year, the land is to be left free to restore its fertility. The sacred soil of God's sanctuary is rejuvenated by rest as an integral component of its life.

The fertility of the land is dependent on both the faithfulness of the tenants and the rejuvenating presence of God. Total rest/Sabbath for the land is specifically termed a "Sabbath for the Lord." The two activities are intimately interconnected. As God rested on the seventh day, so God rests and thereby rejuvenates the soil every seven years. The rejuvenating presence of God is in the soil, and by implication, in all creation.

In the Sabbath year, the land returns to its owner, to the Lord. By returning the land to its owner, the Israelites recognize their absolute dependency on the goodness and authority of the landowner. In that year, the land—and presumably the owner—enjoys complete rest from the turmoil of tenants at work. It is as if God, and indeed the people of God, identify with the land in that year. The Sabbath year thereby highlights the sacredness of the soil.

II: Soil is to be rejuvenated by activating the life-giving power of rest emanating from God's presence in nature.

According to the Sabbath principle, it is vital for the rejuvenation of the sanctuary of Earth by the presence of God that Sabbath time be dedicated by human tenants on God's planet. Or in more contemporary ecological terms, it is vital for the internal rejuvenation/restoration of the domains of nature, including soil, that there be adequate rest time without the influence of external human forces that may continue to deplete a given domain.

8.3 PRINCIPLE THREE: SOIL IS TO BE CELEBRATED AS ALIVE

If we return to the Soil Myth of Genesis 2–4, we can also appreciate that the ground is the source of many forms of life emanating from the soil. The forest of Eden emanates from the soil and flourishes as trees of great beauty and trees with fine fruit. The animals likewise are formed from the soil and enjoy the forest of Eden as partners with the primal human pair. The soil is alive with all the potential flora and fauna of nature.

The sacred dimension of soil, however, only becomes fully apparent in the story of Cain and Abel. Cain is the farmer who serves/tills the soil and brings its produce to God as an act of thanksgiving. After Cain kills Abel, it is the soil that becomes a mediator who enables Abel's voice to be heard by God. The soil amplifies the voice of her child. Then God confronts Cain with the message that "Your brother's blood is crying out to me through the soil" (Genesis 4.10). The soil is sensitive to the blood of Abel and all who suffer unjustly (Habel 2007).

The full force of the sacred nature of the soil only becomes apparent when Cain responds to the decree of God that Cain will become a fugitive. Cain complains that his punishment is more than he can bear and that "Today you have driven me away from the face of the soil, and so I shall be hidden from your face" (Genesis 4.14). For Cain, and presumably all who live on/from the soil in this tradition, the presence of God is not associated with the sky, the temple, or some high mountain, but with the soil. The soil is sacred by virtue of God's presence, a presence that in other traditions may be associated with a specific sanctuary. In this tradition, the soil is a sanctuary.

Many Indigenous Australians also believe that the soil/land is sacred and filled with divine presence. The Rainbow Spirit Elders contend not only that the Creator Spirit emerges from the land and returns to the land where the Spirit's power is eternally present but also that the land/soil is alive within, filled with the life forces of all the species on Earth (Rainbow Spirit Elders 2007). The soil is alive with the Spirit (Habel 2007, pp. 30–32).

A comparable understanding of the land/soil is found in Psalm 104.29–30, which reads:

When you hide your face/presence, they are dismayed,
When you take away your breath/spirit they die
And return to the dust/soil.
When you send forth your breath/spirit they are created
And you renew the face of the ground/soil.

A full appreciation of this pivotal text depends on a understanding of the Hebrew term *ruach*, which depending on context may be translated as "breath," "wind," "air," "atmosphere," or "spirit." All these nuances of the word may be implied in the present text (Hiebert 2008). In this biblical tradition, the atmosphere we breathe is God's breath, God's spirit, that which animates all life.

In this Psalm passage, this life-giving breath/atmosphere/spirit is that which animates all life forms that emanate from the soil, everything from worms to elephants, from humus to humans. But just as significantly, the ground itself or, in the language of the text, "the face/surface of the soil" is renewed, activated, brought to life by the atmosphere or breath of God. The soil is sacred, activated by the atmosphere.

In many Christian churches, worshippers celebrate a harvest festival, thanking God for the fruits of the soil—everything from potatoes to passion fruit. In recent years, with the introduction of *The Season of Creation*, all the domains of nature may become the focus of celebration on specific Sundays. On a Sunday such as Planet Earth Sunday, a bowl of soil may be the central symbol of worship and the living presence of sacred soil celebrated as worshippers celebrate with creation. See www.seasonofcreation.com.

III: Soil is to be celebrated as alive, a sacred domain animated by the spirit/breath of God rather than a resource dismissed as meaningless matter.

In the context of the current environmental crisis, it would seem wise to acknowledge whether through worship, reflection, or personal contact with soil in field, forest, or garden, that soil is indeed alive. It is time, I would argue, to recognize that the so-called "ground on which we walk" is indeed sacred soil pulsing with life. We do not need a voice form a burning bush to remind us that the atmosphere, the very breath of God, renews the life of the soil we share.

8.4 PRINCIPLE FOUR: SOIL IS TO BE HEARD AND HEEDED

Once we recognize that the soil is alive, a sacred force in the cosmos, we can appreciate anew those many passages where Earth/ground/nature laments or praises, passages that we have in the past dismissed as mere poetry. Joel, Hosea, Jeremiah, and other prophets hear the land mourning and crying to God because of the desolation it has suffered. Jeremiah wonders how long the land will mourn and the grass of the field/soil wither (Jeremiah 12.4). Jeremiah even hears the land mourning directly to God (Jeremiah 12.11).

In the dramatic text of Hosea 4, God is holding a court case against the people of the land in which the land and its inhabitants testify against the people.

Hear the word of the Lord, O people of Israel;
For the Lord has a case against the inhabitants of *the land*.
There is no faithfulness and loyalty,
No knowledge of God in *the land!*
Swearing, lying and murder
And stealing and adultery break out;
Bloodshed follows blood shed!
Therefore *the land* mourns
And all who live in it languish;
Together with all the wild animals
And the birds of the air;
Even the fish of the sea are perishing. (Hosea 4.1–3)

Here as elsewhere the domains of nature are genuine participants in God's covenant with God's people. Through their mourning, the land, the soil, and all its inhabitants testify before God in God's covenant court trial with Israel. Do we hear the mourning of the land today? Do we empathize with the cries of a soil suffering from excessive salinity, toxicity, or pollution?

In these passages, the prophets hear the cries of the land, the soil, and the vegetation in pain because of what humans have done and they give voice to that anguish. We may well ask who the contemporary prophets are that speak for the soil of Earth and demand justice. In the context of modern ecology, the soil also has rights—the right to be heard, to be heeded, and to be recognized as a living presence in our contemporary cosmology.

Joel, who has also heard the soil mourning and the crying in pain (Joel 1.10), finally promises:

Do not fear, O soil;
Be glad and rejoice;
For the Lord has done great things.
Do not fear, you animals of the field,
For the pastures of the wilderness
Will be green. (Joel 2.21–22)

In the Psalms, we frequently hear the call for all domains of nature to praise their creator (e.g., Psalm 148). In Psalm 96, the Psalmist cries

Let the skies be glad
And the Earth rejoice,
Let the sea roar
And all that fills it.
Let the field/soil exult
And everything in it.
Then shall the trees of the forest
Sing for Joy. (Psalm 96.1–12)

For us to be sensitive to the feelings of the soil, we need to find ways of making more immediate contact with the soil. As a boy on the farm, it was commonplace for me to see my father handling the freshly ploughed soil and exclaiming, "Good Earth! Sweet soil!" For most people today, the urban landscape has overpowered the sacred soil beneath. We walk on crude concrete not holy ground. We meander through massive shopping centers, not expanses of sacred soil. Most of the time, modern society is detached from the soil even if our greens are grown on a distant farm. Even the parklands and golf courses we may traverse are covered with grass and other vegetation that prevent immediate contact with the soil. The presence of sacred soil in our modern world has almost become a secret that needs to be uncovered.

IV: Soil is to be heard and heeded, seen in person and experienced through our senses, if we are to have any empathy with soil as sacred and enable its voice to be heard.

8.5 CONCLUSION

As we reflect on the principles enunciated above, we might well situate ourselves in a location where rich loose soil is accessible and taking handfuls of soil allow it to run through our fingers. As we contemplate the soil flowing from our hands, we may well wonder about the mystery of this rich reality and appreciate that in many traditions it has been considered sacred. In the current environmental climate, we may also wonder how we might sustain the soil that is so vital to our survival and whether an appreciation of its sacred nature may well be a positive course of action.

In the biblical traditions outlined above, we are led to appreciate that soil is a sacred domain of creation, that we are commissioned to serve rather than subdue, that soil is to be rejuvenated by the life-giving power of rest activated by the spiritual in the soil, that soil is a living reality continuously brought to life by the atmosphere, the sustaining breath of God, and that soil has a voice to be heard if we are to appreciate her rights, rhythms, and rituals in the sanctuary we call Earth.

EDITORS' NOTE

The "adamah" terminology has more recently moved into the pop culture world, with the character Commander Adama in the American science fiction television series *Battlestar Galactica* (1978–1980; 2004–2009). In this series, Commander Adama (William Adama in 2004) leads the human refugees of a galactic war from the "12 colonies" to the "13th colony"—a legendary planet called Earth.

REFERENCES

Habel, N.C. 1995. *The Land Is Mine. Six Biblical Land Ideologies*. Minneapolis: Fortress Press.

Habel, N.C. 2007. The beginning of violence: An ecological reading of Genesis 4. In *Ecumenics from the Rim: Explorations in Honour of John D'Arcy May*, eds. J. O'Grady and P. Scherle, 79–86. London: Transaction Publishers.

Habel, N.C. 2008. Introducing ecological hermeneutics. In *Exploring Ecological Hermeneutics. SBL Symposium Volume XX*, eds. N. Habel and P. Trudinger, 1–8. Atlanta: SBL.

Habel, N.C. 2009. *An Inconvenient Text. Is a Green Reading of the Bible Possible?* Adelaide: ATF Press.

Habel, N.C. 2011. *The Birth, the Curse and the Greening of Earth. An Ecological Reading of Genesis 1–11*. Sheffield: Sheffield Phoenix.

Hiebert, T. 2008. Air, the first sacred thing: The conception of *ruach* in the Hebrew scriptures. In *Exploring Ecological Hermeneutics. SBL Symposium Volume X*, eds. N. Habel and P. Trudinger, 9–19. Atlanta: SBL.

Rainbow Spirit Elders. 2007. *Rainbow Spirit Theology. Towards an Australian Aboriginal Theology*. Adelaide: ATF Press.

Ray, D.K. 2006. It's about time. In *Ecology, Economy and God. Theology That Matters*, ed. D.K. Ray, 154–171. Minneapolis: Augsburg Fortress.

9 Indigenous Māori Values, Perspectives, and Knowledge of Soils in Aotearoa-New Zealand

Beliefs and Concepts of Soils, the Environment, and Land

Garth Harmsworth and Nick Roskruge*

CONTENTS

* Landcare Research NZ Ltd – Manaaki Whenua, Private Bag 11 052, Palmerston North 4442, New Zealand; Email: HarmsworthG@LandcareResearch.co.nz

9.1 INTRODUCTION

Whatungarongaro te tangata, toitū te whenua
(People come and go, but the land endures).

Indigenous Māori in Aotearoa-New Zealand have had a long history and close interdependent relationship with the natural environment particularly the soil resources. People from northern Polynesia migrated to Aotearoa-NZ about 1000 years ago (McKinnon et al. 1997; King 2003), and it was in this new country that Māori culture developed and flourished (Best 1924a; Buck 1950), drawing on the early Polynesian cultural beliefs, customs, language, and philosophies. At present, indigenous Māori make up around 15% of the total population of just over 4 million in a largely multicultural mixed society based on relatively high inter-marriage between Māori and Europeans and other smaller ethnic groups. Around 85% of all Māori now live in urban environments following large migration shifts to cities for employment predominantly after the World War II, from the early 1950s. This Māori migration (~post-1990) is now global—mainly to Australia. This society is very different from when Europeans first colonized New Zealand in the mid-nineteenth century, when there were two distinct and separate cultures, one-Māori, one-English (King 2003). The first contact Europeans had with Māori is generally accepted in 1769. Māori culture since the arrival of the Europeans has gone from being strong and vibrant, through a long period of being at risk from the pressures of colonization, to a newfound Māori cultural renaissance that has progressively grown in the latter half of the twentieth century to the present (Durie 1998; Walker 2004). However, traditional beliefs, values, and cultural perspectives still resonate strongly in a contemporary society, and have taken on new importance through a resurgence of interest in cultural identity and philosophies, and the preservation, development, and use of indigenous language and traditional knowledge (Durie 1998; Walker 2004).

Ancestral lineage (*whakapapa*) links Māori to each other and to the natural environment. This intricate genealogical web, which provides the basis for Māori societal structure, divides Māori into three main hierarchical levels: *iwi* (tribal regional), *hapū* (district, local, sub-tribal), and *whānau* (extended family). Different tribes have authority over distinct geographic areas, although decisions on tribal areas do not always fall under the purview of a single *iwi* or *hapū* group. The basic tenets of traditional Māori society remain strong and influence the way Māori construct tribal status and authority, manage their lands and resources, and relate to other agencies and government. The connection between the land (*whenua*), ancestry, and people is well reflected in Māori proverbs (*whakatauki*) such as the following:

Ko nga mana ko nga mauri o te whenua kei i raro iho i nga tikanga a o tatou tupuna.
(The prestige and life force of the land is enhanced beneath the mantle of our ancestral traditions.)

9.2 TRADITIONAL MĀORI BELIEF SYSTEM

Māori beliefs, custom, and values are derived from a mixture of cosmogony, mythology, religion, and anthropology (Best 1924b; Buck 1950; Marsden 1988; Barlow 1993). Within this complex and evolutionary belief system are the stories of the origin of the universe and of Māori people—the sources of knowledge and wisdom that have fashioned the concepts and relationship Māori have with the environment today. From a Māori perspective, the origin of the universe and the world can be traced through a series of ordered genealogical webs that go back hundreds of generations to the beginning. Figure 9.1 is a very simplified version of this Māori belief system.

This genealogical sequence is referred to as *whakapapa*, and places human beings (e.g., Māori) in an environmental context with all other flora and fauna and natural resources as part of a hierarchical genetic assemblage with identifiable and established bonds. *Whakapapa* (Figures 9.1 and 9.2) follows a sequence beginning with the nothingness, the void, the darkness, a supreme god (Io-matua-kore), emerging light, through to the creation of the tangible world, the creation of two primeval parents,

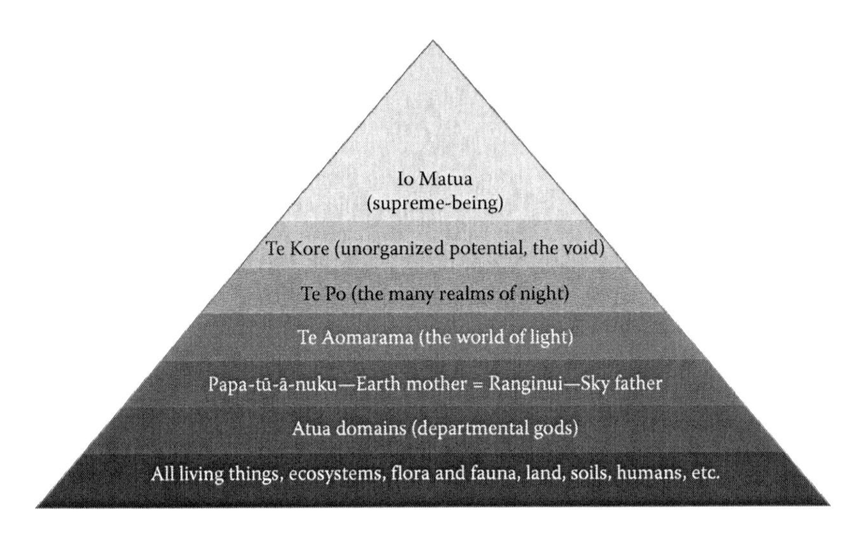

FIGURE 9.1 Te Timatanga—Māori creationist theory from the beginning.

Ranginui—the sky father, and Papa-tū-ā-nuku—the earth mother, the birth of their children, such as the wind, the forest and plants, the sea, the rivers, the animals, through to the creation of mankind. The two primeval parents, once inseparable, had many children, often termed departmental *Ātua*, deities, or Māori gods (about 100 departmental gods), each with supernatural powers, who preside over different domains (Best 1924b; Buck 1950; Marsden 1988; Barlow 1993). In a plan carried out by the children to create light and flourish, the parents were prised apart. The separation of the parents led Ranginui the sky father to form the atmosphere and continually weep with rain for his departed wife, and Papa-tū-ā-nuku the earth mother to form the land and provide sustained nourishment for all her children. As part of this ancestry, a large number of responsibilities and obligations were conferred on Māori to sustain and maintain the well-being of people, communities, and natural resources.

It is within this context that Māori commonly develop frameworks for resource management based on *whakapapa* to help make sense and explain their environments.

Two primeval parents	
Papa-tū-ā-nuku—Earth mother = Ranginui—Sky father	

Departmental Atua (Children)	
Tangaroa	The god of oceans, seas, rivers, lakes, and all life within them (and reptiles, fish, amphibians) and Tū-te wehiwehi (grandson of Tangaroa and also referred to as the father of reptiles, lakes, rivers, freshwater)
Tāne-mahuta	The god of the forests and all living things within them
Tāwhiri-mātea	The god of winds and storms
Rongo-mā-Tāne	The god of cultivated foods (e.g., kūmara-sweet potato), also god of peace
Haumia-tiketike	The god of fern roots and other wild foods
Rūaumoko	The god of earthquakes and volcanoes
Tū-mata-uenga	The god of man and war
Whiro	The god of evil, the domain of darkness and death

FIGURE 9.2 The main Atua or departmental gods, children of Papa-tū-ā-nuku and Ranginui.

9.2.1 Māori Ancestral Links to the Soil

In Māori tradition, the link between Māori and the soil was strong and reciprocal, stretching back to the time of creation. All flora and fauna were the grandchildren (*mokopuna*) of Papa-tū-ā-nuku. In many stories the departmental god Tāne Mahuta formed the first woman called Hine ahuone (woman made from earth) from soil before breathing life into her (Buck 1950; Rangitāne o Wairarapa Inc. 2006; Keane 2011b). In other tribal stories, it was a man Tiki-āhua, who was formed from soil by Tāne-mahuta (Best 1924b; Buck 1950; Rangitāne o Wairarapa Inc. 2006; Keane 2011b). After the creation story came the attainment of knowledge, "to help humankind acquire knowledge." This knowledge (wisdom) was special and pertained to the earth, land, water, flora and fauna, and everything else to enable long-term survival. This deep-rooted ancestral connection between land, soils, living organisms, and people, means there are several words that are the same when expressing characteristics about land and living things. *Papa-tū-ā-nuku* was often referred as a protective cloak, a skin covering the earth. Many words were used for parts of the landscape and interchangeably for humans. *Whenua*, for example, is used to mean both placenta and land; *rae* is used either as forehead or a land promontory; *iwi* refers equally to a nation of people, a tribe, or to bones (as *ko-iwi*); *hapū* can denote pregnancy or sub-tribe; *whānau* means birth or family. In terms of narratives, the complete life cycle starting, with birth and ending in death, was frequently acknowledged. When the demigod Māui failed to convince Hine-nui-te-pō, goddess of the underworld, to let humans die like the moon (die and return) she told him, "*Me matemate-a-one*" (let man die and become like soil).

9.2.2 Māori Traditional Stories

Hawaiki and Rangi-atea are two distant ancestral islands from which early Polynesian migrated, and many ancestral names repeatedly occur in many of the old manuscripts and oral translations (JPS; JPS 1913, 1915). Early Māori explorers who arrived on canoes from Polynesia were often interested in the agricultural qualities of soils in New Zealand (NZ). After landing in NZ, the great navigator Kupe returned to Hawaiki and told of the sweet-scented and rich soils of New Zealand (JPS; Keane 2011b). Turi, captain of the Aotea canoe, decided to settle at Pātea after smelling its fertile soil and "whilst at Patea he distinguished the soil qualities through his senses, and found it to be para-umu—a rich black soil—and sweet scented" (JPS 1913 22(87): 128; Keane 2011b; Roskruge 2011). The sons of Whātonga from the Kurahaupō canoe decided to move to Matiu island in Wellington Harbour because the soil was good and the island was easy to defend. In a number of traditions, Māori explorers from voyaging canoes (*waka*) expressed their views and knowledge of Aotearoa and of soils at various locations around the NZ coast (JPS; JPS 1913 22(87): 129–130; Keane 2011b).

Another tribal story from about 1350 AD, describes the Tākitimu canoe landing at Whangaparāoa, in the eastern Bay of Plenty, North Island, NZ. Its captain, Tamatea-ariki, found that Hoturoa (chief of the *iwi* Tainui) and Ngātoroirangi, a *tohunga* (high priest) of the Arawa canoe, were already there, and he asked them "What kind of land is this?" Ngātoroirangi replied:

"It is good. Some parts are limestone, some are sandy soil, others rich soil, others friable soil, black soil, sand, pumiceous soil, and light sandy soil, red volcanic soil, some parts are gravelly, stony, and some are very loose soils" (JPS 1915, 24(93), 1).

9.2.3 Traditional Environmental Concepts

9.2.3.1 Mātauranga Māori and Māori Values

Mātauranga Māori is wisdom, how to do something and why, and therefore includes all Māori knowledge systems or ways of knowing. Knowledge provides the basis for the Māori worldview or perspective encompassing all aspects of Māori knowledge from philosophy, cosmology to practice. Māori retained, organized, and imparted knowledge orally rather than in any written form. Huge quantities of specialized (especially ancestral and traditional) knowledge were memorized and retained by people such as *tohunga* (priests, specialists), *rangatira* (chiefs), *kaumātua* (elders), *kuia* (elderly female), and *pakeke* (adults). Parts of this knowledge were documented mainly by early Europeans, such as explorers and missionaries, (since about 1769), and from the late nineteenth century, by European and Māori scholars and authors. In moving beyond the traditional *mātauranga Māori*, *mātauranga Māori* has grown into many contemporary forms (e.g., historic, local indigenous knowledge, Māori perspectives) that are complementary to western science knowledge; a view consistent with many recent Māori authors who regard Māori knowledge as a dynamic and evolving knowledge system that represents more than the past (Durie 1998; Harmsworth et al. 2002; Raskin 2009; Roskruge 2012).

Māori values (Marsden 1988; Barlow 1993; Mead 2003) are derived from traditional belief systems and are part of the wider Māori knowledge system (*mātauranga Māori*). Values can be defined as instruments through which Māori make sense of, experience, and interpret their environment (Marsden 1988). They form the basis for the Māori worldview (*Te Ao Māori*), and provide the concepts, principles, and lore Māori use to varying degrees in everyday life, and often form ethics and principles. This can govern their responsibilities and relationship with the environment. Important Māori values (see glossary) include: *tikanga* (customary practice, values, protocols); *whakapapa* (relationships, traditions, connections); *tino rangatiratanga* (self-determination); *mana whenua* (authority over land and resources); *whānaungatanga* (family connections); *kaitiakitanga* (environmental guardianship); *manaakitanga* (acts of giving and caring for); *whakakotahitanga* (consensus, respect for individual differences, and participatory inclusion for decision making); *arohatanga* (the notion of care, respect, love, compassion); *wairuatanga* (a spiritual dimension). Māori values can therefore be translated into, and represented as to what is valued, for example, natural resources, soils, significant cultural sites, significant biodiversity habitats and species, iconic cultural flora and fauna species. *Tikanga* is one of the most encompassing values, and within the context of soils can be used to define customary and best management practices for each *iwi/hapū* group and embodies many of the other values such as *whakapapa* and *kaitiakitanga*.

9.2.3.2 Traditional Māori Environmental Perspectives

Māori sought to understand the total environment or system, its connections through *whakapapa*, not just a part of it and their perspective as today is very holistic and integrated. Māori values form the basis for a number of key Māori concepts for resource management (Marsden 1988; Barlow 1993; Durie 1994; Harmsworth et al. 2002; Mead 2003, Roskruge 2011) and some of the most important environmental concepts are:

- *Kaitiakitanga*—stewardship or guardianship of the environment, an active rather than passive relationship.
- *Ki uta ki tai*—a whole of landscape approach, understanding and managing inter-connected resources and systems from the mountains to the sea (the Māori concept of integrated catchment management).
- *Mauri*—an internal energy or life force derived from *whakapapa*, an essential essence or element sustaining all forms of life, denoting a health and spirit, which permeates through all living and nonliving things. All plants, animals, water, and soil possess *mauri*. Damage or contamination to the environment is therefore damage to or loss of *mauri*.
- *Ritenga*—the area of customs, protocols, laws that regulate actions and behaviors related to the physical environment and people. *Ritenga* includes concepts such as *tapu*, *rahui*, and *noa*, which were practical rules to sustain the wellbeing of people, communities and natural resources. Everything was balanced between regulated and deregulated states where *tapu* was sacred, *rahui* was restricted and *noa* was relaxed access or unrestricted.
- *Wairuatanga*—the spiritual dimension, a spiritual energy and dimension as a concept for Māori wellbeing.

9.2.4 Traditional Māori and the Soil Resource

9.2.4.1 Māori Terms for Landscapes

Much of the heritage of Māori resides in the landscape. Māori had names for almost every part of the landscape and these expressions of place were linked back to *whakapapa* through stories and became increasingly descriptive locally. Examples in Tables 9.1 through 9.3, show that Māori were very

TABLE 9.1
Māori Terms for Landforms, Features, and Descriptors

Māori Expression	English Landform Term Eq.	Māori Expression	English Landform Term Eq.
Papa-tū-ā-nuku	Earth Mother	*whenua*	Land, terrain, placenta
Wāhi	Place, site, location	*haumako, mōmona, ngotongoto*	Flat, fertile, good fertile land
roa	Long or high	*whenuapai*	Highly productive, good land
ika whenua, tuatua, tuarā maunga	Range, axial range	*awa*	River
pou karangaranga	Landmarks, significant land features	*ma, manga*	Stream, tributary, creek
tāhinga	Slope	*kautawa*	Tributary
papa, papatahi, raorao, mānia, papatika, papatairite (level flat)	Flat, plains, broad flat ground	*kekee*	Creek
raorao, mānia, papatika	Flats, plains	*korere*	Channel
tāwakawaka	Undulating	*pūau, huihuinga wai*	Confluence
pāraharaha	Rolling	*parehua*	River terrace
papa	Terraces	*taha*	Bank of riverside
puke	Hill	*waha*	Mouth of river
rapaki, puketai	Hillside, rolling and steep hills	*wahapū*	Entrance, mouth of river
poupou, tūparipari	Steep	*parahua*	Alluvial areas, wet, muddy
paeroa	Low range, steep hills	*repo*	Swamp
pae	Ridge, step, resting place	*roto*	Lake
		moana	Extensive lake area
taumata	Brow of a hill	*wahapū*	Estuary
tau	Ridge of a hill	*whanga*	Bay, harbor, or inlet
tihi, tautara, taumata	Summit, peak	*one, tāhuna*	Beach
tara	Peak	*takutai*	Coast, sea coast
taukahiwi, kahiwi, ranga	Ridge	*mata*	Point of land, headland, headwaters, face, cliff
maunga	Mountain	*rae, mata*	Headland, head of river
riu, awaawa, whārua	Valley	*koutu*	Promontory
pari	Cliff		
āpiti, kapiti, tawhārua	Gorge	*kūrae*	Headland, promontory
tākau	Escarpment	*moutere, motu*	Island
horo	Landslide, landslip, to swallow	*taha moana, tātahi, takutai, tai (tidal)*	Coastal, sea coast
pakohu, awa, kopia, parari	Gully	*whenua kāki, whenua kuti*	Isthmus
ana	Cave, cavern		
waiariki	Hot springs, geothermal		
puia	volcano		
takiwā-waiariki	Hydrothermal, geothermal district		

Source: Adapted from Williams, H. W. 1975. *A Dictionary of the Māori Language*. 7th ed. Wellington, New Zealand, A. R. Shearer, Government Printer. 507pp.; Ngata, H. M. 1993. *English-Māori Dictionary*. Wellington: Learning Media. 559pp.

TABLE 9.2
Rocks

Māori Term	English Equivalent	Māori Term	English Equivalent
toka, kāmaka, kōhatu, pōhatu	Rock, stone	*tāhoata, pungapunga*	Pumice
pōhatuhatu	Rocky	*pounamu*	Greenstone
pōhatu, kāmaka, kōhatu	Stone	*kairangi*	Quality, finest greenstone
horo kōhatu	Crumbling stone	*auhunga*	Pale greenstone
matā, kiripaka, takawai	Quartz, flint	*tongarewa*	Semi-transparent greenstone
omata	Place of quartz	*kutukutu*	Speckled greenstone
matā, mātara, tūhua	Obsidian	*tōtōeka*	Streaked greenstone
karā, ōnewa	Basalt	*Tangiwai*	Transparent greenstone
rino	Iron	*Inanga*	Whitish greenstone
		Pākeho	Limestone

Source: Adapted from Williams, H. W. 1975. *A Dictionary of the Māori Language*. 7th ed. Wellington, New Zealand, A. R. Shearer, Government Printer. 507pp.; Ngata, H. M. 1993. *English-Māori Dictionary*. Wellington: Learning Media. 559pp.

sophisticated in their use of *te reo Māori* (Māori language) to describe accurately different parts of the natural environment and its characteristics and condition, using root words, aggregation of terms, categories, subdivision, and descriptors (Williams 1975; Ngata 1993). These were an early form of oral classification for Māori and through the language these classifications became increasingly complex, structured, sophisticated, and both generic and local. Being based on *whakapapa* there was always a hierarchical assemblage and inter-connection within classifications, linking to them to each

TABLE 9.3
Māori Terms for Degrees of Wetness on Land

Māori Term	English Equivalent	Māori Term	English Equivalent
wai	Water	*para*	Muddy, sediment
waipara	Stream across plain, river with thick muddy sediment	*Paretai*	Bank of a river
repo, ngaere	Swamp	*Pipiwai*	Damp or swampy
mākū	Wet, wet place	*Reporoa*	Long swamp
waipuke	Flood	*Mangareporepo*	Muddy creek
puaha, ngutu awa	The mouth of a river	*Mimi*	Stream or creek
pukaha	Spongy or swampy	*Oaro*	Bog, aro is bog, a boggy place
pukaki	Head of creek, or where the stream meets tidal waters, pu is heaped or bunched up, and kaki means neck	*Opara*	Muddy place
te ngae	Swamp	*pākihi*	Opening, clearing, flat land dried up and poor, swampy, low fertility
putarepo	A place at the end of a swamp that can be crossed	*Manga*	Tributary, drain, creek
whakatahe(a)	Drained	*Puna*	Spring of water

Source: Adapted from Williams, H. W. 1975. *A Dictionary of the Māori Language*. 7th ed. Wellington, New Zealand, A. R. Shearer, Government Printer. 507pp.; Ngata, H. M. 1993. *English-Māori Dictionary*. Wellington: Learning Media. 559pp.

other and enabling everything to be traced back to a common ancestor or *Ātua*, and then back to the stories describing in order the origin of all life, the earth, plants, animals and humans, and so on.

The links between people, the land, and flora and fauna are well recorded in all tribal narratives (Buck 1950; Park 1995). Ancestral stories, *mātauranga Māori*, and oral traditions, were attached to every part of the landscape and its associated features linking people and culture to place, establishing spiritual and ancestral significance, serving to locate points in tribal history and therefore relating the occupation of the land and the way it became part of a tribe. Landmarks such as *maunga* (mountains) and *awa* (rivers) are considered to be *iwi/hapū* (tribal) elements and identifiers when recited in *pepeha* (recitations). Other examples refer to the *mauri* of the soil and sacred earth which was brought from traditional homelands to NZ by Polynesian ancestors and placed on gardens and at other sacred sites to reaffirm connection and whakapapa (JPS 1913, 1915; Keane 2011a,b).

Terminology was therefore contextual and descriptive and combined root words, location, and landscape setting, with more descriptive terms from adjectives such as color, texture, wetness. In most cases, Māori descriptions of soils and land can be used generically, while more specific detail is added at the *iwi/hapū* level denoting more specific knowledge and description locally. Hundreds of terms described different parts of the landscape into elements and attributes of the natural environment. The more general Māori terms have been organized into classifications (Tables 9.1 through 9.3).

9.2.4.2 Māori Terms for Soils

Māori identified at least 50 different types of soils in various locations around Aotearoa-NZ. *Oneone* (pronounced awneyawney) was the general term used to describe soil or earth, and "*one*" was often used as a prefix for the names of different soil types in different landform and ecosystem locations (e.g., sand country–dunes, swamps, alluvial flats, terraces). For example, *one-pū* was sand, *one-parakiwai* was silt, *one-nui* was rich soil made of clay, sand and decayed organic matter, *one-matua* was loam. "*Kere*" was used as a prefix for some types of clay, including *keretū*, *kerematua*, and *kerewhenua*. Other terms included references to animals, birds, fish, and so on, to improve description and indicate relationships, for example, *tenga kākāriki* (parakeet's crop), was a white volcanic sand in the Bay of Plenty (Keane 2011b), and was named in this way because it resembled the rough inner surface of a parakeet's crop (a pouch near the throat).

9.3 HISTORICAL MĀORI SOCIETY AND LAND

9.3.1 Traditional Land Tenure

From mid-eighteenth to the mid-nineteenth centuries, many dominant *iwi/hapū* collectives with distinct *whakapapa* and origins, had well-defined territories marked by distinct places and land features such as *awa* and *manga* (rivers and streams), *maunga* (mountains and axial ranges), rocks, and coastlines. Over time, and following much conflict Māori internal migration and settlement patterns became entrenched in certain geographical locations, with growth in tribal dominance and inter-marriage. Increasingly, Māori started to live and work together in small, geographically distinct groups and settlements as part of larger hierarchical tribal structures (*iwi, hapū, whānau*) while Māori beliefs and values (*tikanga*) continued to be practised under local systems. From the seventeenth century, there is evidence of a well-developed communal land tenure system in many areas in NZ, where resources were shared collectively within and between tribes. This gave rise to a communal society where Māori lived and worked together to survive, adapted to change, and cared for each other in extended families (*whānau*) and sub-tribes (*hapū*), while sustaining and managing natural resources locally and collectively.

The notion of how central land and soil was, and still is to Māori, was described by Asher and Naulls (1987, p. 5): "To the early Māori, land was everything. Bound up with it was survival, politics, myth, and religion. It was not part of life but life itself." Taking culture in its widest context, there was no part of early Māori culture that was not touched by the physical environment, land in particular.

9.3.2 The Treaty of Waitangi

The Treaty of Waitangi (Te Tiriti o Waitangi), signed on February 6, 1840, was enacted between Māori (~540 Māori chiefs) and the Crown (British Government) and conferred responsibilities and obligations on subsequent NZ Governments (representing the Crown) to uphold rights for Māori as British subjects and NZ citizens, to protect their land, estates, water, forests, and other resources or treasures (*taonga*). The Treaty has been the source of arduous debate and interpretation by European and Māori ever since but is enshrined into NZ legislation at many levels. The Waitangi Tribunal was established in 1975 to hear Māori land grievances under modern legislation to address land alienation, confiscation, and breaches of the Treaty.

9.4 CONTEMPORARY MĀORI SOCIETY AND LAND

9.4.1 Māori Land

Through *whakapapa* (ancestral lineage) Māori continue to be affiliated with hierarchical groupings such as *iwi*, *hapū*, and *whānau*. The basic unit of Māori society is still the *whānau*, the extended family. Today, the *hapū* or *iwi* and urban Māori are the main groupings involved in pooling resources for economic, business, social, health, education, employment, and housing programs, and many are involved in environmental and resource management. *Whānau* and *hapū* groupings still provide the basic unit for decision making on specific blocks of land, local business activities, coastal and fisheries resource management, and utilizing specific natural and human capital. Traditional and historical Māori social structure has been an enduring feature of new Māori business and governance models.

Māori today only have a fraction of the land and natural resources to which they once had rights or title (Durie 1998; Harmsworth and Roskruge 2014) and live in a more fragmented, modern, free-market society. In 1840, most land in NZ was under Māori control and tribal ownership but with the onset of British colonization, a raft of Crown–Government interventions increasingly alienated Māori from their collective land resource base. After 1840, and the signing of the Treaty of Waitangi, Māori land drastically reduced in total area (Table 9.4), with a large decrease in area due to the Native

TABLE 9.4
Patterns of Māori Land Ownership from 1840 to 2000

Year	Hectares
1840 (customary land)	27,053,400
1852	15,300,000
1860	9,630,000
1891	4,985,000
1911	3,211,000
1920	2,154,000
1939	1,813,000
1975	1,350,000
1986	1,181,740
1996	1,515,071
2000	1,515,071

Source: Adapted from Durie, M. 1998. *Te Mana, Te Kawanatanga: The Politics of Māori Self-Determination*. Auckland: Oxford University Press. 280pp. and from the TPK Maori land database; Te Puni Kōkiri (TPK) 1998. *Māori Land Information Database* MLIB 1998–2002. Te Puni Kōkiri (TPK) Wellington, New Zealand: Ministry of Māori Development. http://www.tpk.govt.nz/en/services/land/ (accessed 2000–2002).

Land Act in 1860–1862, which individualized the title on Māori land and increasingly opened it up to sale under colonial legislation to direct it away from collective ownership and tribal authority. Māori access to land, forests, and coastal and marine (e.g., fisheries) resources rapidly diminished. Māori gradually lost control, retention, and access to the best soil resources in NZ, and were reduced to the present Māori land area that is mainly hilly and steep with marginal characteristics (Table 9.5). By 1998 Māori collectively owned "freehold" land had diminished to only 6% of the total NZ land area or some 1.5 million hectares (Table 9.4; Durie 1998; TPK 1998; Harmsworth 2003; Hui Taumata 2006; MAF 2011). Most land now under Māori collective ownership comes under a Māori Land Act, Te Ture Whenua Act 1993, that focuses on retaining Māori land, acknowledging rights of *whakapapa*, allowing multiple-ownership, and facilitating and promoting effective use, management, and development (Maughan and Kingi 1998; Durie 1998). Commonly termed Māori freehold land, this land represents a collectively owned land asset base and is held under different types of Māori governance, typically trusts, to protect owners' rights and promote participatory decision making. Māori land analyses intersecting Māori freehold land data with environmental datasets (e.g., Harmsworth 2003; Harmsworth

TABLE 9.5
Land Use Capability LUC 1–8 (Areas and % for Māori Land (TPK1998) Compared with New Zealand National Averages)

Potential of Māori Land by NZLRI Land Use Capability (LUC)				
Land Use Capability Class	**% of Total New Zealand Land**	**Māori Land Area (ha)**	**% of Māori Land**	**Description of Land-Use Capability**
1	0.7%	7514.76	0.50%	Most versatile multiple-use land—virtually no limitations to arable use
2	4.55%	43,733.59	2.89%	Good land with slight limitations to arable use
3	9.22%	85,534.33	5.65%	Moderate limitations to arable use restricting crops able to be grown
4	10.5%	153,972.29	10.16%	Severe limitations to arable use. More suitable to pastoral and forestry
5	0.8%	6883.47	0.45%	Unsuitable for cropping—pastoral or forestry
6	28.1%	507,706.36	33.51%	Nonarable land. Moderate limitations and hazards when under a perennial vegetation cover
7	21.4%	469,830.47	31.01%	With few exceptions can only support extensive grazing or erosion control forestry
8	21.8%	230,142.75	15.19%	Very severe limitations or hazards for any agricultural use
Other	2.97%	9752.96	0.64%	Nonarable land. Moderate limitations and hazards when under a perennial vegetation cover
Total	100.00% (26,930,100 ha)	1,515,071.00	100.00%	

Source: From the New Zealand Institute of Economic Research (NZIER). 2003. *Māori Economic Development: Te Ōhanga Whanaketanga Māori.* Wellington, NZ: New Zealand Institute of Economic Research. 116pp; Landcare Research NZ Ltd GIS 2012. (accessed Feb to Dec 2012); Harmsworth, G. 2003. Māori perspectives on Kyoto Policy: Interim Results. Reducing Greenhouse Gas Emissions from the Terrestrial Biosphere (C09X0212). *Landcare Research Report LC0203/084. Discussion paper for policy agencies (Climate Change Office, MfE, MAF, TPK).* Updated November 2003 GIS tables and statistics. 30pp; Harmsworth, G. R., C. Insley, and M. Tahi. 2010. *Climate change business opportunities for Māori land and Māori organisations*: FRST Contract: C09X0901. Prepared for MAF Wellington, under the Sustainable Land Management Mitigation and Adaptation to Climate Change (SLMACC). Landcare Research report LC0910/157. 59pp.

TABLE 9.6
Landcover Class (LCDBv2) Comparisons for Māori Land and New Zealand Land at 2011

Landcover	New Zealand (LCDBv2)		Māori Land (MLIB)	
Landcover Class	**Area (ha)**	**Area (%)**	**Area (ha)**	**Area (%)**
Indigenous Forest	7,109,546.4	26.4	586,332.5	38.7
Scrub	1,804,316.7	6.7	212,109.9	14.0
Planted exotic forest	1,965,897.3	7.3	206,049.6	13.6
Pastoral (grassland)	10,583,529.3	39.3	401,493.8	26.5
Horticultural	430,881.6	1.6	12,120.6	0.8
Inland water and wetlands	807,903.0	3.0	31,816.5	2.1
Other (e.g., mines, tussock, bare ground)	4,228,025.7	15.7	65,148.1	4.3
Total	26,930,100	100	1,515,071.00	100

Source: Adapted from Landcare Research NZ Ltd GIS 2012. (accessed Feb to Dec 2012); Harmsworth, G. R., C. Insley, and M. Tahi. 2010. *Climate change business opportunities for Māori land and Māori organisations*: FRST Contract: C09X0901. Prepared for MAF Wellington, under the Sustainable Land Management Mitigation and Adaptation to Climate Change (SLMACC). Landcare Research report LC0910/157. 59pp.

et al. 2010; MAF 2011) give statistics for national Land Use Capability (LUC; Table 9.5) and vegetative cover (Table 9.6) on Māori land compared with national NZ land data. Table 9.5 shows that at 2010 <4% (<50,000 ha) of Māori freehold land is classified as LUC 1 and 2—the most highly versatile multiple use land in NZ, containing large areas of high class soils—while a further ~16% or 239,000 ha may be suited to horticulture and cropping. Over 65% of Māori land is classified as steep, hilly, and mountainous (Table 9.5) and as a result much remains in indigenous forest and scrub (Table 9.6).

In 1975, a Waitangi Tribunal was established to hear Treaty of Waitangi land grievance claims by Māori claimant groups through a lengthy and costly legal process—largely *iwi/hapū* claims from all over NZ—responding to land loss due to illegal sale, forfeiture, and colonial confiscation. The number of land claims for property, resources, and proprietary rights lodged with the Tribunal at 2012 number well over 1900. Since the establishment of the Waitangi Tribunal, the resource base to which Māori have access and control is slowly increasing again as a result of continual and constructive redress. With increasing settlement, many tribal organizations are now positioning themselves to manage hundreds of millions of dollars of assets, while other Māori business and enterprise have been highly successful and have flourished outside the Treaty process in the past 20–30 years. Recognized indigenous or aboriginal rights under the 1840 Treaty of Waitangi and in NZ legislation mean that Māori are widely included in decision making for lands, water, coastal, and marine environments, especially where Māori can demonstrate a long-standing cultural or customary tribal relationship with specific natural resource areas.

9.4.2 Māori Economic Development

Māori increasingly moved from rural to urban settlements after the World War II, around 1950. Today, Māori social structure is complex, and fragmented, and most Māori live outside the ancestral lands that they once occupied. Modern Māori businesses take many forms; including *whānau*- and *hapū*-(family) based trusts, incorporations, *rūnanga* (councils), through to limited liability companies and privately owned businesses/enterprises. A large majority of businesses have key characteristics that include a distinctly Māori style of governance, management structure, entrepreneurship, strategic planning, networks, many reflecting Māori values, a cultural dimension, and cultural drivers (e.g., history, land ownership, resources, ancestral connections, cultural values) (Harmsworth 2009).

The term "the Māori economy" has increasingly been used since the late 1990s to indicate a Māori dimension within the NZ economy that is largely culturally and ancestrally based (TPK 2002; NZIER 2003; Whitehead and Annesley 2005; TPK 2007a,b; Harmsworth 2009). The term was defined broadly, and generally referred to total assets owned and income earned by collectives such as Māori-owned trusts and incorporations, Māori-owned businesses, and service providers (Whitehead and Annesley 2005; TPK 2007b). However it has been very difficult to distinguish and quantify the Māori economy as a separate entity from the wider economy as the two are interconnected. Many reports have assessed Māori involvement and contribution to the NZ economy, and this provides a snapshot in time and a baseline to model future scenarios and trends (BERL 2011). In 2011, the 2010 Māori asset base of the Māori economy was estimated at $37 billion, based on a large range of Māori businesses, entities, employers, and household incomes (BERL 2011). In 2010, Māori industry and enterprise in the primary sector—mainly made up of agriculture, farming, forestry, and fishing—contributed approximately to $10.6 billion (or ~30%) of this total Māori asset base. In addition, Māori continue to have major interests and rights in water, land, soil, and geothermal, mineral, and petroleum resources. These figures, however, show that Māori have contributed much value to the NZ economy through a very low land resource base.

At 2012, very little Māori land is in horticulture or cropping (~<2%); most tends to be in pastoralism, mainly sheep and beef (~20%), a reasonable proportion in dairying (about 6% Māori land), and larger areas in production forestry—almost 14% of all Māori land (Table 9.6). In 2003, it was estimated that the annual agricultural and forestry production from Māori communally owned land assets was approximately $750 million per annum, and contributed 7.5% of NZ's total annual agricultural outputs (NZIER 2003). In 2006 (Ahie 2006; MAF 2011), it was estimated that Māori freehold land provided about $633 million to the national GDP. In the first decade of the twenty-first century, more than 15% of the country's sheep and beef exports have come from Māori farming interests, and Māori were farming 720,000 hectares in 2003—mainly in sheep, beef, and dairy. Māori freehold land now carries about 10–15% of the national sheep and beef stock units (MAF 2011). General estimates are that at 2011 Māori agribusiness enterprises provide about 8–10% of the national milk solids production in NZ (MAF 2011), and own around $NZ 100 million worth of shares in Fonterra, the largest dairy company in NZ. Further development of Māori land and its economic base has the potential to produce significant gains.

Because of the limiting agricultural characteristics of the present 2012 Māori land base, many authors continue to describe Māori land as underperforming, and common perceptions are that about 50% of the present Māori land base is underutilized or undeveloped, although these figures are difficult to confirm. MAF (2011) stated that on average, Māori land production was about 60–70% of the national average, but this figure cannot be substantiated. The Māori contribution to NZ's farming economy remains significant and productive Māori land constitutes a large fraction of the national Māori asset base. As the Māori asset base grows, so does its contribution to local, regional, and national economies. At 2012, the Māori asset base is still largely concentrated in export-orientated areas of primary production and processing—mainly agriculture, farming, forestry, and fisheries—with approximately $10.6 billion worth of collective assets recorded in 2010 (BERL 2011).

9.5 CONCLUSIONS

This chapter describes the intricate traditional belief system of the indigenous Māori that links humans to land, soils, ecosystems, and flora and fauna. Many of these beliefs and values influence the way Māori now see the world and make decisions about soils and land environments. The Māori relationship and interdependence with the soil resources of Aotearoa-NZ can be traced back to the time of creation. The primal parent, Papa-tū-ā-nuku, was the earth mother, and the first human was formed from the soil that cloaked Papa-tū-ā-nuku. All life depended on her for its well-being. Traditional beliefs stated that "people have the option of caring for Papa-tū-ā-nuku to maintain their own health, or abandoning her to concentrate on their own short-term personal needs," and "ultimately an unhealthy Papa-tū-ā-nuku will lead to unhealthy people" (Rangitāne o Wairarapa Inc., 2006). This interwoven relationship with land (*whenua*) and soil resources give Māori a personal affinity with land and soils in tribal areas, through their ancestral lineage, values, stories, and language; an affinity that was further developed through a long history of interdependence with natural resources, and a growing interest in horticulture and soil management which gave rise to Māori naming and classifying many parts of the landscape and soils.

The present Māori land base is a fraction of what it once was. However, Māori have large areas under Treaty of Waitangi claims, and are progressively acquiring additional resources and assets through claim settlement processes that have continued since 1840 and, since 1975, through the Waitangi Tribunal legal process between the Crown and *iwi/hapū*. With this growing economic base, Māori are becoming significant partners, investors, co-investors, and owners of land and soil resources. The Māori economy depends largely on the primary sector, and agriculture is a significant portion of this. As the Māori contribution to New Zealand's agricultural and farming economy is significant, Māori rely heavily on sustaining the national soil resource base, which underlies the Māori and NZ economy and its ability to prosper. Modern Māori now seek to achieve the challenging balance between cultural, social, and environmental aspirations through concepts such as *kaitiakitanga* (guardianship of the environment with strong cultural and social objectives), and *tino rangatiratanga* (self-sufficiency, self-determination) combined with increasing economic demands for land development and land management, particularly in the areas of pastoral farming, agriculture, horticulture, forestry, and associated industries. As shown in this chapter, traditional beliefs, concepts, and values still play a significant and integral role in Māori advancement, business, social and cultural development, and economics.

ACKNOWLEDGMENTS

Mātauranga Māori on soils currently resides with a few knowledgeable practitioners (*kaumātua*, *kuia*, and *tohunga*) in different parts of New Zealand, but is now sporadically recorded and most is retained in old documents and some refereed publications. An extensive review has been carried out as part of this chapter. The authors have worked extensively with *iwi*, *hapū*, and *marae* groups throughout New Zealand in soils, land classification, land resource assessment, horticulture, and cropping. We thank those people who have contributed knowledge from past projects that is now used in this chapter. Dr. Nick Roskruge has been a leading authority on preserving knowledge on Māori traditional crops, through collectives such as Tahuri Whenua and the national Māori Vegetable Growers collective, keeping this knowledge alive and active through crop planting programs, use, and application. Many ancestral narratives have come from a variety of publications such as *The Journal of the Polynesian Society*.

GLOSSARY

Aotearoa	Māori name for New Zealand, land/world (ao) of the long (roa) white (tea) cloud
awhinatanga	assist, care for
hapū	pregnant, sub-tribe
iwi	tribe, bones
kaumātua	elderly respected male, one with knowledge and wisdom
kuia	elderly respected female, one with knowledge and wisdom
kaitiakitanga	the ethos of sustainable resource management, guardianship
mahinga kai	food gathering area
manaaki	provide hospitality
manaakitanga	reciprocal and unqualified acts of giving, caring, and hospitality

mana whenua	rights of self-governance, rights to authority over traditional tribal land and resources
mātauranga Māori	Māori knowledge and philosophy
mauri	an energy, a sustaining life force or spirit, a soul, in all living and nonliving things
oneone	soil
Papa-tū-ā-nuku	Earth mother
pepeha	recitations linking people to place
Ranginui	Sky father
taonga	treasure
taonga tuku iho	treasured possessions passed through generations
te ao Māori	Māori worldview
te ao Pākehā	Pākehā worldview
te reo	Māori language, voice
tikanga	customary practice, protocol, values
tino rangatiratanga	self-determination, independence, or inter-dependence
tohunga	knowledge expert, specialist, priest
wāhi tapu	sacred site
wāhi taonga	heritage site
whakatauki	Māori proverb
whakakoha	the act of giving
whakapapa	ancestral lineage, ancestral connections, genealogical relationships
whānau	family, extended family (incl. cousins, twice, thrice over, etc.)
whānaungatanga	family connections and family relationships
whenua	placenta, land
wairua	the spiritual dimension to life

REFERENCES

Ahie M. 2006. Application of technologies to Māori agribusiness, report prepared for Te Puni Kokiri Wellington.

Asher, G. and D. Naulls. 1987. *Māori land. Planning Paper No. 29.* Wellington: NZ Planning Council.

Barlow, C. 1993. *Tikanga Whakaaro: Key Concepts in Māori Culture.* Auckland: Oxford University Press. 187pp.

Best, E. 1924a. *The Māori As He Was.* 2nd edition. Wellington: Dominion Museum.

Best, E. 1924b. Māori religion and mythology. *Dominion Museum Bulletin No. 10.* Wellington: Board of Māori Ethnological Research and Dominion Museum.

Buck, P. (Te Rangi Hiroa) 1950. *The Coming of the Māori.* 2nd edition. Wellington: Māori Purposes Fund Board/Whitcombe & Tombs. 551pp.

Business and Economic Research Ltd (BERL) 2011. *The Asset Base Income, Expenditure and GDP of the 2010 Māori Economy.* Wellington, New Zealand: BERL and Māori Economic Taskforce. 44pp.

Durie, M. 1994. *Whaiora:. Māori Health Development.* Auckland: Oxford University Press. 238pp.

Durie, M. 1998. *Te Mana, Te Kawanatanga: The Politics of Māori Self-Determination.* Auckland: Oxford University Press. 280pp.

Harmsworth, G., T. Warmenhoven, P. Pohatu, and M. Page. 2002. *Waiapu Catchment Technical Report: Māori Community Goals for Enhancing Ecosystem Health.* Foundation for Research, Science, and Technology (FRST) contract TWWX0001. Landcare Research report LC0102/100 for Te Whare Wananga o Ngati Porou, Ruatorea (unpublished). 185pp.

Harmsworth, G. 2003. Māori perspectives on Kyoto Policy: Interim Results. Reducing Greenhouse Gas Emissions from the Terrestrial Biosphere (C09X0212). *Landcare Research Report LC0203/084. Discussion Paper for Policy Agencies (Climate Change Office, MfE, MAF, TPK).* Updated November 2003 GIS tables and statistics. 30pp.

Harmsworth, G. 2009. Sustainability and Māori business. In *Hatched: The Capacity for Sustainable Development,* eds. R. Frame, R. Gordon, and C. Mortimer, 95–108. http://www.landcareresearch.co.nz/resources/business/hatched

Harmsworth, G. R., C. Insley, and M. Tahi. 2010. *Climate Change Business Opportunities for Māori Land and Māori Organisations*: FRST Contract: C09X0901. Prepared for MAF Wellington, under the Sustainable Land Management Mitigation and Adaptation to Climate Change (SLMACC). Landcare Research report LC0910/157. 59pp.

Harmsworth, G. R. and N. Roskruge. 2014. Indigenous Māori values, perspectives, and knowledge of soils in Aotearoa–New Zealand: B. Māori use and knowledge of soils over time. In *The Soil Underfoot*, eds. G.T. Churchman and E. R. Landa, 257–268, Boca Raton: CRC Press.

Hui Taumata 2006. Review of Māori land tenure: A report prepared by the Māori Land Tenure Review Committee for the Hui Taumata, Wellington, New Zealand. June 2006 (unpublished).

Journal of the Polynesian Society (JPS). Vols. 1–119 http://www.thepolynesiansociety.org/journal.html; http://www.jps.auckland.ac.nz/browse.php (accessed February–December 2012)

Journal of the Polynesian Society (JPS) 1913. Whatahoro, H. T. 1913. (Percy-Smith, S. trans.). The lore of the whare wānanga. Part II Te Kauwae-Raro, things terrestrial. Chapter III Discovery of New Zealand by Kupe as related by Te Matorohanga. *Journal of the Polynesian Society* 22(87): 129–130.

Journal of the Polynesian Society (JPS) 1915. Wahi II Te Kauwae-Raro, Ara: Ngā Korero ta Tai o Nehe a ngā Ruanuku o te Whare Wānanga o Te Tai-Rawhiti. Upoko X. Te Haerenga mai o "Takitimu" ki Aotearoa (Te Roanga). *Journal of the Polynesian Society* 24(93): 1. Told by Te Matorohanga, written by H. T. Whatahoro, and translated by S. Percy-Smith, http://www.jps.auckland.ac.nz/document/Volume_24_1915 (accessed September-December 2012).

http://www.jps.auckland.ac.nz/document/Volume_22_1913

http://www.jps.auckland.ac.nz/document/Volume_22_1913/Volume_22%2C_No._87/The_lore_of_the_whare_wananga._Wahi_II._Te_Kauwae-Raro%2C_Upoko_III._by_Te_Matorohanga%2C_p_107–133?action=null (accessed September-December 2012).

Keane, B. 2011a. Oneone—soils—Uses of soils. *Te Ara—The Encyclopaedia of New Zealand*, URL: http://www.teara.govt.nz/en/oneone-soils/1; (accessed February–December 2012).

Keane, B. 2011b. Oneone—soils—Soil in Māori tradition. *Te Ara—The Encyclopaedia of New Zealand*, URL: http://www.teara.govt.nz/en/oneone-soils/2; (accessed February–December 2012).

King, M. 2003. *The Penguin History of New Zealand*. Wellington: Penguin Books. 570pp.

Landcare Research NZ Ltd GIS 2012. (accessed Feb to Dec 2012).

Marsden, M. 1988. The natural world and natural resources. Māori value systems and perspectives. *Resource Management Law Reform Working Paper 29. Part A*. Wellington, Ministry for the Environment.

Maughan, C. W. and T. T. Kingi. 1998. Te ture whenua Māori: Retention and development. *The New Zealand Law Journal* (January): 27–31.

McKinnon, M., B. Bradley, and R. Kirkpatrick. 1997. *The New Zealand Historical Atlas: Visualising New Zealand. Ko Papa-tū-ā-nuku e Takoto Nei*. Auckland: David Bateman & Department of Internal Affairs New Zealand. 290pp.

Mead, H. 2003. *Tikanga Māori: Living by Māori Values*. Wellington: Huia Publishers and Te Whare Wananga o Awanuiarangi. 398pp.

Ministry of Agriculture and Forestry (MAF). 2011. *Māori Agribusiness in New Zealand. A Study of the Māori Freehold Land Resource*. Wellington, Ministry of Agriculture and Forestry. 42p.

Ngata, H. M. 1993. *English–Māori Dictionary*. Wellington: Learning Media. 559pp.

New Zealand Institute of Economic Research (NZIER). 2003. *Māori Economic Development: Te Ōhanga Whanaketanga Māori*. Wellington, NZ: New Zealand Institute of Economic Research. 116pp.

Park, G. 1995. *Ngā Uruora: The Groves of Life. Ecology and History in a New Zealand Landscape*. Wellington: Victoria University Press. 376pp.

Rangitāne o Wairarapa Inc. 2006. *Ngāti Hāmua Environmental Education Sheets*. Greater Wellington Regional Council. Masterton: Lamb-Peters Print. 39pp.

Raskin, B. 2009. *The Adoption and Use of Technology in Food Production: Māori Experiences and Perspectives. Literature Review Report*. Wellington: Institute of Environmental Science and Research. 39pp.

Roskruge, N. 2009. Hokia ki te Whenua: Return to the land, unpublished PhD thesis, Massey University, Palmerston North, New Zealand.

Roskruge, N. 2011. Traditional Māori horticultural and ethnopedological praxis in the New Zealand landscape. *Management of Environmental Quality* 22(2): 200–212.

Roskruge, N. 2012. *Tahua-roa: Food for your Visitors. Korare: Māori Green Vegetables their History and Tips on Their Use*. Palmerston North: Institute of Natural Resources, Massey University. 110pp.

Te Puni Kōkiri (TPK) 1998. *Māori Land Information Database* MLIB 1998–2002. Te Puni Kōkiri (TPK) Wellington, New Zealand: Ministry of Māori Development. http://www.tpk.govt.nz/en/services/land/ (accessed 2000–2002).

Te Puni Kōkiri (TPK). 2002. *Māori in the NZ Economy*. 3rd edition. Wellington: Te Puni Kōkiri.
Te Puni Kōkiri (TPK). 2007a. *Historical Influences: Māori and the Economy*. Wellington: Te Puni Kōkiri. 12pp.
Te Puni Kōkiri (TPK) 2007b. *The Māori Commercial Asset Base*. Wellington: Te Puni Kōkiri. 20pp.
Walker R. 2004. *Ka Whawhai Tonu Matou: Struggle without End*. Wellington: Penguin Books. 462pp.
Whitehead, J. and B. Annesley, 2005. The context for Māori economic development: A background paper for the 2005 Hui Taumata, February 2005. Wellington: The Treasury. 33pp.
Williams, H. W. 1975. *A Dictionary of the Māori Language*. 7th edition. Wellington, New Zealand, A. R. Shearer, Government Printer. 507pp.

10 Integrative Development between Soil Science and Confucius' Philosophy

*Xinhua Peng**

CONTENTS

10.1 INTRODUCTION

Confucianism is a profound school of learning founded by the ancient Chinese thinker Confucius (551–478 BC) in the fifth century BC. By the second century AD, thanks to the advocacy by influential officials and intellectuals of the Han Dynasty (206 BC–AD 220), such as Dong Zhongshu, Confucianism became central to national ideology. During the historical and social developments of later times, the school would have a profound impact on the Chinese people in terms of their belief, values, culture, economy, and politics.

Although it first appeared as a school of learning, Confucianism as an institutionalized belief system for the Chinese people's ultimate concern is also known as "the Confucian religion" (*rujiao*). Since the Eastern Han Dynasty (25–220), there have been "three religions" in China—Confucianism, Taoism, and Buddhism—that influence each other and together form the spiritual framework of China. Although sometimes the relationship among them incurred intervention by the state, they have enjoyed a peaceful coexistence most of the time. Common Chinese people are inclined toward an eclectic belief system. As a result, the idea of the integration of the three religions became ever more popular, so much so that some members of the literati also advocated their complete integration. All Confucian scholars during and after the Song Dynasty (960–1279), whether they were for the exclusion of Buddhism and Taoism or for "the unity of the three religions," learned immensely from the other two faiths. However, a comparison between Confucianism and the other two clearly shows its major characteristics—placing more emphasis on the actual existence of humans rather

* State Key Laboratory of Soil and Sustainable Agriculture, Institute of Soil Science, CAS, Nanjing, 210008, P.R. CHINA; Email: xhpeng@issas.ac.cn

than descriptions of a world beyond human life. For this reason, we regard Confucianism as humanistic and, in a sense, atheistic. It is the emphasis on the actual existence of humans that makes Confucianism the only belief system among the "three religions" that requires its followers to assume direct responsibility for state governance.

In the Confucian cosmology, the world has two poles, *yin* and *yang*, which work together to create everything. *Yang* stands for a lively masculine force, and functions as the primordial drive; *yin*, a quiet, yet necessary soft force, gives life to living things. In the *Book of Changes*, which is one of the Confucian classics, *yang* is represented by the symbol *qian* (— — —), and *yin* by the symbol *kun* (— —). Both symbols are called *yao*. A combination of three *yao*, one on top of another, is a *gua*. There are altogether eight *gua*, each made up of three *yao*. Any two of them can combine into a *guaxiang* comprising six *yao*, so that there are altogether 64 *guaxiang*. These are used in the *Book of Changes* to explain the nature and structure of the universe, relations among things in the human world, and the relationship between humans and nature. In this all-inclusive cosmic system, the most fundamental concept is the unity of opposites—heaven and earth, *yin* and *yang*. According to the book, *qian* stands for heaven and father, while *kun* stands for earth and mother.

10.2 ORIGIN AND MEANING OF THE CHARACTER *TU* (土)

What is *tu* ("earth"), and how did the character come into being? As early as the Zhou Dynasty (1046–256 BC), people realized the nature of *tu* from experience; according to *Rites of Zhou*, "*Tu* is that from which everything grows." That is to say, there is *tu* wherever plants grow, and vice versa. This indicates the scientific meaning of *tu*. A more specific description was provided in the later times.

According to *Shuowen Jiezi* (explaining and analyzing characters), "*Tu* is what can sprout living things (i.e., green plants) on the earth. The horizontal strokes (=) stand for the surface soil and subsoil horizon, respectively. The vertical stroke (|) symbolizes the growth of things (e.g., plants)." This is a specific explanation of the character's origin, meaning, and form, in which it is recognized that *tu* is what produces living things (vegetation). That is to say, *tu* is understood in terms of the growth of plants (the natural aspect). In other words, there is *tu* wherever there are plants growing, and vice versa, in the natural course of events. From this, it can be seen that ancient Chinese people had a scientific understanding of *tu*. The explanation in *Shuowen Jiezi* reveals the relationship between *tu* and vegetation. It can be said that the understanding of *tu* among working people of the time when the book was written already involved vague ideas and imagination about the name of soils and its relation to crop production.

In symbolizing the "surface and middle of earth," the horizontal strokes (=) mean the upper and lower levels of *tu*. They give a graphic representation of the relationship between *tu* and the growth of plants, which are partly above the surface of the earth and partly beneath it. Moreover, the two horizontal strokes stand for surface soil and substratum. The difference in length between them implies the rough ratio of surface soil to substratum, which will vary with soil type (Figure 10.1).

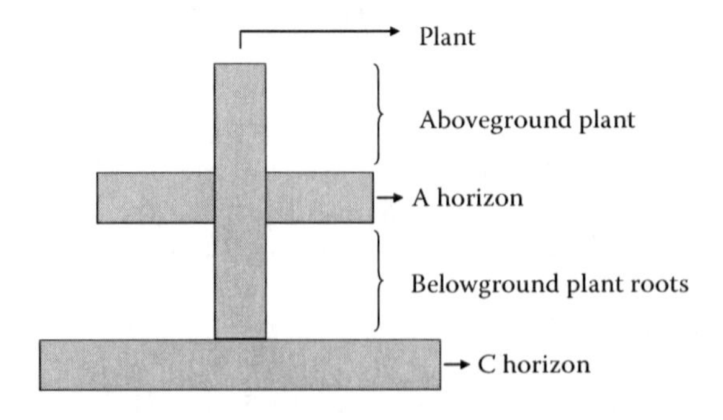

FIGURE 10.1 Soil in Chinese word. (Modified from Wang, Y. 1980. *Soil Science in Ancient China*. Science Press, Beijing.)

The vertical stroke (丨) means that plants grow out of *tu*—usually upward and upright. Similarly, the character *zhi* (植, "plant") comprises 直 (straight) and 木 (wood). The components of the character *tu* provide a graphic representation of the mutual reliance between soil and plants, and illustrate the character's origin and scientific meaning (Figure 10.1). This is thought provoking in terms of the biological perspective on soil.

10.3 ORIGIN AND MEANING OF THE CHARACTER *RANG* (壤)

What is *rang* (soil in English), and what is the meaning of the character of soil? As early as the Zhou Dynasty (1046–256 BC), it was noted in *Rites of Zhou* that "*rang* refers to that which is tilled and on which plants are grown and cultivated." This is an insightful description of the formation of *rang* and its scientific meaning. According to *Shuowen Jiezi*, "*Rang* means soft soil without lumps." Ma Rong's commentary on *Yu Gong* (Old Chinese geography) states, "*Rang* is gentle and good by nature." Zheng Xuan's commentary on a chapter in *Rites of Zhou* states, "*Rang* is soft and mild in appearance." All three explanations indicate the physical features of *rang*. They are in keeping with each other in terms of the understanding of *rang*. *Rang* is more productive and fertile than *tu* because it is neither hardened nor too loose, but soft and gentle, suitable for the growth of all kinds of crops.

But how does *rang* come into being? According to Zheng Xuan, "*Rang* is also *tu*, a variation of the latter." This is an insightful generalization. Such "variation" results from the human activities of "tilling and cultivation." That is to say, through the human initiatives of fertilization and intensive cultivation, *tu* can mature into *rang*. Therefore, *rang* is essentially the same as *tu*, but it has acquired what is beyond the original properties of *tu*, thanks to the human act of tillage. The natural qualities of *tu* have changed to some degree. In scientific terms, "tillage and cultivation" will change the physical, chemical, and biological properties of *tu* to some degree and increase its productivity. In other words, *rang* is a more fertile form of *tu*. This is also apparent from the composition of the character *rang* (壤), which can be broken into 土 (earth) and 襄 (help, assistance). It is obvious that the human factor is essential to the formation of *rang*. This is the ancient Chinese people's understanding of *rang*. Today, it has acquired new meanings, which are in keeping with nature. It is also thought provoking in terms of the cultivation of (agricultural) soil.

In *Nong Ya* (Agronomy in China), it is said that *rang* means fertility. That is to say, *rang* refers to fertile *tu*. This can serve as another note on Zheng Xuan's remark that "*Rang* is also *tu*, a variation of the latter."

Both the explanations in *Shuowen Jiezi* ("*rang* means soft soil") and in *Yu Gong* ("*rang* is gentle and good by nature") suggest that *rang* is more productive and fertile than *tu*, because it is neither hardened nor too loose, but soft and gentle, suitable for the growth of all kinds of crops.

10.4 COMPARISON BETWEEN *TU* AND *RANG*

The fundamental difference between *tu* and *rang* is that the former is naturally formed while the latter is formed by cultivation. Yet, there is no strict criterion for telling one from the other. For instance, black *tu* (black earth in English) and yellow *tu* (loess in English) can also refer to agricultural soil, while red *rang* (red soil in English) can be a general term for soil in south China. Natural soil can be turned into cultivated soil, and cultivated soil may also turn back into natural soil after a long term without cultivation. So, the Chinese word for "soil," *turang*, is a combination of *tu* and *rang* reflecting the harmonious unity of humans and nature.

10.5 IMPLICATIONS OF SOIL IN ANCIENT CHINA AND CONFUCIAN IDEAS

10.5.1 Maternal Implications of Soil

Earth has always been honored as the mother of humankind, as apparent from the common expression in Chinese "earth is our mother." The virtues of earth are fully reflected in humans' moral activities. Regarding earth as their "mother," people are generated, supported, and nurtured by earth, and try to imitate the virtues of earth. Such virtues are often associated with women (the *yin* sex), wife (birth and tenderness), and subjects (submission). It is said in *Yijing* (*Book of Changes*) that "the Way of the earth is the Way of wives and subjects." This provides a perspective for our analysis of some elements in Confucian ethics, such as "loyalty, filial piety, integrity, and righteousness." In the monarchial times of China (221 BC–AD 1911), all the desired characters of subjects contained elements of the docility and submission of soil.

With earth regarded as the mother of all living things, there is the sense of soil being regarded as the skin of the earth, which bears a certain similarity to the title of the soil science journal *Geoderma*. For instance, there is the following remark, "For land, famous mountains are its assistants, rocks are its bones, rivers are its veins, plants are its hair, and earth is its flesh. What is above three *chi (Chinese length unit, 3 chi = 1 meter)* is *fertile*, and what is below three *chi* is *infertile*." This indicates the unity of and consistency between *yin*, *kun*, earth, and the virtues (i.e., essential characteristics) of soil. There are many other Chinese words referring to soil (e.g., 田, which often refers to the shape of arable lands on the field; 壤, which usually refers to loose and soft surface soil). In Chinese classics, *di* (地, earth or land) is often used in the same sense as *tu*; in both ancient Chinese and modern Chinese, the two are often joined to refer to land in its contemporary common sense.

10.5.2 Harmonious Nature of Soil

The Confucians incorporated the theory of *Wuxing* (the five elements, also translated as phases/movements/stages) from the *Yin-Yang* School. The theory holds that the universe is composed of five elements—metal, wood, water, fire, and earth (*tu*), which "generate" and "restrict" each other. These are also five kinds of power that determine the destiny of humankind. Among the five elements, earth functions as an important basis. In the "generation cycle," fire can generate earth. Here, "fire" can be interpreted as various kinds of weathering. In *Zhou Yi Can Tong Qi* (a comparative study of the Zhou [Dynasty] *Book of Changes*), Wei Boyang of the Eastern Han Dynasty also said that "fire can turn into earth." The real meaning is that weathering can turn rocks into earth (*tu*). Such "earth" needs to be fertile and able to nurture plants. Thus, in the "restriction cycle" of *Wuxing*, wood can "restrict" earth. In modern terms, this means that plants absorb all kinds of nutrition from soil. Conversely, only the kind of soil that is fertile and able to support plants can count as *tu*. This is only a primitive understanding of how soil is formed, but its scientific basis and far-reaching significance remains relevant today as the world's first theory on the formation of soil.

10.5.3 Soil as a Symbol of Power

In Confucian cosmology, like any other living beings, humans are generated by *yin* and *yang*, *qian* and *kun*, and heaven and earth. Thus, humans are both a component of, and a phenomenon in, the universe. However, heaven and earth are personified to some degree and worshipped accordingly. Confucians of the Han Dynasty were convinced that human faults would infuriate heaven and earth and incur punishment, which would appear as natural disasters such as earthquakes and thunderstorms. In the later period of imperial China, state ceremonies for making offerings to earth were held in the Temple of Earth and personally presided over by the emperor. The purpose was to communicate with the earth and pray for protection. In feudal times, making offerings to heaven was a privilege reserved for the emperor, who was called the Son of heaven. Thus, heaven was separated from common people by a long distance. Ceremonies for making offerings to earth could be held in either local areas or communities. The place for doing so was called *she* (alter in English), and the gathering for the purpose was called *shehui*. Later, *shehui* would become an equivalent to the English word "society," suggesting the significant role played by sacrifices to earth in Chinese people's existence in groups. In communities, earth was worshipped as a local natural god in charge of that particular place.

FIGURE 10.2 The altar of land and grain in Beijing.

Earth was so important that it gradually evolved into a symbol of power. During the Zhou Dynasty, at a ceremony of sacrifice, the feudal lord would receive a lump of earth wrapped in white thatch grass. He would carry it to his fief and build a temple to enshrine it. In 1421, in the reign of the Yongle Emperor of the Ming Dynasty (1368–1644), the Altar of Land and Grain (what is now Zhongshan Park) was built in Beijing. From then it was used by emperors of Ming Dynasty and Qing Dynasty (1644–1911) to offer sacrifices to the God of Earth and the God of the Five Grains. At the center of the altar is an earthen terrace with bluish-green earth in the east, red earth in the south, white earth in the west, black earth in the north, and yellow earth at the center. A stone column called *Shezhu* (power in English) stands at the center of the terrace. The earth in five colors in the altar stands for China's vast and colorful soils, and symbolizes the traditional idea that "all under heaven belongs to the monarch" and the everlasting endurance of the dynasty. The location of five soils is also in agreement with the geographic distribution of the main soil types in China. Moreover, they stand for the five basic elements in traditional Chinese astrology—metal, wood, water, fire, and water (Figure 10.2).

10.5.4 Emotional Elements of Soil

The Chinese are deeply attached to their native places and believe that each place nurtures a unique group of people. When a traveler is unable to adapt to the climate and dietary habit of a distant place, that is called "failure to acclimate oneself to the water and soil." In imperial times, as soon as man had succeeded in the imperial examination, he would go home, report to his parents, and make offerings to his ancestors. It was customary for a retired official to return to his hometown and purchase land to become a landlord. When someone was dying, he would usually expect to be buried in the clan cemetery in his hometown, which would be his proper final resting place. When folks traveled far from his hometown, they would often carry a handful of soil from his hometown as a memento. Nowadays, many successful overseas Chinese take delight in doing something for their hometowns, for they believe that they should eventually return, like fallen leaves returning to the roots. In a word, such love of one's hometown and native place is a part of traditional Chinese culture and can be traced back to the Chinese people's age-old attachment to land.

10.6 SOIL TILLAGE AND THE CONFUCIAN IDEA OF THE GOLDEN MEAN

The Way of the Golden Mean, which is the quintessence of Confucianism, means an impartial and moderate attitude that one should assume when dealing with others. This is an important guideline for soil cultivation. According to *Lü Shi Chun Qiu* (*The Annals of Lü Buwei*), "What is hard desires to be soft, and vice versa; what is at rest desires to work, and vice versa; what is barren desires to be fertile, and vice versa; what is quick desires to be slow, and vice versa; what is damp desires to be dry, and vice versa." These five major principles reflect the Way of the Golden Mean in soil cultivation, as follows:

1. "What is hard" refers to hard and compact soil, and what is "soft" means loose and soft soil. "What is hard desires to be soft" means that soil that is too hard should be made softer, "and vice versa" means that soft soil should be made more compact. In other words, soil cultivation is supposed to improve the structure of soil so that it is neither too hard nor too soft and is in the best condition in terms of water, air, heat, and fertility.
2. "What is at rest" means fallow, and "what is at work" means land in continuous tillage. "What is at rest desires to work" means tillage after fallow time, "and vice versa" means that land should be fallow after continuous tillage. In other words, the maintenance of soil fertility should be achieved by soil cultivation between tillage and fallow.
3. "What is barren desires to be fertile, and vice versa"—This means that the fertility of soil should be adjusted through soil cultivation. Barren soil is to be made more fertile, by cultivation, whereas soil that is already fertile should be kept from intensive cultivation.
4. "What is quick desires to be slow, and vice versa"—This involves soil's ability to maintain its fertility and the speed with which such fertility is released. The kind of soil that releases its fertility too quickly, resulting in the excessive growth of crops followed by slow growth, should be enabled to maintain its fertility better and release it more steadily by cultivation; conversely, the kind of soil that releases its fertility too slowly, resulting in the slow growth of crops at first, should be enabled to release its fertility more quickly through cultivation methods.
5. "What is wet desires to be dry, and vice versa"—This means that the water content of soil should be adjusted through cultivation. "What is wet" is low lying and excessively wet land, and the opposite is high and excessively dry land. Wet soil is made drier by deep drainage and dry soil is made wet by irrigation. These principles indicate that Chinese people were already highly experienced in soil cultivation by the spring and autumn period and the Warring States period.

Fundamental to soil tillage is the principle of "three adaptations"—adaption to the time, the place, and the physical conditions. This is more specifically expressed by maxims such as "choose the proper time and suit measures to the conditions of the land," "adapt to the climatic conditions and consider the geographical conditions," "adapt to the climatic conditions and recognize what is suitable for the land," "comply with what is suitable for the climatic conditions, geographical position, and the nature of things," and "make use of the geographical and climatic conditions." The Chinese theory of "climatic, geographical and human conditions" requires people to recognize, respect, and follow objective laws. As pointed out by Jia Sixie in *Qi Min Yao Shu* (Main Techniques for the Welfare of the People), "If one can adapt to the climatic conditions and consider the geographical conditions, he will achieve greater success with less effort; if he indulges his will and goes against the proper way, he will be laboring in vain."

10.7 END NOTE

I do not profess to be an expert on this subject and have found that little has been written on it. So, this is merely a humble introduction offered in the hope that someone might be inspired to continue to delve into the relationship between soil science and Confucianism in China and, in doing so, help to disseminate Chinese culture. Most of the data for this chapter come from the references listed below, and I have merely edited and translated these data into English for international readers. The topic deserves more study.

FURTHER READING

Chen, F. 2009. A study of soil improvement technology in ancient China. MS thesis, Nanjing Agricultural University, Nanjing.

Gong, Z., Zhang, X., Chen, J., and Zhang, G. 2003. Origin and development of soil science in ancient China. *Geoderma* 115: 3–13.

Lahmar, R. and Ribaut, J.-P. (eds.). 2005. *Sols Et Sociétés. Regards Pluriculturels.* Beijing: The Commercial Press.

Lin, P. 1996. *Soil Classification and Land Use in Ancient China.* Beijing: Science Press.

Wang, Y. 1980. *Soil Science in Ancient China.* Beijing: Science Press.

Zeng, X. 2008. *A History of Agronomy in China.* Fuzhou: Fujian People's Press.

11 Soil

Natural Capital Supplying Valuable Ecosystem Services

Brent Clothier and Mary Beth Kirkham*

CONTENTS

11.1 INTRODUCTION

The natural capital concept attempts to integrate economic thinking with ecological principles by considering nature's stocks of materials and energy as capital. In economics, interest, or rents, flow from financial or built capital, and so by analogy in nature, ecosystem services, which benefit mankind, flow from natural capital stocks. Nature comprises an assemblage of natural capital stocks, and they, in sum, form our ecological infrastructure. Marchant et al. (2013) consider ecological infrastructure to be how natural capital is arranged. Bristow et al. (2010) noted that soil is a key component of ecological infrastructure, and that it is a prime natural capital stock that provides valuable ecosystem services.

Around the turn of the century, frameworks for quantifying natural capital and valuing ecosystem services were proposed (Costanza et al. 1997; Daily 1997). In 2001, the United Nations sponsored a major initiative called the millennium ecosystem assessment to determine the state of the world's ecosystem services and to consider the consequences of the diminution of ecosystem services as a result in the degradation of the world's ecological infrastructure. Their findings (Millennium Ecosystem Assessment 2005) classified ecosystem services into four broad, and often overlapping, types:

- *Supporting services*: Those necessary for production of the other services, such as soil formation.
- *Provisioning services*: Production of food, fuel, and fiber.
- *Regulating services*: The buffering and filtering of water, gases, and chemicals.
- *Cultural services*: Heritage, recreation, aesthetics, and spiritual well-being.

Since these seminal studies, there has been a burgeoning growth in studies on ecological infrastructure, natural capital, and ecosystem services. And furthermore, many of these have focused on

* Systems Modelling, The New Zealand Institute for Plant & Food Research Limited, Palmerston North, New Zealand; Email: brent.clothier@plantandfood.co.nz

the ecosystem services that flow from the natural capital stocks of soil, a key element of our ecological infrastructure (Daily et al. 1997; Bristow et al. 2010; Dominati et al. 2010a; Haygarth and Ritz 2010; Jury et al. 2011; Robinson et al. 2013a,b). The goal of these works has been to develop new frameworks to assess the state of the soil's ecosystem services, so that they can be used to inform the development of policies and practices that will ensure the sustainability of the natural capital stocks of the world's soils.

So, what is a good time stamp for soil sustainability? Four thousand years? Franklin Hiram King, the world's first professor of agricultural physics, who was appointed by the University of Wisconsin in 1888 (Tanner and Simonson 1993), visited Korea, Japan, and China in 1909. In his book, "Farmers of Forty Centuries" (King 1911; see also Landa 2014), he describes how he set out to "... consider the practices of some five hundred millions of people who have an unimpaired inheritance acquired through four thousand years." His book is a prescient forerunner of ecosystem services thinking. King (1911) wrote that "... we were amazed at how these nations have been and are conserving and utilising their natural resources and surprised at the magnitude of the returns they were getting from their fields. It is evident that these people, centuries ago, came to appreciate the value of [natural resources] in crop production." This language mimics that of the current research into in natural capital and ecosystem services.

King (1911) found that "... judicious and rational methods of fertilisation are everywhere practiced. Lumber and herbage for green manure and compost, and ash of the fuel and lumber used at home finds its way ultimately to the field as fertilizer. Manure of all kinds, human and animal, is religiously saved and applied to the fields." The *supporting services* of nutrient cycling can be seen in this excerpt from King's (1911) writing. He observed that "... much intelligence and the highest skill are exhibited by the old-world farmers in the use of their wastes," and he concluded that because of this "... it is possible after even 40 centuries for their soils to produce sufficiently for the maintenance of such dense populations." Here, King (1911) is describing the *provisioning services* of the oriental soils that have been sustained for 40 centuries. King (1911) goes on to say "... that these people, centuries ago, came to appreciate the value of water in crop production. They have adapted conditions to crops and crops to conditions until with rice they have a cereal which ensures maximum yields against both drought and flood." King (1911) is referring here to the *regulating services* provided by soils. Given the return of human and animal waste to the soil to maintain *supporting services*, King (1911) noted that "... the drinking of boiled water is universally adopted as an individually available and thoroughly efficient safeguard against a class of deadly disease germs." He concluded that "... the cultivation of tea in China and Japan is another of the great industries of these nations and it plays an important part in the welfare of the people [and] there is little reason to doubt that this industry has its foundation in the need of something to render boiled water palatable for drinking purposes." So, soil management over 40 centuries has led to the *cultural service* of having a "cuppa" and participation in tea ceremonies.

Franklin King's perceptions of a century ago provide a pertinent and prescient underpinning to this chapter on our contemporary assessment of the flow of valuable ecosystem services from the natural capital stocks of the world's soils. But "... much has changed in the last century" write Kubiszewski and Costanza (2012). They add that "... when the world was still relatively empty of humans and their built infrastructure, natural resources were abundant [but now] the human footprint has grown so large that in many cases real progress is constrained more by limits on natural resources and ecosystem services than by limits on built capital infrastructure."

11.2 NATURAL CAPITAL, ECOLOGICAL INFRASTRUCTURE, AND ECOSYSTEM SERVICES

Natural capital is defined as nature's stocks of materials and energy. These natural capital stocks are arranged and connected to comprise landscape elements such as rivers, soils, aquifers, wetlands, and biota (Bristow et al. 2010). These form ecosystems from which flow the ecosystem services that

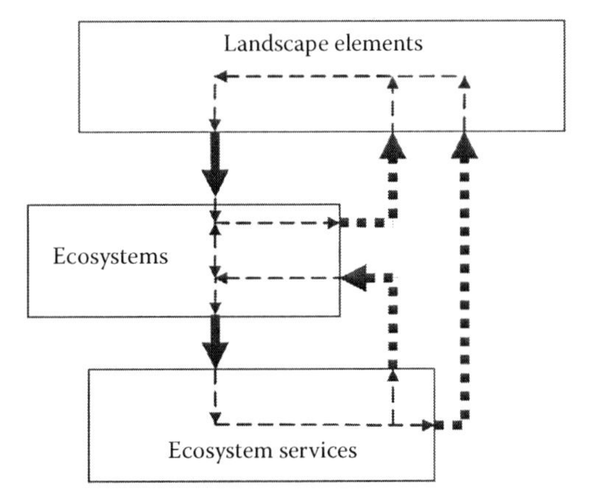

FIGURE 11.1 Ecological infrastructure consists of landscape elements, of which soil is one element, ecosystems including the soil ecosystem, and ecosystem services with soil ecosystem services being one category of these services. Interconnections within and between these are shown. (From Bristow, K.L. et al. 2010. *Proceedings of the 19th World Congress of Soil Science,* Brisbane, Australia, August 1–7, 2010 (on CD).)

benefit mankind, either directly or indirectly (Figure 11.1—reproduced from Bristow et al. 2010). They describe ecological infrastructure as being how natural capital is arranged, for it is the connectedness within and between the landscape elements, ecosystems and ecosystem services that is critical for the continuing delivery of ecosystem services. They add that it is "... the spatially and temporally interconnected natural elements, ecosystems and ecosystem processes and functions, which maintain the continued regeneration and evolution of the biosphere."

Soil is a natural capital element, a critical component of ecological infrastructure, which supplies valuable ecosystem services.

Soon after the pioneering research by Costanza et al. (1997) and Daily (1997), the Millennium Ecosystem Assessment (2005) developed a classification of the ecosystem services that flow from our natural assets, and they depicted how these sustain human well-being. Their classification comprised four services: supporting, provisioning, regulating, and cultural flows. The supporting services, such as soil formation and nutrient cycling, underpin the three other services. These latter three services provide the constituents of well-being through providing security, basic materials, health and social relations, as well as ensuring freedom of choice and action. This four-way service classification, along with the valuation results from Costanza et al. (1997), resulted in a significant change in our way of thinking about soils and nature. The focus in the latter part of the twentieth century was on the green revolution, and increasing crop yields through new breeding techniques and better soil management—a focus on only the provisioning services provided by soils and climate.

The growth in ecosystem services research has seen a reassessment of the fourfold classification of the millennium ecosystem assessment (Robinson et al. 2009; Dominati et al. 2010a,b). Robinson and coworkers developed a definition of soil natural capital by consideration of mass, energy, and organization, and they distinguished between two aspects of natural capital: quantity and quality. Their analysis enabled them to show soil moisture, temperature, and soil structure as the valuable stocks, rather than those traditionally associated just with provisioning services, namely inorganic and organic materials. Dominati et al. (2010a) then developed a framework that enables exploration of how and why soil natural capital stocks change and they defined the nature and role of soil processes. They were able to link the external drivers of change to their impacts on stocks and services. They distinguished between processes and services. The framework of Dominati et al. (2010a) is reproduced here as Figure 11.2.

It consists of five elements: soils as natural capital; the formation, maintenance, and degradation of soil natural capital; the drivers of soil processes and impact on soil properties; the flow from the

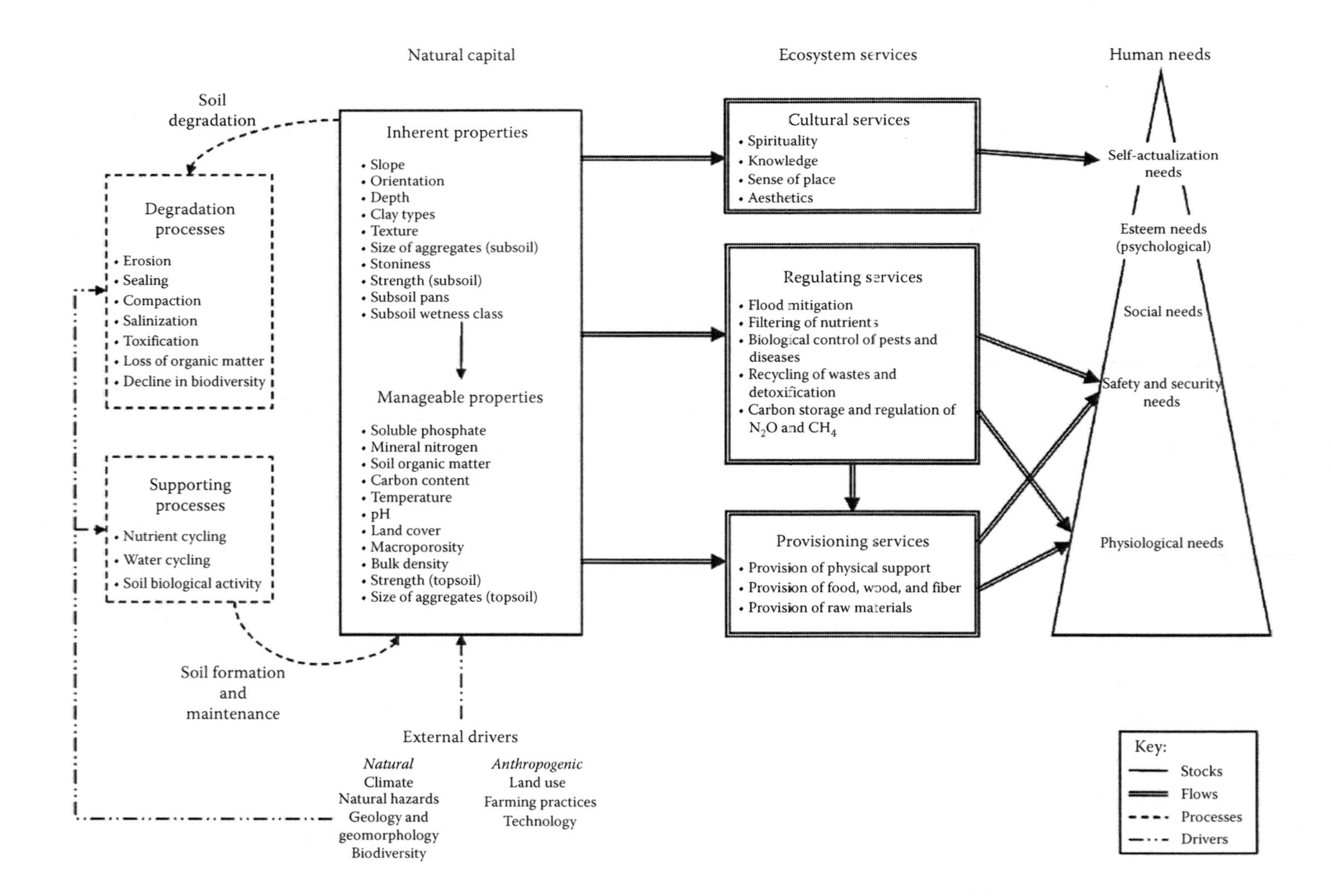

FIGURE 11.2 The framework proposed by Dominati et al. (2010a) for the provision of ecosystem services from soil natural capital. (From Dominati, E., Patterson, M., and Mackay, A. 2010a. *Ecological Economics* 69: 1858–1868.)

stocks of the three services of regulation, provision, and cultural; and how these meet human needs. The left-hand side of Figure 11.2 shows what comprises soil natural capital, both the inherent properties and the manageable characteristics, and furthermore how natural and anthropogenic drivers can affect these properties. These drivers, and the changes they impart on these properties, can lead to either, or both, degradation processes and supporting processes. These led to further soil formation, or not, depending on the balance between degradation and supporting processes.

The framework of Dominati and coworkers for soil natural capital underpins one landscape element of ecological infrastructure, for, as Bristow et al. (2010) note, soil is just one key and valuable component of ecological infrastructure. Soil supplies valuable ecosystem services, both directly and indirectly through its interconnectedness with other landscape elements.

Bristow et al. (2010) note that "... the recent financial crisis has led to massively increased investments in built infrastructure as a means of rapidly stabilising and reinvigorating economies." One of these researchers wryly laments that "... given the worsening water and food crises and increasing population pressures, one wonders why larger investments are not being made to ensure that our ecological infrastructure has the capacity to continue to produce sufficient flows of ecosystem services to satisfy the world's future needs" (Marchant, pers. comm., 2013). This has extended ecosystem services thinking beyond mere classification and valuation to consideration of the value of investing in ecological infrastructure.

We will now consider the value of investment into ecological infrastructure in relation to supporting processes, plus some consideration of regulating and cultural services. These are the "hidden" ecosystem services, relative to the provisioning services which are more easily quantified through standard economic assessments. Finally, we will discuss how the value of land and the price of land are out of kilter. We show how, through a judicial hearing process, it is possible to use a holistic ecosystem services approach to maintain the value of ecological infrastructure through limiting peri-urban expansion onto prime soils.

11.3 SUPPORTING PROCESSES AND SOIL PROPERTIES

Soils and plants are interlinked landscape elements, and their interconnectedness is critical for the supporting processes that underpin the value of soil natural capital. Deurer et al. (2010) reported on an assessment of soil carbon stocks (SCS) in two kiwifruit orchards in the Bay of Plenty of New Zealand (Figure 11.3). The orchards were adjacent to each other and on the same soil type. However, whereas one block had been planted 25 years ago, the other was just 10 years old. Soil carbon was

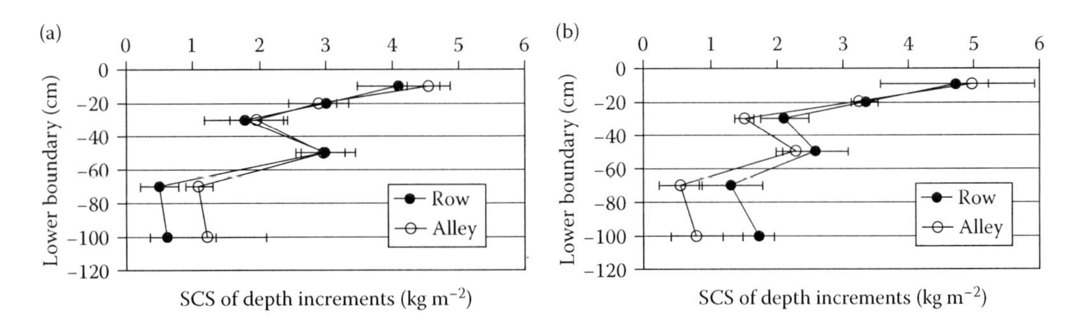

FIGURE 11.3 SCS in "Hort16A" kiwifruit orchards of different ages. (a) "Young" block. The total SCS to 1 m depth are 13 ± 2.1 kg m^{-2} in the row and 14.7 ± 0.5 kg m^{-2} in the alley. (b) "Old" block. The total SCS to 1 m depth are 15.7 ± 0.8 kg m^{-2} in the row and 13.3 ± 0.3 kg m^{-2} in the alley. (From Deurer, M. et al. 2010. In *Proceedings of the Workshop: Farming's Future: Minimising footprints and maximising margins*, 445–459. Fertiliser & Lime Research Centre, Massey University, February 10–11, 2010.)

sampled in layers down to 1 m in each orchard, with four profiles sampled along the row, and four along the alley. The soil carbon stocks (SCS, kg-C m^{-2}, or kilograms of carbon per square meter) were calculated for each of the layers.

There was no difference between the SCS summed to 0.5 m between the row and the alley in either orchard. Furthermore, there was no difference in the SCS to 0.5 m between the "young" and the "old" orchard. Nor was there any difference in the SCS between 0.5 and 1 m in the alleys of the different-aged orchards. However, there was a significant difference in the 0.5–1 m SCS in the row under the vines of the "young" and "old" orchards. So it can be seen that the growing kiwifruit roots are building up the levels of soil carbon at depth. In sum down to 1 m, the SCS in the "young" orchard was 139 tonnes of carbon per hectare (t-C ha^{-1}), whereas it was significantly higher at 145 tonnes of carbon per hectare in the "old" orchard. This bears out the findings of Kong and Six (2010) that roots are effective at storing recalcitrant carbon in soil. The results set out in Figure 11.3 show that the interconnectedness of the soil with the vine is actually leading to soil formation by "growing" the depth of the root zone. So through the anthropogenic driver of land-use change, supporting processes have led to soil formation at depth, and the maintenance of soil processes throughout (Figure 11.2). The land-use practice of kiwifruit growing is investing carbon into the soil as an element of ecological infrastructure and "growing" the soil.

Soil carbon content is a manageable property of natural capital, and a change in the carbon content of soil can bring with it attendant benefits or losses—depending on the sign of the change.

Kim et al. (2008) quantified the carbon status of two neighboring apple orchards in the Hawke's Bay of New Zealand. The orchards had both been planted at the same time some 10 years earlier. Whereas one orchard embarked on a certified organic program of orchard management, the other employed certified integrated fruit production (IFP) practices. In the organic orchard some 5–10 tonnes per hectare of compost were applied annually to provide for the trees' nutrient requirements, whereas small amounts of inorganic fertilizer were used in the other. As a result of these quite different carbon-management processes, the SCSs of the top 0.1 m in the tree rows of the two orchards had become quite different: 3.8 kilograms of carbon per square meter for the organic orchard of which 110 grams of carbon per square meter was the labile carbon found using hot-water extraction; and 2.6 kilograms of carbon per square meter for the IFP orchard, of which 60 grams of carbon per square meter was hot-water extractable. There was variability within and between these orchards in terms of the hot-water extractable carbon. Kim et al. (2008) obtained a number of samples from both orchards and carried out 40-day incubations to determine the nitrogen mineralization rate in the soil. Their results are reproduced here in Figure 11.4.

There was found to be a strong and positive relation between the soil's hot-water carbon (HWC) content and the rate of nitrogen mineralization. Here then, the investment of compost into the ecological infrastructure of the soil has led to a change in the soil's manageable property of its carbon content and this has enhanced the supporting process of the nutrient cycling of nitrogen. The value of the supporting processes coming from the soil's natural capital has been enhanced through the investment of carbon into the soil's ecological infrastructure.

But this carbon investment provides other returns. Deurer et al. (2009) and Clothier et al. (2011) report on the impact on the soils' macropore structure of the different soil carbon management practices between these two apple orchards. The depthwise profiles of macroporosity in the surface soil of the two orchards are shown in Figure 11.5 (from Clothier et al. 2011). The porous structure of the 43 mm long soil cores from these two orchards was resolved using 3D x-ray-computed tomography, and macropores were defined as being pores of diameter greater than 0.3 mm. For the two cores shown in Figure 11.5, the volumetric average macroporosity in the core from the organic orchard was 9.4 (±1.7)%, whereas it was just 3.1 (±1.4)% in that from the IFP orchard. Deurer et al. (2009) present the results as the average of three cores taken from each of the orchards. Whereas the mean macropore radius was essentially the same for the cores from both orchards, namely 0.4 mm, the volumetric macroporosity was 7.5 (±2.1)% for the organic cores and 2.4 (±0.5)% for the IFP ones. As well, Deurer et al. (2009) inferred the Euler–Poincaré number for the macroporous networks

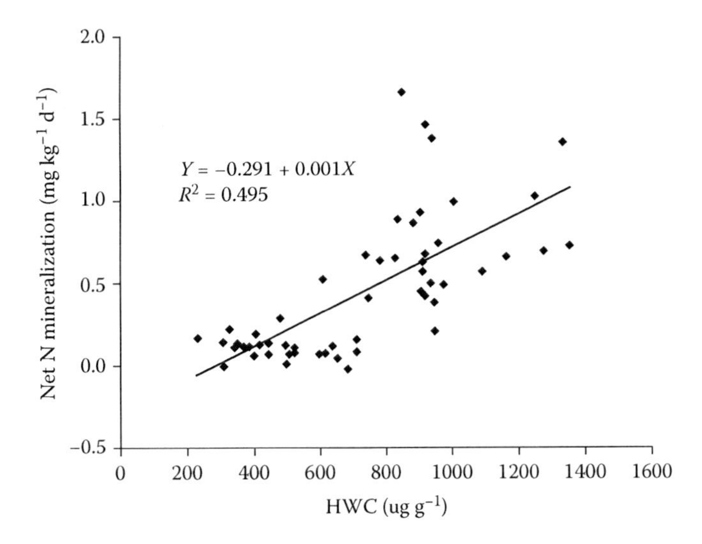

FIGURE 11.4 The correlation between HWC content of the SCS and the ecosystem services of net nitrogen-mineralization. (From Kim, I. et al. 2008. In *Proceedings of the Workshop: Carbon and Nutrient Management in Agriculture*, 67–75, Fertiliser & Lime Research Centre, Massey University, February 13–14, 2008.)

using the computed tomography. The Euler–Poincaré number provides a measure of the connectedness of the macropores, such that the smaller the number, the less connected are the macropores. For the surface soil of the organic orchard, the Euler–Poincaré number was found to be 0.0196 (±0.0048) per cubic millimeter, which was significantly greater than that found for the soil of the IFP orchard at 0.0066 (±0.0029) per cubic millimeter. Thus, the anthropogenic driver of different carbon investment strategies between the two orchards has not only had an impact on the soil's manageable properties of its macroporosity, but also the connectedness of its macroporous network

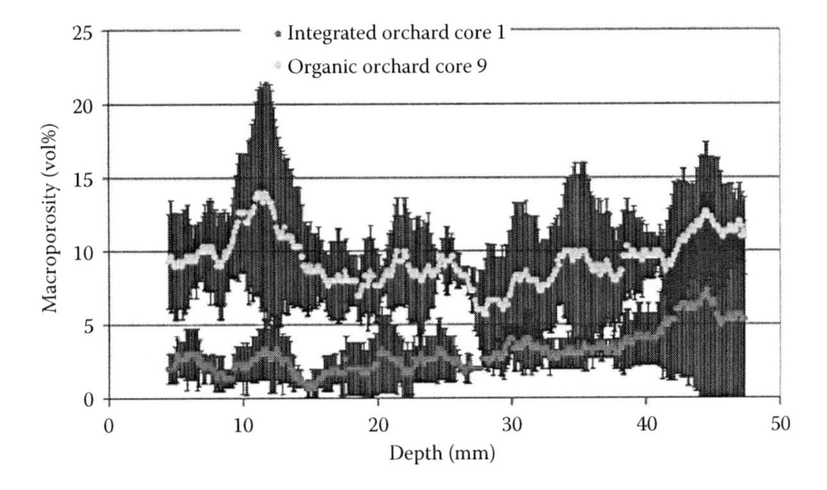

FIGURE 11.5 The macroporosity (pores > 0.3 mm diameter) as a function of depth below the soil surface in the tree row of an organic (core 9) and integrated (core 1) apple–orchard system. On average the volumetric macroporosity of core 1 was 3.1% ± 1.4 and of core 9 it was 9.4% ± 1.7. For each depth, and core, the average and standard deviation of three subcolumns is shown, where the dimension of the subcolumns was 43 × 20 × 17 mm, such that the volume of the subcolumns was 14620 cubic millimeters. (From Clothier, B.E. et al. 2011. In *Sustaining Soil Productivity & Climate Change: Science, Policy and Ethics*, 115–137. Chichester, Wiley-Blackwell.)

(Figure 11.2). These have affected the natural capital value of the two soils through altering their respective supporting processes.

Deurer et al. (2009) reported on other characteristic differences between the two soils. Every month during their study, Deurer et al. (2009) measured the fresh weight of the deep-burrowing anecic worms. The fresh weight of worms was significantly greater in the organic orchard soil, at 154 (±47) grams per square meter, than in the IFP orchard soil where it was 85 (±53) grams per square meter. Therefore, the different orchard practices had led to greater densities of anecic earthworms, which would have led to the greater volume of macropores in the organic orchard (Figure 11.5). But as Deurer et al. (2009) commented, macroporosity is not just about the creation of the pores. They also need to be stable. Deurer et al. (2009) found that the mean weighted diameter of aggregates, a measure of the stability of aggregates to water disruption—a soil property required to maintain macroporosity—was greater in the organic orchard (1.82 mm) than in the IFP orchard (1.33 mm). Here then, orchard practices have affected another manageable property of the soil's natural capital—the size of the aggregates (Figure 11.2). Deurer et al. (2009) also found that the microbial biomass carbon was significantly different between the two soils. It was 143 (±40) grams per square meter in the soil of the organic orchard, and just 73 (±16) grams per square meter for the IFP one. So the soil's supporting processes of biological activity have been affected by the driver of orchard management in relation to carbon investment (Figure 11.2).

A decade earlier these two soils would have been identical, being the same genoform (Droogers and Bouma 1997). However, through different carbon investment practices, they have changed natural capital properties (Figure 11.2) and are now different phenoforms of the same genoform (Droogers and Bouma 1997; Pulleman et al. 2000). The soils' supporting processes are now different (Figure 11.2), as different carbon investment practices have altered their ecological infrastructures so that the value of the contemporary ecosystem services that flow from them is different (Figure 11.1).

11.4 REGULATING SERVICES

As a result of the different ecological infrastructures of these two neighboring apple orchards, there are different regulating services that flow from their soils' natural capital (Figures 11.1, 11.2, and 11.5). The different porous networks of the two phenoforms of the soils from these two orchards now provide different regulating services in terms of gaseous exchange and nutrient filtering.

By using the approach of Vogel et al. (2002), Deurer et al. (2009) simulated gaseous diffusion through the respective porous networks of the two soils from these neighboring apple orchards. Their analysis considered flow by gaseous diffusion between subcolumns along the core each of width 4.3 mm. They calculated the relative apparent diffusion coefficient, D_r, as being the diffusive flow through the porous networks of the subcolumns, relative to free-air diffusion along that same path length. The relative diffusion coefficient at the whole column scale was $D_r = 0.0239$ (±0.0079) for the macroporous organic orchard soil, and just 0.0056 (±0.0009) for the integrated cores. Indeed for the integrated soil, 42% of all segments had a D_r of zero as there was a complete lack of macropore continuity. So the soil of the organic orchard 'breathes' more easily than that in the integrated orchard (*cf.* Figure 11.5). Deurer et al. (2009) concluded that the regulating service for gaseous diffusion in the organic soil would be valuable since the greater aeration provided by that soil's connected macroporosity (Figure 11.5) would indicate less favorable conditions for N_2O production and emission. The results of van der Weerden et al. (2012) confirm this, for they found increased pore continuity reduced the duration of anaerobicity leading to lower emissions. They found that D_r could explain nearly 60% of the variability in their experiments with two soils. Indeed, considering the two regression equations for their two soils and the D_r values above would suggest that N_2O emissions from the organic orchard would be 15–35 times lower than those in the integrated one. Avoiding the generation and emission of such a powerful greenhouse gas as N_2O (Tate and Theng 2014) thus ensures that this organic soil would be providing a valuable regulating service for the atmosphere.

These orchard soils, with their high macroporosity, also provide valuable nutrient regulation services, as described in more detail in Robinson et al. (2013b). From the data of Kim et al. (2008), presented above in Figure 11.4, it can be calculated that about 100 kg of nitrogen per hectare per year is being mineralized each year in the surface soils of these orchards by the conversion of organic nitrogen into ammonium. Green et al. (2010) measured the leaching of nitrogen under apple orchards using fluxmeters, and they found the annual leaching of nitrogen below the rootzone was just 8–14 kg of nitrogen per hectare per year, despite some 700 mm of drainage. So only 8–13% of this endogenously generated nitrogen is lost as leachate, and so only a small fraction of the mineralized nutrient is being dispatched to receiving groundwaters, thereby limiting nonpoint source pollution. The macropores of the soil have performed a valuable regulating service since they have isolated the source of the nitrogen in the soil's matrix by directing bypass and preferential transport of the incident rainfall through the rootzone. Thanks to the macropores, the mineralized nitrogen is contained within the soil's matrix, where the trees' roots are able to access it as a plant nutrient. Macropores perform an important regulating service here by buffering the endogenously mineralized nitrogen in the soil. Clothier et al. (2008) estimated the annual global value of the services provided by macropores in soil to be worth US$304 billion.

The prime motivation in the kiwifruit study of Deurer et al. (2010) (Figure 11.3) was to determine whether perennial fruit crops can maintain or enhance soil carbon storage, so that these results could be incorporated in carbon-footprinting protocols designed for greater differentiation of New Zealand's fruit products in environmentally concerned markets such as Europe. The rate of carbon sequestration was calculated from the data in Figure 11.3 to be about 0.4 tonnes of carbon per hectare per year by considering SCS down to 1 m. Holmes et al. (2012) extended the work of Deurer et al. (2010) by sampling down to 9 m in a 30 year-old kiwifruit orchard, as well as in a neighboring pasture, which would have been the antecedent land use before the orcharding. By considering the difference in the 9 m SCS, they calculated that the kiwifruit vines were sequestering carbon at a rate of 6.3 tonnes of carbon per hectare per year, which is what you would obtain by extrapolating the 1 m results of Deurer et al. (2010) down to 9 m. This rate of carbon sequestration, if accounted for in a life cycle assessment of the carbon footprint of export kiwifruit, would essentially halve the current value of the carbon footprint calculated for a New Zealand kiwifruit that landed on a shelf in a European supermarket. This reinforces the finding of Figure 11.3 that kiwifruit vines are increasing the value of the soil's natural capital by "growing" soil at depth. But furthermore the vines are also engendering a valuable regulating service of significant carbon storage in the soil, thereby lessening the flow of carbon to the atmosphere (Figure 11.2).

11.5 TERROIR VALUE AND CULTURAL SERVICES

The ecosystem services provided by soil, climate, and local landscape, coupled with grape variety and the skills of the viticulturalists and oenologists, confer on wine the valuable concept of terroir. Vineyards cover just 33,600 ha of New Zealand, yet they generate NZ$1.1 billion of export revenue. The NZ$ was worth US$0.86 in March 2013. The provisioning value of the terroir that derives from the ecological infrastructures of vineyards is therefore easily quantified, in sum. Clothier et al. (2011) noted that soil and weather ecosystem services can explain most of the terroir effect. Conradie et al. (2002) and Bodin and Morlat (2006) found that vine water supply, soil depth, and potential vine vigor were the major contributing factors to terroir. Clothier et al. (2011) quantified the role of the supporting processes of soil-water storage and nitrogen mineralization on the natural capital value of terroir, and we review their findings here.

Clothier et al. (2011) developed a natural capital valuation model for terroir that contains four submodels of a plant-growth model based on weather and soil, soil and water management practices, environmental impacts, and economic valuations. They defined a temporal trend from bud break to harvest in both the soil water content of the rootzone and soil nitrogen storage that they considered would confer the maximum value to terroir. Freely available water is considered best right up until flowering to ensure maximum fruit set and optimum nitrogen storage in the roots, vine, shoots, and leaves. After flowering through until veraison—the onset of berry ripening—it is best if there is a reduction in soil water and nitrogen, so as to limit vegetative vigor. Following veraison, lower levels of water and nitrogen are considered ideal to limit vegetative vigor, enhance light penetration to the bunches, and encourage a rise in sugar levels in the grape berries. The economic modeling by Clothier et al. (2011) penalized any deviation from these ideal time courses. On the basis of the value of the provisioning service of wine production, they took the maximum terroir value of terroir to be NZ$25,000 ha^{-1}.

The model could be run to provide irrigation as, and only as, needed. Clothier et al. (2011) first ran their model considering irrigation water to be cost-free, but accounted for the costs of pumping. The value of terroir in this case is shown in Figure 11.6a (top curve) as a function of soil depth. For vineyards on shallow soils, there is a need to irrigate more frequently, and the operational cost of doing so reduces net terroir value. For deeper soils, the larger water-holding capacity and greater provision of nitrogen through mineralization result in greater deviations away from the sought-after decline in water and nitrogen stocks. There are additional costs associated with leaf plucking and pruning. The optimum depth to maximize terroir is around 1 m. Clothier et al. (2011) reran their model with the price for water at NZ$4 per cubic meter, being the cost of domestic water in some cities. The return on investment into the soil's natural capital changes (Figure 11.6a, bottom curve). It is now less because of the additional costs associated with using irrigation to meet water demands over and above those that cannot be supplied by the supporting services from the soil and climate. For the shallowest soil of 200 mm, the inability to store sufficient water and the additional costs of supplying water for the vine's needs mean that terroir has virtually no value. In this case, where water is a valuable commodity, deeper soils will have a much higher natural capital value.

Soil carbon is a manageable property of soil natural capital and it plays a key role in supporting processes and regulating services (Figure 11.2). Soil carbon can be changed through land management practices, as we have discussed above (Figures 11.4 through 11.6). Clothier et al. (2011) compared the terroir value of a genoform with its soil carbon content of 1% (Figure 11.6b, top curve) with that of a phenoform having a soil carbon content of 1.75% (Figure 11.6b, bottom curve).

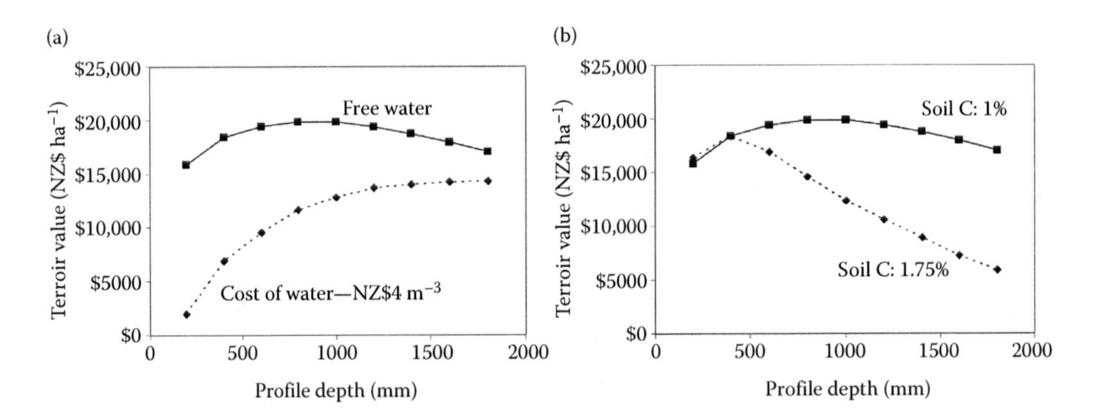

FIGURE 11.6 (a) The pattern of terroir value as a function of soil depth calculated for a viticultural soil in Marlborough, New Zealand. The maximum terroir value was set at NZ$25,000 ha^{-1} and penalties were accrued throughout the season as rootzone conditions deviated from the ideal time courses for water and nitrogen stocks. For the upper curve (solid line) there was no cost for the water, whereas for the lower curve (dotted line) water was priced at NZ$4 m^{-3}. (b) The pattern of terroir value as a function of soil depth calculated for a viticultural soil in Marlborough, New Zealand. For the upper curve the soil had a soil carbon content of 1%, whereas for the lower curve (dotted line) the carbon content of soil was increased to 1.75% and there was enhanced soil water storage and nitrogen mineralization. (From Clothier, B.E. 2011. In *Sustaining Soil Productivity & Climate Change: Science, Policy and Ethics*, 115–137. Chichester Wiley-Blackwell.)

FIGURE 11.7 **(See color insert.)** The ecological infrastructures of the vineyards of Marlborough, New Zealand, also provide valuable cultural services in relation to aesthetics, tourism, recreation, fashion, music, social needs, and cuisine. Each year the Marlborough Wine & Food Festival celebrates the terroir provided by the local viticultural infrastructures.

They altered the water and nitrogen supporting processes of the soil in relation to carbon using the soil-water storage changes as suggested by Rawls et al. (2003), and for nitrogen mineralization by Kim et al. (2008). The increase in the water and nutrient supporting processes provides a marginal increase in the terroir value for very shallow soils, but once the soil depth increases beyond about 500 mm when soil carbon is at 1.75%, the additional water and nitrogen results in vegetative vigor. This is because the soil water does not track down the sought-after declining late-season pattern of soil water content and nitrogen storage because of the soil's enhanced water storage and greater mineralization. For this reason, greater terroir value is associated with soils having high stone contents and low carbon levels, such as New Zealand's Gimblett Gravels. However, the optimum is for a zone of higher soil carbon to just a depth of between 200 and 400 mm (Figure 11.6, right, bottom curve). This would suggest there is terroir value from using shallow organic mulches made from composted prunings or the crushed grape skins left over from winemaking which is called marc. This would, inter alia, provide some enhanced water storage and requisite nitrogen mineralization, without inducing vigor and a subsequent loss in terroir value.

Vineyards, wines, and viticultural landscapes also provide a range of cultural services, ranging from movies such as *Sideways*, through wine tourism, to vineyard festivals. Every year in New Zealand there are a number of cultural festivals held in one, or across many, vineyards (see http://www.marlboroughwinefestival.co.nz/index.htm) and these provide for the cultural services of recreation, fashion, aesthetics, cuisine, and music (Figure 11.7). The soils of vineyards are valuable natural capital stocks.

11.6 LAND PRICE, NATURAL CAPITAL, AND THE LAW

The horticultural industries of New Zealand annually generate NZ$3.5 billion of export revenues and contribute another NZ$1.5 billion to the domestic economy (www.freshfacts.co.nz). This $5 billion industry covers just 70,000 hectares of land, and it is often prime land with high-class and versatile soils on the periphery of cities. This small area of prime natural capital stocks not only provides this provisioning ecosystem service worth NZ$5 billion, but also it provides other valuable supporting, regulating, and cultural services. New Zealand's urban areas and built infrastructures cover nearly 1 million hectares of land, and every year there is a loss of 40,000 hectares of productive lands to peri-urban expansion (Mackay et al. 2011). Being on the periphery, these horticultural lands are in the front line of the battle for peri-urban expansion. For rather than infill within the extant city boundary, housing, industry, and commercial enterprises are seeking to move outwards from the city center. This makes good economic sense, since the price of land rapidly drops away with distance from the central business district (CBD). In the Hawke's Bay region of New Zealand, the so-called 'fruitbowl' of New Zealand since it has 30% of New Zealand horticultural planted area, inner-city land sells for between NZ$4 and NZ$7 million per hectare (Figure 11.8). Yet on the rural fringe which abuts the horticultural lands, the price of property is just NZ$60,000 ha^{-1}. Thus, land price and the natural capital value of soil are strongly related, but sadly in an inverse way!

Robinson et al. (2013b) detail one attempt of peri-urban expansion of a city onto prime horticultural land. They discuss the ensuing judicial processes that sought to protect the valuable land on the rural fringe of the Hawke's Bay City of Hastings. We summarize these here. The hardware retailer Bunnings purchased 4 ha of orchard land on the outskirts of Hastings and sought resource consent to build a large-format big-box store there. This was a smart economic decision given the price differentials highlighted in Figure 11.8. The resource consent was, however, declined.

Bunnings appealed that decision and the Environment Court hearing was held during March 2011. One of us acted as an expert witness for the respondent, the Hastings District Council (HDC) (Clothier 2011). Clothier (2011) argued that "... we cannot afford to lose such valuable natural capital assets, whose presence is needed for their ecosystem services, and whose use will be needed to enable the horticultural industries to realise their strategic goals, and whose functioning will continue to enhance the life-supporting capacities of the Heretaunga Plains", as required by the HDC's District Plan for the Heretaunga Plains.

Clothier (2011) also noted that "several key ecosystem services are provided by the soil of this site: primary production, nutrient cycling, water storage, platform, and water supply regulation" for the vadose zone of unsaturated soil to which the nearby Karamu Stream is linked. He added

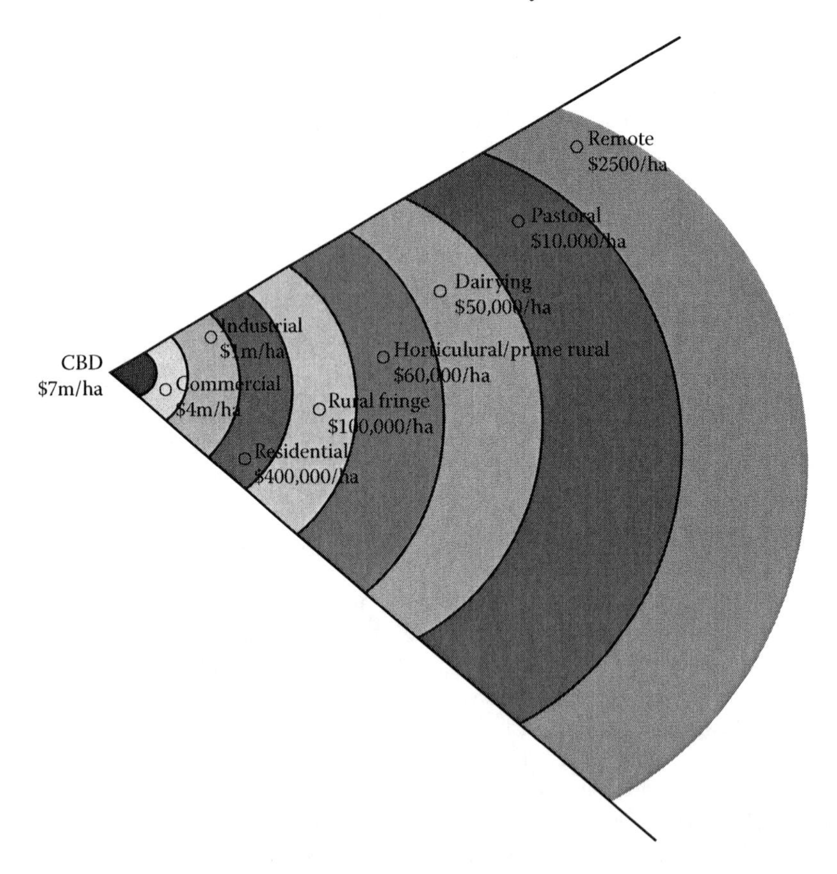

FIGURE 11.8 Reduction in land values (in NZ$ = US$0.0.86) in the Hawke's Bay of New Zealand as distance increases away from the CBD and the land-use changes. (From Mackay, A.D. et al. 2011. *New Zealand Science Review* 68(2): 67–71.)

"that this deep soil has no impeding layers of low conductivity which means that it can provide the ecosystem service of water supply regulation" to the Karamu Stream, which a hard, impermeable surface of the roof of a large-format store and its parking lot could not.

Bunnings' lawyer in his closing address considered that "... there is no quantitative or qualitative analysis of the ecosystem services at the site other than in relation to food production."

The judgment (Dwyer 2011) was cautious and noted that "... we do not propose to enter that [natural capital] debate... but it seemed to us that Dr Clothier took a somewhat more holistic approach to assessment of the value of the soils of the site." The appeal was declined, and costs awarded to the HDC. In his costs decision, Judge Dwyer (Dwyer 2012) stated clearly that "... in reaching [our] decision we emphasised the importance of the District Plan to protect the rural resource." The latter term is, in essence, natural capital. That seems to infer that the rural resource needs to be protected to ensure the continued flow of ecosystem services from it.

Although Judge Dwyer and his two commissioners did not directly buy into a natural capital argument, they did note a holistic view was needed. Holism, it seems, is an ecosystem services approach in principle.

Since we are witnessing a greater use of the ecosystem services approach in policy and regulatory development, it will be increasingly used in judicial proceedings that seek to protect the valuable ecosystem services that flow from the interconnected natural capital stocks of our ecological infrastructures.

11.7 CONCLUSIONS

Natural capital is a stock of natural material or energy. These stocks sum to be the constituent parts of our landscapes—our rivers, aquifers, lakes, and vegetation. The critical link between all of these elements is the soil. These interconnected elements in total form our ecosystems from which flow the valuable, and indispensable, ecosystem services that benefit and sustain mankind.

Soil is a valuable natural capital asset, a critical component of ecological infrastructure. Soil supplies valuable ecosystem services, from the supporting processes that deliver provisioning services which are patent, to the regulating and cultural services which tend to be latent.

Land is a finite resource, and soil formation proceeds only slowly. Thus, we must ensure that the inventory value of the world's soil stocks is not diminished. Indeed, through investment in their ecological infrastructure we can grow their asset value so that we can ensure an enhanced flow of ecosystem services from our soils to meet the demands of a growing population with greater aspirations of prosperity.

Mankind's history in relation to investment into the ecological infrastructure of soil provides salutary lessons. The over exploitation of the soils in the Midwest of the United States in the 1930s was the prime cause that led to the disastrous era of the Dust Bowl with its attendant socioeconomic upheavals. In setting up an investment scheme into the ecological infrastructure of the US soils, Franklin D. Roosevelt said "... the nation that destroys its soil destroys itself." By way of contrast, the soil investment schemes practiced by Asian farmers have enabled them to farm the same land for 40 centuries. F.H. King (1911) noted that all the services flowing from their soils had been sustained for 4000 years, and that the provisioning service was undiminished since "... the yield is the product of brain, brawn and utilised waste." A beacon.

REFERENCES

Bodin, F. and R. Morlat. 2006. Characterisation of viticultural terroirs using a simple field model based on soil depth I: Validation of the water supply regime, phenology and vine vigour, in the Anjou vineyard (France). *Plant and Soil* 281: 37–54.

Bristow, K.L., S.M. Marchant, M. Deurer, and B.E. Clothier. 2010. Soil—A key component of ecological infrastructure. *Proceedings of the 19th World Congress of Soil Science,* Brisbane, Australia, August 1–7, 2010 (on CD).

Clothier, B.E., S.R. Green, and M. Deurer. 2008. Preferential flow and transport in soil: Progress and prognosis. *European Journal of Soil Science* 59: 2–13.

Clothier, B.E., A.J. Hall, M. Deurer, S.R. Green, and A.D. Mackay. 2011. Soil ecosystem services: Sustaining returns on investment into natural capital. In *Sustaining Soil Productivity & Climate Change: Science, Policy and Ethics*, eds. T.J. Sauer et al., 115–137. Chichester, Wiley-Blackwell.

Clothier, B.E. 2011. Statement of Expert Evidence for Hastings District Council. In the matter of Appeal ENV-2009-WLG-0182, Environment Court, Wellington, New Zealand.

Conradie, W. J., V.A. Carey, V. Bonnardot, D. Saayman, and L.H. van Schoot. 2002. Effect of different environmental factors on the performance of sauvignon blanc grapevines in the Stellenboch/Durbanville districts of South Africa. I. Geology, soil, climate, phenology and grape composition. *South African Journal of Enology and Viticulture* 23(2): 78–91.

Costanza, R., R. d'Arge, R. de Groot, S. Farber, M. Grasso, B. Hannon, K. Limburg et al. 1997. The value of the world's ecosystem services and natural capital. *Nature* 387: 253–260.

Daily, G.C. 1997. *Nature's Services: Societal Dependence on Natural Ecosystems*. Island Press, Washington DC.

Daily, G.C., P.A. Matson, and P.M. Vitousek. 1997. Ecosystem services supplied by soils. In *Nature's Services: Societal Dependence on Natural Ecosystems*, ed. G.C. Daily, 113–132. Washington, DC: Island Press.

Deurer, M., D. Grinev, I. Young, B.E. Clothier, and K. Müller. 2009. The impact of soil carbon management on soil macro-pore structure: A comparison of two apple orchard systems in New Zealand. *European Journal of Soil Science* 60: 945–955.

Deurer, M., H. Rahman, A. Holmes, S. Saunders, B. Clothier, and A. Mowat 2010. Quantifying soil carbon sequestration in kiwifruit orchards. Development of a sampling strategy. In *Proceedings of the Workshop: Farming's Future: Minimising footprints and maximising margins*, 445–459. L. Currie and C.L. Christiansen (Eds), Fertiliser & Lime Research Centre, Massey University, February 10–11, 2010.

Dominati, E., M. Patterson, and A. Mackay, 2010a. A draft framework for classifying and measuring soil nature capital the ecosystem services. *Ecological Economics* 69: 1858–1868.

Dominati, E., M. Patterson, and A. Mackay 2010b. Response to Robinson and Lebron—Learning from complementary approaches to soil natural capital and ecosystem services. *Ecological Economics* 70: 139–140.

Droogers, P. and J. Bouma. 1997. Soil survey input in exploratory modelling of sustainable soil management practices. *Soil Science Society America Journal* 61: 1704–1710.

Dwyer, B.P. 2011. Decision. In the matter of Appeal ENV-2009-WLG-0182, Environment Court, Wellington, New Zealand.

Dwyer, B.P. 2012. Costs Decision. In the matter of Appeal ENV-2009-WLG-0182, Environment Court, Wellington, New Zealand.

Green, S., M. Deurer, B. Clothier, S. Andrews, K. Cauldwell, A. Roberts, M. Wellwood, and P. Thomas 2010. Water and nitrogen movement under agricultural and horticultural land. In *Proceedings of the Workshop: Farming's Future: Minimising footprints and maximising margins*, 79–88, L. Currie and C.L. Christiansen (Eds), Fertiliser & Lime Research Centre, Massey University, February 10–11, 2010.

Haygarth, P.M. and K. Ritz. 2010. The future of soils and land use in the UK: Soil systems for the provision of land-based ecosystem services. 2010. *Land Use Policy* 265: S187–S197.

Holmes, A., K. Müller, and B.E. Clothier. 2012. Carbon storage in kiwifruit orchards to mitigate and adapt to climate change. Final Report SFF C09/20 August 2012, Ministry of Primary Industries, Wellington, 72 p.

Jury, W.A., D. Or, Y. Pachepsky, H. Vereecken, J.W. Hopmans, L.R. Ahuja, B.E. Clothier et al. 2011. Kirkham's legacy and contemporary challenges in soil physics research. *Soil Science Society of America Journal* 75: 1589–1601, doi:10.2136/sssaj2011.0115.

Kim, I., M. Deurer, S. Sivakumaran, K.Y. Huh, S. Green, and B. Clothier. 2008. N-mineralisation in two apple orchards in Hawke's Bay: The impact of soil carbon management and environmental conditions. In *Proceedings of the Workshop: Carbon and Nutrient Management in Agriculture*, 67–75, L. Currie and L.J. Yates (Eds), Fertiliser & Lime Research Centre, Massey University, February 13–14, 2008.

King, F.H. 1911. *Farmers of Forty Centuries*. Rodale Press, Inc., Emmaus, Pennsylvania. 441 p.

Kong A.Y.Y. and J. Six. 2010. Tracing root vs. residue carbon into soils from conventional and alternative cropping systems. *Soil Science Society America Journal* 74: 1201–1210.

Kubiszewski, I. and R. Costanza. 2012. Ecosystem services for sustainable prosperity. In *State of the World 2012*, ed. L. Starke, 177–182. Washington, D.C.: Island Press, Center for Resource Economics.

Landa, E.R. 2014. Potash, passion, and a president: Early 20th century debates on soil fertility in the United States. In *The Soil Underfoot*, eds. G.J. Churchman, and E.R. Landa. 269–288, Boca Raton: CRC Press.

Mackay, A.D., S. Stokes, M. Penrose, B. Clothier, S.L. Goldson, and J.S. Rowarth. 2011. Land: Competition for future use. *New Zealand Science Review* 68(2): 67–71.

Millennium Ecosystem Assessment. 2005. Living Beyond Our Means: Natural Assets and Human Well-Being. A Statement from the Board, pp. 28 (http://www.maweb.org/documents/document.429.aspx.pdf, (accessed 18 November, 2012).

Pulleman, M.M., J. Bouma, E.A. van Essen, and E.W. Meijles. 2000. Soil organic matter content as a function of different land use history. *Soil Science Society America Journal* 64: 689–694.

Rawls, W.J., Y.A. Pachepsky, J.C. Ritchie, T.M. Sobecki, and H. Bloodworth. 2003. Effect of soil organic matter on soil water retention. *Geoderma* 116: 61–76.

Robinson, D.A., I. Lebron, and H. Vereecken. 2009. On the definition of the natural capital of soils: A framework for description, evaluation, and monitoring *Soil Science Society of America Journal* 73(6): 1904–1911.

Robinson, D.A., N. Hockley, D.M. Cooper, B.A. Emmett, A.M. Keith, I. Lebron, B. Reynolds et al. 2013a. Natural capital and ecosystem services: Developing an appropriate soils framework as a basis for valuation. *Soil Biology and Biochemistry* 57: 1023–1033.

Robinson, D.A., B.M. Jackson, B.E. Clothier, E.J. Dominati, S.C. Marchant, D.M. Cooper, and K.L. Bristow. 2013b. Advances in soil ecosystem service concepts, models and applications for Earth System Life Support. *Vadose Zone Journal,* 12(4), doi:10.2136/vzj2013.01.0027.

Tanner, C.B. and R.W. Simonson. 1993. Franklin Hiram King—Pioneer Scientist. *Soil Science Society America Journal* 57: 286–292.

Tate, K.R. and B.K.G. Theng. 2014. Climate change—An underfoot perspective? In *The Soil Underfoot*, eds. G.J. Churchman and E.R. Landa. 3–16, Boca Raton: CRC Press.

Van der Weerden, T.J., F.M. Kelliher, and C.A.M. de Klein. 2012. Influence of pore size distribution and soil water content on nitrous oxide emissions. *Soil Research* 50: 125–135.

Vogel, H.-J., J. Cousin, and K. Roth. 2002. Quantification of pore structure and gas diffusion as a function of scale. *European Journal of Soil Science* 53: 465–473.

Section III

Culture and History

"Soils have their own histories, both natural and human." J.R. McNeill and Verena Winiwarter (From Soils, soil knowledge and environmental history: An introduction, 1–6, p. 3) in *Soils and Societies: Perspectives from environmental history* ed. J.R. McNeill and V. Winiwarter. Isle of Harris, UK: The White Horse Press (2010).

12 Bread and Soil in Ancient Rome

A Vision of Abundance and an Ideal of Order Based on Wheat, Grapes, and Olives

Bruce R. James, Winfried E. H. Blum, and Carmelo Dazzi*

CONTENTS

12.1 INTRODUCTION

In antiquity, food was power.

—Peter Garnsey (1999)

Throughout Roman history land remained the major, and indeed the only respectable, form of investment. The entire administrative structure of the Empire rested on the foundation of an agricultural surplus.

—K. D. White (1970)

. . .If the theory of the insidious decline of the soil as the result of human cultivation is correct, would humanity not have met its demise a long time ago? Is humanity's survival for thousands of years and its enormous growth in numbers since the invention of agriculture not proof enough that there must be elements of sustainability not accounted for by this theory?

—Joachim Radkau (2008)

* Department of Environmental Science and Technology, University of Maryland, College Park, Maryland, USA; Email: brjames@umd.edu

All members of soil communities are conditioned, within the limits of adaptation to them, by their soil...something mortal happens to the spirit of civilizations when their relationship with soil becomes one of exploitation.

—Edward Hyams (1976)

Thus far the tilth of fields and stars of heaven;
Now will I sing thee, Bacchus, and, with thee,
The forest's young plantations and the fruit
Of slow-maturing olive...
Earth of herself, with hooked fang laid bare,
Yields moisture for the plants, and heavy fruit,
The ploughshare aiding; therewithal thou'lt rear
The olive's fatness well-beloved of Peace.

—Virgil's *Georgics* (II)

The words of the environmental historians Peter Garnsey, K. D. White, Edward Hyams, and Joachim Radkau speak cogently to the underlying, mysterious roles of soil in the sustainability, resilience, and continuity of human civilizations based on agriculture, in particular, that of ancient Rome. Those of Virgil complement the modern historians and speak to us of the ancient reverence for the bounty of the soil, particularly the olive as a symbol of peace. The soils of the Mediterranean Basin were a source of political power, economic wealth, and surplus food; all of which were essential for the growth and evolution of the Roman civilization over millennia.

A common refrain written by many over centuries since 500 CE is that the ancient Romans misused, exploited, or abused soils on erodible landscapes of fragile Mediterranean ecosystems. As a result, the Roman Empire as a "monster state" and "vicious and ridiculous system" (Hyams 1976) ultimately fell, and the civilization of the western Mediterranean region collapsed in the fifth century CE into a society controlled by barbarians (Tainter 1988). A "ruined landscape" remains today with low soil fertility, eroded soils, and little native vegetation characteristic of the bioregion (Montgomery 2007; Ibáñez et al. 2013). Have we overlooked a narrative of sustainability in Roman times by focusing too much on the oft-told, historical drama of the fifth century CE? What can we learn from events in earlier centuries of Roman history that speak to the rise and resilience of their culture, and not just their presumed demise based on soil exploitation?

A close reading of the environmental and agricultural histories of ancient Rome, linked to modern soil science and interpreted with new theories of the dynamics of ecosystem disturbance and recovery, leads to a more heuristic narrative of cultural and ecological change in Roman history. This is particularly germane to our understanding of the dramatic transition period of approximately 240 years of Roman Antiquity starting with the onset of the Punic Wars in 264 BCE. This time of cataclysmic social change encompassed the years of the Crisis of the Republic (133–44 BCE), including civil wars and the assassination of Julius Caesar (44 BCE). It ultimately set the stage for the two centuries of *Pax Romana* (from ~27 BCE) that gave birth to the nascent Roman Empire under the first emperor, Caesar Augustus, who ruled from 27 BCE to 14 CE (Gibbon 1776; Baker 2006).

We explore this particular period of Roman history as a time of dynamic cultural evolution based on a mosaic of resilient ecosystems supported by soils that were used and managed for agriculture. We focus on the peninsula of Italy (Figure 12.1), the island of Sicily, and the Mediterranean coastal zones of North Africa where wheat for bread, grapes for wine, and olives for oil were produced. This unusual triad of foods that was and still is the core of the Mediterranean diet (Estruch et al. 2013) based on the seeds of a non-native, annual grass and the fruits of two native, woody perennials was produced on a challenging landscape that required considerable creativity to provide enough food to feed burgeoning metropolises, especially Rome, and a mighty military machine spread over a vast geographic region. Land stewardship practices varied regionally and changed temporally from small farms owned and worked by peasants to huge estate farms

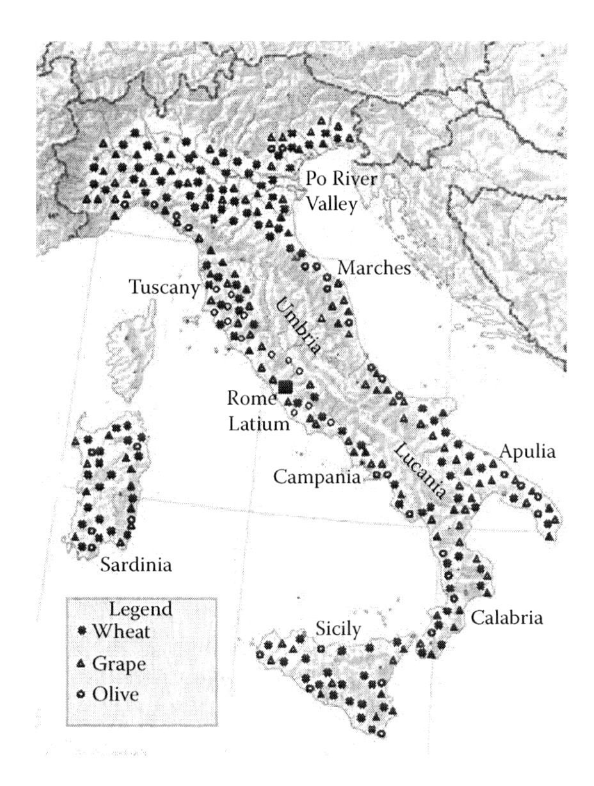

FIGURE 12.1 Map of the Italian peninsula with named political regions and cropping areas for wheat, grapes, and olives.

(*latifundia*) controlled by absentee owners and supported by an abusive, but profitable, system of slave labor (Weber 1998).

Did this evolving system of Roman land use and food production from 264 to 27 BCE create a vulnerable civilization that sowed the seeds of its own destruction centuries later, or did it comprise creative responses to challenges? Is the tale of the exploitation of soil by humans in ancient Rome a prophecy of Cassandra calling us to beware and learn from the missteps of history? Or, is she exhorting us to learn from the Roman's new, effective human–soil relationships and to develop land stewardship practices that sustain our agriculturally based society? The answers are nuanced and manifold, and they provide intellectually challenging ways to view this seminal period of the environmental, cultural, and soil history of the Western world.

12.2 TESTING A SOIL-BASED HYPOTHESIS FRAMED BY AN ETHOS OF CIVILIZATION AND THE ECOLOGY OF DISTURBANCE

We hypothesize that the soils of ancient Rome and its colonies, and the way they were managed for food production over the centuries of transition from the Republic to *Pax Romana*, fostered sustainability in the sense of enabling enhanced ecosystem stability, cultural resilience, and economic vitality (Brundtland 1987). Their agricultural production systems were ideally suited to the sloping land and erodible soils of the Mediterranean region with hot, dry summers and cool, moist winters. We further hypothesized that the simple diet of the Romans, principally based on bread from leavened flour of wheat, wine from fermented juice of grapes, and oil from pressed olives, was a key to their success. It sustained all social classes during the eventful years of cultural transition from the Late Roman Republic to the beginning of the Empire. Wine, wheat, and olive oil all can be stored and transported without spoilage, and these qualities were essential for their trade and consumption in ancient Rome.

This two-part hypothesis based on Mediterranean soils and the diet of the ancient Romans can be further elucidated by examining modern historical and cultural tenets underpinning the concept of human civilizations. Filipe Fernández-Armesto (2001) describes a "civilizing tradition" for human culture as one based on long-term continuity and an ethos that comprises city life maintained by an ideal of physical and behavioral order, agriculture grounded by a vision of abundance, and writing emerging as a synthesis of symbolic imagination and abstract thought.

The historian, Arnold Toynbee (1946) provides a complementary, dynamic notion of civilizations. He posits that they arise from natural environments that challenge people, but not too severely, and only when the people are ready to respond to such challenges through lifestyle change and habitat modification. Furthermore, he argues that civilizations decline when the people lose their creativity of response to challenges via a lack of innovative leadership and the breakdown of unity of society as a whole.

These theoretical, historical models underlying the development and decline of civilizations as the ultimate human, complex societies based on food production have remarkable parallels with new ecological thinking about the stability and resilience of natural ecosystems of which humans are a part (Pickett et al. 1988; Reice 1994; Paine et al. 1998; Scoones 1999; Collins et al. 2000). This new thinking has superseded older, traditional ideas related to the diversity and stability of climax ecological communities as the ultimate biomes in a region determined by physical, biological, and edaphic conditions, and emerging from less stable and less diverse pioneer ecological communities (Odum 1969). Experimental work on change and disturbance in ecology has led to the following observations and conclusions about how ecosystems actually remain stable and resilient in time:

1. Ecosystem stability is a function of constant recovery from disturbances of intermediate severity, rather than being based on an unchanging climax community of a region.
2. "Preserving" undisturbed climax communities may be counterproductive for the maintenance of biodiversity and the continuance of a stable ecosystem, a counterintuitive idea based on observations of disturbed ecosystems and experimental work on biodiversity.
3. Spatial and temporal patchiness create opportunities for resilience and recolonization, particularly through regrowth of organisms that survive the disturbance and those that migrate into patches, or are recruited from outside the system, due to the changed environmental conditions within patches.
4. The absence of predictable disturbances may have greater impacts than their occurrence, reflecting the idea that populations and ecological communities become adapted to regular disturbances, such as soil plowing, nutrient additions, snow accumulation, or seasonal light and temperature changes (Reice 1994).

These tenets of the disturbance–recovery dynamics of ecosystems suggest that the continuity and ultimate stability of Roman civilization during the tumultuous transition from Republic to Empire were maintained by regular disturbances of soils and land of the Mediterranean biome used for agriculture, horticulture, and viticulture. The disturbances were intermediate in severity and predictable in time and space, and they created agricultural and urban "patches" on the landscape where human land management emulated, to a degree, natural ecological plant communities and seasonal variability of the Mediterranean vegetative communities. These are dominated by evergreen, scrub oak communities known as *macchia* in Italian, *maquis* in French, and *chaparral* in English, derived from the Spanish word for the scrub oak, *chaparro*. Furthermore, creative responses in soil use led to a system of physical and behavioral order based on cities, large estate farms, slavery, and a crop surplus. The social changes that took place in these ecosystems in Roman Antiquity continue to have significant impacts on the political, cultural, and economic developments of countries and human societies around the world.

12.3 MYSTERY OF MEDITERRANEAN SOILS UNDERFOOT: ENDURING AND CHANGING ROLES AS A KEY NATURAL BODY AND RESOURCE FOR ROMAN CIVILIZATION

What is the "soil underfoot," and what do we know about it as a natural body and resource underpinning ancient Roman civilization for millennia in the Mediterranean region? Answers to this question emerge most clearly when the perspectives on the nature of soil, as envisioned by soil scientists, are complemented by those from the fields of history (comprising environmental history, anthropology, and archaeology) and ecology. Taken together, these perspectives of the soil, particularly in the Mediterranean region, provide a heuristic way of exploring the mystery of the soil's role in the continuity and cyclical nature of Roman civilization during our period of study.

Richmond Bartlett, late professor of soil chemistry at the University of Vermont in the United States, used to describe soils as organized natural bodies on the landscape in a "remarkably stable state of nonequilibrium," as predicted by thermodynamics but governed by kinetics, and maintained in a metastable condition by living and nonliving processes, including inputs and outputs of gases, solutes, solids, and energy (R. J. Bartlett, personal communications, 1976–2005). The development and very existence of soil as the uppermost, thin skin of the earth's crust (from several centimeters to many meters deep; depending on the combined effects of climate, topography, organisms, time, and parent material as soil forming factors) is evidence of a lack of chemical equilibrium through the maintenance of the low entropy in its biotic and abiotic systems (Schrödinger 1945; Smeck et al. 1983). Yet, remarkably, soil is a robust, steady-state system that has supported human cultures for millennia since we emerged from African forests onto the savanna as hunting and gathering, bipedal hominids. Dynamic processes govern soil change and stability on time scales from nanoseconds to millennia, and on spatial scales from the Ångstrom level of molecules to the global scale of thousands of kilometers. They have endured dynamically and resiliently; acting as sinks and sources of nutrients, carbon, pollutants, and wastes; and as essential sources of food, feed, and fiber for humans. As a result, soils are the basis for sustaining human societies and the environment (Blum 2005).

There is cultural evidence regarding the role of soil throughout history, as captured in the customs, folklore, and traditions of various populations in several countries. In ancient societies, the soil always had a privileged status by virtue of its fundamental role in providing foodstuffs. Indeed, it shaped the lifestyle and ways of thinking in these societies, and led to the emergence of the modern subdiscipline of soil science, ethnopedology, which explores the perceptions and practices of ancient peoples and their descendants in how they use, manage, protect, and revere the soil. Recognition of these functions emerges not only from the many proverbs found throughout the world in folklore referring to the soil as a source of life and wealth, but also from the etymological ties between "soil" and "man." In ancient Hebrew, *adamat*, the word for soil, comes from the same root as Adam, the first man. In Latin, *humus*, one of the main soil constituents, shares the same root with *homo*, for man (Hillel 1991).

Walter Clay Lowdermilk, an American soil conservationist, traveled the Mediterranean region and China in the 1930s after the Dust Bowl years in the Great Plains of the United States to learn about soil erosion and its role in the persistence and decline of human civilization. He wrote in *Conquest of the Land through 7,000 Years* (1953), his "Eleventh Commandment," exhorting humans to take care of soils as key natural resources:

> Thou shalt safeguard thy fields from soil erosion, thy living waters from drying up, thy forests from desolation, and protect thy hills from overgrazing by thy herds, that thy descendants may have abundance forever.

He identified sloping land and soil erosion in a Mediterranean climate with dry summers and wet winters as key challenges for the establishment of enduring culture and a permanent agriculture, and wrote further:

> I have laid special emphasis on saving the physical body of soil resources rather than their fertility. Maintaining fertility falls properly to the farmer himself. Conserving the physical integrity of the soil resource falls to the Nation as well as to the farmer and landowner, in order to save the people's heritage and safeguard the national welfare...If the soil is destroyed, then our liberty of choice and action is gone, condemning this and future generations to needless privations and dangers.

He exhorted late twentieth-century humans to avoid the catastrophes of history; to preserve and conserve the remarkably stable physical state of soils created over millennia.

Archaeologists (Butzer 1996, 2005, 2012), agricultural and landscape historians (White 1970; McNeill 1992), and plant biogeographers (Grove and Rackham 2003) with expertise in the Mediterranean region provide new data, information, and narratives on soil, plant communities, and land use changes of the past. Their perspectives enhance those of soil chemists, ethnopedologists, and soil conservationists. Butzer (1996) describes the Mediterranean system of soil and land management as sustainable over a period of 5–10 millennia, and that complex systems of land management allowed urban societies to flourish for 4000 years. White (1970) documents the extensive knowledge that the ancient Romans had for assessing soil quality for wheat, grape, and olive production, along with advice on animal husbandry and pastoralism. Saltini (1984) and Sirago (1995) provide further insight into the agronomic practices of the Romans.

McNeill (1992) describes changes in the land in five mountainous regions of the Mediterranean, and he supports the concept that sustainable soil and land use could only persist in a narrow range of human population density consisting of farmers and pastoralists who worked the land and were supported by its bounty. Large populations led to poor soil management, excessive use, and increased soil erosion, while underpopulation led to a lack of attention to land terracing, erosion control, and soil fertility management. He makes the point that mountainous soils and land in the Mediterranean exist in a tenuous balance in an environment where it is either too wet or too dry for agriculture and human settlements, and that the key variable for human survival and impact is vegetation. The complementarity of soils and vegetation on sloping uplands used for pastoralism and relatively flat lowlands used for the triad of grapes, olives, and wheat was key for the persistence of Roman cultures for centuries (Figures 12.1 and 12.2).

Grove and Rackham (2003) refer to and disagree with a "Ruined Landscape Hypothesis" for the Mediterranean in Antiquity amid concerns about modern climate change and desertification. The extant vegetative *maquis* community (often growing on siliceous, acid soils) and *garrigue* or *phrygana* (short, shrub plants on calcium-rich soils with higher pH than those of *maquis*) is not degraded, according to their biogeographical assessment, but is natural, in light of soil conditions and climate of the region. They take issue with the value-laden notion of "land degradation" of the region, and describe ways that Mediterranean plant communities are adapted to fire, drought, seasonal temperature fluctuations, and ways of recovering and being resilient.

Radkau (2008) writes that "an impartial environmental history does not recount how humanity has violated pure nature; rather, it recounts the processes of organization, self-organization, and decay in hybrid human-nature combinations... The solutions to environmental problems are often hidden within social and cultural history, and it is there that we must first decipher them." This insight is germane to the ancient Roman cultural perceptions of the soil. He claims and cites the work of Grove and Rackham (2003) in support of the idea that the soils and land of the Mediterranean were not degraded and deforested in ancient times, but were so changed in the last 200 years by human development pressures and abuse. The need to build sailing ships in Roman times, especially during and after the first Punic War (264–241 BCE), consumed much timber from these landscapes, but did not exceed the renewable rate of tree regrowth. In contrast, the coming of steam-powered, iron-clad ships in the nineteenth century resulted in the consumption of vast amounts of wood and charcoal before the use of coal and oil for fuel in steam engines. Free-roaming goats in the age of steamships then could denude the forested landscape since there was little economic incentive to protect the trees that were once managed carefully for the building of

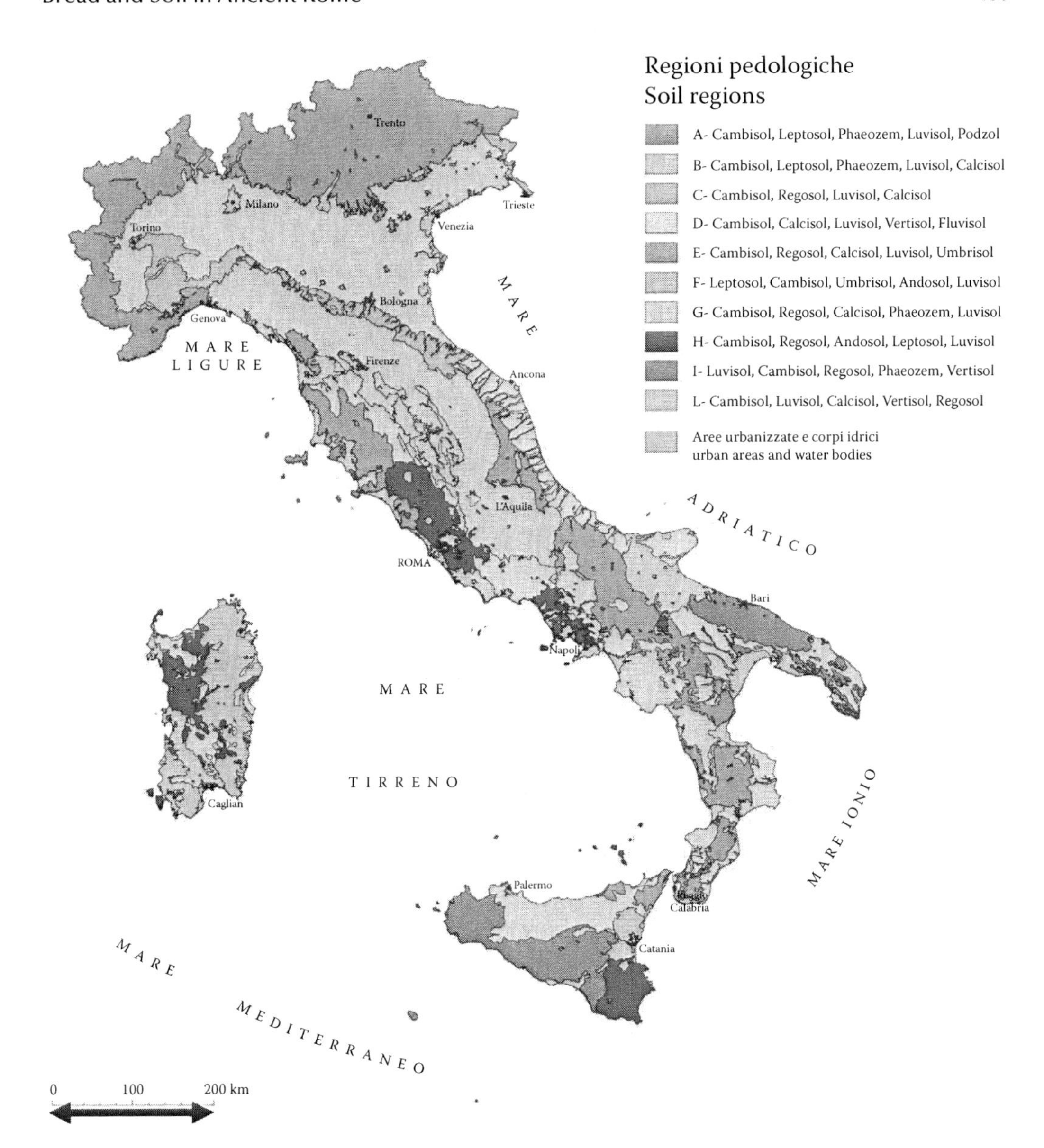

FIGURE 12.2 **(See color insert.)** Soil map of Italy (Costantini et al. 2012). Soils are classified according to the World Reference Base (IUSS Working Group WRB 2006). The combination of different soils inside each area indicates the complexity and the spatial variability of Italian soils.

wooden ships (McNeill 1992; Radkau 2008). Is this, then, how the Mediterranean landscape came to look as it does today, or is there a longer historical picture that shows Roman misuse of the land?

Butzer (2003, 2005) provides the apposite insight for understanding land–soil–human interactions over long periods of history, "that historical ecology is never self-evident, that causation is elusive, and that human perceptions and behavior are integral to an understanding of 'change'." He analyzes and critiques the divergent viewpoints on "ruined landscapes," soil degradation, and cultural change—three phenomena often assumed to be in a cause–effect relationship, simply because there is correlation among them in time and space. Butzer (2005) emphasizes that historical disturbances of land in the Mediterranean region did not necessarily lead to soil degradation as grapes and olives were introduced as important crops in the region beginning in the Early and Middle

Bronze Ages (2600–1600 BCE). Instead, he suggests that major soil erosion periods coincided with the arrival in these regions of foreign inhabitants unaccustomed to living in a biome with a significant summer drought and used for olive and grape production on steep slopes. Based on empirical studies in the region, he concludes that a "long-term ecological transformation and periodic disequilibrium" resulted from human land use decisions and experience. In this way, humans had significant influences on the Mediterranean landscape, but he disagrees with the conclusions of Grove and Rackham (2003) that humans had only minimal effects on the soils and plant communities.

12.4 WARS, WEALTH, AND LAND USE CHANGE LEADING TO AN ENDURING *PAX ROMANA*

The essential human perceptions and behavior of the ancient Romans with regard to their land and soils were studied and interpreted in the late nineteenth century by Max Weber, the father of modern sociology, writing on the agrarian sociology of ancient civilizations (English translation: Weber 1998). Further insight has been provided by his biographer, environmental historian, Joachim Radkau (2008). The perspectives of sociology and political economy from Weber, coupled to the ethical and historical insights of Radkau, clarify these more empirical perspectives on ancient Roman soils and landscapes by soil scientists, biogeographers, and archaeologists.

Weber (1998) traces the evolution of wealth from maritime trade to land ownership on the Italian peninsula in the early Roman Republic years (from approximately 500–30 BCE). Land ownership at this time replaced wealth from seafaring as the basis of the military and financial power of the Roman state, and remained so for centuries. Voting rights and citizenship were given only to landowners, principally peasants in the early Republican years. Land ownership also subjected the owners and their sons to provide military service when called, and as a result, small farms were considered "nurseries" for raising soldiers who were healthier and stronger than men from the city where less physical labor was performed and human disease was rampant (Dale and Carter 1955; Scheidel 2003).

Conquering land in military campaigns in Italy and overseas became a key incentive to serve willingly in the army and to sustain a fighting spirit in battle. Newly acquired land was divided up and awarded as booty among peasants serving as foot soldiers in the army, their city-based officers, and aristocrats in power in Rome. As a result, land ownership and military service became mutually reinforcing throughout the time of extensive Roman colonization during the Republican years. This led to land ownership patterns and laws based on individual rights and duties, rather than being oriented to the village and grazing commons as units of social organization and land management, as was the case in the *poleis* of Greece (Weber 1998). These developments were especially dramatic during and after the three Punic Wars between Rome and Carthage (264–146 BCE).

In 264 BCE, the peninsula-based, Roman legions marched into the First Punic War with the thalassocratic (from Greek, rule of the sea), coastal city of Carthage in North Africa. The war began over disputes surrounding the island of Sicily between Africa and Italy. Rome emerged as the controlling, maritime, political power in the entire Mediterranean basin 118 years later, at the end of Third Punic War. Sicily was only its first of many overseas colonies, and it became a bountiful Roman breadbasket at the end of the First Punic War. Following the Third Punic War, a razed Carthage ultimately was rebuilt by the Romans as a key source of grain, olive oil, and wine for Rome and its military juggernaut (Baker 2006). The lands of the growing list of colonies (Salmon 1970) became essential for wheat production, the main food source for Roman citizens and military forces.

Rome's success in the First Punic War resulted from the rapid development and expansion of a navy and new military tactics that allowed ship-borne Roman soldiers to attack and board Carthaginian ships. The navy protected Italian lands and maritime trade routes, and it invaded foreign lands in the Mediterranean region. With silver flowing into Rome from defeated Carthage as war indemnity beginning in 241 BCE, and with foodstuffs from two other conquered islands, Sardinia and Corsica, Rome began the transition from small farms owned and worked by peasant families on the Italian peninsula to the *latifundia* dependent on slavery. This transition occurred first in southern Italy, Sicily, Sardinia, and Corsica, and then in the northernmost, coastal region of North Africa with a Mediterranean climate and soils. The *latifundia* emerged from the confiscation and consolidation (legal and illegal) of the small holdings previously owned and farmed by peasant families (agronomic enterprises that may have become unprofitable or fallen into disuse during the wars when the young men were conscripted). The *latifundia* were also established by distribution of conquered lands, described above, as a much coveted prize for military service and as a key investment of the patrician class in Rome. They were often parceled out in units of 2–200 *iugera* (approximately 0.5–50 ha), depending on the new owner's years of military service or status in the Roman elite.

Human–soil–land relations changed drastically as a result of the establishment of the *latifundia*. The new owners typically lived in the city, and a steady, seemingly endless supply of slaves worked the land and grew the crops. Slaves were another one of the prizes of military conquest by Rome, and hundreds of thousands of them were captured, sold, and bought to sustain the new, large-scale farms producing essential foodstuffs (Weber 1998). Cheap labor was a key resource for the profitable production of wheat, wine, and olive oil. Small farms as communally owned land were at a disadvantage compared to the large estate farms (Weber 1998). The *latifundia* and the growing urban centers were mutually supportive in that the large farms could supply the necessary food surplus cheaply, and their efficiency drove small farmers out of business and into the growing ranks of the poor *plebeian* (landless) class in the cities, thereby increasing the demand for food from the *latifundia* (Morley 2001). The use of slave labor on the *latifundia* led to unsuccessful insurrections by slaves in three Servile Wars in southern Italy and on Sicily during the years 135–132, 104–100, and 73–71 BCE. The most famous was the Third Servile War led by the slave and former gladiator, Spartacus, when more than 120,000 slaves revolted. These occurred as olive and grape productions were becoming dominant on the *latifundia*, and were centrally important for tax revenue for the *fiscus* (central treasury) in Rome.

There were two unexpected outcomes of the conquest of Carthage by Rome in the Punic Wars. The first was that the Romans realized the value and importance of the fertile soils of the Carthage region and the Bagradus River valley (now known as the Medjerda, the only river that flows year-round in what is now the northern part of Tunisia). The second outcome was that the Romans learned indirectly the value of plantation-type farming that had been started by the Phoenicians, who established Carthage as a trading port and agricultural region during the 1st millennium BCE (White 1970). After the Second Punic War and the destruction of the Italian countryside by the Carthaginian general Hannibal, this type of intensive farming became particularly important for the food supply in Rome, as described above.

The Bagradus River valley is an extensive region of flat, fertile land suitable for grain, grape, and olive production, and it was to this region that Rome sent early *coloni* (farmers or sharecroppers) from whom *conductores* (overseers) collected wheat, taxes-in-kind, and rent paid by these small holders to the *fiscus* in Rome (Kehoe 1988). This system morphed into the *latifundia* from which the annual harvest (*annona*) of crops and payments was derived. In addition to the Mediterranean triad of crops, others included barley (*Hordeum vulgare*) as an alternative to wheat (*Triticum* sp.), beans (*Phaseolus* sp.) as a good protein source, and tree fruits, such as figs (*Ficus carica*). Honey was an additional agricultural product from this region.

This is a region of North Africa receiving only 500–700 mm of rain annually today, and it perhaps received 100 mm less in Roman times (Kehoe 1988). Despite this, the Mediterranean climate and soil conditions made it a prime area for wheat, wine, and olive oil production. The latter was sold all over the western Mediterranean region, and wheat was shipped to Rome in large quantities, perhaps as much as 50,000 tonnes per year soon after the time of Julius Caesar, who laid the plans for rebuilding Carthage in 49–44 BCE before his assassination. The plans were executed by Caesar Augustus in the first century CE (Lendering 2013; World Heritage Convention 2013). Roman Carthage, named *Colonia Julia Carthago* by Caesar Augustus, became a key port from which these huge quantities of grain were shipped. By the first and second centuries CE, the new Carthage was the second largest city in the Mediterranean region with a population of perhaps 500,000 people, and may have supplied as much as two-thirds of Rome's wheat at the time of the *Pax Romana* in 27 BCE.

The demands of the Roman *fiscus* for tax revenue and rents put intense pressure on the small farms and *latifundia* established in this river valley. A large *fundus* of 1600 ha in the region might have had within it as many as 12 decentralized small farms, and the combination was important for a resilient grain production system in the region. Kehoe (1988) estimated that small farmers pressed into the service of the *fundus* increased by 60% the area of land under cultivation. The farmers also eliminated the traditional biennial fallow year and began to cultivate pastureland to meet the grain and tax demands of Rome. In this way, the short-term grain and fiscal demands of Rome overshadowed traditional, long-term, land management practices used by the non-Roman, native peoples of the region.

With complete control of the Mediterranean region by 167 BCE and the elimination of Carthage in 146, Rome plunged into the Crisis of the Republic (Baker 2006). Ironically, it was also a time of massive amounts of war treasure flowing into Rome from Carthage, Corinth, Sardinia, and Corsica. The water supply of Rome doubled, industry flourished, and incomes were rising fast, albeit unevenly (Baker 2006). There was much inequality in wealth and living conditions, and only the aristocrats benefitted from the surge of new wealth from the conquests.

Key contentious issues between rich and poor Romans at the time were land ownership and military service (Baker 2006)—issues that were linked, as detailed above. The displacement of peasants from their land resulted in intense resentment of a burgeoning class of urban poor who could not return to work their land after military campaigns, even as skilled hired hands in war-related industries in the cities. Owing to the glut of slaves holding those positions between 140 and 133 BCE, unemployment and unrest became significant social and economic problems. The First Servile War, a grain shortage in Rome, and a massive desertion by 20,000 soldiers in Spain due to long terms of combat duty, all contributed to the social instability. The price of grain doubled twice in 133 BCE, further exacerbating the anger of the restive, *plebeian* population in Rome. Legislation introduced into the Roman Senate at that time by Tiberius Gracchus (older brother of Gaius), on behalf of the common people, limited the acquisition of public land by senators to 125 ha (500 *iugera*). Tiberius Gracchus was ultimately murdered as a result of this contentious legislation, but it brought to the fore the importance of land use and ownership in the Late Republic (Baker 2006).

The land ownership requirement for military service was abolished in 107 BCE, but so were most opportunities to receive a parcel of land as a reward for military service due to the now-dominant *latifundia*. Civil war erupted in 70 BCE between forces loyal to Julius Caesar, who had famously conquered Gaul in northern Europe, and those loyal to Pompey the Great, who eliminated piracy of essential grain supplies on the Mediterranean in a mere 3 months. The ultimate victory by Julius Caesar established his power over Rome and its far-flung colonies, and this began the long period of more than four centuries of the Roman Empire.

Caesar Augustus, the founder of the new Carthage and the first Roman emperor, was deified by the Roman people. He ushered in the *Pax Romana* during which there was relatively little military activity, except at the frontiers of the nascent Empire. By 17 BCE, Egypt's Nile River valley became the primary breadbasket for Rome and was a key conquest by Caesar Augustus following the military defeat of Marcus Antonius (commonly known in English as Mark Antony) in the Battle of Actium, the final war of the Roman Republic. Augustus renewed the worship of earthly gods, not from the underworld, and rekindled interest in the goddess Diana (for hunters, woodlands, and childbirth) and the goddess Mother Earth (for vegetation, regrowth, and bountiful production). The god Apollo (originally Greek) represented peace and art, while Jupiter became the patron god of Rome. When a plague hit Rome in 19 BCE, followed by a grain shortage, Augustus's links to these gods were expected to address these social crises (Baker 2006).

Through this period of transition from the Punic Wars to the *Pax Romana*, land use and soil management changes figured prominently in the key cultural and agricultural transitions of the time. They affected the powerful, rich patricians who owned the *latifundia*, as well as the poor *plebeian* class displaced from their farms and crowded into cities following military service far from their home lands. This was a time of considerable social and economic change, but it eventually led to an extraordinary period of relative peace. Seen in this light, it was a time of challenges, creative responses, and considerable resilience in Roman society as it expanded and diversified (Toynbee 1946).

12.5 ROMAN RESPONSE TO CHALLENGE IN THE POST-PUNIC WAR YEARS AS A LITERATE "COSMOTROPH": WRITING ABOUT AND PRODUCING WHEAT, GRAPES, AND OLIVES

Rome became a huge "cosmotrophus" ("nurturer of the world") and cosmopolis ("city of the world") as the first Western city to exceed 500,000 inhabitants in the Late Republic, and it ultimately approached one million: "*Romanus spatium est urbis et orbis idem*." "The world and the city of Rome occupy the same space." In Ovid's words (*Fasti* 2.684): "There is no part of the world which is not also Rome. To be a citizen of Rome was thus to be a citizen of the world" (Edwards and Woolf 2003).

Rome was without rivals in the Mediterranean region at a time when most cultural and commercial centers were much smaller with populations of around 20,000 to 50,000, with some cities reaching hundreds of thousands (e.g., Alexandria). Morley (2001) has estimated that from 175 to 28 BCE, the total population of Italy increased from 4.5 to 12 million people, with a free (nonslave) urban count of 400,000 to 1.5–1.6 million. The fraction of the population working on farms may have been as high as 14% by 28 BCE (up from approximately 8% in 225 BCE). The nonagricultural population was living in Rome and in one of the more than 400 urban centers of Italy. As a result, Rome was feeding upon and dependent on its colonies for wheat, wine, and olives; thus, it could be described in ecological terms, as a "cosmoheterotroph" consuming the carbon captured by photosynthesis within Mediterranean agro-ecosystems for energy, and producing, by simple processing, nutrients for human consumption in the forms of starch in wheat flour, alcohol in wine, and fat in olive oil.

Some of the earliest Latin documents were written in support of the essential agricultural production systems of the region during this time; in fact, White (1970) states that "Cato [the Elder]... taught agriculture to speak Latin." Cato's (234–149 BCE) *De Agri Cultura* (written in ~160 BCE) was the first treatise to provide agronomic recommendations for farming the land, and it explicitly assumed that slaves were being used on *latifundia* to produce wine and olive oil (Weber 1998). It provided advice (assuming it would be read aloud on farms) on slave management and recommendations that combined proven practices recommended by Cato with exhortations to try new methods for vines and fruit trees. Virgil's (70–19 BCE) *Georgics* (~29 BCE) is a poem published as four books (I–IV) on the subject of farming (White 1970). By the time of Columella's (~4–70 CE) *De Re Rustica* (12 systematic volumes) and his earlier, two-volume *De Arboribus*; grape production methods dominated all others in his advice to the owners of *latifundia*. Despite the dominance of estate farming, small-scale farms persisted in an extensive patchwork of large and small units in southern Italy and in heavily wooded Sicily, where *latifundia* were most common (White 1970), but almost all written advice was aimed at wine and olive oil production. This group of Roman agronomists, including Cato the Elder, Varro, Virgil, and Columella wrote some of their early Latin prose and poetry during the Punic Wars and the transition to the *Pax Romana*, the key period of challenge and response in Roman political, cultural, and environmental history related to soils and land.

A key question surrounding the cultural changes from the Punic Wars to *Pax Romana* is: how did the unusual diet of the Romans based on bread from wheat, wine from grapes, and oil from olives contribute to and support such remarkable change in human history? Closely linked is the question of the role of these Latin authors who advised on how to produce them economically. A look at the soils, terrestrial ecology, and biogeography of the Mediterranean region helps to frame an answer, especially when coupled with historical knowledge about agronomic practices for these crops and modern understanding of a healthy human diet. Their writings provide us a source of "data," in a scientific sense, derived from and supporting "experimental agriculture" in the form of slave-dominated *latifundia*. From this perspective, we present and interpret their findings, and integrate them with modern understandings of the ecology of the Mediterranean region.

The geomorphology and plant biogeography of the Mediterranean basin are governed by a xeric climate with striking seasonality of precipitation in a region dominated by mountains at the interface of land and sea (European Environmental Agency 2008). Indeed, McNeill (1992) describes the Mediterranean not as the "sea between the lands," but as the "sea among the mountains." Most of the annual 1 m of precipitation falls between November and April, with considerable amounts of snow in the mountains. As a result, the sloping land, coupled to heavy winter rainfall or spring snowmelt water is erosive, and this is made worse if the soil is not vegetated (Dazzi and Lo Papa 2013), or if it consists of heavy loam or clay loam, mostly developed on marls (Blum and Gomer 1996).

The native vegetation (especially the *garrigue*) contains fragrant, flammable oils (e.g., from lavender, thyme, sage, and rosemary) that make the ecosystem prone to low-level, regular fires during the dry season. If there is suppression of such fires (e.g., by humans extinguishing them), then the winter growth of woody vegetation leads to a build-up of flammable, oil-rich fuel and ultimately fires much larger and hotter than those that occur in the absence of fire suppression. Many seeds of these types of plants require intense heating by fire to germinate, so it is clear that the biome is naturally maintained by fire, is constantly recovering from disturbance due to burning, and is adapted to the arid summer and wet winter (Grove and Rackham 2003). The characteristic open canopy of the flammable vegetation allows light penetration to the soil surface and the growth of grasses and other herbaceous vegetation among the woody plants. This gives the landscape its characteristic savanna-like look that is so pleasing and welcoming to humans (Grove and Rackham 2003). These attributes of the vegetation, climate, and topography make the Mediterranean-type ecosystem prone to regular disturbances (that may or may not be of intermediate severity), and such disturbances define an ecosystem that is resilient and constantly being recolonized following fires, floods, and drought. How did the Roman agronomists describe such landscapes and make recommendations for farming on estate farms of this period of Roman history?

In the writings of the ancient Latin authors, there is clear evidence that the soil was considered an essential resource for plant growth, food production, and a profitable agriculture. Cato the Elder provided advice on buying a farm by distinguishing, in descending order of quality, nine types of plots of land: the first was a good vineyard (*vinea est prima*), the last a grove (*nono glandaria silva*) where animals could graze freely. Between these two extremes, there was the irrigated garden (*hortus inriguus*), olive grove (*quarto oleum*), the lawn (*quinto pratum*), and the arable land (*sexto campus frumentarius*).

Varro, about 100 years later, renewed the importance of soil evaluations made by Cato on the basis of their suitability for different crops, and he proposed a theory according to which soils are different because they originate from the blending of 11 types of substances. Varro defined the soil as "the element in which the seeds are sown and germinate" stating the importance of determining if the soils are rich, poor, or discrete: the rich soils support all types of plants, can be easily worked and provide good yields.

A century later, Columella drew up what can be defined as the first attempt to classify soils for agricultural purposes. In the second book of the *De Re Rustica*, he proposed a separation of soils into three different types based on their morphology: soils of plains, hills, and mountains. Each was in turn divided into six classes according to their quality: poor, rich, loose, strong, damp, and dry. Columella explained how, by means of some "perceptions" that are still used today, it is possible to appreciate properties of the soil, such as texture and structure. *De Re Rustica* became a reference book for soil management throughout the Middle Ages, and remained so until the advent of the modern era (Saltini 1984), based on its systematic agronomic framework and for the detailed advice provided.

Virgil proposed that physical and chemical soil analysis (II, 230–250) was very simple and effective. For wheat cultivation, Virgil states the importance of an adequately prepared, compact soil. But he maintains that black, fat, and sticky soils (clayey) (II, 204) can be suitable. In fact, he adds that, with plowing, compact soil becomes soft. Columella (II, 9, 3) and Varro (I, 23) confirm these recommendations, stating that wheat also grows in compact, wet, and heavy soils. Varro (I, 9) states that in rather moist (*umidior*) and clayey soil, it is better to sow durum wheat (*Triticum turgidum durum*), a rustic, high-protein grain, rather than the more "delicate" bread wheat *(Triticum aestivum)*.

Virgil suggests a simple analysis to understand if the soil is suitable for wheat or another crop. He writes (II, 230): "dig a hole in the soil and then put the whole earth removed into the hole and tamp it down with your feet. If the removed soil completely fills the hole, it is a loose soil" and therefore it is more suitable for the vine and for the meadow, and not for wheat. "If, on the other hand, it will not be able to refill completely the hole, and some soil remains, this is a clayey soil (and therefore suitable for wheat). With plowing you should obtain hard clods and compact fields."

Such a soil test was accepted and repeated by Columella (II, 2, 19), the most rigorous among the ancient Latin agronomists, who in his *De Re Rustica* writes:

> *Nam perexigua conspergitur aqua glaeba, manuque subigitur ac si glutinosa est, quamvis levissimo tactu pressa inhaerescit, et picis in morem ad digitos lentescit habendo, . . . quae res admonet nos inesse tali materiae naturalem succum et pinguitudinem. Sed si velis scrobibus egestam humum recondere et recalcare, cum aliquo quasi fermento abundaverit, certum erit esse eam pinguem; cum defuerit, exilem; cum aequaverit, mediocrem.*

"Moisten with a little water a small clod of soil and work it with your hand. If it is sticky, and even as a result of the lightest pressure of the fingers, it thereby adheres, then this indicates to us that it contains a natural juice and is fat. After digging a pit, thou shalt lay the soil which thou hast quarried back in the pit and press it down. If it will overfill the pit, it contains a ferment, and this is proof that the soil is fertile. If it does not fill the pit, the soil will be poor, and if it will just fill the pit, it will be mediocre." The word "ferment" suggests that the soil contained some sort of liquid between or within soil crumbs that acted like a nutritive juice.

For the cultivation of barley, Columella (II, 9, 14) and Varro (I, 9) suggest dry and loose soil. For the cultivation of the vine Virgil suggests (II, 229) loose soil, while for the olive tree (II, 179), the poorest soils are suitable, either gravelly or clayey.

The Roman agriculture writers do not limit their suggestions to indicating the most suitable crops for different kinds of soils; they also include in their books, short treatises on soil science or pedology, in which a system of soil classification is developed. They pay particular attention to the classification of soils in relation to humidity (wet or dry), and they emphasize soil texture (the proportions of sand, silt, and clay) and soil structure (the aggregation of the sand, silt, and clay particles) as loose soils and compact soils. These features had intermediate grades and various combinations. For example, Varro (I, 9) described different types of soils, on the basis of the prevalence of pebbles, gravel, sand, clay, humus (*carbunculus*):

> *In illa enim cum sint dissimili vi ac potestate partes permultae, in quis lapis, marmor, rudus, arena, sabulo, argilla, rubrica, pulvis, creta, glarea, carbunculus, id est quae sole perferve ita fit, ut radices satorum comburat, ab iis quae proprio nomine dicitur terra, cum est admixta ex iis generi bus aliqua re, cum dicitur aut cretosa sic ab illis generum discrimini bus mixta.*

"Therefore, there are several substances in the soil, with various consistencies and strength, such as rock, marble, rubble, sand, silt, clay, reddle [red ochre or iron oxides], dust, limestone, ash, and charcoal. Charcoal under the sun heats up so much that will burn the roots of plants; and if the soil is made with some part of these substances, it is called, for example, calcareous, or it is defined as mixed, according to the presence of other substances."

The Roman agronomists suggest several other ways to distinguish and classify soils. First, they considered how they appear visually (color, darkness, etc.), how they feel between the fingers, and the perception of density or weight. Virgil wrote (II, 248ss): "We can realize which soil is fertile and fat (*pinguis*) in this simple manner: when you rub the soil between your hands it doesn't shatter (*fatiscit*), but it will stick to the fingers as tar (*in morem picis*)...it is easy to see black soils and any other colors identify and distinguish soils." Virgil also refers (II, 249) to the analysis of soils through the native vegetation growing on them, and it should be carefully observed before clearing the land for agriculture by tree cutting and burning: "On wet soils grow quite high herbs (*maiores*). But damp cold earth (*scelleratum frigus*) is difficult to distinguish (*exquirere difficile est*); sometimes, we have evidence for it (*pandunt vestigio*) only through the presence of the brown ivy, the poisonous yew, and fir trees."

With reference to viticulture, Virgil writes that ferns grow in soils with southern exposure, and these are suitable for planting the vine (II, 186). Thorns, oleaster (wild olive), and other plants with wild berries grow in soils that are difficult to work, but these are suitable for olive trees (II, 179). Virgil, as already mentioned, suggests some simple analyses of the soil fertility. If we wish to know if our soil is salty (due to sodium chloride) or bitter (due to magnesium sulfate) and therefore unsuitable for any type of crop, he recommends the following test: "Untie from the smoke house ceiling a wicker basket, or take a filter from a winepress; then you press up to the brim of the basket the soil and add fresh water from a spring. The water naturally will percolate through the soil, and large droplets will stream from the wicker basket. The taste of the water will give a reliable clue, and the bitter taste [typical taste of sulfate salts] will twist the mouth to a pucker by those who taste the filtered water from the soil."

These texts by the Roman agronomists testify to the importance of soil fertility management for successful and profitable production of crops, particularly grapes and olives after the spread of *latifundia* following the Second Punic War. At that time, foreign trade was restricted and the Italian countryside was recovering from the devastation wrought by Hannibal in the Second Punic War. Despite the priority given to *latifundia* by these writers, small farms, orchards, and market gardens were interspersed within the large estate farms (White 1970).

The Roman writers recognized various effects of climate and soil properties on crop quality in different regions of the Italian peninsula; from the relatively flat coastal regions to the much steeper Apennine Mountains regions. The parent materials from which the soils had developed were recognized as important in distinguishing soils of the Italian peninsula for the purposes of crop production (White 1970; Figures 12.1 and 12.2). Volcanic ash was responsible for the high natural soil fertility in Etruria, Latium, Sicily, and Campania: "when after an eruption, the burning ashes have caused temporary damage, they fertilize the country for years, and render the soil good for the vine and very strong for other produce" (Strabo, the Greek historian and geographer, VI.2.3 cited in White 1970). Campania was renowned for its soil fertility, and its high-quality emmer wheat (*Triticum dicoccum*; known as *farro* in Italian, Latin, and in the culinary world) from the plains was complemented by the finest wines and oil in all of Italy.

The soil parent materials comprise a spatially diverse array, including limestone and marl, sandstone, and marine-derived clays (Bini 2013). Extensive, level land for agriculture was rare on the Italian peninsula, with the exception of the Po River Valley in northern Italy; indeed, four-fifths of the peninsula is classified as hilly and mountainous. The more level land was used for grain production, the foothills were used for vines and orchards, and the mountainous regions were reserved for forestry and transhumance, principally with sheep. Etruria had the highest bread wheat yields, Apulia the best quality, and the best emmer wheat came from Campania (White 1970). The fertile soils of Sicily largely fed Rome and the army, while grain production in Italy fed most of the rest of the peninsula. Demand for wheat was high enough that production on the lands in Sicily, Italy, and North African maintained an adequate supply from diverse sources.

Soil erosion has been identified as a major modern concern in Italy (Dazzi and Lo Papa 2013), especially in the absence of vegetation or in agronomic or horticultural production systems using clean cultivation (without weeds). How much the ancient Romans contributed to the soil properties and conditions of today (Figure 12.2) has been a controversial topic (Grove and Rackham 2003; Corti et al. 2013). As Corti et al. (2013) stress, the Romans used crop rotations to avoid a decline in soil fertility and introduced legumes into the rotation. They were well aware of the value of manure and legumes, as well as fallow, for the maintenance of soil fertility and soil structure (the aggregation of sand, silt, and clay into various larger crumbs or blocks of soil). Maintenance of soil structure through the use of the added organic matter from crop residues and animal manures helped to control erosion. As discussed above, Butzer (2005) points out that the ancient Romans surely would have learned from any mistakes made that resulted in soil erosion or declining crops yields. In the process, they would have learned by experience and developed more conservation-minded practices. His empirical studies point to anthropogenic transformations of soils, but probably not degradation of them or long-lasting disturbance of the ecosystem.

Lin Foxhall (1996) argues that terracing of the land was not extensively practiced in ancient Roman times to prevent erosion; instead, she posits that many existing terraces have been built since the Middle Ages. She cites ancient authors who describe a process of trenching around olive trees and vines to a depth of 1–1.5 m. Such trenching captured and stored winter rain water for use by the plants in the arid summer, controlled weed growth, and incorporated organic residues and other soil amendments. Trenching thereby reduced runoff of excess water on sloping land, but was labor-intensive and could be maintained by only the wealthiest landlords using slave labor.

The extent to which ancient Roman farming practices caused severe soil erosion remains uncertain and contentious, but it is clear that ancient Roman soil management was based on creative practices. Examples are animal manuring, trenching, crop rotation, legume production, and fallowing; each of which helped to minimize soil loss and maintain soil fertility. Such practices can be viewed as ones that were learned through experience (i.e., "adaptive management" in the modern language of sustainability) and they simulated to some degree the natural processes of seasonal growth of native vegetation.

These soil management practices for the production of wheat, grapes, and olives were used for centuries, and they were particularly important in the transition from the Republic to the Empire as the *latifundium* emerged and supplanted or blended with smaller-scale, traditional farms in Italy, Sicily, and North Africa. They also supported a diet that was surprisingly simple, but nutritious. Of the nearly 1 million residents of Rome at the time of the Late Republic after the Third Punic War and before *Pax Romana* (147–27 BCE), Garnsey (1998) estimates that the majority were poor, and as many as 150,000 may have received the *plebeian frumentatio* (poor people's food ration). Gaius Gracchus initiated the *frumentatio* in 123 BCE, and 5 *modii* of unmilled grain per month (~33 kg), principally wheat, was available at low cost (free after 58 BCE). The lists of those to receive the *frumentatio*, however, did not include noncitizens, foreigners, slaves, women, and girls. If these individuals were not able to share the food ration with a family member on the list, then they would have to purchase grain on the open market. Without such access, hunger and malnutrition would be common (Garnsey 1998).

Garnsey (1998) estimates that such a ration provided double the daily need for calories and sufficient protein (if eaten as whole wheat containing 10–14% protein), and wheat-based food contributed approximately 75% of food energy. If eaten in the form of yeast bread, wheat provided a full complement of amino acids, important because wheat grain alone is low in lysine and threonine. The yeast also enhanced intestinal uptake of calcium, iron, zinc, and other essential elements in the human diet. The wheat grain and milled flour available to the poor of Rome probably was of low quality and contained variable amounts of weed seeds and other impurities. The wealthy Romans preferred white flour and bread due to it being cleaner, but ironically, it was less nutritious than whole wheat flour. In contrast, whole grain flour contains phytic acid, which binds with calcium, iron, and zinc, making them less bioavailable in the intestinal tract than in the absence of phytic acid, for example, in white flour from which the wheat bran and germ have been removed. So, in a Roman diet dominated by coarsely milled, whole wheat eaten as unleavened bread and other dishes, and without other sources of these essential micronutrients, nutrient deficiencies probably occurred, especially among the poor (Garnsey 1998). Given the importance of adequate dietary iron and calcium for pregnant and lactating women, plus young women of child-bearing age; the quantity, kind, and form of consumed wheat could have had significant effects on maternal health.

Puls was a grain-based staple food dish prepared by boiling wheat flour in water or milk, followed by mixing with cheese, as described by Cato the Elder; and boiled pulses (seeds of legumes, e.g., beans, lentils, and chickpeas) or vegetables (*holera*) were served as a side dish (*pulmentarium*) to *puls*. Yeast bread from wheat, coupled with olive oil, provided an almost complete complement of protein, carbohydrate, and fat, especially when supplemented by the addition of pulses (Garnsey 1999). Known as "poor man's meat," broad beans and other pulses were commonly eaten as a protein source (sometimes mixed into wheat bread), and other legumes were interplanted in olive plantations and vineyards as a source of nitrogen for the perennial crops for example, lupin (*Lupinus* sp.) and vetch (*Vicia* sp.); now called winter annual cover crops. The Romans, especially the lower classes, ate little meat, in part because it is far more efficient and far less costly for humans to eat plant protein and carbohydrates directly than to feed the grain to animals to produce meat. In addition, the lack of extensive grazing land and productive pastures in the arid summer for meat-producing animals to supply a metropolis like Rome meant that the soil and land resources could not support such a food system based on meat.

Wheat and barley were the two dominant grains in the Roman diet, with two varieties of wheat being grown the most: durum wheat, now principally used for pasta, and bread wheats used for unleavened and raised breads that were much prized in the Roman diet. Although durum wheat is harder than bread wheat and higher in protein, it is low in gluten, a necessary protein found in much higher levels in bread wheat and rye, the two principal grains used for leavened breads. Durum and bread wheats are also the only free-threshing, naked types of wheat in which the wheat seeds are not tightly-encased in a glume, as in the hulled types [emmer, einkorn, and spelt wheats (*Triticum spelta*)], making durum and bread wheats easy to separate from the grain head or spikelet and mill into flour.

Durum wheat is tetraploid, having four sets of chromosomes in a cell, whereas bread wheat is hexaploid. Their wild ancestors (einkorn and goatgrass, *Aegilops cylindrica*, are both diploid). The higher number of chromosomes in durum and bread wheats than in their diploid parents imparts greater adaptability and vigor under challenging weather conditions. This change in the genetic makeup of wheat during its early domestication in Mesopotamia and the Mediterranean may have been an important, if serendipitous, development that made wheat such a successful crop for the Mediterranean region. Planted in the fall and harvested in the early summer, wheat production schedules and soil management needs complemented those associated with the harvest and processing of olives and wine in the fall, but there were regional differences due to climate, elevation, and aspect.

Wine was the principal source of essential water in the Roman diet, and the quality ranged widely in accordance with vineyard soil properties, climate, elevation, and type of grape. It was drunk with all meals, and it was considered "civilized" to dilute it with water. Those who drank it neat (undiluted), such as the Celts and Gauls, were seen as "barbaric" for adopting such a practice. The Gauls prized wine from Rome, and accepted it readily as a bribe to maintain peace with the Romans. Soldiers in the Roman army drank a wine akin to vinegar in quality, diluted with water. The Romans for a time forbade the cultivation of vines outside of the Italian peninsula to protect their own use, consumption, and profits from vine production. Grapes were harvested in the early fall for wine making, and the techniques used are similar to those of today. Wine yields were extremely high (e.g., average of 3000–6000 liters/ha), and it was a very profitable, even if labor- and capital-intensive to produce.

Table grapes were grown mainly in the area around a town. There were several kinds used by the Romans. Examples are *Duracina*, mentioned by Cato the Elder for its fleshy pulp; *Ambrosia*, for long-term preservation; *Dactylus*, with its elongated grapes; *Forensis*, the most popular in the markets; and *Visullae* for its delicious taste. Among wine grapes, *Spianta* was widespread in the area of Ravenna; *Murgentina* in the area of the Vesuvius; *Helvanaca,* which produced a wine of poor durability; *Argitis*, known for its productivity; *Psithia*, that Virgil considered suitable to produce *passito* wines (made from raisins or partially dried grapes); and *Falernae* that produced the wine most appreciated by the Romans (Forni 2002).

Olives (*Olea europaea*) were grown on the poorest soils, and fertilization was not recommended for the highest yields and best-quality olive oil. The most suitable soil for olives is clay mixed with sand, overlying stones. It also does well in compact soils, if irrigated and fertile. Columella provided a series of precepts for evaluation of the environmental and soil conditions suitable for the cultivation of olives. Although considered "first among all trees" by him (Mattingly 1996), the olive is temperamental: it requires a temperature range of 16–22°C for successful growth, and it does not tolerate prolonged, severe drought. Interestingly, however, it can thrive with as little as 150 mm of annual precipitation, far less than required for grapes and cereal grains, and it does not grow well in depressed or high areas; it prefers intermediate elevations. Columella argues that it is important to plow the soil under the olive tree at least twice annually, coupled to cultivating deeply with a two-tooth hoe. After the autumnal equinox, the farmer should undermine the plants (hoe around the trunk of the tree), and add animal manure at least every 3 years. Recommending the use of animal manure seems to contradict the advice to avoid fertilizing olive trees, but doing so infrequently may have contributed to higher yields in the long term.

As a woody perennial, the olive is long-lived (300–500 years on average) and remarkably resilient when grown spaced out on the landscape, much like the native vegetation; it is able to regrow by putting up adventitious shoots from the roots following damage to the top by burning, frost, disease, cutting, or other disturbance (much like the native vegetation). It does not bear fruit for up to 7 years after planting, but continues to produce good crops for decades, even centuries. Like wine, olive oil was and is a profitable crop, and in Roman times, it was a key component of the diet (Garnsey 1999).

Olive oil was used as a skin emollient and cleanser in public baths, as a combustible fuel for light and heat, for perfumes and cosmetics, and, of course, for human consumption. Mattingly (1996) estimates that the Romans consumed 10–50 L of olive oil per person per year. It is 100% digestible, contains little protein, provides all essential lipids, is rich in vitamins A and E, and is high calorie (approximately 18 times the caloric content as wine per milliliter). In the Roman diet, it supplied 80% of the necessary fats, 12% of calories, but only 2% of the mass of food consumed. The solid residue derived from the pressing of olives for the oil was used for fertilizer, animal feed, and fuel.

This "eternal trinity" of crops (Garnsey 1999) on which the Roman civilization was based takes on an almost sacred quality in light of the soil, landscape, and climatic conditions under which the three grew successfully. When we analyze the transition of the Roman civilization from the Late Republic to the beginning of the *Pax Romana*, it is evident that the diversity of plants involved, the range of soils supporting their growth, and the spatial array of landscapes and climatic conditions of the region combined to allow for a continuously growing human population centered on a few cosmopolitan centers, principally Rome. The spatial proximity and diversity of landscapes also provided a resilient system of interacting components that avoided the situation in which the whole region experienced the same effects of extreme weather events (e.g., drought or heavy rain) at the same time. Tainter (1988) identifies such asynchronous variability as a beneficial situation that allowed for economic, ecological, and cultural resilience through trade in well-developed networks, as existed in the Mediterranean region. To quote Radkau (2008):

> The olive tree and grape have shaped the landscape and the way of life in the Mediterranean region since prehistoric times. Still today, one can see the 'age-old marriage of oil and vine' along many Mediterranean coastlines, a sign that it must have developed ecological stability over thousands of years...[they] have also a modernizing side, containing from early on an impetus to trade and to move beyond the subsistence economy.

12.6 SOILS AND CREATIVE SUSTAINABILITY IN ROMAN ANTIQUITY: TRANSDISCIPLINARY HYPOTHESIS TESTING AND CASSANDRA'S HEURISTIC LESSONS

In ancient Greek mythology, the beautiful Cassandra was known for crying out her dire prophecies and warning the people to heed her predictions, to no avail. The god, Apollo, cursed her with this insight for spurning his sexual advances, and refused her the power to alter what she knew was to come or to convince others to believe her. In contrast, we can learn how to respond creatively to modern-day Cassandras and challenges related to soil, water, food, and their critical roles in sustaining our modern, urban-based, agriculturally dependent civilization. In contrast to the common, pessimistic narratives about the rise and fall of the Roman Empire as a warning to twenty-first century people, we have shown that new learning may emerge by focusing on an earlier time period of dramatic change in Roman Antiquity. Using transdisciplinary knowledge, methods, and evaluations from environmental history, systems ecology, and soil and agricultural sciences, we have demonstrated the possibility of viewing human–nature relationships and environmental change in soil–plant systems in ways that may lead to creative responses to such challenges.

The transition from the Roman Republic to *Pax Romana* took place in an ecosystem sustained by regular disturbances due to fire, flood, and seasonality in temperature and moisture. The soils and biogeography of the region are adapted to such changes, and they provided a terrestrial resource base that was ideal for the production of wheat, grapes, and olives. These crops constituted the core of the Mediterranean diet that has sustained inhabitants of the region for many centuries, and remains so today (Estruch et al. 2013). The changes that took place in the region over the more than 200 years of our study were dramatic, and they altered soil landscapes and Roman culture in significant ways that probably did not result in soil degradation. In contrast, we find that changes in human–soil relationships of the time are examples of creative responses to challenges of intermediate severity, as Toynbee (1946) has theorized. Long, devastating wars, imposition of autocratic rule, displacement of small farmers by absentee land owners of *latifundia*, and widespread use of slavery may be morally reprehensible, economically contraindicated, or politically unacceptable when seen through our modern lenses. Nevertheless, the ultimate two centuries of relative peace following the Crisis of the Republic were certainly a remarkable time, both in the long history of warfare in Roman Antiquity leading up to it, and in our modern era of militarism in the Western countries and cultures descended from the Romans.

In a sense, the information and knowledge we have used from plant ecology, soil science, and environmental history were set out for us on the land and in written records centuries-to-millennia ago, and we have collected and interpreted them in a balanced way today. We also have used the three-component method of Radkau (2008) in environmental history based on "the geographic reach of environmental problems, the level of social authority that deals with them, and the type of knowledge that is employed in the process."

Based on these methods of investigation, we have reached several conclusions and propose new learning as a result:

1. The changes in Roman society in this period occurred in a climatic region that allowed for the production and shipment of foods that emulated the natural plant communities of the region. The distances between where the crops were produced and where they were consumed were short, and there existed well-developed systems for trade over those distances. The Roman civilization of the time likely persisted because of the closely linked subregions and the possibility for coordination of production and consumption if one region had crop failures.
2. The changes in land use that followed the challenges of the Punic Wars were dramatic and creative in displacing locally based, small-scale farmers by city-dwelling land owners of large *latifundia* worked by imported slaves.

3. The diet of the Romans based almost exclusively on bread, wine, and olive oil is simple and nutritionally sound, and it was a stable base for sustaining the population through the centuries of change.

The methods we have used in making our evaluation may be summarized with the overlapping circles representative of a Venn diagram (Figure 12.3). In so doing, we have shown the enhanced value of a "disciplinary triangulation" method for interpreting data and information from each of the fields of study individually. The starred zone in the diagram illustrates where our predictions and learning may be most effective and reliable. In trying to assess the "sustainability" of modern practices and predictions, we can use such a diagram and approach to recognize the dynamic qualities of learning from environmental history, soil science, and ecology. The parallels are both striking and reassuring as a learning tool, and they provide a way to respond to Cassandra in heuristic and positive ways.

In looking anew at this time of crisis in Roman history, we see a period of considerable creativity that was in line with the civilizing ethos of Fernández-Armesto (2001) and the framework of Toynbee (1946) for viewing the dynamics of civilizations. Fernández-Armesto's civilizing ethos is based on an ideal of order, a vision of abundance, and writing as symbolic imagination. The Romans of our period of study expected a future harvest of grains, olives, and grapes, either from the Roman peninsula or the island and coastal zones of the colonies. These products could be stored and shipped, thereby enabling a continuous, year-round source of food from local and regional sources. The cities that grew tremendously during the period had restive populations of unemployed, landless *proletarians*, but physical and behavioral order was maintained in resilient ways of governance and social organization. And finally, the ethos of the Romans was based on writing in Latin, which was refined and put to practical use during this period by Roman agronomists, among others (White 1970). A sense of history and a worldview emerged from these writings, both of which still speak to us today in soil science, environmental history, and ecology.

The questions we asked and hypotheses that we framed rhetorically have been addressed in the course of our investigation of Roman history from the beginning of the Punic Wars to the emergence of the Empire. Human–soil–land relationships were indeed crucial for the resilience of the Romans in that time. It is not evident that they sowed the seeds for the ultimate fall of the

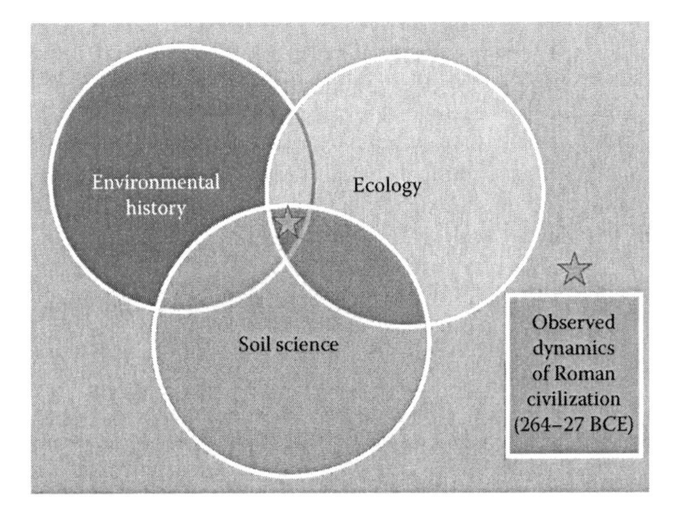

FIGURE 12.3 Disciplinary triangulation for answering "why" questions about soils and sustainability in Roman Antiquity and in testing interdisciplinary hypotheses from the perspectives of environmental history, ecology, and soil science.

Empire; indeed, the changes we describe and evaluate supported continuity of the Roman civilization (Fernández-Armesto 2001) during the next 500 years. It is also not evident that the Romans of this period created a degraded landscape and eroded soils. The climate, topography, and diversity of the biogeography of the region appear to have contributed a resilient resource and biophysical system that was supportive of the Romans, and that was resilient in the face of natural and cultural change over centuries.

REFERENCES

Baker, S. 2006. *Ancient Rome: The Rise and Fall of an Empire*. London: BBC Books.

Bini, C. 2013. Geology and geomorphology. In *The Soils of Italy*, eds. E.A.C. Costantini and C. Dazzi, 39–56. Dordrecht: Springer Science + Business Media.

Blum, W.E.H. 2005. Functions of soil for society and the environment. *Reviews in Environmental Science and Biotechnology* 4: 75–79.

Blum, W.E.H. and D. Gomer. 1996. Runoff from soils on marls under semi-arid Mediterranean conditions. *International Agrophysics* 10: 1–10.

Brundtland, G.H. 1987. *Our Common Future: Report of the World Commission on Environment and Development*. United Nations. Annex to document A/42/427.

Butzer, K.W. 1996. Ecology in the long view: Settlement histories, agrosystemic strategies, and ecological performance. *Journal of Field Archaeology* 23: 141–150.

Butzer, K.W. 2003. The nature of Mediterranean Europe: An ecological history. Book review. *Annals of the Association of American Geographers* 93: 494–530.

Butzer, K.W. 2005. Environmental history in the Mediterranean world: Cross-disciplinary investigation of cause-and-effect for degradation and soil erosion. *Journal of Archaeological Science* 32: 1773–1800.

Butzer, K.W. 2012. Collapse, environment, and society. *Proceedings of the National Academy of Sciences* 109: 3632–3639.

Collins, J.P., A. Kinzig, N.B. Grimm et al. 2000. A new urban ecology. *American Scientist* 88: 416–425.

Corti, G., S. Cocco, G. Brecciaroli, A. Agnelli, and G. Seddalu. 2013. Italian soil management from antiquity to nowadays. In *The Soils of Italy*, eds. E.A.C. Costantini and C. Dazzi, 247–293. Dordrecht: Springer Science + Business Media.

Costantini, E.A.C., G. L'Abate, R. Barbetti, M. Fantappiè, R. Lorenzetti, and S. Magini. 2012. Carta dei suola d'Italia, scala 1:1.000.000 (*Soil map of Italy Scale 1:1.000.000*)—Consiglio per la ricerca e la sperimentazione in agricoltura—S.EL.CA. Firenze, Italia.

Dale, T. and V.G. Carter 1955. *Topsoil and Civilization*. Norman: University of Oklahoma Press.

Dazzi, C. and G. Lo Papa. 2013. Soil threats. In *The Soils of Italy*, eds. E.A.C. Constantini and C. Dazzi, 205–245. Dordrecht: Springer Science + Business Media.

Edwards, C. and G. Woolf. 2003. Cosmopolis: Rome as world city. In *Rome the Cosmopolis*, eds. C. Edwards and G. Woolf, 1–20. Cambridge: Cambridge University Press.

Estruch, R., E. Ros, J. Salas-Salvadó et al. 2013. Primary prevention of cardiovascular disease with a Mediterranean diet. *The New England Journal of Medicine*. 368: 1279–1290.

European Environment Agency. 2008. *Biogeographical Regions: Europe 2001*. http://www.eea.europa.eu/data-and-maps/data/ds_resolveuid/3B2B7DDA-4584-4963-8D93-A533D3C830DF (accessed March 13, 2013).

Fernández-Armesto, F. 2001. *Civilizations: Culture, Ambition, and the Transformation of Nature*. New York: The Free Press.

Forni, G. 2002. La produttività. In *Storia dell'agricoltura Italiana, l'età antica*. 431–440. Firenze: Accademia dei Georgofili. Edizioni Polistampa.

Foxhall, L. 1996. Feeling the earth move: Cultivation techniques on steep slopes in Classical Antiquity. In *Human Landscapes in Classical Antiquity*, eds. G. Shipley and J. Salmon, 44–67. London: Routledge.

Garnsey, P. 1998. *Cities, Peasants, and Food in Classical Antiquity*. Cambridge: Cambridge University Press.

Garnsey, P. 1999. *Food and Society in Classical Antiquity*. Cambridge: Cambridge University Press.

Gibbon, E. 1776. *The History of the Decline and Fall of the Roman Empire: Abridged Edition*. London: Penguin Books.

Grove, A.T. and O. Rackham. 2003. *The Nature of Mediterranean Europe: An Ecological History*. New Haven: Yale University Press.

Hillel, D. 1991. *Out of the Earth: Civilization and the Life of the Soil*. Berkeley: University California Press.

Hyams, E. 1976. *Soil and Civilization*. New York: Harper Colophon Books.
Ibáñez, J.J., J.A. Zinck, and C. Dazzi. 2013. Soil geography and diversity of the European biogeographical regions. *Geoderma* 192: 142–153.
IUSS Working Group WRB. 2006. International Union of Soil Sciences. *World Reference Base for Soil Resources 2006: A Framework for International Classification, Correlation and Communication*. World Soil Resources Reports No. 103. Rome: Food and Agriculture Organization.
Kehoe, D.P. 1988. *The Economics of Agriculture on Roman Imperial Estates in North Africa*. Gottingen: Vanderhoeckt & Ruprecht.
Lendering, J. "Carthage." *Livius: Articles on Ancient History*. 11 January 2013. Livius.org. http://www.livius.org/cao-caz/carthage/carthage.html (accessed January 21, 2013)
Lowdermilk, W.C. 1953. *Conquest of the Land through 7,000 Years*. Agric. Info. Bull. no. 99. Washington: US Department of Agriculture, Soil Conservation Service.
Mattingly, D.J. 1996. First fruit? The olive in the Roman world. In *Human Landscapes in Classical Antiquity*, eds. G. Shipley and J. Salmon, 213–253. London: Routledge.
McNeill, J.R. 1992. *The Mountains of the Mediterranean World*. Cambridge: Cambridge University Press.
Montgomery, D.R. 2007. *Dirt: The Erosion of Civilizations*. Berkeley: University of California Press.
Morley, N. 2001. The transformation of Italy, 225–28 B.C. *Journal of Roman Studies* 91: 50–62.
Odum, E.P. 1969. The strategy of ecosystem development. *Science* 164: 262–270.
Paine, R.T., M.J. Tegner, and E.A. Johnson. 1998. Compounded perturbations yield ecological surprises. *Ecosystems* 1: 535–545.
Pickett, S.T.A., J. Kolasa, J.J. Armesto, and S.L. Collins. 1988. The ecological concept of disturbance and its expression at various hierarchical levels. *Oikos* 54: 129–136.
Radkau, J. 2008. *Nature and Power: A Global History of the Environment*. Cambridge: Cambridge University Press.
Reice, S. 1994. Nonequilibrium determinants of biological communities. *American Scientist* 82: 424–435.
Salmon, E.T. 1970. *Roman Colonization under the Republic*. Ithaca: Cornell University Press.
Saltini, A. 1984. *Storia delle scienze agrarie: 1. Dalle origini al rinascimento*. Milan: Edagricole.
Scheidel, W. 2003. Germs for Rome. In *Rome the Cosmopolis,* eds. C. Edwards and G. Woolf, 158–176. Cambridge: Cambridge University Press.
Schrödinger, E. 1945. *What Is Life? The Physical Aspect of the Living Cell*. Cambridge: Cambridge University Press.
Scoones, I. 1999. New ecology and the social sciences: What prospects for a fruitful engagement? *Annual Review of Anthropology* 28: 479–507.
Sirago, V.A. 1995. *Storia agraria romana. 1*. Naples: Liguori editore.
Smeck N.E., E.C.A. Runge, and E.E. MacIntosh. 1983. Dynamics and genetic modelling of the soil systems. In *Pedogenesis and Soil Taxonomy I. Concepts and Interactions*, eds. L. Wilding, N. Smeck, and G.F. Hall, Vol. II, Part A, 51–81. Amsterdam: Elsevier.
Tainter, J.A. 1988. *The Collapse of Complex Societies*. Cambridge: Cambridge University Press.
Toynbee, A.J. 1946. *A Study of History*. New York: Oxford University Press.
Weber, M. 1998. *The Agrarian Sociology of Ancient Civilizations*. London: Verso Classics.
White, K.D. 1970. *Roman Farming*. London: Thames & Hudson.
World Heritage Convention/UNESCO. Archaeological site of Carthage. UNESCO. 2012. http://whc.unesco.org/en/list/37 (accessed January 21, 2013).

13 The Anatolian Soil Concept of the Past and Today

Erhan Akça[*] and Selim Kapur

CONTENTS

13.1 INTRODUCTION

Anatolia,[†] also called Asia Minor, has an area of about 500,000 km^2, is located at the crossroads of Europe, Africa, and Asia (Figure 13.1). Several civilizations namely Hittites, Greeks, Urartians, Persians, Romans, Byzantines, Arabs, Seljuks, Moguls, and Ottomans occupied Anatolia owing to its favorable climate and rich natural resources, including its soils. These various civilizations in turn strongly modified both the cultural and physical landscapes of Anatolia. The story of why they settled in these lands and their effects upon them reflect the special features of its geological and landscape histories. To a very large extent, these are expressed in the soils of Anatolia, which gave them the materials for their nutrition, fuel, clothing, and shelter. In time, the interplay of these products of the soils with the ingenuity of its different inhabitants was to bring Anatolia through the agricultural revolution. Archaeological evidence outlined in this chapter suggests that it may even have provided the location of the very beginnings of the agricultural revolution: Anatolia may have been the crucible of this vital change in mankind's evolution. This proposition is examined in this chapter, which also describes the state of soils in Anatolia today.

13.2 SOILS AND THEIR ORIGIN IN ANATOLIA

Anatolia is characterized by a complex geology consisting of several continental fragments joined together into a single landmass in the Late Tertiary, about 30 million years ago (Okay 2008). This led to the formation of many sloping (Figure 13.2) and mountainous landscapes, with an average elevation of 1100 m.

[*] Adıyaman University, Technical Programs, Altınşehir Kampüsü, 02040, Adıyaman, Turkey; Email: erakca@gmail.com

[†] The use of the term "Anatolia" in the chapter refers to a geographic and historic *context*, rather than being a political recognition as the well-established national *concept* of Turkey in 1923 (establishment of the Turkish Republic) that has been adopted since the mid-Ottoman period.

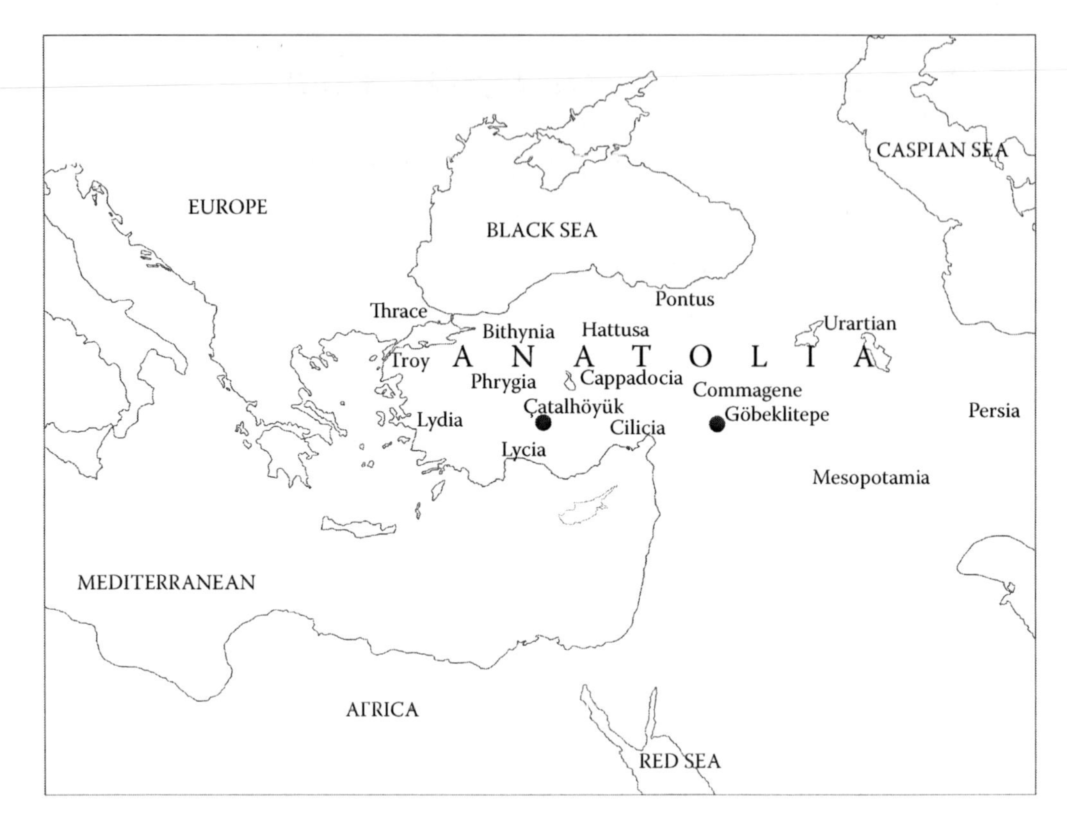

FIGURE 13.1 Civilizations of Anatolia.

FIGURE 13.2 Steep sloping limestone mountains elongated from the west to the east of Anatolia.

FIGURE 13.3 The shallow soils (Leptosols/Entisols) of the Mediterranean region of Anatolia.

The majority of the Anatolian soils have resulted from the vigorous earth movements of mountains ("neotectonics") that have taken place since the Miocene, that is, approximately 5 million years ago (see Figure 13.3 for an example). The subsequent development of the steep slopes inevitably led to the mass transport of soils even at 60 km distances and the continuous destruction of the landscape (Aksoy et al. 2010). This transportation is a Pleistocene phenomenon that was induced by earth movement and especially by the wet pluvials–glacials of the Mediterranean belt, south of the Alps. Thus, shallow soils (Leptosols/Entisols) formed on the steep slopes (Dinç et al. 2005).

The next most dominant soils of Anatolia are located in the drier parts of the country, having developed on ancient lake basins, and mudflow was induced by the tectonic movements responsible for the formation of the terraces that have given rise to tectonically induced terraces of the Quaternary, that is, within the 500,000 to 100,000 BP (Dinç et al. 1997). The next most dominant soils of Anatolia, the red Mediterranean soils, are located in the drier parts of the country, having developed on ancient lake basins and mudflow deposits. Other widely distributed soils occur throughout Anatolia along river valleys and lake basins that are covered by the materials transported following the neotectonic activities. These particular soils have been subjected to a long history of exploitation since the Neolithic period, which is the New Stone Age, a period in the development of human technology, beginning about 12,500 BP, in some parts of the Middle East (especially in Anatolia), and later in other parts of the world and ending between 4500 and 2000 BC.

13.3 EARLY CIVILIZATIONS

Anatolia has developed more than 13,000 plant species due to its desirable soil and climate conditions. Prior to the agricultural revolution, Anatolian settlers enjoyed a wide variety of plants in their diets, including *Triticum aestivum* (common wheat or bread wheat), *T. dicoccoides* (wild emmer), *T. urartu*, and *T. monococcum* ssp. *Boeoticu* (einkorn). These were/are the main cultivars of today's cereals.

The inhabitants of Anatolia became excellent terrace constructors, particularly in the mountainous Mediterranean, Black Sea, and Eastern Anatolian regions. The Urartian terraces built on the sloping landscapes of Lake Van with a slope angle of 27° (Akça et al. 2008) are an excellent example of this. These terraces are still intact at several locations around Lake Van, and some are successfully managed by the local communities for nonirrigated cereal and vegetable cultivation (Figure 13.4). However, some of these unique historic terraces are threatened by inappropriate

FIGURE 13.4 Ancient Urartu terraces (2800 years BP) with a 27° slope.

modern agricultural technology (machinery, chemicals, and so on) and especially heavy cultivation which all degrade the quality of soils such as their aggregation and organic matter content and induce erosion.

13.4 EARLY SETTLEMENTS

The prime soils of Anatolia are/were found on the plains of Cilicia, Lydia, Thrace, Lycia, Mesopotamia, and Central Anatolia (Figure 13.1). The first settlements discovered in the Neolithic of Anatolia were mostly located around the fertile soils and water resources, such as Çayönü (adjacent to the Tigris and Euphrates) and Çatalhöyük (Central Anatolia); they date from around 9000 BP (Braidwood et al. 1981; Fairbairn 2005). These soils were mostly cultivated for cereals on shared properties, and extensive grazing of small ruminants was also a common practice on the indigenous small ruminant pathways that were the long-standing routes followed by the Central Anatolian shepherds for their large herds roaming from the south of Central Anatolia to the north. These pathways were well-established paths developed by the trampling action of the sheep until the Bronze Age. Today, some are even used by locals. The Çatalhöyük site is considered to be one of the earliest sites to flourish in the agricultural revolution. The settlers' perception in selecting an appropriate settlement site revealed their knowledge of a good environment for agriculture and for hunters. The agricultural technology of the Neolithic has not been well documented except for some sickle blades and wall paintings in Çatalhöyük that revealed probable cultivation activities (Fairbairn 2005; Rosen and Roberts 2005; Hodder 2006) (Figure 13.5).

13.5 THE FIRST FARMERS?

Anatolia's widespread Neolithic landscapes have attracted national and international experts in the social and natural sciences for decades, and, especially today, in studies aimed at ascertaining if the hunter-gatherers who constructed the site of Göbekli Tepe (Figure 13.6), 7000 years before Stonehenge and the Great Pyramids of Egypt were built, were also the first people to start cultivating the soil (Schmidt 2010).

Unfortunately, no single consensus as to whether or not this was the first farming society has yet been formed despite the discovery revealing the consumption of pistachio, almonds, peas, wild

FIGURE 13.5 Çatalhöyük (9000 years BP) excavation site.

barley, wild emmer, and lentils by the inhabitants (Hirst 2009). However, that Göbekli Tepe is a good candidate for being that site is evidenced by the presence there of the first humanly modified grain—the einkorn wheat (single grain wheat) in soil. The grain is now becoming more closely linked to one of archaeology's major puzzles—the mysterious and ancient site of Göbekli Tepe and its limestone megaliths decorated with bas-reliefs of animals. Nobody has been able to satisfactorily interpret their full significance. The region of the contemporary Göbekli Tepe has a dry Mediterranean climate, and dryland farming has been practised throughout the historical ages on typical Mediterranean soils. The area was semi-arid, as indicated by the widespread limestones and calcretes (secondary, rock-like accumulations of calcium carbonate formed by soil processes, hence "pedogenic," Figure 13.7) of Early Holocene and Late Pleistocene age.

FIGURE 13.6 Göbekli Tepe (12,000 years BP) excavation site.

FIGURE 13.7 Pedogenic calcrete surface and profile.

These are overlain by long-cultivated soils commonly known as *terra rossas,* red and reddish-brown soils of the Mediterranean climatic area. The soils with calcrete formed in this climate favored the cultivation of wheat and the Mediterranean vegetation of today (especially Mediterranean shrubs, olives, pistachio, and figs) owing to their porous character and good structure suitable for dry farming. Terra Rossas were also widespread on stone terraces constructed near the settlement sites which were composed of grazing belts, pottery production, churches, administrative compounds, irrigation canals, and cisterns of the mid- to late Roman period coastal areas (the Roman to the Byzantine period). Olives can be easily dated back to the early Roman period, which suggests the consistency of a Mediterranean type of climate prevailing in the area since then. Mid-Ottoman (about 1500 AD) grafted wild olive trees still cover vast areas along the southern slopes of the Taurus Mountains running parallel to the Mediterranean Sea coast. The major plant genetic material obtained from the excavation area (a well-known olive production site in Turkey) was the olive stones that were determined to be an extinct species of an olive stock. Although local, the species cultivated today do not resemble the stone characteristics of the disclosed species from the excavation site.

13.6 LATER CIVILIZATIONS AND ATTITUDES TO SOIL

Seeking to protect the soils, the people of the Phrygian kingdom, who reigned in Anatolia from c.1200 to 700 BC enacted laws bearing punishment for the misuse of the land (Kealhofer 2005). In the Bronze Age, the Hittite kings, rulers of the first known empire of Anatolia (Akurgal 1962), granted the privilege to cultivate the land to the landlords who were responsible for supporting the imperial army as well as peasants working as corvée (unfree labor). Monuments expressing the blessing of the people to the gods for the productivity of soils have been found in several parts of Anatolia (Figure 13.8). These powerful monuments to crop production suggest a view of soils as indispensable resources responsible for the survival of the empire. Several clay tablets have been found, which record land endowment by the Hittite emperors to their citizens. Erkut and

FIGURE 13.8 Hittite İvriz relief carving in Central Anatolia showing Tarhundas, the Storm God holding bunches of grapes and cereal, and King Varpalavas expressing his gratitude.

Reyhan (2012) write that the tablets bear the seal of the king and guarantee the donation of land in the form of inheritance to the children and grandchildren of the person concerned, who may be defined as vassals.

13.7 ANATOLIAN SOILS TODAY

The land of Anatolia has a rich diversity of landscapes and soils, and a semi-arid climate with an average precipitation of 650 mm (Kapur et al. 2002). The region is largely mountainous, and plains comprise only 10% of the total area. Thus, steep slopes along with high temperatures and a low humidity cause erosion and rapid decomposition of organic matter, yielding soils with low productivity. High yielding soils are quite limited, mostly being found on riverbanks and alluvial plains, which makes these soils invaluable assets of the society. This is well defined by several poets and artists who have displayed their faithfulness to soil through the history of Anatolia. The thirteenth century philosopher Mevlana quotes that "One should behave like the soil in humility and selflessness" and the twentieth century poet Asik Veysel expressed his emotions via these words:

> I ripped your stomach with shovel and hoes
> I tore your face with tooth and nail
> You still welcomed me with a rose
> My faithful love is the black soil.

The soils with limited production capacity have been well maintained by Anatolian settlers since the Neolithic period by supplying manure, building terraces for conserving the soils against erosion, by water harvesting (the accumulation of rain water for reuse as in cisterns or under-tree pools, etc.) (Kapur et al. 2003), and by employing legumes in crop rotations. The conserved soils of the natural/reworked-leveled and man-made terraces have most probably been the favored media for water harvesting, and for delineating specific stripes, that is, also sectors of estate value. This

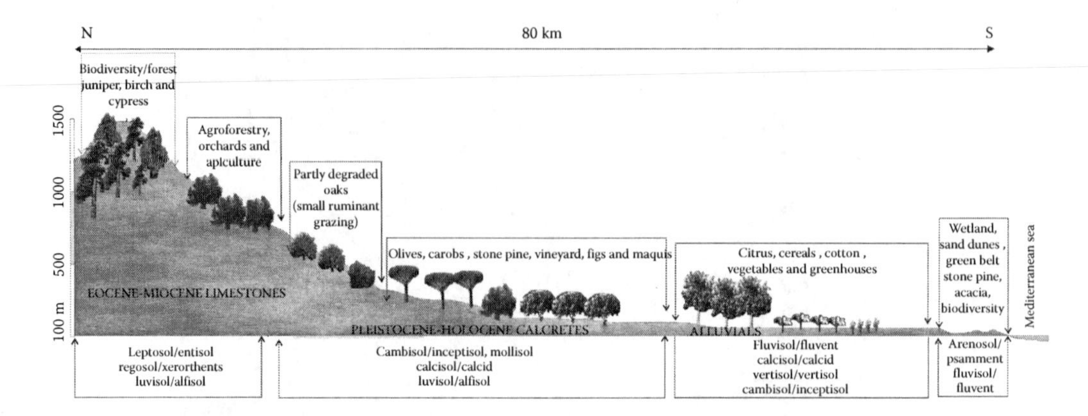

FIGURE 13.9 Mediterranean anthroscape showing dominant land uses (top) and classifications of soils, according to IUSS Working Group WRB (2006)—given first, and Soil Survey Staff (2010)—given second, in each case.

type of unique olive/carob/vine, pistachio/olive/vine, and stone pine/olive/carob Mediterranean anthroscapes (landscapes reshaped by humans in the course of history) are found throughout the coasts of Anatolia (Eswaran et al. 2011) (Figure 13.9). These are comparable to similar landscape developments in other parts of the world, such as the organic material rich "Plaggen Boden" of Northern Europe, and the "Satoyama" landscapes of Japan where the historical paddy fields on sloping land harmoniously integrate the highlands and downstream lands.

13.8 SOIL THREATS IN ANATOLIA

Although soils were regarded as precious and have been a limited resource in Anatolia throughout history, the misuse of the soil dates back to the Roman period, and has caused severe erosion due to deforestation of the highlands of Anatolia (Bal et al. 2003). The increasing population pressure on prime soils of Anatolia since the seventeenth century, coupled with the contemporary migration to this area has caused degradation of soils (Kapur et al. 2006).

Soil sealing (especially urbanization on productive agricultural lands), desertification and soil erosion are the most important land problems and threats to soil in present-day Anatolia. Customary crop production (rain fed) is being abandoned in many parts of the plains with shifts occurring particularly to irrigated cash crops, such as cotton, maize, canola, and soybeans, all of which demand an excessive use of chemicals and water. With irrigation drastic changes have occurred in the cropping pattern of the plains. Because of the mismanagement of the land, some of the main types of degradation in Anatolia are erosion by water or wind, soil salinization and alkalization, soil structure destruction and compaction, biological degradation, and soil pollution. The salinization potential in irrigated plains of Anatolia, such as Harran (SE Anatolia), Çukurova (S. Anatolia), and Menderes (W. Anatolia), poses a risk to increases in the welfare of farmers in the future.

Approximately 86% of the land is suffering from significant soil erosion because of the climatic and topographic conditions of the country (Aksoy et al. 2010). Another important land problem for Turkey is the fractured structure of land ownership. Cultivated land sizes are very small and the ownership of the land is fragmented due to heritage laws. Private farms less than 5 ha constitute 65% of the total arable land which exerts an intense pressure on soils for high-income generation activities via excess irrigation and use of agrochemicals. The latest law on soil protection (No. 5403) in Turkey, which sought to conserve the soils of Anatolia that have been degraded since millennia, was empowered on July 19, 2005.

13.9 CONCLUSIONS

Soils are essential components of the earth and play an integral part in the evolutionary progress of humanity. They have been the major components of the cultural landscapes in the ancient and prehistoric settlements of Anatolia. The first settlers of Anatolia in the Neolithic period built their houses in wetlands, on riverbanks, lakeshores, and sea sides to secure their food, fiber, and fuel resources. There is archaeological evidence for the first farms in the agricultural revolution occurring in Anatolia. In the course of history, several monuments, reliefs, orthostats (an architectural component of the Asyrians and Hittites, a block of rock with relief used as the facade of the buildings particularly temples) and decorations were built to respect the productivity of the fertile soils. Terraces, aqueducts, dams, and irrigation networks, along with cisterns were built in Anatolia for protecting soils, and collecting and conveying water.

However, the harsh topography and uneven distribution of precipitation, along with strong winds, has always threatened the soil in Anatolia; erosion remains a major issue for Anatolian soils. Population increases and migrations through history and until today have also led to the sealing of soils, particularly on fertile plains in the recent millennia. This has caused the irreversible loss of soils.

As in the past, the soils of Anatolia will continue to be an integral source for global food security due to their productivity. Thus, maintaining and improving natural resource management sites for sustainable land management necessitates strong consideration of the environmental friendly past land use practices of Anatolia in order to make recommendations for contemporary conservation management and planning.

There is an increasing demand by stakeholders for detailed soil surveys in Anatolia for sustainable soil and water management, urban development, renovation of cadastral (land registration) plans and construction of infra and ultra structures such as the petroleum pipelines crossing the whole of Anatolia. Raw material/soil excavation for roads, buildings, and brick manufacturing along with the highly developed ceramic industries of Anatolia are other prospective fields that require considerable technical input from detailed state-of-the-art soil survey reports.

REFERENCES

Akça, E., K.M. Çimrin, J. Ryan, T. Nagano, M. Topaksu, and S. Kapur. 2008. Differentiating the natural and man-made terraces of Lake Van, Eastern Anatolia, utilizing earth science methods. *Lakes & Reservoirs: Research and Management* 13: 83–93.

Aksoy, E., P. Panagos, L. Montanarella, and A. Jones. 2010. *Integration of the Soil Database of Turkey into European Soil Database 1:1.000.000.* Milan, Italy: European Commission Joint Research Centre, Institute for Environment and Sustainability.

Akurgal, E. 1962. *The Art of the Hittites.* London: Thames and Hudson.

Bal, Y., G. Kelling, S. Kapur, E. Akça, H. Çetin, and O. Erol. 2003. An improved method for determination of Holocene Coastline changes around two ancient settlements in southern Anatolia: A geoarchaeological approach to historical land degradation studies. *Land Degradation & Deevlopment* 14: 363–376.

Braidwood, L.S., H. Çambel, and W. Schirmer. 1981. Beginnings of village-farming communities in Southeastern Turkey: Çayönü Tepesi, 1978 and 1979. *Journal of Field Archaeology* 8: 249–258.

Dinç, U., S. Şenol, S. Kapur, İ. Atalay, and C. Cangir. 1997. *Soils of Turkey.* Adana, Turkey: University of Çukurova Publication.

Dinc, U., S. Senol, C. Cangir et al. 2005. *Soil Survey and Soil Database of Turkey.* In *Soil Resources of Europe,* 371–375. Ispra, Italy: JRC-IES European Soil Bureau Research Report.

Erkut, S. and E. Reyhan. 2012. Land acquisition and land donation documents in Hittite. *ACTA TURCICA, Thematic Journal of Turkic Studies,* IV(1): 80–86.

Eswaran, H., S. Berberoğlu, C. Cangir et al. 2011. The anthroscape approach in sustainable land use. In *Sustainable Land Management,* eds. S. Kapur, H. Eswaran, and W.E.H. Blum, 1–50. Berlin: Springer.

Fairbairn, A. 2005. A history of agricultural production at Neolithic Çatalhöyük East, Turkey. *World Archaeology* 37: 197–210.

Hirst, K. K. 2009. *The Archaeologist's Book of Quotations*. Walnut Creek, California: Left Coast Press.
Hodder, I. 2006. *Çatalhöyük: The Leopard's Tale*. London: Thames & Hudson.
IUSS Working Group WRB. 2006. *World Reference Base for Soil Resources*. World Soil Resources Reports. Rome: FAO.
Kapur, S., E. Akça, D.M. Özden et al. 2002. Land degradation in Turkey. In *Land Degradation in Central and Eastern Europe,* eds. J.A. Robert and L. Montanarella, 303–318. Luxembourg: European Soil Bureau Research Report Office for Official Publications of the European Communities.
Kapur, S., P. Zdruli, E. Akça, and B. Kapur. 2003. *Natural and Man-Made Agroscapes of Turkey: Sites of Indigenous Sustainable Land Management (SLM)*. Briefing Papers of the second SCAPE Workshop in Cinque Terre (IT), 13–15 April 2004.
Kapur, S., E. Akça, B. Kapur, and A. Öztürk. 2006. Migration: An irreversible impact of land degradation in Turkey. In *Desertification in the Mediterranean Region. A Security Issue*, eds. W.G. Kepner, J.L. Rubio, D.A. Mouat, and F. Pedrazzini, 291–301. NATO Security Through Science Series Volume 3, Dordrecht Netherlands: Springer.
Kealhofer, L. 2005. *The Archaeology of Midas and the Phrygians*. Philadelphia: University of Pennsylvania Museum Press.
Okay, A.R. 2008. Geology of Turkey: A synopsis. *Anschnitt* 21: 19–42.
Rosen, A., and Roberts, N. 2005. The nature of Çatalhöyük: People and their changing environments on the Konya Plain. In *Çatalhöyük Perspectives: Reports from the 1995–1999 Seasons*, ed. I. Hodder, 39–54. Cambridge: McDonald Institute and British Institute of Archaeology at Ankara Monograph.
Schmidt, K. 2010. Göbekli Tepe—The stone age sanctuaries: New results of ongoing excavations with a special focus on sculptures and high reliefs. *Documenta Praehistorica* 37: 239–256.
Soil Survey Staff. 2010. *Keys to Soil Taxonomy*, 11th edition. Washington, DC: USDA-Natural Resources Conservation Service.

14 Deconstructing the Leipsokouki

A Million Years (or so) of Soils and Sediments in Rural Greece

Richard B. Doyle and Mary E. Savina*

CONTENTS

14.1 INTRODUCTION

In the summer, this is a landscape of light grays, tans, and greens (Figure 14.1). Fields of wheat stubble cover all the flatter parts of the landscape, on the uplands and terraces. The moderately dense oak woodlands up-valley, near Rodia, (Figure 14.1) transition to scattered oak trees down-valley, near Mega Sirini (Figure 14.2). Where there is no green oak or tan straw, there is light gray mudstone—the gray bedrock seems close to the surface everywhere, and it is exposed in incised side stream valleys and slope breaks. There is little water in the stream and the summer heat is only occasionally relieved by a passing thunderstorm. This is the late summer view of the Leipsokouki Valley, 20.5 km^2 (square kilometers), in the prefecture of Grevena, western Macedonia, Greece.

Just about every centimeter of reasonably flat and gently sloping ground is used for cropping, and that is because the soils are fertile, especially for dryland crops such as wheat, rich in minerals such as calcium, magnesium, and potassium. Beneath the highest flat surfaces, near the drainage divide, the soils are deep and rich in clays (particles about a thousandth of a millimeter in size), calcium carbonate has leached (washed out) of their upper layers and redeposited lower down as hard carbonate nodules—all signs of lengthy soil-forming processes. On the prominent gently sloping surfaces partway down the slope, many soils are somewhat less clay-rich, more calcareous throughout, and—importantly—have in places been covered by material moved down the steep slopes, on

* Tasmanian Institute of Agriculture, School of Land and Food, University of Tasmania Private Bag 54, Hobart TAS 7001 Australia; Email: richard.doyle@utas.edu.au

FIGURE 14.1 **(See color insert.)** View of upper Leipsokouki catchment looking northwest. Photo taken after wheat harvest; fields are covered in light tan wheat stubble. Gray areas are exposed mudstone bedrock. Most trees in view are oaks (*Quercus* sp.). Note narrow valley floor.

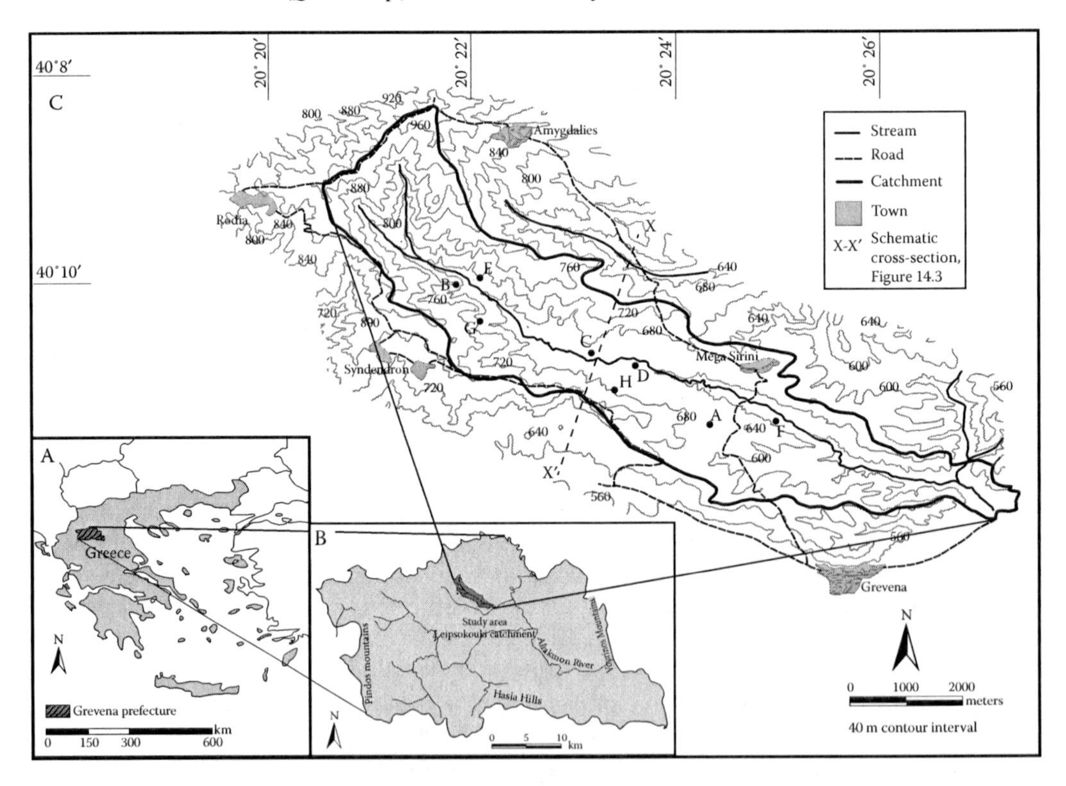

FIGURE 14.2 Index map of the Leipsokouki catchment. Inset A shows location of Grevena prefecture within Greece. Inset B shows the location of the Leipsokouki catchment within Grevena prefecture. A–H: Sites mentioned in text. Sites A through H correspond to the following locations from Doyle, 2005 (available at http://eprints.utas.edu.au/2346/): A = P7a; B = C6; C = C14, P30; D = P33; E = C13; F = P37; G = P47; H = P60. Dashed line X–X′: approximate location of schematic cross section (Figure 14.3).

which younger soils are now forming. On these surfaces, then, is a mosaic of many soil types. The soils on the flat surfaces just above the stream channel are young and not well developed—similar to the poorly developed soils of the steeper hillslopes.

Grevena is landlocked: it has none of the harbors and beaches characteristic of most of Greece and no olive trees, either. Even the goldsmiths in the Macedonian capitals of Pella and Vergina crafted their victory wreaths with acorns and oak leaves, not olives and laurels, back in the time of Philip II, Alexander's father. There are no monumental ancient structures: no theatres, no palaces, no elaborate citadels. In the mid-1980s, when we started our work, Greveniots asked us why we were there—"nothing interesting happens here now, and nothing interesting has happened in the past—and no one comes here." Grevena invites us to study an evolving rural landscape where the soil resource is as crucial in the present as in the past, because rural areas such as Grevena feed cities. As our anthropologists relay the stories they hear, we realize that we are learning a story comparable to those we know from the post-European rural histories of our home countries of Australia, New Zealand, and mid-America, a story of changing climates and farming technologies, changing expectations of the hinterland, and changing soils and landscapes. The search for the causes of these changes becomes a massive "whodunit"—is this mosaic of "fresh" gullies and valley fills, with remnant deep soils on ridgelines, controlled primarily by climate? By tectonic activity? By ancient, even geologic history? By human actions? By some combination? The story in the Leipsokouki goes back at least 14,000 years, maybe more, in contrast to the few centuries of intensive post-European-settlement use of the land near our homes. In an age of changing climates, with more intensive agriculture globally, perhaps figuring out the past history of a place such as the Leipsokouki will be not only an interesting academic exercise but also a map of a possible future.

Actually, Grevena has each summer attracted one group of foreigners since the late 1960s: earth scientists interested in on-land evidence for crustal plate movements. Two huge slices of ocean crust and mantle flank Grevena on the east and west: the Vourinos and Pindos ophiolites. All you need to know about ophiolites for this story is (1) that their rocks make for odd soils (too rich in some elements and too poor in others), but when they are weathered and mixed with other materials, they provide some essential soil mineral nutrients and (2) that you can walk more easily to the Moho (the crust/mantle boundary) in Grevena than practically anyplace else in the world—and certainly infinitely easier than via a 10 km deep borehole through an ocean basin (Rassios 2004).

An additional complication in the Leipsokouki also makes deciphering its history an interesting challenge. Not only are there no ancient structures, there are also no places in the valley or on the slopes where continuous, multiple layers of material have accumulated. What we have are remnants, preserved accidentally—a few ancient soils in the uplands, some pockets of transported slope materials (colluvial fills), and some discontinuous deposits exposed along the stream valley and its tributaries (valley fills and fans). Everything, of course, is covered with soil, but the nature of that soil varies over short distances, and does not necessarily reflect the soil factors from the place where it is found—because it used to be somewhere else (Doyle 1990, 2005). Imagine putting together a puzzle from 10% of the parts, most of which have a good bit of the pattern scratched off. Reconstructing this history means finding, describing, and interpreting these scattered remnants individually and together.

Why so little evidence? To a geomorphologist studying how landscapes form, this one is incising, that is, the main stream and its tributaries are cutting down through the bedrock (Savina 1993). Hence, the exposures of the mudstone on the slopes, in the ravines, and in the valley bottoms. There is little room for anything in those valley bottoms, so it is no wonder that the roads and settlements, ancient and modern, follow the uplands (Figure 14.2). Not only do the uplands have the broadest flat areas, covered by the deepest young sediments, but they also have the deepest soils and springs that are the water sources for the villages. We think of this as an "upside-down" landscape—the broadest, flattest areas are at the higher elevations, where in other places, they might be in wide valley floors adjacent to stream channels.

Two mechanisms have moved sediment down the hillslopes and streams in the Leipsokouki catchment: mass movement (gravity) and running water. Mass movement has left hillslope material (colluvium) on the slopes, on fan-shaped deposits at the base of slopes, and in the valley floor. Running water has mostly contributed bedded gravels and finer material (alluvium) to the valley fills but on some slopes has also produced thin layers of fine-grained material (sheetwash deposits—formed by thin layers of water moving sediment). In a stream basin such as the Leipsokouki, without continuous water flow, these processes and deposits are linked. We recognize alluvium by its characteristic bedding, moderate grain sorting, and grain-to-grain support. Some alluvial beds can be traced along the stream, at least over short distances. These characteristics allow us to say with some confidence that these deposits formed by running water coming down the center of the valley. Most likely, the material was transported and deposited during prolonged wet weather with moderate rainfall intensities (of the sort that Grevena experiences in the spring and autumn). Other stream valley deposits, however, have little obvious bedding. They are poorly sorted, with a matrix of fine material that supports the larger grains. These deposits have come down hillslopes and steep side valleys into the valley bottom by mass movements such as landslides and debris flows. Probably, these slope failures happened during intense rainfalls. The volume of the slope deposits exceeded the capacity of the main stream to move them very far, as might happen during an isolated, large summer thunderstorm when there was little water in the main channel.

The two kinds of deposits, formed by running water (alluvium) and mass movement (colluvium), constitute the valley fills (Figure 14.3). Certain valley fills in the Leipsokouki, particularly the older ones and those in the central and upper parts of the catchment, contain a high percent of slope

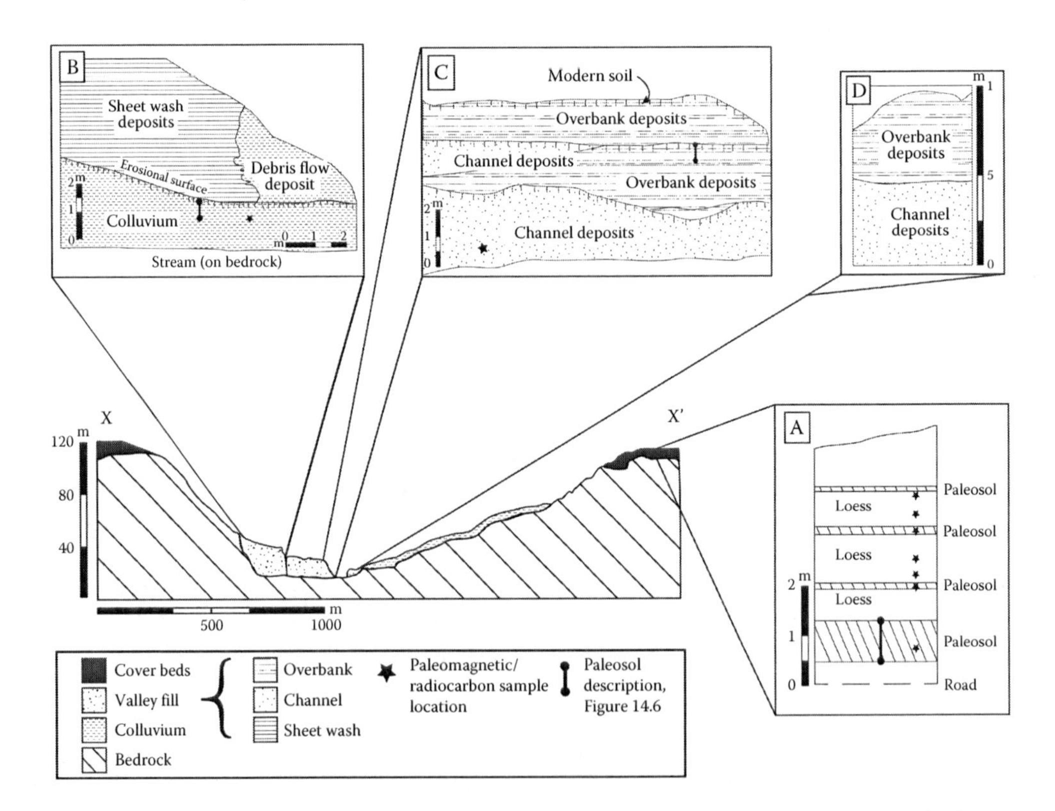

FIGURE 14.3 Schematic mid-catchment cross section looking downstream. Insets A, B, C, and D show stratigraphy at sites A, B, C, and D, respectively, projected onto the cross section at appropriate locations. Dates and descriptions of sections: Site A: Plio/Pleistocene loess/paleosol sequence: All samples are paleomagnetically reversed. Site B: Syndendron valley fill deposits on pre-Syndendron colluvium: 14,750 ± 1000 cal years BP (Wk 9816). Site C: Sirini valley fill deposits: 4150 ± 500 cal years BP (Wk 9919). Site D: Leipsokouki valley fill deposits: 140 ± 130 cal years BP (Wk 9926) (date on colluvium above floodplain deposits in a nearby section (C18 of Doyle 2005)).

deposits, compared to "regular" stream alluvium. The geometry of the catchment, in addition to variations in precipitation, must be one of the reasons. The Leipsokouki catchment is elongated in the NW-SE direction, with a ratio of length to width-at-mid-catchment of about 5:1. Tributaries to the axial (main) stream are short and steep—just the kind of geometry that promotes mass movements such as debris flows, which then dump material straight into the main valley, forming a highly variable suite of fans and other valley floor deposits.

The slope deposits now in valley fills must have come from somewhere. In the Leipsokouki, the two most likely sources are directly from the steep valley sides or through one of the tributary systems. In either case, the "hole" left by the slope movement is filled in over time with new material moving in from the hilltops and valley sides. These are the colluvial fills on the hillslopes. Thus, stream deposition and hillslope deposition are sometimes out of sync in the Leipsokouki, with deposition and soil development on hillslopes occurring slightly later than stream deposits. One can think about the cycle this way: When the main stream incises (as is the case now), the short and steep valley side streams and the hillside gullies become connected to the long main channel. The drainage is integrated, and a particle of sediment has only to reach the side channels to have a chance of making it to the main valley bottom. However, when enough deposits block the main channel and the stream junctions, sediment can no longer get into the main stream. Incision temporarily ceases, depressions and channel heads fill with colluvium, and soils begin to form on the new deposits. The landscape is temporarily stable, awaiting the next phase of incision, which might be touched off by an earthquake, by collapse of thick, weak material, by a climate change (for instance, more torrential rains), by vegetation clearing, or even by something more random. As if this was not complicated enough, different places in the catchment will react differently.

In order to decipher "what happened when," geomorphologists and soil scientists working on materials less than 40,000 years old rely on soil development and radiocarbon dates. In areas with an archaeological record, such as Grevena, we can also use the presence or absence of archaeological materials in the deposits as a dating tool—if the materials are diagnostic of particular periods in an archaeological chronology.[1] Grevena has a thin record of settlement from many periods (Figure 14.4), but fortunately, many of the scraps of pottery from its deposits are recognizable. If a deposit contains fragments that are recognized as Hellenistic, for instance, that deposit has to be of Hellenistic age—or younger. If a deposit contains fragments from a variety of periods, it must be younger (or the same age) as the youngest of those periods.

To understand how soil development and archaeology help us date valley and hillslope deposits, it is useful to go outside the Leipsokouki, to the Xeropotamos on the east side of the Aliakmon, about 14 km northeast of Mega Sirini. (Xeropotamos—what a great name!—much more interesting than "dry river"). Soil development is a little clearer there than in the Leipsokouki, because of a less complex parent material—river sands and gravels. Of the three river deposits in the Xeropotamos (Figure 14.5), the two youngest can be dated archaeologically. The youngest, only about 2 m above the modern stream, contains Hellenistic and Roman pottery fragments and must be younger than 2500 years old. Soil has just started to form on this deposit. The second fill, approximately 3.2 m above the modern stream, contains abundant Early Neolithic pottery fragments. From their large size, angular corners and relative fragility, we know they cannot have traveled very far. In fact, a series of post holes associated with the pottery indicates to us that the material is *in situ*: this is an occupation site. The soil formed in and around the Neolithic fragments, therefore, tells us what a soil about 8000 years old looks like. It is much more developed than the younger soil; it is browner and clay has been relocated from the upper part of the profile (the vertical sequence of soil layers) to lower down. Above these two deposits, about 20 m above the modern stream, a redder soil with even more clay has formed in much older material. Unfortunately, there are neither archaeological materials nor charcoal to allow us to date this oldest soil—but the color, degree of development, and position in the landscape indicate that this soil is at least 15,000 years old. When we see reddish soils with marked clay concentrations in other parts of Grevena, we can be confident that they are older than Holocene (the last 11,700 years of geologic time). A group of soils differing in age but

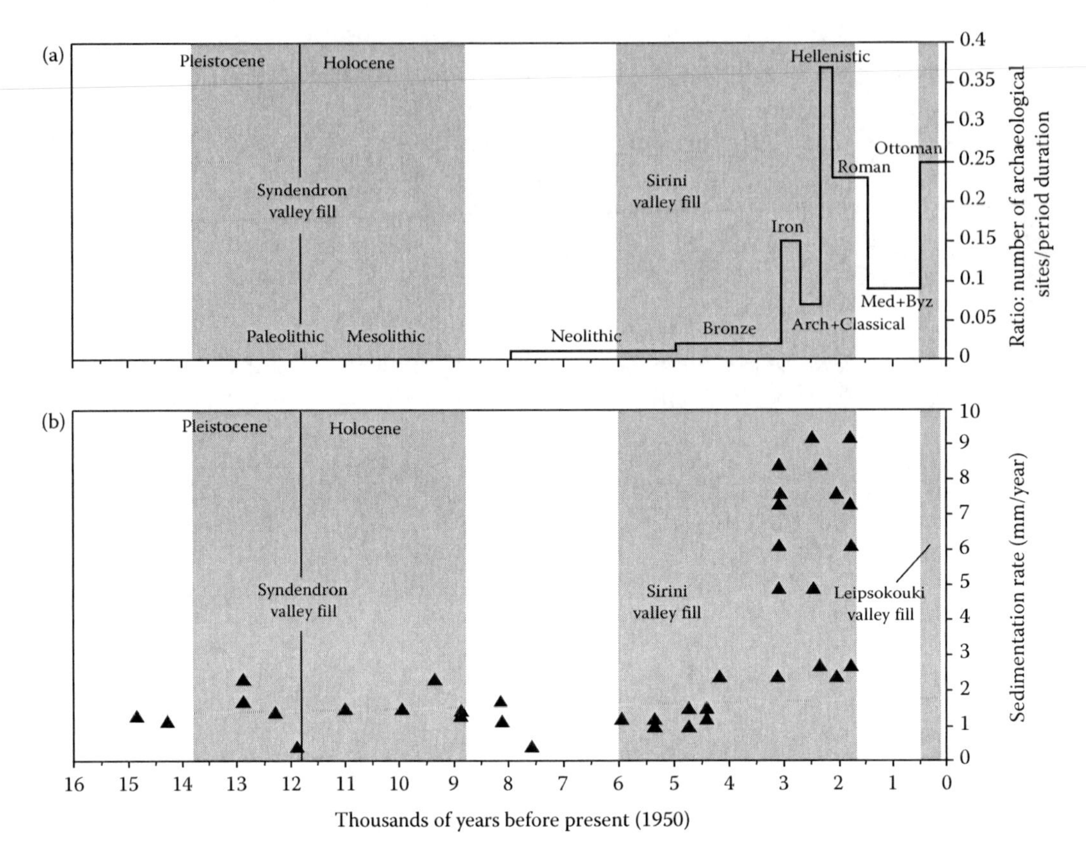

FIGURE 14.4 (a) Bar graph showing number of sites recorded by Grevena Project in the prefecture (1986–1990) for each archaeological period divided by the duration of the period (site numbers ÷ time). Height of each bar is a rough measure of the intensity of settlement during that period. (b) Sedimentation rates (mm/year) for valley fill exposures calculated from radiocarbon dates. Sedimentation rate = thickness of sediment ÷ duration of sedimentation. Shaded vertical columns show the approximate dates and durations of the Syndendron, Sirini, and Leipsokouki valley fills. See Note 1 for discussion of the archaeological and geological periods.

similar in all other respects is called a "chronosequence"—and we are very lucky to have one in Grevena. Similar schemes that relate soil characteristics to ages of soil development have also been developed in other areas of Greece (Runnels and Van Andel 2003).

14.2 AS IT HAPPENS: THE PRESENT DAY

We will tell the story of the Leipsokouki from several key exposures. As a first step, it might be helpful to figure out what is going on now.

Underneath the wheat stubble shown in Figure 14.1, the modern soil on these flatter areas is thick, black, uniform vertically, and clay-rich. During summers and other prolonged dry periods, it is cracked at the surface—and the cracks can extend more than a meter below the surface. It is nearly impossible to shovel out a soil pit because the soil is so hard and compact when it is dry. And when we have finally exposed the soil, the clayey surfaces that break off are shiny and covered with faint streaks. Put all of these characteristics together and we have a Vertisol: a soil with such a high concentration of clays, particularly ones that swell when wet and shrink when dry, that it naturally turns itself over (see Figure 14.6d for a modern Vertisol profile at Site H). The polished surfaces and streaks ("slickensides") show where masses of soil have rubbed against each other while swelling and shrinking.

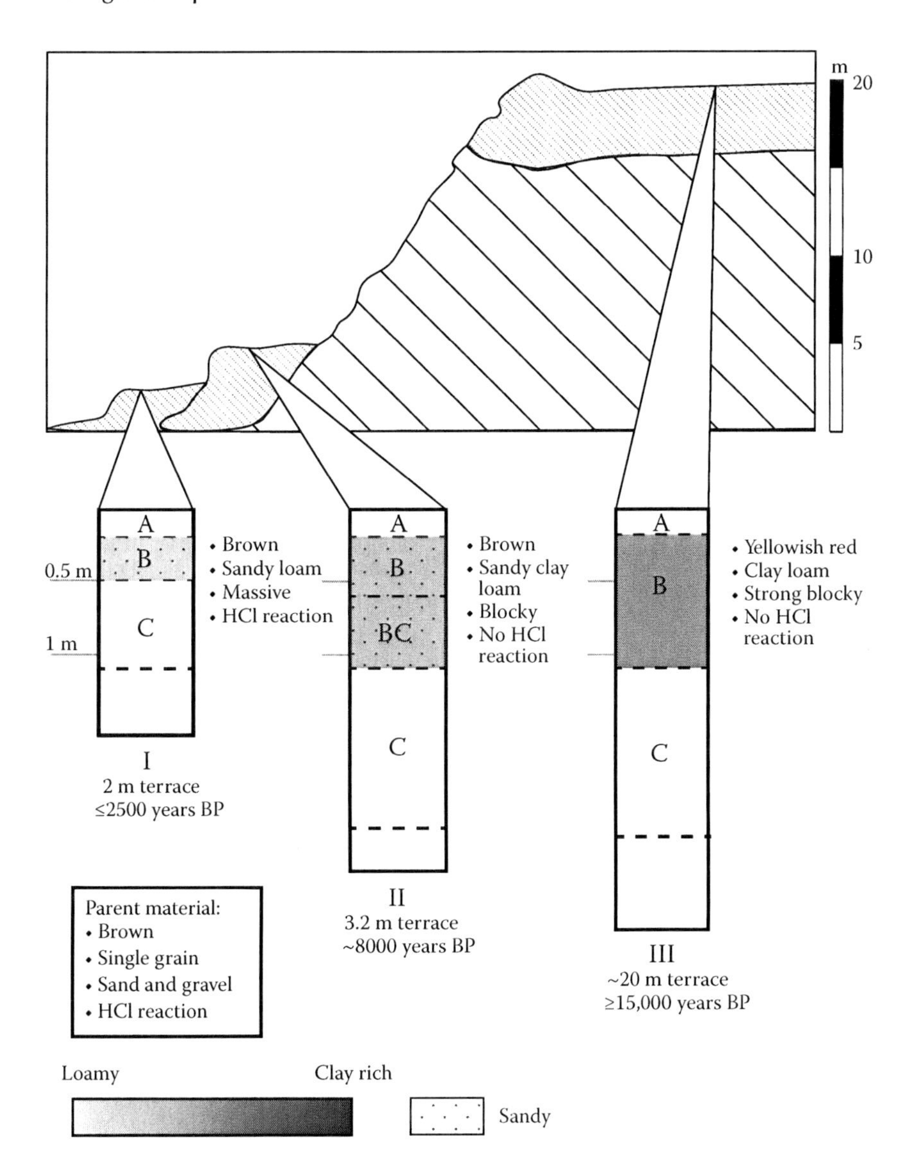

FIGURE 14.5 Soil properties at Xeropotamos chronosequence. Numbers on the left of profiles are depths below ground surface. On the soil profiles, letters A, B, and C refer to particular horizons. In the text, A horizons are referred to as "topsoil." B horizons are referred to as "subsoil." C horizons are slightly altered soil parent materials. The figures for the soils include the following properties from standard soil horizon descriptions. They are included here to highlight the contrasts between horizons and profiles: Color (from Munsell soil color chart); texture (grain size); structure (type and degree of aggregation); reaction to hydrochloric acid (a measure of calcium carbonate content).

These Vertisols are both good and bad for farming. On the plus side, they are rich in the mineral nutrients plants need to grow. Ultimately, this is thanks to eroded material from the ophiolites, which retains high levels of calcium and magnesium even when weathered and transported. The climate also helps. Rainfall in this region is low enough that the mineral nutrients do not leach out of the soil. And the self-turning properties of the Vertisols (they "inVert" or "self-mulch") means that nutrients and organic carbon are relatively evenly distributed through the upper part of the soil profile. No matter how much material is removed from these soils by erosion, they will still maintain a reasonable fertility, especially for dryland grains.

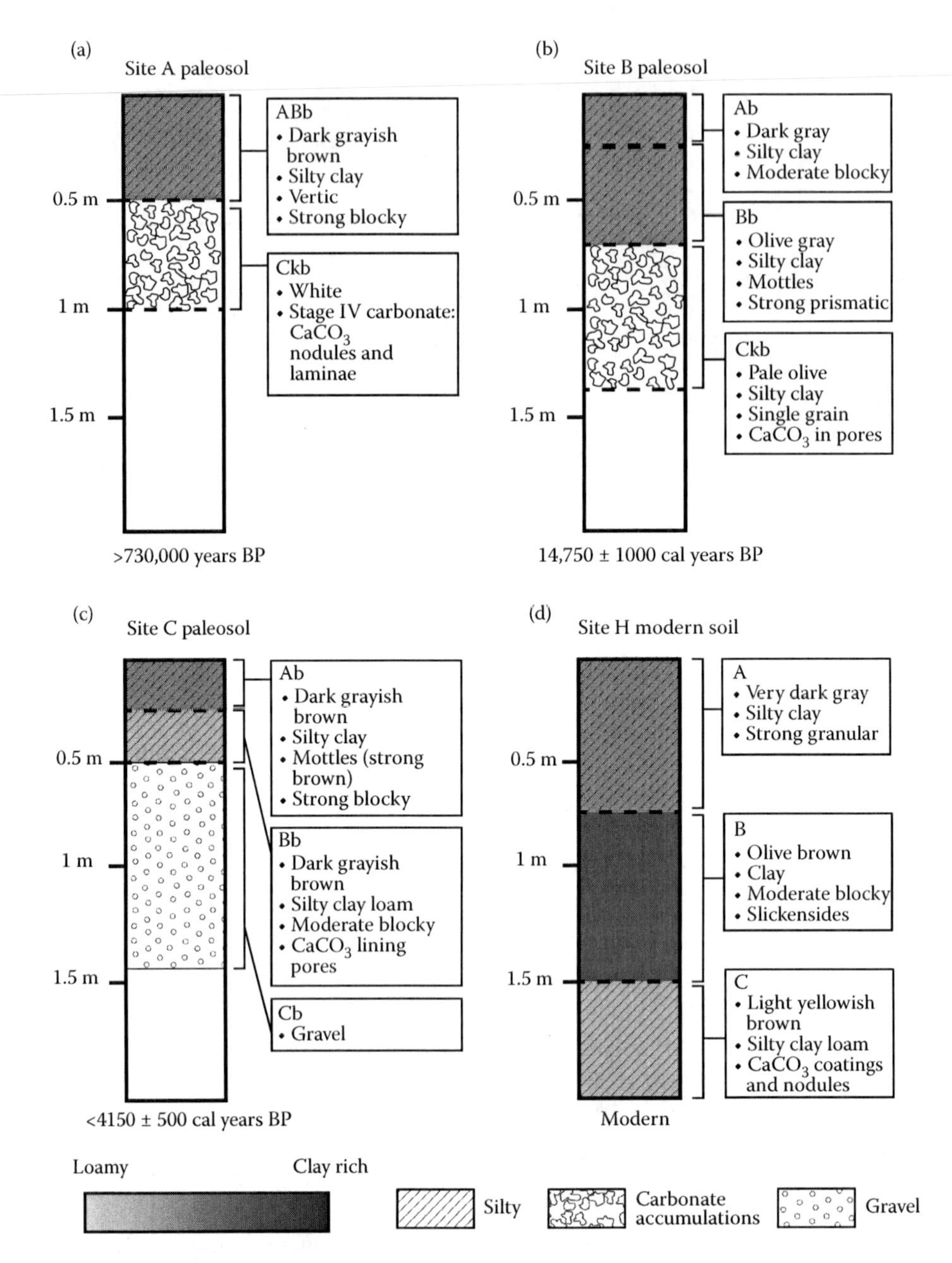

FIGURE 14.6 Soil and paleosol properties at selected sites: (a) Site A paleosol in Plio-Pleistocene loess/paleosol sequence. (b) Site B paleosol below Syndendron valley fill (date of 14,750 ± 1000 cal years BP (Wk 9816) from top of this paleosol). (c) Site C paleosol within Sirini valley fill (date of 4150 ± 500 cal years BP (Wk 9919) from this section). (d) Site H modern (strongly developed) Vertisol profile. Numbers on the left of profiles are depths below ground surface in (d) and depth below the top of the buried soil in (a), (b), and (c). Because of erosion, the buried soils were probably thicker at one time. On the soil profiles, letters A, B, C refer to particular horizons. In the text, A horizons are referred to as "topsoil." B horizons are referred to as "subsoil." C horizons are slightly altered soil parent materials. "b" means "buried." "k" means horizon contains accumulated calcium carbonates. The figures for the soils include the following properties from standard soil horizon descriptions. They are included here to highlight the contrasts between horizons and profiles: Color (from Munsell soil color chart); texture (grain size); structure (type and degree of aggregation); mottles, slickensides, and other vertic features, and description of calcium carbonate aggregates in profiles where these features are present.

However, the high clay contents make for soils that are challenging to cultivate. In late fall through spring, the predictable rainy season, the soil becomes a mass of wet, heavy, sticky clay. You can double the weight of your shoes by hiking through this soil for only a few minutes. In mid-late summer, before the rains start again, the soil is so hard and dry that you can barely scratch the surface, unless you have a big, powerful tractor and a strong plow. Before mechanized agriculture arrived in Grevena in the 1960s or so, cultivating and cropping these soils would have been a real challenge. And one downside of big machines is that they tend to tip over when operating on sloping ground along the contours. So the tractors and harvesters run up and down the slopes. Many current farmers burn and till their fields after harvest, creating bare ground with newly broken up soil and furrows running up and down the slopes, just in time for the rains to erode the soil down the slopes.

In this broken-up landscape, the more steeply sloping areas are not now cultivated. Many are gray, the color of the bedrock mudstones and the thin soils that overlie them (Figure 14.1). These are also the parts of the landscape with trees, some of which were cleared from the gently sloping areas as mechanized farming took hold and small fields merged into larger ones. When crops are growing, these are also the only parts of the landscape where sheep and goats from village flocks can graze.

Grevena has always been a rural area, and the mix of landscape uses we see now is probably a decent representation of life in more ancient times, judging from the types of artifacts found in some of the colluvial deposits (see discussion of Site G, Section 14.3.5). However, the pattern of cultivation on the landscape may have looked quite different in antiquity, because of the challenges of cultivating the thick, clayey soils on the more gentle slopes without machinery. It is speculation, but perhaps farmers in antiquity cultivated on the slopes, just as much as on the flat areas—and, of course, gravity being what it is, these steeper slopes will erode. Another possibility is that farmers in antiquity may have made use of the cracking and self-mulching properties of the Vertisols and broadcast grain seeds by hand without tilling (Limbrey 1990). Grains falling into the cracks would be watered by the initial rainfalls in autumn and still have access to nutrients that are distributed through the entire soil depth because of the shrink-swell process.

During our field work, completed over a period from 1986 to 2001, we also saw the effects in Grevena of population and agricultural trends elsewhere. If the present is the key to the past, as earth scientists are fond of saying, then these observations may help us as we explain trends of the last 15,000 years or so. Extreme weather events, namely drought in the late 1980s, led to out-migration from the area, as farming became more difficult. Countering this trend were the effects of Greek integration into the European Community; money flowing into the Greek agricultural sector supported traditional grazing and promoted experimental cropping of unusual crops such as sunflowers and (in Crete) bananas. North of Greece, two countries, Yugoslavia and Albania, lost their totalitarian governments in short order during this period. Yugoslavia literally fell apart, as the death of Marshall Tito dissolved the tenuous glue that held together its pieces and this part of the Balkans was at war for about a decade. Albania's collapse was less violent, but no less radical: an impoverished country that for years had been completely isolated suddenly opened up, and much of its male population migrated away—into Greece—for better opportunities. Many of these Albanians found work as farm hands because rural areas of Greece such as Grevena became relatively depopulated in the late twentieth century.

What do these population trends mean for land use? First of all, because tractors can do the work of many people and mules, it is likely that the same amount of land, or even more, is now cultivated in central Grevena, including the Leipsokouki, even if there are fewer residents. (However, this is not true for the whole Grevena prefecture; in the Pindos foothills to the west, fields that were cultivated earlier in the twentieth century have now been abandoned and are used for pasture.) Other traditional land uses have also changed. Mechanized agriculture means farmers need not keep horses and mules. Transhumant grazers, who move long distances between summer pastures in the Pindos Mountains and winter pastures at lower elevations (famously described by Wace and Thompson in 1914), may have declined in numbers, although oral tradition and recent field observations conflict on this point (Chang and Tourtellotte 1993). Within the Leipsokouki, numbers of sheep and goats kept in village flocks remained relatively constant in the twentieth century (Doyle 1990), meaning continuing grazing pressure on the land, particularly noncultivated areas, including the need to access and harvest oak branches for winter feed. In parts of Grevena, trees have regrown, because gun emplacements, installed during World War II and the subsequent Greek Civil War, now look into thick forests and would be of little use today unless those trees were gone, as they must have been in the 1940s and the 1950s.

At first blush, the gullies exposing fresh bedrock look as if they have been there forever, and that they are still actively altering the landscape (Figure 14.1). It turns out, though, that healed gullies, partly covered with oak trees, lie adjacent to the very fresh ones. And trees are slowly stabilizing some vertical faces of even the fresh-looking ones (Oliver Rackham, pers. comm.). If the gullies were more active in the past, then we cannot attribute their formation to the modern, mechanized agriculture that probably began to arrive in Grevena in the 1960s, though it is likely that the cropping system (including the yearly burning of wheat stubble, fall plowing, and restriction of sheep and goats during the growing season to the very ravines and slopes that are most erodible) is not ameliorating erosion. Did the landscape, with its gullies, always look like this? Was the 10% or so of the Leipsokouki with bare rock exposures usable in the past? The answers to these questions are in remnant deposits of stream sediments and hillslope debris.

14.3 WORKING BACKWARDS: THE REST OF THE HISTORY

14.3.1 At the Catchment Divide

A few million years ago, the Leipsokouki catchment looked nothing at all like it does today. The gray mudstones had been deposited, uplifted, and tilted. No longer under the ocean, these rocks were now exposed at the surface. Naturally, they eroded—but eventually new kinds of material began to accumulate in a basin on top of them. In parts of the Leipsokouki, at least 120 m thickness of this material is preserved; in other locations in Grevena, up to 200 m are preserved (Chatzipetros et al. 1998), part of a wedge of sediment that thickens to the south to up to 600 m (Doutsos and Kokkalas 2001). First, braided streams deposited beds of gravel, now exposed in the lower reaches of the Leipsokouki valley and elsewhere. Later, finer grained material—relatively uniform in grain size (silt, fine sand, and a little clay)—was deposited on top of the gravels (Doyle 2005). Deposition was not continuous; during periods when it slowed, the land surface was sufficiently stable for soils to form. Today, the remnants of these fine-grained deposits are exposed near the broad drainage divides in the Leipsokouki (Figure 14.2, Site A; Figure 14.3, inset A). They (and the gravels, for that matter) have been completely removed from the center of the Leipsokouki basin by the down-cutting of the main stream and its tributaries.

Earth scientists reading about these fine-grained deposits will, by now, have lightbulbs going off in their heads. These catchment divides in Grevena are just a small area of global deposits of wind-blown fine sediment, called "loess." Sequences of loess deposits, with intervening soils, characterize many continental interiors, especially those that had ice sheets nearby during the last several million years. The material that forms the loess originates as rock flour, produced as stones carried by glaciers grind against the glacial bed. This material is carried by running water under or near the glacier and deposited in streams and lakes out in front of the glacier. From there, it is easily picked up by the wind.

Thick sequences of loess deposits in places such as China, New Zealand, and Alaska have been used to determine the timing of glacial advances and other climate changes. The old soils that separate the loess deposits formed in periods generally warmer and wetter, when the ground surface was stable enough to allow soil-forming processes to outstrip deposition.

The section at Site A consists of five units of loess—they also occur at the top and bottom of the sequence—separated by four buried soils (paleosols) (Figures 14.3 and 14.6a). These "old soils" are key to understanding how long these deposits took to form.

We are so used to the look of the soil under our feet that it is hard to imagine what the ground looked like before soil started to form. At the Xeropotamos site, mentioned above, the soil formed in loose gravelly and sandy river deposits. At Site A, now the boundary of the Leipsokouki catchment, the ground surface at first was the loose dust of the loess. Once a few plants sprouted, their roots grew, holding the soil in place, and their leaves fell and decomposed, adding organic matter to the ground surface and below. The color changed from light tan to brown. An early indication that soil is starting to form is the darkening of the topsoil, related to organic matter accumulation, decomposition, and mixing.

Vegetation helped stabilize the ground surface and soil continued to form. Small mineral particles forming the loess started to decompose and the loess compacted. Released nutrients, such as potassium, helped the plants grow, and what was left behind from the original minerals became clays, dominated by aluminum, iron, silicon and oxygen. Helped along by water, the clay accumulated in the subsoil, which also became redder. Finally, calcium, released by weathering of the mineral grains or blown in with wind and rain, accumulated in the subsoil as calcium carbonate cement.

The differing color, clay concentration, and thickness of the surface soils at the Xeropotamos showed us how to recognize and differentiate surface soils of Holocene age (less than 11,700 years old) from older soils. But the soils at Site A ("paleosols" or "old soils") are buried, so what can we say about them?

Let us focus on the deepest of the exposed soils (Figure 14.6a). In addition to a remarkable concentration of clay, a dark color, and thickness, all of which suggest great age, it also has a horizon (a "soil layer") that is completely cemented by calcium carbonate. In fine-grained deposits such as loess, such a layer takes hundreds of thousands of years to form (Machette 1985). Although the three soils overlying it at Site A are less well developed, they still record thousands of years of soil formation. Moreover, at least 16 more ancient soils are exposed in the top 70 m of these deposits along a new, longer road cutting nearby. When we tally up the amount of time represented by this exposed sequence of loess deposits and paleosols, based on soil development alone, the duration represented amounts to between 2.1 and 6.9 million years. If we make some reasonable assumptions about how many buried soils might be below the exposed sections, the total age might be between 2.4 and 7.7 million years.

Other indications of the antiquity of these materials come from the fossils in them and from a peculiar feature of natural materials—their ability to retain the direction of the Earth's magnetic field at the time they were deposited. Everyone knows that elephants no longer roam through Europe and everyone knows that the magnetic field of the earth points north. As recently as 730,000 years ago, though, the magnetic field of the earth was reversed—and elephantimorphs (creatures related to elephants) roamed through Grevena, as recently as 160,000 years ago and possibly as long ago as 4–5 million years.

The elephantimorph and other vertebrate remains found in deposits comparable to those of the Leipsokouki catchment divide are three distinct ages. The oldest are skeletal remains of *Mammut borsoni*, a mastodont, found in gravels in Milia just 9 km northeast of Mega Sirini (Tsoukala 2000), thought to be between 3.6 and 5 million years old (Pliocene epoch, for those keeping geologic time). Several types of mammals, including deer, a giraffe, horse, bear, and wolf, found in a deposit near Dafnero, 10 km further east, date to between about 1.8 and 2.6 million years ago (Koufos 2006) and a skeleton of *Elephas antiquus* from Grevena city, near the mouth of the Leipsokouki is dated to 160,000–170,000 years ago (Tsoukala and Lister 1998; Tsoukala et al. 2010).

The Earth's magnetic field flips every now and then between its current position (N magnetic pole near the N geographic pole) and the reverse (N magnetic pole near the S geographic pole). The record of those inversions can be used to help date rocks and sediments all over the world. The last flip—from a reversed position to today's "normal" position—happened about 730,000 years ago, after the sediments now exposed along the divide accumulated (all the upper sediments at Site A are reversely magnetized) (Doyle 1990, 2005). These paleomagnetic results tell us that the top of the section at Site A is at least 730,000 years old—Pleistocene in age.

Most scientists believe that loess deposits and their intervening paleosols record climate changes—such as from glacial to interglacial—in the regions where they are now found (e.g., Marković et al. 2011). Even though there were no glaciers in central Grevena, near the Leipsokouki, glaciers did form in the nearby Pindos Mountains to the west (Figure 14.2), where the oldest preserved deposits are more than 350,000 years old (Hughes et al. 2006). Cold temperatures during glacial advances would have inhibited plant growth, so there was not much to stop wind from blowing dust from these glaciated areas, and ones even further away, into the Leipsokouki area to accumulate as loess. The paleosols probably formed during warmer and wetter interglacials.

The time period in which these Plio-Pleistocene beds formed may overlap with the earliest evidence of hominins [the group consisting of modern humans, extinct humans, and their direct ancestors; hominid is a more general term that also includes the other great apes and their ancestors] in the Grevena area (Harvati et al. 2008). Some hominins were probably in Greece perhaps as much as 500,000 years ago (Tourloukis and Karkanas 2012). Artifacts as old as 200,000 BP have been discovered south and west of Grevena (e.g., Runnels 1995; Runnels and Van Andel 2003). These early humans were certainly affected by the climate changes that ultimately controlled deposits like those in the Leipsokouki; most workers (e.g., Roebroeks and Villa 2011) believe early humans had little influence on the landscape.

There is nothing left of these Plio-Pleistocene loess and paleosol deposits on lower slopes or in the valleys—at least, not in place—so either these deposits never formed where the valley is now, or (more likely) they have been thoroughly eroded away. As the Leipsokouki main stream and its side tributaries have elongated headward and cut downward with time, more and more of the drainage divide deposits have eroded away. The mass of fine-grained material still at the drainage divides is like a loaded gun, ready to erode upslope and deposit downslope once the trigger has been pulled. That trigger is renewed down-cutting of the stream and its tributaries. Near the drainage divides the landscape is still stable, whereas the areas downslope and downstream seem to be in an endless cycle of erosion, valley filling, and (temporary) stabilization.

We do not know exactly when the last loess was deposited and the down-cutting of the Leipsokouki started; it probably happened at the same time as the down-cutting of the Aliakmon River, because erosional remnants of the Plio-Pleistocene beds are stranded along terraces of the Aliakmon (Harvati et al. 2008). The transition from sediment accumulation to uplift and incision may be related to large-scale earth movements ("tectonics") in Grevena, which is in a part of the Aegean that is pulling apart, the visible consequences of which are faults (earth fractures that move in earthquakes) and isolated basins, many with lakes (Chatzipetros et al. 1998; Goldsworthy and Jackson 2001). A fault at Site A that displaces the loess–paleosol sequence (and therefore must be younger than these deposits) is one local indication of tectonics as was a damaging 6.6 magnitude earthquake in 1995 centered in Grevena SE of the Leipsokouki (Stiros 1998). In fact, the incision of the Aliakmon probably touched off the down-cutting of all its tributaries, because it changed the level to which all streams above it flow, just as a change in sea level will change the behavior of rivers that flow to a coast. Whenever it began, this down-cutting pulled the first trigger, touching off the waves of instability in the Leipsokouki catchment that persist even today.

14.3.2 The Buried Soil at Site B

We know that the Leipsokouki reached its present level, about 120 m below the drainage divide, before the end of the Pleistocene (11,700 years ago) because of valley-bottom deposits that are older than that. At Site B (Figure 14.2 for location; also Figures 14.3, inset B and 14.6b), soil properties, archaeology, and radiocarbon dating help us to sketch in the next stages of the stream's history.

At Site B, the oldest material is an unsorted mix of sand, silt, and small pebbles. This mix of grain sizes indicates that the material probably moved down a hillslope or small stream valley as a mass movement. In fact, it is the lowest unit of a fan-shaped deposit that emerges into the main valley from the west side. For such a fan to form, the steep slopes that characterize the Leipsokouki catchment today must have already existed. Also, it is no accident that this deposit both formed and is preserved at a tributary junction: this is the type of place where sediment carried down the smaller, steep side valleys will tend to accumulate because the wider main valley has limited water flow and a gentler gradient.

Where did this material come from? In contrast to some of the younger colluvial (mass movement) deposits, this soil parent material is largely fine-grained, with only a few bedrock fragments. So we need to look for a source of this fine-grained material upslope. The loess deposits and paleosols described in the previous section fit the bill nicely.

A soil has formed on this material (Figure 14.6b), so this particular surface must have been stable long enough to produce its distinctive weathering features: a 40-cm-thick clay-rich subsoil horizon with strongly developed soil aggregates (structure) above a horizon with carbonate accumulation (the calcium carbonate has been leached from the top 55 cm of the soil and redeposited in this horizon). Another feature of this soil is the abundant ~1 cm diameter charcoal fragments in the uppermost horizon. These have been dated to 14,750 ± 1000 cal years BP (calendar years "before present," where present is 1950[1]), a date that gives a maximum age for the overlying material.

Although affected by mass movement, the landscape around Site B must then have been stable for a long enough time for the soil to develop. This well-developed soil less than 2 m above the modern stream bed tells us that the valley floor reached its present level at least 15,000 years or so ago.

At three other sites in the Leipsokouki catchment, similarly well-developed soils on colluvium are also exposed near the stream channel and covered with younger material. At these sites, soils formed on slopes that most soil scientists would consider too steep to be stable. Perhaps it is their position near the base of slopes and at tributary junctions that has preserved them.

Charcoal dates these soils from 11,000 to 14,800 cal years BP. Although we have not examined the charcoals closely enough to determine what kind of plants they are derived from, pollen records from sites of comparable age in northern Greece indicate that vegetation at this time was mostly steppe (grassland) with scattered oaks and pines (Bottema 2003). The fires that produced the charcoal may have been ignited from lightning strikes similar to those associated today with summer thunderstorms. These charcoal dates from the Leipsokouki fall squarely into a gap in the archaeological record, one that persists throughout Greece despite intensive survey and excavation (Perlès 2001). Most Paleolithic (300,000 to 13,000 year ago) sites in Greece were abandoned before the Pleistocene/Holocene boundary (11,700 cal years BP) and Mesolithic sites (ca. 9000 to 11,000 cal years BP) are exceedingly rare. Most sites with deposits spanning these periods have sharp breaks between Paleolithic and Mesolithic layers and between Mesolithic and Neolithic. Exactly what these breaks mean is a subject of hot debate in the archaeological community. Further, with one exception (Theopetra Cave), Greek Mesolithic sites are coastal (Perlès 2001; Runnels and Van Andel 2003; Runnels et al. 2009; Strasser et al. 2010). The evidence for human occupation strengthens only with the advent of Early Neolithic, starting approximately 9000 cal years BP in Greece with an Initial or "Aceramic" phase before "classic" Early Neolithic at about 8450 cal years BP (Perlès 2001, 2003). (See the Note (1) at the end for discussion of archaeological and geological time scales; also see Figure 14.4).

14.3.3 The Syndendron Valley Fill: Instability before Human Settlement

Multiple exposures in the Leipsokouki highlight a phase of accelerated erosion on hillslopes and deposition in valleys that occurred roughly between 13,800 and 8800 cal years BP. The present thickness of this Syndendron valley fill ranges from 1.6 to 8 m (Doyle 2005), but it is probable these deposits were once thicker. At several exposures, we have found enough charcoal below, within, and above the Syndendron valley fill deposits to constrain its age. The oldest bounding dates are 14,800 to 11,000 cal years BP and the youngest bounding dates are 9900 to 7500 cal years BP. At each exposure, we can define an approximate apparent duration of deposition; these range from 1000 to about 6500 years. This range in ages and apparent durations is expected, for several reasons. First, the bounding dates are just that. At Site B, for instance, there is no upper bounding date and the buried paleosol contains charcoal dated at 14,750 cal years BP. We know the Syndendron deposit above it must be younger—but we do not know how much younger. Second, all radiocarbon ages have uncertainties (1000 years in the case of Site B). Third, even in a relatively small catchment such as the Leipsokouki, it is likely that slope failures and valley filling occurred at different times up and down the catchment (just as a few gullies are now active while others are healing). Finally, we are looking at scattered remains of this valley fill. We cannot trace the fill with confidence through the entire valley based on the existing exposures—so we do the best we can. The range of dates gives us a picture of a relatively long period—perhaps as much as 5000 years—characterized by slope failures and valley filling at different spots in the catchment.

At one site (E, Figure 14.2), we can get a better idea of the timing and some insights into the cause of the erosion and deposition. Here, the old paleosol has been eroded—and the sediment just above that surface of erosion is loaded with puffy ash and charcoal that looks as if it has not traveled very far. The ash dates to 12,250 ± 700 cal years BP and is covered by thin beds of fine-grained material. Because Site E is on a slope, it is likely that sheetwash is responsible. Moreover, the lack of a sharp break between the ashy deposit and the sheetwash deposits suggests that the sheetwash closely followed a fire that would have burnt vegetation holding the soil in place.

The deposits at Site B also show some of the characteristic features of this valley fill (Figure 14.3, inset B). The well-developed paleosol, discussed in Section 14.3.2, is cut by a sharp erosional contact. Above it, two deposits interfinger with each other: another unsorted mass of mixed grain sizes, including large fragments of bedrock, against which several meters of thin beds of silt and sand, have backed up. The unsorted material probably reached the valley through a mass movement and may have blocked the stream channel, forming a temporary barrier and short-lived lake in which the bedded, finer-grained sediments accumulated.

Other Syndendron valley fill exposures also have a mix of deposits, including unsorted debris flow sediment with blocks of bedrock and thin layers of finer sediment. Higher on the slopes, colluvium that correlates with Syndendron valley fill either covers well-developed soils or forms in depressions cut into well-developed soils, reinforcing the idea that there was a period of landscape stability preceding the Syndendron events. Also, like the valley fills, many of the colluvial deposits are "multi-media"—they contain both unsorted material and thinly bedded material. On hillslopes, this thinly bedded material likely formed through repeated sheetwash, a process that is most active on unvegetated slopes. However, the hillside colluvial deposits also include some things not found in the valley floor deposits—blocks of previously weathered soil and nodules of calcium carbonate.

These blocks and nodules are likely transported into the colluvial fills by gravity (mass movement) and they raise an interesting question about defining soil development. In the previous examples we have talked about, the soil has formed on deposits that either were certainly not soil to begin with—river material in the case of the Xeropotamos—or else are old enough for soil development to obscure any inherited weathering. However, if a "real" soil erodes and then is deposited somewhere else, the "new" soil has a head start, because some of the weathering has already happened. We can get confused and assign an age that is too old to a young deposit. The blocks of soil and the carbonate nodules (suggesting thousands of years of soil formation) in the colluvial fill are extreme and helpful examples—they provide a clue to the process but are themselves easy to figure out. In other cases, it is not so easy to tell if a soil was entirely formed *in situ* (in the place where we find it) or had a previous history of soil development. In the Leipsokouki, many younger valley deposits alternate between channel gravels and finer-grained overbank deposits that contain blacker, soil-like layers. Given what we know about soil development, it is not reasonable to think that the color and other properties of these "little soils" developed in the short intervals between coarser-grained deposits. Most likely, the materials were partly weathered before being transported to the exposures where we find them. We might expect, then, to find better developed soils where older soils have been redeposited. As we will see in Section 14.3.5, this is exactly what the youngest hillslope colluvial deposits show.

Even though only a tiny fraction remains of the Syndendron sediments in the Leipsokouki stream and on the hillsides, it speaks to a set of profound landscape changes: erosion of older deposits (including parts of older soils), and deposition of up to 8 m of valley fill followed by renewed erosion.

14.3.4 When and Why?

As we have seen, the period of five thousand years from about 13,800 to 8800 cal years BP, when the Syndendron valley fill and correlative slope deposits formed, was preceded by a period of landscape stability of unknown length. It was also followed by a period of relative stability when the main stream and its tributaries slowly incised through the Syndendron deposits.

Few, if any, large valley fills are reported from this five thousand year period in other parts of Greece and the eastern Mediterranean (Dusar et al. 2011). What caused such a uniquely large volume valley fill in the Leipsokouki? The Syndendron deposits date to a period with little evidence for human presence in Greece, let alone human modification of the landscape (Perlès 2001, 2003; Gkiasta et al. 2003) including, specifically, an absence of evidence for human-manipulated fire regimes in Europe between about 70,000 and 10,000 cal years BP (Daniau et al. 2010). This period does, however, straddle the boundary between the Pleistocene and Holocene (11,700 cal years BP), a period when glaciers had retreated in the Northern Hemisphere, and sea levels and temperatures were also rising. Two interesting phenomena associated with this transition might have affected the Grevena area. First, perhaps the climate near the end of the Pleistocene did not go initially from cold and dry to warm and wet. Instead, the increase in rainfall may have lagged behind the temperature increase (Bottema 2003). Because vegetation changes inherently lag behind climate changes, the Greek landscape may still have been covered with grassy, steppe-like vegetation and open oak forests (perhaps less effective than thick forest at retarding erosion) when the rain finally did increase. Second, several short-lived climate swings across Eurasia and North Africa are associated with pulses of fresh water entering the North Atlantic from waning glacial ice sheets. Each time one of these pulses arrived, it triggered massive changes in ocean and atmospheric circulation, changes large enough to affect Asian monsoons and rainfall patterns over Europe. The three most profound changes seem to have happened at about 12,900–11,800 (the Younger Dryas), 11,250–11,050 (the Pre-Boreal Oscillation), and at 8200 cal years BP (the "8200 year event") (Magny and Bégeot 2004).

Note that the landscape instability that resulted in the Syndendron valley fill and colluvium overlaps with the Younger Dryas and the Pre-Boreal Oscillation events. The 8200 year event, however, occurs during an episode of relative landscape stability in the Leipsokouki. Coincidence in time, like all correlations, does not imply causation. Still, the pattern of climate changes resulting from these short-term events may help explain how a particular area, like the Leipsokouki, might have responded. For instance, climate reconstructions for the 8200 year event show a belt of cooler, wetter weather that stretched across Europe and the Near East (Magny et al. 2003; Berger and Guilaine 2009; c.f. their Figure 6). Further to the north and south, the prevailing climate associated with this event was dry: cold and dry to the north in places like Iceland, warm and dry to the south in North Africa and much of the Middle East. Between the arid belts and the wetter belt, narrower transitional regions may have fluctuated between very dry, windy and warm summers and very wet winters (Berger and Guilaine 2009). The narrowness of these transitional belts implies that small changes in the climate drivers could produce great changes in transitional zones, such as the one near Grevena. Perhaps this short-term climate variability helps explain why these very-late-glacial and early Holocene climate oscillations seem to have caused so much havoc in the Leipsokouki when there is little evidence of massive deposition during the last glacial maximum. This is a mystery that still puzzles us.

Although there is no evidence that humans affected the Leipsokouki landscape at this time, we found a stone tool, some enigmatic groupings of rocks and soil bricks and a hearth feature at one site (F; Figure 14.2) associated with a paleosol within the Syndendron valley fill, dated at 9350 ± 600 cal years BP. Because fine-grained Syndendron sediments continued to be deposited after the paleosol formed, we suspect that the hearth feature is close to the age of the paleosol.

14.3.5 The Next Big Thing—People Influence the Landscape: The Sirini Valley Fill

The earliest evidence for human settlement in Grevena falls into the period of relative landscape stability in the Leipsokouki from about 8800 to 6000 years ago that followed the Syndendron episode. The nearly 20 Early Neolithic sites found so far in the prefecture are on flat areas above floodplains, near the larger perennial streams (Wilkie and Savina 1997). Many sites have mudbrick, large pottery fragments, and stone tools, remains of places where people lived, not simply passed through. We do not have radiocarbon dates on these sites, so we have to do our best with dates of sites in nearby regions. The Grevena sites are probably younger than the 8450 cal years BP Early Neolithic start (Perlès 2001, 2003; Nancy Wilkie, pers. comm.). Some authors (e.g., Weninger et al. 2006) propose that the climate change associated with the 8200 BP event was the driving force behind Neolithic migration into Greece.

In the Leipsokouki, soils began to form on flatter areas (for instance, on the surface of the Syndendron valley fill) and stream reincision began in the lower valley, working its way up into the headwaters and slopes over a period of a few thousand years. The next phase of deposition in the valleys started about 6000 years ago and ramped up (at least in terms of sedimentation rate) at about 3000 years ago (Figure 14.4).

The Sirini valley fill deposits are complex. At Site C (Figure 14.3), for instance, at least 4 m of fill sit above a wavy, nearly horizontal surface eroded into mudstone bedrock. The two discrete "packages" of gravels topped by fine-grained sediments, one above the other, are separated by a soil, probably formed from partly weathered material deposited with the fine-grained sediments. Deposits that correlate with the lower of the two channel gravels yield a date of 4150 ± 500 cal years BP. Unlike the older Syndendron valley fill, with its debris flow and sheetwash deposits deposited on slopes, Site C and other Sirini sites have deposits one might reasonably expect to find in valley bottoms, transported there by stream flow: channel gravels with mudstone pebbles capped by thin, fine-grained deposits that look like alluvium deposited during floods on floodplains (overbank deposits). This change is significant because it means that for the first time in the long erosional history of the Leipsokouki, mudstone bedrock was available to be eroded by running water that flowed in channels on the hillslopes (gullies). Sheetwash is perfectly capable of moving silt and fine sand grains, but deeper, faster water is needed to erode and transport bedrock fragments.

The soil formed on the older overbank deposits is consistent with the radiocarbon date and the implication that the deposit is mid-Holocene (Figure 14.6c). The buried topsoil and subsoil horizons had barely started developing. Some of the original calcium carbonate had leached out of the topsoil into the subsoil, where it accumulated as soft coatings on the pores in the soil (contrast this with the completely carbonate-cemented horizon at Site A). In addition, the soil has reddish surface coatings, which also developed as the soil formed. Before this soil had the chance to develop further, however, it was eroded and another package of channel gravels and overbank sediments deposited on top. The modern soil forming at the present ground surface is less developed than the buried one. For instance, no leaching of carbonate has occurred.

Sirini valley fills range in thickness from 2–10 m. This range is similar to that of the Syndendron valley fill; however, in sites where both are found, the Sirini deposits are inset into the Syndendron fill (Figure 14.3). Thus, we are confident that the stream eroded through the Syndendron deposit back to bedrock before the Sirini valley fill was deposited.

The Sirini valley fill and its correlative hillslope colluvial deposits span the period from the Bronze Age through Roman times (see Figure 14.4). In addition to radiocarbon, many of these deposits can be roughly dated based on pottery fragments contained in them; the deposit must be the same age or younger than the youngest artifact found in it. A colluvial fill at Site G (Figure 14.7b) illustrates this principle well. Four distinct layers of colluvium occupy an erosional hollow cut into mudstone bedrock, perhaps originally the head of a gully or a landslide scar. The colluvial layers are unsorted and unbedded—something we would expect in a mass movement deposit on a hillside. The lowest colluvial layer has charcoal dated at 2175 ± 175 cal years BP. It and the colluvial layers above contain pottery that the archaeologists working with us date confidently to Hellenistic, a period (dated precisely by historic convention) that begins with the death of Alexander the Great in 323 BCE (2273 BP) and ends 177 years later (2096 BP), when the Romans arrived in Greece. Considering the vagaries of all these dating techniques, this sort of agreement is pretty amazing! The two soils in this section—one a buried soil formed on the third youngest of the four colluvial layer and the other the modern soil on top—are poorly developed even by comparison to the buried soil at Site C, described above. Perhaps 500 years or so of development has occurred in each of the buried soils in the colluvial sequence at Site G.

Site G is one of several colluvial deposits preserved on slopes in the Leipsokouki mid-catchment that correlate with the Sirini valley fill. The material filling these depressions varies depending on what is locally available and on the age of the deposit (and these two factors are related). Many fills with pottery fragments dating from the Bronze Age or Early Iron Age have rather uniform dark earthy material that seems to have been topsoil when it arrived in the colluvial fill. Other deposits, such as Site G, are lighter in color and contain bedrock fragments. Figure 14.7a schematically shows an intermediate stage in the process of converting a landscape from one covered with thick materials, such as loess/paleosol sequences or well-developed soils, into one covered mostly by bedrock. Perhaps, also, the "soil-like" deposits reach their final destinations by a process of soil creep while the thicker

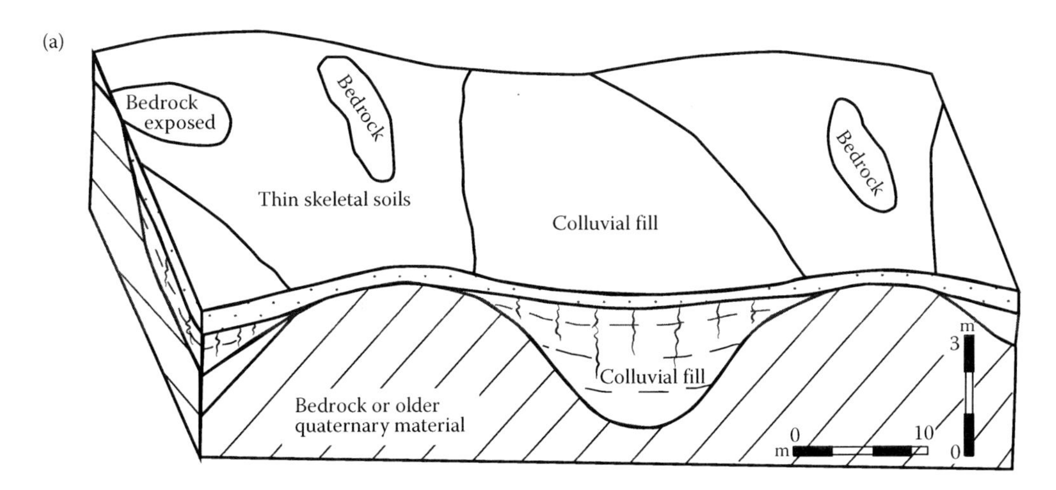

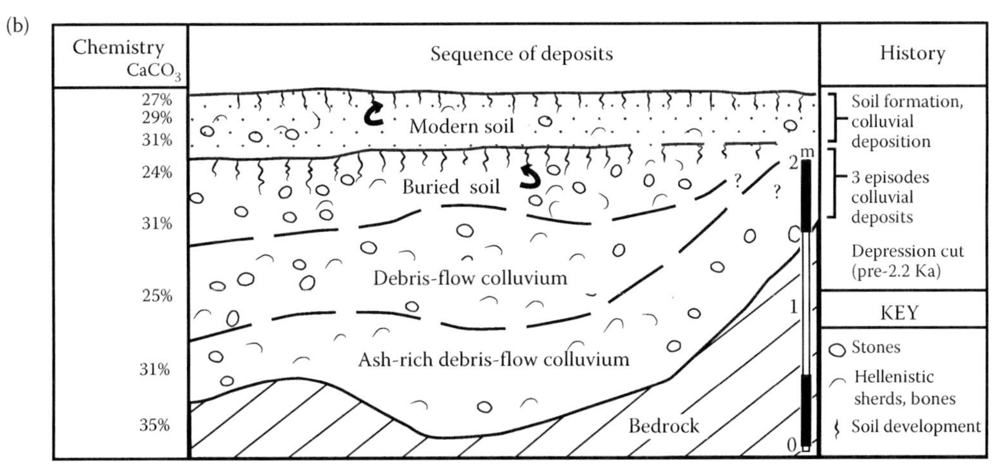

FIGURE 14.7 (a) Block diagram showing typical setting of Late Holocene hillslope colluvial fills. (b) Sketch of colluvial fill at site G. $CaCO_3$ percent shows leaching of modern and buried soils relative to the concentration in the colluvial fills the soils are developed on; mudstone bedrock has the highest $CaCO_3$ percent of all materials in this section.

colluvial deposits like those at Site G form by mass movements capable of transporting rocks (Doyle 1990, 2005). By the stage represented in the figure, the deep deposits have been mostly eroded. What is left are small patches of bedrock, thin "skeletal" soils, and colluvial deposits derived in part from the very old soils that are otherwise gone downslope and down valley. These materials are variously distributed on the landscape; whatever is nearby makes it into the colluvium. As long as deep, weathered soils are available on the landscape, the colluvium will contain them—and, as noted in Section 14.3.3, new soils will have a head start on further weathering as a result. But as erosion exposes more and more bedrock, that rock will become part of the mix that is eroded into (and preserved by) the colluvial fills. Eventually, all the reserves of soil will be gone and the bedrock will be widely exposed.

Perhaps this scenario sounds extreme, but it is common throughout central and eastern Grevena. Artifacts and other remains of homes and villages move into colluvial fills along with the soil. No one purposely builds a home in a gully (or directly on mudstone bedrock for that matter); yet in the Leipsokouki and elsewhere in the prefecture, the only remains of some settlements are the fragments that have been transported into and preserved in colluvial deposits such as at Site G. Like the soil materials in the deposit, the habitation site must originally have been up the hill—where now there are only bedrock exposures. Dates on the colluvial deposits tell us that the present landscape, with its mix of exposed mudrock, deep soils on the uplands, and variable thicknesses of soils on the slopes must have been in existence for about 2500–4000 years; it is not a twentieth-century phenomenon. Before 4000 years ago, farmers and grazers may have had access to an entire landscape of relatively gentle slopes mantled by thick, fertile soils. Now much of that resource has been lost.

Continuing the cycle, another period of erosion followed the end of Sirini deposition, lasting for approximately a thousand years, starting at about 1700 years ago (Figure 14.4). The final depositional interval, called the Leipsokouki, created volumetrically smaller deposits than either the Syndendron or the Sirini; most are less than 2 m above the channel (Figure 14.3, Site D). Like the Sirini deposits, these youngest valley fills contain distinct coarse channel and finer overbank deposits. Because of negligible soil development, young radiocarbon dates, and Ottoman-age pottery in the sediment, we think this valley filling occurred mainly within the last millennium.

Given the evidence for human activity and climate change during the Sirini period in the Leipsokouki, we think it likely that both affected the landscape, probably simultaneously. There is no question that people used the landscape in the Leipsokouki and much of Grevena, during the periods when the Sirini and Leipsokouki valley fills (and their correlative hillslope deposits) were forming (Figure 14.4). Evidence is particularly strong for human occupation during the Early Iron Age—and then more-or-less continuously from the Hellenistic through the successive Roman, Ottoman, and modern periods, with a dip in site numbers during the Medieval and Byzantine periods. We know a little bit about what these people were doing, based on the artifacts found in places like Site G. Some Hellenistic and Roman sites have loom weights—pyramid-shaped pieces of fired clay hung at the ends of the warp threads to keep them taut. At a few sites, these loom weights are so numerous as to make us think these sites might have been used for large-scale, perhaps commercial weaving. Some sites have bones of sheep or goats; others have seeds of wheat or barley. Pollen from cereal grains is also found in the oldest recovered deposits in a wetland near Mega Sirini; these probably date to 2780–2580 cal years BP (Chester 1998). The picture that emerges is somewhat similar to the village-based enterprises today: cereal grain farming, local flocks of sheep and goats, with the addition of textile manufacture, something not found in these villages today.

In the greater world outside Grevena, three or four empires were rising and falling during this time. Alexander the Great, from Greek Macedonia himself, consolidated an empire that stretched from the Mediterranean through western India. Roman rule took over more or less where the Hellenistic left off; with the breakup of the Roman empire, Greece became administratively attached to Anatolia, first as part of the Byzantine empire and then as part of the Ottoman empire. (The tautness of the "attachment" varied a lot). People living in Grevena have the resources to be self-sufficient at most times. When part of an empire, though, Grevena likely served larger functions too—perhaps as a supplier of grain, meat and cloth, and as a receiver of goods from elsewhere (Harold Koster, pers.

comm., 1988). The European Union's efforts to promote Greek sunflower oil and Greek bananas, and support Greek shepherds, so as not to have to import these commodities from outside the Union, might not be too different philosophically from the situation of Grevena during these empires. So not only is there archaeological evidence for a more peopled landscape during the time of the Sirini valley filling, we infer that these people supplied raw materials to a larger, market economy. From our experience today, we know that agriculture, both for self-sufficiency and for larger markets, can be done carefully or carelessly with respect to long-term protection of soil resources. The scale is not as important as the practices—and also what is happening with the climate. In a time of drought or prolonged cold, for instance, it might be relatively easy for a group of villagers to move (after all, they are the only ones depending on their efforts), or they might weather that drought more easily because of the diversity of their household operations. In a market economy, though, if one's livelihood depends on payments needed to buy essentials that are no longer being locally grown, then there may be more of a temptation to keep trying to farm and graze, even at the risk of soil loss and degradation.

The rules of the landscape stability game change once humans are in the picture, as we know from the record of the last two centuries. In the United States, we do not have to look further than the disastrous "Dust Bowl" period of the 1930s for evidence of the synergistic effects of climate change and human action. The drought that marked the Dust Bowl period, including high temperatures and large magnitude (but very infrequent) erosive rainstorms, was neither confined to the "classic" Dust Bowl area of the Southern Great Plains (climate records all over the continental United States show the change in weather patterns) nor was it unprecedented. Several droughts in the preceding few centuries were of equal or greater magnitude and scope as the one in the 1930s (Woodhouse and Overpeck 1998). Looking even further back, Holocene windblown sand and dust underlie areas in the Great Plains that were cultivated during the early twentieth century. When these areas were settled during a wetter period in the nineteenth century and planted with wheat to supply an overseas war, high commodity prices encouraged extensive tillage. When prices dropped, even more marginal land came into production to compensate for the lower prices (Opie 1989). Historians and climatologists still argue about the Dust Bowl, but almost everyone agrees that it was the combination of climate change and a particular kind of agriculture that caused the disaster (c.f. Cook et al. 2009). In the Leipsokouki, a similar synergistic mechanism involving climate and human action may have produced the Sirini and Leipsokouki erosional and depositional episodes of the late Holocene. If the change in sediment type from the Syndendron valley fill (debris flows and sheetwash, predominantly fine deposits) to the Sirini and Leipsokouki valley fills (channel gravels and overbank sediments) marks an increase in the importance of gullies for instance, then human action, in clearing and tilling ground, may be a triggering factor. Although much fine material remained on the landscape after the Syndendron valley-filling episode (and, indeed the loess/paleosol sequences are still exposed near the drainage divides), removal of so much of that material over time may have altered hillslope hydrology by reducing infiltration, in turn promoting gully-type erosion over sheetwash. We still have work to do connecting the human activities and landscape change.

In a comprehensive review of climate proxy records from 6000 BP around the Mediterranean, Finné et al. (2011) note that regional climate was wetter from 6000 to 4500 BP and that the following 3000 years were drier. In the Leipsokouki, the Sirini valley fill and colluvium formed during this relatively dry period. One would expect a drier climate to result in more natural fires and certainly with more people around, there would be more opportunities for human-started intentional and unintentional blazes. If followed by intense rainfall before vegetation regrows to cover surfaces and strengthen soils with roots, fires can result in serious hillslope erosion. Fires might be expected to have different effects on hillslopes, depending on the material available at the surface. If fine-grained, loose material, easily transported by sheetwash, is at the surface, for instance, postfire erosion may move more material than if the ground surface is cohesive and rough. Once more cohesive materials, such as bedrock or Vertisols, are exposed, however (Figure 14.7), infiltration rates may have decreased, leading to more gullying. In fact, across the Mediterranean, postfire erosion may not be as severe now as it was before about 2500 BP simply because less erodible material is now at the ground surface (Shakesby 2011). We continue to try to resolve how fire affected landscape change at different periods in the Leipsokouki.

Superimposed on top of the overall trends of moisture and temperature are shorter-term climatic "events," like the three that ended the Late Glacial and opened the Holocene, but with less profound effects (the ice sheets, the climatic elephants in the room, so to speak, were gone). Various researchers have described events at about 5300 (Magny and Haas 2004); 4300–3800 (Magny et al. 2009); 2800 (Chambers et al. 2007; Plunkett and Swindles 2008); 1000–700 (the Medieval Climate Anomaly); and ~500–150 cal years BP (the Little Ice Age) (Roberts et al. 2012). Different parts of the Mediterranean responded in a variety of ways to these changes. Though a few Sirini valley fill deposits are older, most formed in the 3000-year-long drier interval that includes the 4300–3800 and the 2800 cal year BP events. The Medieval Climate Anomaly (MCA) coincides with the most recent phase of relative landscape stability in the Leipsokouki and also a smaller number of archaeological sites in Grevena (Figure 14.4). The Leipsokouki valley fills seem to be younger than the MCA and to overlap with at least the first part of the Little Ice Age. Figuring out what these climate events may have done in Grevena—and how humans responded to them—is part of the work we have yet to do.

14.4 SUMMARY/CONCLUSIONS

By studying the soils and sediments preserved in the Leipsokouki, we have been able to reconstruct its landscape history. The conditions for the cycles of erosion and deposition over the last 15,000 years or so were primed by the overall tectonics of the area (which creates the slope angles) and by the accumulation of fine-grained deposits of loess during successive glacial periods. Downcutting of the streams set off the valley and hillslope erosion, which in turn provided the materials that accumulated in valley fills and on hillslopes. In stable periods, soils formed on these deposits. It seems likely that the deposition and subsequent erosion of the older valley fills were triggered by local responses to climate change. When more people arrived, however, sedimentation rates increased (Figure 14.4) and land use began to play a stronger role in landscape changes. Each successive set of valley fills is lower and smaller than its predecessors.

The landscape history of the Leipsokouki influences its potential for agriculture, now and in the future. Because of steep hillslopes and the absence of soil cover in much of the landscape, the resource available for agriculture is limited. Already some areas of Grevena are too dry for agriculture without irrigation. Erosion rates on European agricultural landscapes are estimated at 3–40 tons per hectare per year, compared to a maximum rate of soil formation of about 1.4 tons per hectare per year (Verheijen et al. 2009). Preliminary modeling studies in the Leipsokouki and surrounding areas, using the Revised Universal Soil Loss equation (RUSLE) found even higher erosion rates, suggesting that the hillslopes and overall landscape are still unstable (Beavis 2011, 2012) and supporting our observations of thin soils covering bedrock on many hillslopes in central Grevena.

The record in the Leipsokouki differs from others in the Eastern Mediterranean in a couple of major ways. First of all, the most voluminous valley fill of all, the Syndendron fill, seems unusual in its timing and the types of sediment it produced. Second, the Leipsokouki (and Grevena in general) has about 4000 years of human settlement before land use strongly affected the landscape. However, the Leipsokouki is the only catchment in Grevena we have studied in such detail. Intriguingly, spot observations we have made in other parts of the prefecture, including in the valleys near the Leipsokouki, suggest that there is more to the history than we know now.

Since Claudio Vita-Finzi published *The Mediterranean Valleys* in 1969, soil scientists, geoarchaeologists, historians, and others have described valley and hillslope deposits across Europe and have sought the causes of the erosion and deposition in some combination of geologic events (tectonic and volcanic), climate change, human action, and the internal workings of complex landscape systems. The professional literature documenting these deposits grows every year; reviewing it comprehensively is not within the scope of this case study of a single, small catchment. Besides, such a study has recently been published. Dusar et al. (2011) conclude that before about

3500 years ago, climate was probably the most important driver of landscape change in the Eastern Mediterranean, largely through its effects on vegetation and rainfall. Effects attributable to humans became more obvious later (and at different times in different parts of the region, depending on when human populations increased and began altering vegetation). In the Leipsokouki, human-affected landscape change seems to have started in the Early Iron Age (see Figure 14.4). Across the Mediterranean, landscape change seems to have reached its peak during the Hellenistic and Roman periods—exactly the time of the most rapid Sirini phase valley filling in the Leipsokouki. Climate change that increased aridity plus the depletion of the soil resource may have limited later erosion and sediment accumulation in post-Roman times (Dusar et al. 2011).

River and hillslope deposits are not limited to the Mediterranean; they have been subjects of study in most parts of the world. Interestingly, explanations for landscape changes have tended to differ from region to region, partly because of truly different geographies and histories, but partly also because of different research traditions. Vita-Finzi suggested that both the older and the younger of his two widespread valley fills were initiated by climate change, in the late-glacial and post-Roman periods, respectively. Many studies of the Mediterranean after Vita-Finzi's work (up until the twenty-first century) pointed the finger at humans as the cause of most of the Holocene land use changes, particularly since about 5000 years ago (e.g., van Andel et al. 1986). Some ecologists argue that the diversity of Mediterranean ecosystems is due, in part, to the long history of interaction with humans (Blondel 2006). With increased recognition of abrupt climate changes in the Holocene and latest Pleistocene, more and more evidence is emerging to support a climatic influence on landscape evolution particularly before about 3500 BP (Dusar et al. 2011). Tectonics has been shown to influence recent landscape evolution as well, but in more limited scope. In North America, on the other hand, scholars tend to attribute almost all Holocene landscape change to climate, at least before the arrival of Europeans in the seventeenth century. Increasingly, North American scholars now recognize that indigenous settlers also affected the landscape, by altering vegetation. In Australia, where evidence of humans goes back 35,000 years or more, archaeologists and geomorphologists recognize the influence of both humans and climate and consider the current vegetation in parts of the continent an adaptation to both. In New Zealand, where humans arrived only in about the thirteenth century, researchers look to a combination of tectonics (not only earthquakes but also tectonic effects on climate and the uplift of large areas of erodible rocks), volcanism, and heavy rainfalls to explain the truly massive scale of landscape change before settlement. Evidence of the landscape effects of Europeans, starting in the nineteenth century, is obvious and it is becoming increasingly clear that Maori also influenced the landscape. Our experiences in these four distinct parts of the world have helped make us skeptical of single explanations for landscape change and, indeed, the Leipsokouki is an excellent example of all of these causes—and the internal dynamics of a complex system. The story of this small catchment has revealed how all these processes interact.

NOTE

1. *Geological and archaeological dating*: Geologists describe the past using both relative and absolute dates. Subdivisions of the geological time scale—such as "Cambrian" and "Cretaceous" are examples of relative dating. Most of the events described in this essay occur in the last few of those geologic subdivisions. These are the Pliocene epoch (from 5.3 to 2.6 million years ago), and the Quaternary system, which is divided into the Pleistocene epoch (from 2.6 million years ago to 11,700 years ago) and the Holocene epoch (the last 11,700 years; Walker et al. 2012). A variety of elemental systems are used for absolute geologic ages, including those that define the geologic subdivisions. The absolute dates in this essay are based on the decay of radioactive ^{14}C ("radiocarbon"), which is produced naturally in the upper atmosphere and incorporated in living organisms, replacing about one in every trillion atoms of the much more abundant but nonradioactive ^{12}C and ^{13}C. Over time, ^{14}C decays to ^{14}N. The more ^{14}C measured in organic matter (such as charcoal

incorporated into a sediment), the younger the organic matter. All radiocarbon dates in this chapter, from our own work or that of others, are reported as calibrated years "before present" ("BP"), where "present" is 1950. Calibration was done in 2005 with the OXCAL computer program (see Stuiver et al. 1998). For each radiocarbon date, we also supply the calibrated error range, which in this case is two standard deviations older and younger than the calculated date (the range of the 95% probability) (Doyle 2005). The radiocarbon dating for our project was done at the University of Waikato Radiocarbon Dating Laboratory, New Zealand, abbreviated "Wk" in figure captions. Unless otherwise specified, other dates (e.g., "5000 BP") refer to calendar dates before present.

Similar to the way geologists use relative ages, archaeologists also have a relative age scale. Early parts of the archaeological time scale refer to the "stone age" or "lithic." From oldest to youngest, these stone age periods are the Paleolithic, Mesolithic, and Neolithic. After the Neolithic, the archaeological time scale moves to the familiar Bronze Age and Iron Age and then to names for periods that reflect the dominant cultural or political influence. The boundary dates for archaeological periods (e.g., "Hellenistic," Figure 14.4) are agreed on by archaeologists working in particular areas and are generally supported by radiometric dating. These archaeologists attribute artifacts to a particular period based on similarities of form, material, and decoration. Figure 14.4 includes the archaeologic periods and boundaries in use for Greece (Nancy Wilkie, pers. comm.).

ACKNOWLEDGMENTS

We thank Nancy Wilkie, Director of the Grevena Project, for her leadership, financial support, and encouragement, for making the work on this project so intellectually stimulating, for keeping us in food and drink while in the field, and for help in more ways than we can properly acknowledge. Field work in Greece was conducted with permission and support of the American School of Classical Studies in Athens, the 15th Ephoreia of Prehistoric and Classical Archaeology in Larissa and the 11th Ephoreia of Byzantine Antiquities at Veroia, and the Greek Institute of Geology and Mineral Exploration (IGME). We thank our three academic institutions, the University of Tasmania and the Tasmanian Institute of Agriculture, Carleton College, and Victoria University of Wellington, for financial support, timely study leaves, and academic schedule adjustments and encouragement. We also thank the editors of this volume, Jock Churchman and Ed Landa, for the invitation to contribute this chapter and for all of their work putting the volume together. We thank Caroline Scheevel for her work on the figures.

REFERENCES

Beavis, M. A. 2011. Soil erosion risk assessment near archaeological sites in Grevena, Northwestern Greece using the Revised Universal Soil Loss equation (RUSLE) and GIS. BA diss., Carleton College.

Beavis, M. A. 2012. Soil erosion risk assessment near archaeological sites in Grevena, Northwestern Greece using the Revised Universal Soil Loss equation (RUSLE) and GIS. In *Computer Applications and Quantitative Methods in Archaeology*. Archaeology Computing Research Group CAA 2012 40th Conference, School of Humanities, University of Southampton.

Berger, J.-F. and J. Guilaine. 2009. The 8200 cal BP abrupt environmental change and the Neolithic transition; A Mediterranean perspective. *Quaternary International* 200: 31–49.

Blondel, J. 2006. The "design" of Mediterranean landscapes: A millennial story of humans and ecological systems during the Historic Period. *Human Ecology* 34: 713–29.

Bottema, S. 2003. The vegetation history of the Greek Mesolithic. In *The Greek Mesolithic: Problems and Perspectives*. *British School at Athens Studies* 10: 33–49.

Chambers, F.M., D. Mauquoy, S. A. Brain, M. Blaauw, and J.R.G. Daniell. 2007. Globally synchronous climate change 2800 years ago: Proxy data from peat in South America. *Earth and Planetary Science Letters* 253: 439–44.

Chang, C. and P. A. Tourtellotte. 1993. Ethnoarchaeological survey of pastoral transhumance sites in the Grevena region, Greece. *Journal of Field Archaeology* 20: 249–64.

Chatzipetros, A. A., S. B. Pavlides, and D. M. Mountrakis. 1998. Understanding the 13 May 1995 Western Macedonia earthquake: A paleoseismological approach. *Journal of Geodynamics* 26: 327–39.
Chester, P. I. 1998. Late Holocene vegetation history of Grevena Province, Northwestern Greece. PhD diss., Victoria University of Wellington.
Cook, B. I., R. L. Miller, R. Seager, and J. E. Hansen 2009. Amplification of the North American "Dust Bowl" drought through human-induced land degradation. *Proceedings of the National Academy of Sciences of the United States of America* 106: 4997–5001.
Daniau, A. L., F. d'Errico, and M. F. S. Goni. 2010. Testing the hypothesis of fire use for ecosystem management by Neanderthal and Upper Palaeolithic modern human populations. *PLOS One* 5: e9157
Doutsos, T., and S. Kokkalas. 2001. Stress and deformation patterns in the Aegean Region. *Journal of Structural Geology* 23: 455–72.
Doyle, R. B. 1990. Soils, geomorphology and erosion history of the Leipsokouki Catchment, Nomos of Grevena, Greece. MSc diss., Victoria University of Wellington.
Doyle, R. B. 2005. Late Quaternary erosion, deposition and soil formation near Grevena, Greece; Chronology, characteristics and causes. PhD diss., University of Tasmania (available at http://eprints.utas.edu.au/2346/).
Dusar, B., G. Verstraeten, B. Notebaert, and J. Bakker. 2011. Holocene environmental change and its impact on sediment dynamics in the Eastern Mediterranean. *Earth-Science Reviews* 108: 137–57.
Finné, M., K. Holmgren, H. S. Sundqvist, E. Weiberg, and M. Lindblom. 2011. Climate in the Eastern Mediterranean, and adjacent regions, during the past 6000 years—A review. *Journal of Archaeological Science* 38: 3153–73.
Gkiasta, M., T. Russell, S. Shennan, and J. Steele. 2003. Neolithic transition in Europe: The radiocarbon record revisited. *Antiquity* 77: 45–62.
Goldsworthy, M. and J. Jackson. 2001. Migration of activity within normal fault systems; Examples from the Quaternary of mainland Greece. *Journal of Structural Geology* 23: 489–506.
Harvati, K., E. Panagopoulou, P. Karkanas, A. Athanassiou, and S. R. Frost. 2008. Preliminary results of the Aliakmon Paleolithic/Paleoanthropological Survey, Greece, 2004–2005. In *The Palaeolithic of the Balkans*, eds. A. Darlas and D. Mihailović, 15–20. Oxford: Archaeopress.
Hughes, P. D., J. C. Woodward, P. L. Gibbard, M. G. Macklin, M. A. Gilmour, and G. R. Smith. 2006. The glacial history of the Pindus Mountains, Greece. *Journal of Geology* 114: 413–34.
Koufos, G. D. 2006. The Neogene mammal localities of Greece: Faunas, chronology and biostratigraphy. *Hellenic Journal of Geosciences* 41: 183–214.
Limbrey, S. 1990. Edaphic opportunism? A discussion of soil factors in relation to the beginning of plant husbandry in south-west Asia. *World Archaeology* 22: 45–52.
Machette, M. N. 1985. Calcic soils of the southwestern United States. *Special Paper—Geological Society of America* 203: 1–21.
Magny, M. and C. Bégeot. 2004. Hydrological changes in the European midlatitudes associated with freshwater outbursts from Lake Agassiz during the Younger Dryas event and the Early Holocene. *Quaternary Research* 61: 181–92.
Magny, M. and J. N. Haas. 2004. A major widespread climatic change around 5300 Cal. Yr BP at the time of the alpine Iceman. *Journal of Quaternary Science* 19: 423–30.
Magny, M., C. Bégeot, J. Guiot, and O. Peyron. 2003. Contrasting patterns of hydrological changes in Europe in response to Holocene climate cooling phases. *Quaternary Science Reviews* 22: 1589–96.
Magny, M., B. Vanniere, G. Zanchetta, E. Fouache, G. Touchais, L. Petrika, C. Coussot, A. V. Walter-Simonnet, and F. Arnaud. 2009. Possible complexity of the climatic event around 4300–3800 Cal. BP in the Central and Western Mediterranean. *Holocene* 19: 823–33.
Marković, S. B., U. Hambach, T. Stevens, G. J. Kukla, F. Heller, W. D. McCoy, E. A. Oches, B. Buggle, and L. Zoller. 2011. The last million years recorded at the Stari Slankamen (Northern Serbia) loess-palaeosol sequence: Revised chronostratigraphy and long-term environmental trends. *Quaternary Science Reviews* 30: 1142–54.
Opie, J. 1989. 100 years of climate risk assessment on the High Plains: Which farm paradigm does irrigation serve? *Agricultural History* 63: 243–69.
Perlès, C. 2001. *The Early Neolithic in Greece: The First Farming Communities in Europe*. Cambridge: Cambridge University Press.
Perlès, C. 2003. An alternate (and old-fashioned) view of neolithisation in Greece. *Documenta Praehistorica Neolithic Studies* 10: 99–113.
Plunkett, G. and G. T. Swindles. 2008. Determining the sun's influence on late glacial and Holocene climates: A focus on climate response to centennial-scale solar forcing at 2800 Cal. BP. *Quaternary Science Reviews* 27: 175–84.

Rassios, A. E. 2004. *A Geologist's Guide to West Macedonia, Greece*. Grevena: The Grevena Development Agency.

Roberts, N., A. Moreno, B. L. Valero-Garces, J. P. Corella, M. Jones, S. Allcock, J. Woodbridge et al. 2012. Palaeolimnological evidence for an east-west climate see-saw in the Mediterranean since AD 900. *Global and Planetary Change* 84–85: 23–34.

Roebroeks, W. and P. Villa. 2011. On the earliest evidence for habitual use of fire in Europe. *Proceedings of the National Academy of Sciences of the United States of America* 108: 5209–14.

Runnels, C. 1995. Review of Aegean prehistory IV: The Stone Age of Greece from the Palaeolithic to the advent of the Neolithic. *American Journal of Archaeology* 99: 699–728.

Runnels, C., M. Korkuti, M. L. Galaty, M. E. Timpson, S. R. Stocker, J. L. Davis, L. Bejko, and S. Mucaj. 2009. Early prehistoric landscape and landuse in the Fier region of Albania. *Journal of Mediterranean Archaeology* 22: 151–82.

Runnels, C. N. and T. H. van Andel. 2003. The Early Stone Age of the Nomos of Preveza: Landscape and settlement. *Hesperia Supplements* 32: 47–134.

Savina, M. E. 1993. Some aspects of the Geomorphology and Quaternary Geology of Grevena Nomos, Western Macedonia, Greece. In *Beiträge Zur Landeskunde Von Griechenland IV, Salzburger Geographische Arbeiten*, ed. H. Reidl, 22: 57–75 Salzburg, Austria: Universität Salzburg.

Shakesby, R. A. 2011. Post-wildfire soil erosion in the Mediterranean: Review and future research directions. *Earth-Science Reviews* 105: 71–100.

Stiros, S. C. 1998. Historical seismicity, palaeoseismicity and seismic risk in Western Macedonia, Northern Greece. *Journal of Geodynamics* 26: 271–87.

Strasser, T. F., E. Panagopoulou, C. N. Runnels, P. M. Murray, N. Thompson, P. Karkanas, F. W. McCoy, and K. W. Wegmann. 2010. Stone Age seafaring in the Mediterranean: Evidence from the Plakias region for Lower Palaeolithic and Mesolithic habitation of Crete. *Hesperia* 79: 145–90.

Stuiver, M., P. J. Reimer, E. Bard, J. W. Beck, G. S. Burr, K. A. Hughen, B. Kromer, F. G. McCormac, J. van der Plicht, and M. Spurk. 1998. INTCAL98 Radiocarbon age calibration 24,000–0 Cal. AD. *Radiocarbon* 40: 1041–83.

Tourloukis, V. and P. Karkanas. 2012. The Middle Pleistocene archaeological record of Greece and the role of the Aegean in hominin dispersals: New data and interpretations. *Quaternary Science Reviews* 43: 1–15.

Tsoukala, E. 2000. Remains of a Pliocene *Mammut borsoni* (Hays, 1834) (Proboscidea, Mammalia), from Milia (Grevena, W. Macedonia, Greece). *Annales de Paleontologie* 86: 165–91.

Tsoukala, E. and A. Lister. 1998. Remains of straight-Tusked Elephant, *Elephas (Palaeoloxodon) antiquus* Falc. & Caut. (1847) ESR-dated to Oxygen Isotope Stage 6 from Grevena (W. Macedonia, Greece). *Bollettino della Societa Paleontologica Italiana* 37: 117–39.

Tsoukala, E., D. Mol, S. Pappa, E. Vlachos, W. van Logchem, M. Vaxevanopoulos, and J. Reumer. 2010. *Elephas antiquus* in Greece: New finds and a reappraisal of older material (Mammalia, Proboscidea, Elephantidae). *Quaternary International* 245: 339–49.

Van Andel, T. H., C. N. Runnels, and K. O. Pope. 1986. Five thousand years of land use and abuse in the Southern Argolid, Greece. *Hesperia* 55: 103–28.

Verheijen, F. G. A., R. J. A. Jones, R. J. Rickson, and C. J. Smith. 2009. Tolerable versus actual soil erosion rates in Europe. *Earth-Science Reviews* 94: 23–38.

Vita-Finzi, C. 1969. *The Mediterranean Valleys: Geological Changes in Historical Times*. Cambridge: Cambridge University Press.

Wace, A. J. B. and M. S. Thompson. 1914. *The Nomads of the Balkans: An Account of Life and Customs among the Vlachs of Northern Pindus*. London: Methuen.

Walker, M. J. C., M. Berkelhammer, S. Björck, L. C. Cwynar, D. A. Fisher, A. J. Long, J. J. Lowe, R. M. Newnham, S. O. Rasmussen, and H. Weiss. 2012. Formal subdivision of the Holocene Series/Epoch: a Discussion Paper by a Working Group of INTIMATE (Integration of ice-core, marine and terrestrial records) and the Subcommission on Quaternary Stratigraphy (International Commission on Stratigraphy). *Journal of Quaternary Science* 27: 649–59.

Weninger, B., E. Alram-Stern, E. Bauer, L. Clare, U. Danzeglocke, O. Joris, K. E. Claudia et al. 2006. Climate forcing due to the 8200 Cal Yr BP Event observed at Early Neolithic sites in the Eastern Mediterranean. *Quaternary Research* 66: 401–20.

Wilkie, N. C. and M. E. Savina. 1997. The earliest farmers in Macedonia. *Antiquity* 71: 201–07.

Woodhouse, C. A. and J. T. Overpeck. 1998. 2000 years of drought variability in the Central United States. *Bulletin of the American Meteorological Society* 79: 2693–714.

15 Knowledge of Soil and Land in Ancient Indian Society

*Pichu Rengasamy**

CONTENTS

Our distant ancestors found their food by hunting and foraging. They indirectly depended on soils that provided plants, but they did not markedly alter soils by their actions. With transition to agriculture, human impact and dependence on soils was inevitable. Development of agricultural technologies during the evolutionary processes of civilization led to the stabilization of human communities through their settlement in fixed locations, rather than being nomadic in search of livelihood.

The Indian civilization, with a recorded history of more than 5000 years, is one of the oldest civilizations in the world. In ancient India, agriculture initially developed in river valleys where water was available throughout all the seasons. Settlements in highly productive agricultural regions led to other developments in trade, art, and literature. Before British colonization, India was never ruled by a single government although the land mass was geographically interconnected. However, different customs developed in different regions, in the north, west, east, and south of India.

Ancient Indian farmers understood their soils in relation to the production of specific crops, and the accumulated knowledge by experience transferred through generations. With the development of art and literature, knowledge was ably molded in the form of proverbs, folksongs, short poems, and literary texts. The development of agriculture occurred through the *Vedic era* (2000–600 BC) in the North of the subcontinent and the *Sangam Era* (200 BC–200 AD) in the South, and agriculture became the main occupation of the Indian community. The prime role of agriculture in society is articulated in *Thirukkural*, a well-known Tamil literature that was written more than 1500 years ago (Rajagopalachari 2009), as follows:

"*Suzhandrum erpinnathu ulagam; adhanal uzhanthum uzhave thalai*" meaning "Many other industries may be taken up, but ultimately the world depends on the plow (agriculture). So, despite its troubles, it is the worthiest occupation."

"*Uzhuthundu vazhvare vazhvar; matrellam thozhuthundu pin selbavar*" meaning "They only live by right that till the soil and raise their food. The rest are parasites."

Similarly, in the *Vedas*, texts of the Sanskrit literature, it is stated "upon this handful of soil our survival depends. Husband it and it will grow our food, our fuel, and our shelter and surround us with beauty. Abuse it and the soil will collapse and die taking man with it." (Lal 2010).

Prasna Upanishad, a companion of *Vedas*, stated "*Kshiti, jal, pawak, gagan, sameera—panch tatva yah adham sharira*" meaning "Soil, water, energy, sky and air are the five elements from which the human body is made" (Lal 2010).

* School of Agriculture, Food and Wine, The University of Adelaide, Adelaide, South Australia; Email: pichu.rengasamy@adelaide.edu.au

In vast areas covering the Indus Valley Civilization (3000–1500 BC), archaeological evidence has shown the protection measures taken against annual flooding, irrigation to secure increased crop yields, and drainage of large alluvial areas as a prerequisite to the valley's prosperity (Abrol 1990). The *Vedic hymns*, specifically those in *Rig Veda,* contain several notes on irrigated agriculture, river courses, water reservoirs, wells, and water-lifting structures.

15.1 ETHNOPEDOLOGY IN ANCIENT INDIA

Ethnopedology is the study of indigenous environmental knowledge, and is a hybrid discipline structured from the combination of natural and social sciences, such as soil science and geopedological surveys, social anthropology, rural geography, agronomy, and agroecology (Barrera-Bassols and Zinck 2003). A classical example of ethnopedology in ancient Tamil country (southern India) is described in the *Sangam literature* comprising *Akattiyam and Tolkappiam* (Balambal 1998). *The Sangam Landscape* (Tamil: *Akathinai*) is the classical love poetry that assigns human experiences related to specific habitats. These poems were categorized into different *tinais* (modes); each *tinai* was closely associated with a particular landscape, and imagery associated with that landscape—its flowers, trees, wildlife, people, climate, soil, and geography—are all woven into the poem in such a way as to convey a mood, associated with one aspect of a romantic relationship. There are seven modes (*Tinais*), five of which are geographical and associated with types of landscapes. These are

Kurinji (derived from the flower of the *kurinji* region: *Strobilanthes kunthiana*; common name: *Neelakurinji*): mountainous regions (mood associated with the union of lovers) with a cool and moist climate, dominated by red and black soils with stones and pebbles and occupied by hill tribes, their occupation being honey gathering.

Mullai (named after the flower of the landscape, viz. *jasmine*): forest and pasture landscapes (mood associated with waiting for lovers) with a warm and cloudy climate, dominated by red soil and occupied by farmers.

Marutham (named after the flower of the region, *Marutham*): agricultural regions of plains and valleys (mood associated with lovers' quarrels) with the combination of all seasons, dominated by alluvial soils and occupied by pastoralists and agriculturists.

Neytal (named after the regional flower water lily): coastal landscapes (mood associated with pining) with the combination of all seasons, dominated by sandy and saline soils and occupied by fishermen.

Paalai (named after the flower *Paalai*): deserts and wastelands (mood associated with elopement or long separation) with hot summers, dominated by salt-affected soils and occupied by bandits.

The *Marutham* region was the fittest for cultivation, with highly fertile soils. As described in *Sangam* literature, the land, in general, was classified according to its fertility, as *Menpulam* (fertile land), *Pinpulam* (dryland), *Vanpulam* (hard land), *Kalarnilam* (sodic land), *Uvarnilam* (alkaline land), and *Uppunilam* (salty land). *Menpulam* yielded richly for a variety of crops, while *Pinpulam* was cultivated with dryland crops, due to a lack of irrigation facilities. The yield from *Vanpulam* was limited. *Kalarnilam, Uvarnilam, and Uppunilam* were considered unfit for cultivation. The soils from *Uvarnilam* (highly alkaline soils), and also known as *usur* in northern India, were used for laundering–steaming the clothes mixed with the soil and steam being generated by boiling the water in mud pots. The term "usa" is used for alkaline earth in *Vedic Samhitas* and *Brahmanas* that recommend the use of this soil for cleaning clothes (Agarwal and Shukla 1984). Now, we know that these high pH soils contain significant amounts of sodium carbonate that is a major component of modern-day washing powders. Laundry workers were able to distinguish the alkaline soils from the saline soils by tasting. The soil types were also commonly known, mainly on the basis of their color or other visible properties, such as *Chemman* (red soil), *Kariman* (black soil), *Vandalman* (alluvial soil), *Chemporai* (laterite soil), and *Manal* (sandy soil).

As mentioned in *Thirukkural* (Rajagopalachari 2009), farmers believed that plowing, manuring, weeding, irrigation, and the protection of crops must be done according to a specific method to obtain a good yield. A wide range of tools necessary for agriculture were manufactured locally and the basic tool was the plow, known as *kalappai* (Cooke et al. 2003). The Tamil poet of ninth century AD, *Auvaiar* says, "*Akala uzhuvathinum azha uzhu*," meaning "it is better to deep plow rather than plowing broader." It appears that the modern soil management of deep ripping was followed by farmers of ancient days. *Thirukkural* states "*Thodippuzhuthi Kaksa unakkin pidithu eruvum vendathu salappadum*," meaning "If the plowed soil is left to dry to a fourth of its bulk, there will be plenteous crop without even a handful of manure being put in." This statement emphasizes that adequate aeration of the soil is necessary for raising a good crop.

15.2 ECONOMICS AND LAND USE

Arthashastra, a treatise in Sanskrit on government, economics, and management in ancient India ascribed to *Kautilya*, the chief advisor to India's first emperor, *Chandragupta Maurya* (315–291 BC) provided detailed descriptions on land and soil classification in relation to productivity, and outlined stringent rules that enabled different assessment of production systems to meet the needs of the state and of the cultivators (Abrol 1990). *Kautilya* emphasized that agriculture should receive policy and administrative support from the king to achieve economic sustainability. He enumerated the interactions among seasons, rainfall, and soil management through manuring in relation to specific crops. He classified the lands based on water regimes and suitability for different crops. On the relative value of different categories of land, he emphasized that land is what is made of it by people.

15.3 CONCLUSION

This chapter provides only a few examples from the literature of the awareness of ancient Indian farmers of the relation between soil properties and crop production, and how agriculture and soil management were linked with social anthropology. There may be many other examples to be found in the literature of a score of Indian languages other than Tamil and Sanskrit, of which the author has no knowledge. While modern soil technologies have evolved through rigorous scientific methods, the ancient community has accumulated the knowledge on land and soils through observation and inference—processes that are essential for any scientific investigation.

REFERENCES

Abrol, I.P. 1990. Fertility management of Indian soils—A historical perspective. *14th International Congress of Soil Science, Transactions* 5: 203–207.

Agarwal, D.K. and Shukla, S.C. 1984. Washerman and washing materials in ancient India. *Indian Journal of History of Science* 19: 314–322.

Balambal, V. 1998. *Studies in the History of the Sangam Age*. Delhi: Kalinga Publications.

Barrera-Bassols, N. and Zinck, J.A. 2003. Ethnopedology: A worldwide view on the soil knowledge of local people. *Geoderma* 111: 171–195.

Cooke, M., Fuller, D.Q., and Rajan, K. 2003. Early historic agriculture in Southern Tamil Nadu. *Proceedings of the European Association for South Asian Archeology Conference,* Bonn, Germany, 341–350.

Lal, R. 2010. Soil quality and ethics: The human dimension. In *Food Security and Soil Quality*, eds., R. Lal and B.A. Stewart, 301–308. Boca Raton, FL: CRC Press.

Rajagopalachari, C. 2009. *Kural—The Great Book of Tiruvalluvar*. Mumbai, India: Bharatiya Vidya Bhavan. 187.

16 The Evolution of Paddy Rice and Upland Cropping in Japan with Reference to Soil Fertility and Taxation

Masanori Okazaki and Koyo Yonebayashi*

CONTENTS

16.1 INTRODUCTION

Farmers have a saying that goes along the lines "Rice production depends on the soil, wheat production depends on the fertilizer." This saying points to the fundamental differences in agricultural production between paddy rice (*Oryza sativa*) and upland crops such as wheat (*Triticum* spp.). People have always understood that if rice is grown in paddy fields, continuous cropping causes no growth penalty; yet, continuous cropping in upland conditions, even of rice, causes a steady decline in yield. In this chapter, we describe the role of soil science in Japanese agriculture historically with reference to soil fertility and its relationship to taxation.

16.2 THE DAWN OF TAXATION

Upland rice cultivation in Japan is believed to have begun about 6400 years ago. Analyses of "plant opals" (phytoliths) from soil and of rice DNA (deoxyribonucleic acid) indicate that rice came to Japan via three different routes (Sato, 2002): from the Yangtze River area of China to northern Kyushu directly, via the Korean Peninsula, and from southeast China to the southwestern islands. A type of shifting cultivation system involving rice, Japanese barnyard millet (*Echinochloa esculenta*), foxtail millet (*Setaria italica*), and a few other cereals prevailed on the flat and gently sloping plains and terraces 5000–7000 years ago. Paddy rice (temperate *japonica* rice) was introduced from the Yangtze River area into Japan in the Yayoi period (BC 1000–AD 250), along with the technology for constructing large irrigation systems (Sato, 2002; Prasad et al., 2006). Paddy field remnants have been found that are dated at 3000 years before present (BP).

* Ishikawa Prefectural University, 1-308, Suematsu, Nonoichi, Ishikawa 921-8836 Japan; Email: okazaki@ishikawa-pu.ac.jp

Paddy rice cultivation in Japan occurred before a centralized administration was established, and during those early times, the only tax that was levied was for a local fund to support the prevailing polytheistic religious practices.

16.3 TAXATION UNDER THE RITSURYO CENTRALIZED ADMINISTRATION AND MANORIAL SYSTEM

A centralized administration system was first established in Japan by the *Taika* Reforms in 645 and reinforced in 701 by the enactment of the Taiho Code (*Taihō-ritsuryō*) at the behest of the Emperor Monmu. The Taiho Code formed a fully fledged law supporting a centralized administration system that continued with little change for the next several centuries. During this period, the combined area of paddy fields and upland cropping has been estimated to be about 0.8 million hectares (ha) (Figure 16.1). Takashima (2012) estimated agricultural land area and crop production in Japan for 3 benchmark years. For 725, he estimated that the total area of paddy rice was 0.59 million ha, and that of upland crops was 0.19 million ha, which yielded 1.38 million tons (t) of dehusked rice and 0.23 million t of upland crops. Under the Taiho Code, a land-holding system based on the centralized administration system used in China was introduced into Japan. In the Japanese land-holding system, the family register was reviewed every 6 years (y), and land allocation was adjusted to the equivalent of 2400 m^2 for each male and 1600 m^2 for each female (age 6 years and over). The Japanese land tax system of the time was based on this land allocation system, and was therefore fundamentally different from the Chinese tax system of the Tang period (618–907), which was based only on male family members (aged 21–60 years). The tax under the Japanese system was 3–10% of the crop yield as payment-in-kind under the system of the assessed annual rated yield of the domain. More productive land (more fertile soil) promised higher yields to farmers and to the tax collectors. A description was provided in the *Harima-no-kuni fudoki* (Harima County Report) of 715 that livestock bedding that had been made of grass cut with a sickle could be added to the soil to improve crop production, which introduced higher yields.

In the middle of the eighth century, the agricultural land allocation system began to break down. The first step was the introduction of the *Sanze-isshin-no-ho* in 723, a law that allowed land cultivators to own their land for a period of three generations as a means of encouraging the cultivation of rice fields. Twenty years later, the *Konden-einen-shizai-ho* was promulgated. This law permitted permanent ownership of newly cultivated land, allowing farmers who cleared new land to own it permanently. The central government, led by Tachibana-no-Moroe, enforced the *Konden-einen-shizai-ho* to promote the reclamation of land for rice fields. This signified an overthrow of the complete public ownership of land, which, being a major premise of the Ritsuryo legal code system,

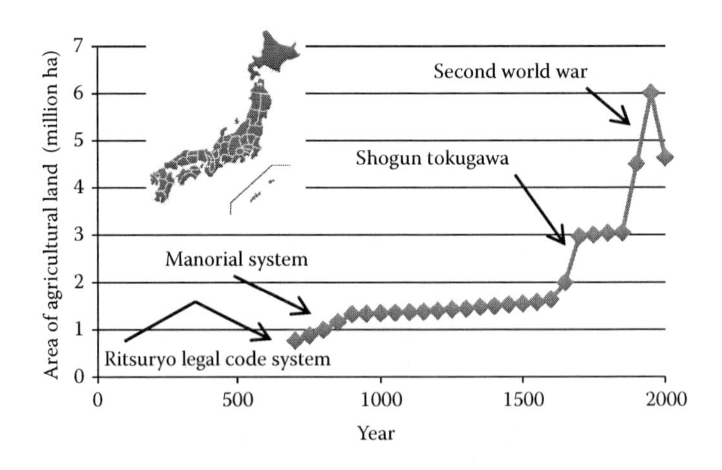

FIGURE 16.1 Development of agricultural land in Japan from the 700s to the present day.

would portend the demise of the Ritsuryo system. By the ninth century, aggregation of land holdings under the system of complete state ownership had ceased and given rise to a new manorial aristocracy who controlled a population of serfs who were responsible for the agricultural production and management of the manor. With the gradual breakdown of the centralized administration system in the middle of the eighth century, the manor-based local governments throughout Japan introduced different systems of tax collection. When the tax burden became unbearable, farmers in the manors fled in the night to new villages, or rose in revolt against their lords. By the tenth century, the centralized Ritsuryo administration system was entirely replaced by the manorial feudal system.

16.4 TAXATION UNDER FEUDALISM

More than 80% of the total population in the feudal period of Japan were farmers. Feudal lords collected taxes from farmers according to the expected productivity of paddy rice. But, they also had to protect the riverbanks from flooding to ensure that irrigation could continue with minimal disruption, and thus provide a stable flow of tax income from the farmers. In 900, the total area of agricultural land in Japan had increased to 1.33 million ha (0.86 million ha of paddy rice and 0.47 million ha of upland crops), which yielded 1.94 million t of dehusked rice and 0.55 million t of upland crops (Takashima, 2012). Around this time, the Emperor Daigo ordered the compilation of new laws. The *Engishiki*, a 50-volume set of detailed laws and regulations, appears to have been started in 905 by Fujiwara no Tokihira, was completed by Fujiwara no Tadahira in 927, and was finally implemented in 967 (Torao, 1995).

By the middle of the twelfth century, the area of paddy fields had increased to 0.96 million ha. But, the area of upland crop fields had not increased at anywhere near the same rate as that of paddy fields (Takashima, 2012), because the feudal lords selectively favored paddy rice cultivation on account of the stable yields, and therefore the tax revenue, it offered. The seventh of 30 volumes of *Seiryoki* (the biography of Seiryo Doi), titled *Shinmin-kangetsu-shu* (Lessons for the People), was an agricultural report written in 1564 (Sasaki, 1981 states that it was published in the 1620s and 1630s) by So-an Matsu-ura (1564), a retainer of Seiryo Doi, the feudal warlord of current-day Ehime Prefecture. It is believed to be the oldest agricultural book in Japan (Doi, 1980). The text discusses the application of fertilizers to agricultural fields on the basis of soil classification, the evaluation of land suitability for crops, and paddy rice cultivation under irrigated or dryland conditions. However, the agricultural description in *Seiryoki* has largely been neglected, because most of the volumes concentrated on tales of warfare. Kanno (1958) found that *Shinmin-kangetsu-shu* used some soil names that we still use today. These include *matsuchi* (loamy soil derived from residual parent material), *onji* (Akahoya ash in southern Kyushu, a volcanic ash derived from the Kikai caldera), and *giro* (clay formed in the Holocene, i.e., the last 11,700 years).

In the late sixteenth century, Hideyoshi Toyotomi, the Chief Adviser to the Emperor, ordered Mitsunari Ishida, the Magistrate of Land Survey, to conduct a land survey as a basis for equitable tax collection from farmers based on the crop yield (this became known as the Taiko Land Survey and Evaluation). From 1582 to 1598, Ishida surveyed 166 prefectures throughout Japan and precisely estimated taxation levies to establish the Toyotomi government. This taxation system was continued by the Shogun Tokugawa throughout the Edo era (1603–1868). Generally, 60% of the total crop yield was paid as tax. Land productivity in each village was estimated by the amount of rice and upland crops produced. Under the Taiko Land Survey and Evaluation, agricultural land used for paddy rice and upland fields was classified into six classes: "extremely fertile," "fertile," "moderately fertile," "low fertility," "very low fertility," and "estimated to be non-fertile." Along rivers, low-lying land that was easily irrigated had been mainly developed for paddy fields, whereas upland crop fields were located around houses on gently convex landforms. This pattern of land use had led to the creation of gleyed paddy soils (U.S. Taxonomy; Haplaquepts) from what had previously been gray lowland soils (Haplaquents). Brown lowland soils (Dystrochrepts) persisted without any change to their pedological features by upland crop cultivation.

The average annual production of paddy rice from "fertile" paddy rice fields was taken by the tax system to be 2.25 t ha^{-1} (dehusked), with productivity decreasing by 0.3 t ha^{-1} with each drop in fertility grade. The crop productivity of "fertile" upland fields was estimated to be 1.8 t ha^{-1} for wheat and other upland crops. By 1600, the total area of agricultural land (paddy fields plus upland fields) had reached 2.05 million ha, and total paddy rice production was 2.70 million t. The *Hyakusho-Denki*, published in the seventeenth century (unknown authorship, 1979), and the *Dosei-Ben*, published by Nobukage Sato in 1729, describe a similar soil classification and agricultural practices based on soil type (Sato, 1729). The *Dosei-Ben* in particular was strongly affected by the *Yukung* (Chinese agricultural book) (Kanno, 1958). It described and classified Japanese agricultural soils into six types (mellow (loamy), clayey, salt accumulated, silty, andic (originated from volcanic ash), and sandy), subdivided them by color, and referred to their distribution (Matsui, 1988). Yasusada Miyazaki (1623–1697) wrote the *Nogyo Zensho* (An Agricultural Book of Universal Applicability), published in 1697, to educate readers and promote the cultivation of many plant species (Miyazaki, 1697). This text was apparently written without knowledge of the *Seiryoki*. With the development of wood-block printing technology, and as judged by the diversity of crops throughout Japan, the *Nogyo Zensho* influenced many farmers. In the Edo era, during which Japan was largely isolated from contact with the outside world, severe famines demanded the cultivation of root crops. Thanks to the work of Konyo Aoki (1698–1769), who in 1735 published the *Banshoko* (*An Incentive for the Production of Sweet Potato*), sweet potato production had spread to volcanic ash soils (Andisols). Andisols are rich in soil organic matter and the highly reactive clay minerals allophane and imogolite, which tightly bind phosphorus (P), and so, are characterized by low levels of available P. They also tend to be soft and easy to till, and are commonly associated with gentle slopes. The application of manure and compost to Andisols is normally recommended to improve soil conditions for the cultivation of root crops. Nagatsune Ohkura (1768–1861) published the *Menpoyomu,* a technical manual for growing cotton (*Gossypium* spp.), in 1833 (Ohkura, 1833) and wrote many agricultural books in his long lifetime. Nobuhiro Sato (1769–1850) proposed agricultural policy and finance to a policy advisor of the Satsuma-clan in his *Nosei Honron* (Takimoto, 1927), and published many other books. He wrote the *Uko Shuran* (*Collected Volumes of Yukung*) (Takimoto, 1927) in 1829 and concluded that Yu, the celebrated engineer and Minister of Public Works of the Chinese Yao dynasty (around 4000 years BP), recognized differences in soil type and established nine classes of soil based on morphological features (Kanno, 1958; Matsui, 1988). The Shogun Tokugawa and feudal lords vigorously promoted the opening of new agricultural land through clearing and reclamation to increase the income tax base from farmers, who comprised 90% of the total population. In general, the only land that was not developed for agricultural fields was that which was difficult to irrigate or was not fertile. The scientific and technological developments of the Edo era (seventeenth through nineteenth centuries) supported the expansion of agricultural land, and by the middle of the eighteenth century, scholars recognized that the productivity of paddy rice fields was relatively uniform despite large variations in soil fertility. This might explain why there seems to be little evidence of farmers complaining about rice taxes during this period. The effort to develop paddy rice cultivation did, however, eventually bring the attention of farmers to soil fertility, and the use of night soil for crop production became common in the suburbs of large cities, with farmers even prepared to pay for the waste because of the lack of manure from animal husbandry. Paddy rice farmers were not anxious about either soil erosion, but upland rice producers on sloping land were. Paddy rice farmers were also unconcerned about salt accumulation (from night soil application). Upland crops were very much secondary agricultural products, resulting in taxation being squarely focused on rice production.

By the early nineteenth century, many Japanese scientists learned Dutch as a means of acquiring western scientific knowledge. As Spain and Portugal were actively seeking opportunities to invade or occupy Japan, the Shogun Tokugawa took the policy of isolationism, maintaining contact only with the Netherlands through the Dutch East India Company from 1641 to 1858, because such trade was not attendant with attempts to propagate Catholicism.

16.5 TAXATION UNDER THE REVISED LAND TAXATION LAW OF THE MEIJI GOVERNMENT

Immediately after the Meiji Restoration in 1868, the Meiji government set out to revise the taxation system and promulgated the Revised Land Taxation Law in 1873. The new law fixed the tax rate at 3% of the land value throughout Japan. The rate was later reduced to 2.5% in 1877 to appease farmers. At the same time, the Meiji Restoration dismissed the feadal warlords and their vassal *samurai* class completely. Land was privatized and taxation was determined not from the expected crop yield, but from the land value, and was paid as money rather than as a portion of the production. Surprisingly, the *samurai* class in the Edo period had been collecting land tax without ever seeing their own estates. They had not been collecting it in the form of rice but instead were getting the salary in the form of gold or silver currency, because under the feudal system, the feudal clan received the taxes directly from the feudal estate rather than from the vassal. Yet, most Japanese agricultural land was in feudal estates that permitted the vassals to collect tax. As a result of this dysfunction in the system, the *samurai* class easily accepted the Revised Land Taxation Law (Isoda, 2003). Around 1900, the total area of agricultural land was estimated at 3.20 million ha, with paddy rice production of 4.80 million t. The Meiji government struggled to collect the taxes. In part for this reason, they invited academics from an assortment of countries to improve scientific knowledge. Japan had close contact with the Netherlands for about 300 years through Dejima, a small island off Nagasaki in Kyushu. Foreign academics from countries such as Great Britain, the United States, France, China, Germany, and the Netherlands were paid high salaries and worked in many scientific fields, including civil engineering and chemistry. Max Fesca (1846–1917) undertook a comprehensive survey of the soils of Japan as a foreign advisor to the Meiji government from 1882 to 1894. He published the soil map of Kai (present-day Yamanashi Prefecture) in 1887 (Fesca, 1887), in which he described soils derived from different parent materials, and which gave rise to a body of important information on agricultural fields for crop cultivation: soils derived from green tuff (a type of pyroclastic rock called green-colored tuff) were suitable for tea (*Camellia sinensis*), *konnyaku* (a tuber used to make a flour used in noodles; *Amorphophallus konjac*), and fiber crops, whereas soils from pyroclastic material (volcanic ash) were best used for sweet potato (Sasaki, 1978; Kyuma, 2009, 2011). Another foreign academic, Oskar Kellner (1851–1911), taught at Komaba Agricultural School in Tokyo and publicized his research into the nutritional analysis of livestock feed. During his 1881–1882 stay in Japan, he conducted research on chemical fertilizers. Japanese agricultural practice at the end of the 1800s focused on application of manure and compost; the use of other organic fertilizers, such as dried sardines and soybean cake, had just started. Experiments on the application of organic fertilizers were performed at agricultural experiment stations throughout Japan.

16.6 AGRICULTURE IN INDUSTRIALIZED JAPAN

Chemical engineering, especially calcium superphosphate manufacturing, developed in the early years of the twentieth century. After this, Japan imported large amounts of ammonium sulfate made by the Haber–Bosch process, resulting in a shift from organic fertilizers to inorganic fertilizers.

With the industrialization of Japan in the late nineteenth and early twentieth centuries, after absorbing technology from the western European countries and the United States, the production of nitrogen (N), P, and potassium (K) fertilizers increased rapidly (Yamada, 1970). The ultimate result of research into the efficiency of N and P uptake from fertilizer applied to rice paddy soil was the proposed reduction in the quantity of fertilizer applied at the time of planting, and the subsequent addition of a top dressing (a surface application of fertilizer which is not plowed in) at the time when the rice was at the heading stage (the emergence of the head in which the grain develops from the leaf sheath). Experiments into the application of inorganic N, P, and K have been conducted at agricultural experimental stations in Japan since 1921. In 1943,

Shioiri (1943) summarized 3 years of data from all the prefectural agricultural experimental stations in Japan and concluded that unfertilized paddy rice fields typically yielded about 30% less grain than fertilized fields (Figure 16.2). Paddy fields can accumulate N, P, and K because of the cyclical management of oxidation (8 months between cropping periods) and reduction (4 months of flooding). In contrast with paddy rice, upland rice yields from unfertilized fields were 40% less than those that received fertilizer. Other upland and tuber crops displayed similar results (Yamane, 1981).

After World War II, the large influx of returnees from former Japanese colonies caused a sharp increase in demand for food and agricultural land. The Natural Resources Service of the General Headquarters Supreme Commander for the Allied Powers (1945–1952) performed a soil survey throughout Japan and classified the soils into eight types: bog soils, gray brown podzolic soils, red and yellow podzolic soils, reddish-brown lateritic soils, ando soils, planosols, alluvial soils, and lithosols (Sasaki, 1978).

In the mid- to late twentieth century, the Japanese government promulgated several laws related to soil fertility. *Kokudo Chosa Ho* (Land Survey Law) in 1951, which aimed to characterize the land and its productivity and *Kodobaiyo Ho* (Law of Improvement of Agricultural Land by Amendments) in 1952, which aimed to improve soil acidity, were proclaimed. In addition, *Chiryoku Zoshin Ho* (Law of Increment of Soil Fertility) in 1984 was promulgated to increase food production. On the basis of these laws, the following projects began: *Teii-Seisannchi-Chosa* (Survey of Low-Productivity Areas) from 1947, *Sehikaizen-Chosa* (Survey for Improvement of Fertilizer Application) from 1953 to carry out soil surveys, test irrigation programs, and determine fertilizer application standards for paddy rice, *Chiryoku-Hozen-Kihon-Chosa* (Basic Survey of Conservation of Soil Fertility), from 1959 to 1980 to improve the soil fertility of upland fields, and *Dojo-Kannkyo-Kiso-Chosa* (Basic Survey of Soil Environment), from 1980 to 2004. These projects were to be carried out by the institutes and experimental stations of the Ministry of Agriculture, Forestry, and Fisheries (Ministry of

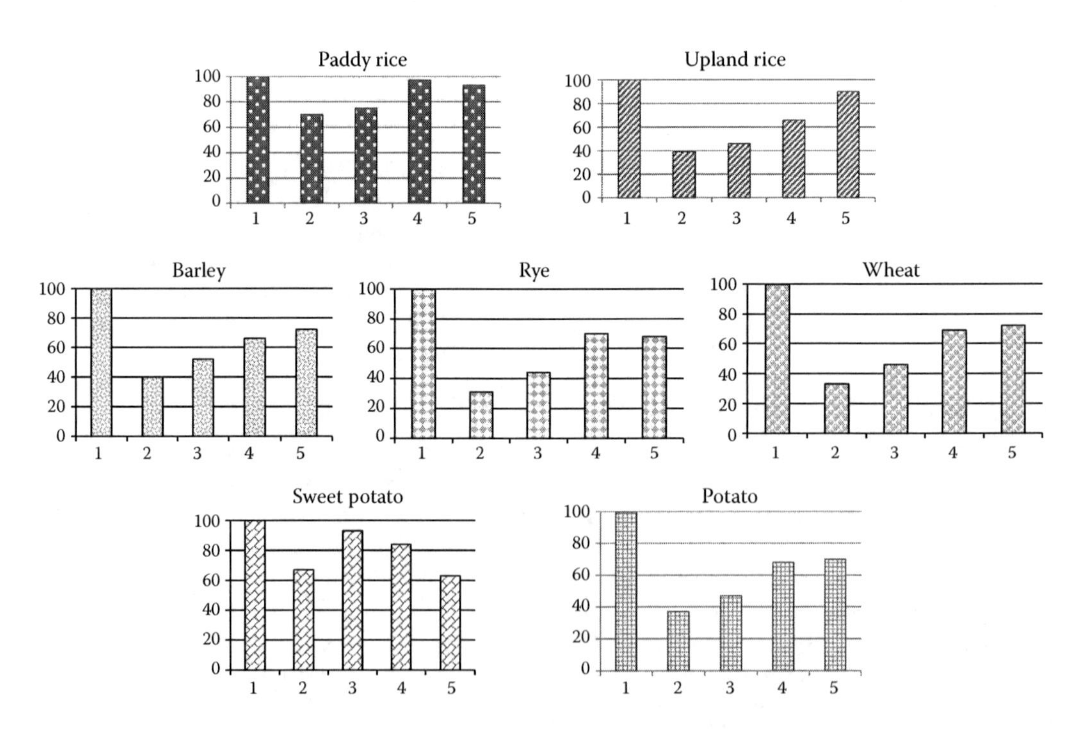

FIGURE 16.2 The effect of N, P, and K application on paddy rice, upland rice, barley, rye, wheat, sweet potato, and potato in Japan (Shioiri, 1943). Key to x-axes: 1. All fertilizers (N, P, and K) added, 2. No fertilizers added, 3. K and P added but not N, 4. K and N added but not P, 5. N and P added but not K.

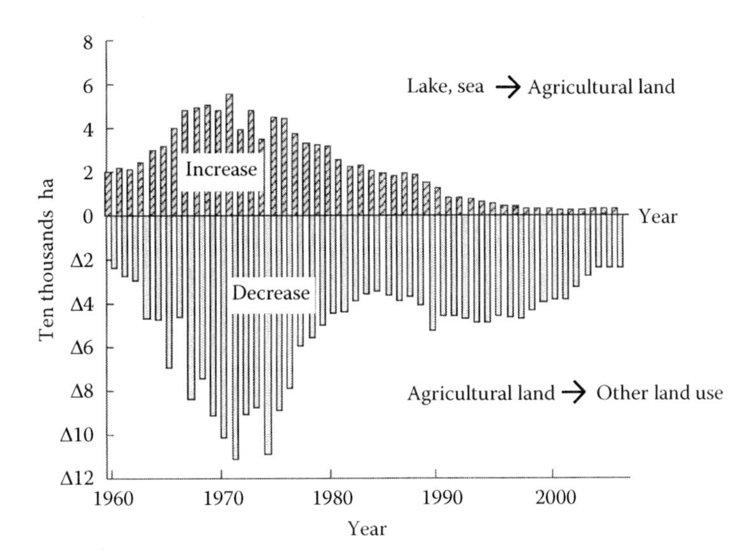

FIGURE 16.3 Changes in the shift of land from reclaimed land to agriculture, and from agriculture to other land uses in Japan in the latter half of the twentieth century. (From Ministry of Agriculture, Forestry, and Fisheries, 2012.)

Agriculture, Forestry and Fisheries, 2008) to stimulate food production, resulting in an increase in agricultural lands up to the 1970s.

A rapid decline in the area of agricultural land, however, began in the 1970s (Figure 16.3) (Ministry of Agriculture, Forestry and Fisheries, 2012). The decline was caused not by land degradation such as soil erosion or salt accumulation, but by the economic depression of farmers and it has directly reduced Japan's self-sufficiency in food production: Japan now produces <40% of the food it consumes. The decline of Japan's agriculture proceeds steadily despite the many national programs of the twentieth century. The stagnation of appropriate agricultural practices leads to more soil degradation. We have to remember that, in agriculture, all roads lead back to the soil (Hambidge, 1938).

REFERENCES

Doi, S. 1980. *Shinmin-Kangetsu, Seiryoki* Vol. 7. In *Seiryoki (Shinmin-Kangetsu), Nojutukan-Seiki and Ashu-Hoppou-Nogyo-Zensho*, Series of Agricultural Books no. 10, eds. T. Yamada, J. Iinuma, M. Oka, and S. Morita, 3–294. Tokyo: Nobunkyo (in Japanese).

Fesca, M. 1887. *Abhandlungen und Erlaeuterungen zur Agronomischen Karte der Provincz Kai*. Tokyo: Kaiserliche Japanische Geologischen Reichsanstalt (in German).

Hambidge, G. 1938. Soils and men—A summary. In *Soils and Men*, Yearbook of Agriculture 1938. USDA, 1–44, Washington, DC: United States Government Printing Office.

Isoda, M. 2003. *Household Account of Samurai. Shincho-Shinsho 005*, 38–41, Tokyo: Shincho-Sha (in Japanese).

Kanno, I. 1958. The germ of soil science in the Orient. On the description of soils written in the "Yukung," the chapter of the "Shooking". *Pedologist* 2: 2–10 (in Japanese).

Kyuma, K. 2009. "Do-Sei" (soils and their productive quality). The origin of the word and changes of its concept as related to early soil surveys in Japan by Max Fesca. *Fertilizer Science* 31: 75–110 (in Japanese).

Kyuma, K. 2011. Greetings from Fesca's writing. *Fertilizer Science* 33: 1–19 (in Japanese).

Matsui, T. 1988. *Soil Geography Prologue*, 1–13. Tokyo: Tsukiji Shokan (in Japanese).

Matsu-ura, S. 1564. *Engishiki*. In *Kotaishiki/Koujinshiki/Engishiki*. Japanese History Series No. 26, eds. K. Kuroita, and the Editorial Committee for Japanese History Series, 1–1032. Tokyo: Yoshikawa Kobunkan (published in 1968) (in Japanese).

Ministry of Agriculture, Forestry and Fisheries. 2008. The Report of Dojo-Hozen-Chosa-Jigyo (The Report of Soil Conservation Survey Program), 1–485. Tokyo: The Ministry of Agriculture, Forestry and Fisheries (in Japanese)

Ministry of Agriculture, Forestry and Fisheries. 2012. http://www.e-stat.go.jp/SG1/estat/List.do?lid=000001061493

Miyazaki, Y. 1697. *Nogyo Zensho*. In Series of Agricultural Books no. 12, eds. T. Yamada, J. Iinuma, M. Oka, and S. Morita, 3–392. Tokyo: Nobunkyo (published in 1978) (in Japanese).

National Institute for Agro-Environmental Sciences. 2012. http://agrimesh.dc.affrc.go.jp/soil_db/ (in Japanese).

Ohkura, N. 1833. *Menpoyomu*. In Series of Agricultural Books no. 15, eds. T. Yamada, J. Iinuma, M. Oka, and S. Morita, 317–411. Tokyo: Nobunkyo (published in 1977) (in Japanese).

Prasad, P. V. V., Boote, K. J., Allen Jr., L. H., Sheehy, J. E., and Thomas, J. M. G. 2006. Species, ecotype and cultivar differences in spikelet fertility and harvest index of rice in response to high temperature stress. *Field Crops Research* 95: 398–411.

Sasaki, J. 1981. *Endogenous Technology and Society in Japan*. HSDRJE-41/UNUP-221, 1–41. Tokyo: The United Nations University, http://d-arch.ide.go.jp/je_archive/pdf/workingpaper/je_unu41.pdf.

Sasaki, S. 1978. Soil survey in Japan. In *Soil Survey Methods*, ed. Editorial Committee for Soil Survey Methods, 3–8, Tokyo: Hakuyusha (in Japanese).

Sato, N. unknown. 1729. *Dosei-Ben*. Enlarged by N. Sato, ed. K. Oda, Tokyo: Rokugo-Kan (published in 1874) (in Japanese), cited from Kyuma, 2009.

Sato, Y. 2002. *History of Rice in Japan*, 104–109. Tokyo: Kadokawa-Shoten (in Japanese).

Shioiri, M. 1943. Basic knowledge of fertilizer application. In *Lectures of Soil and Fertilizer Science*, 133–180. Tokyo: Asakura Shoten (in Japanese)

Takashima, M. 2012. Agricultural production and economic growth in ancient Japan: A quantitative analysis of arable land, land productivity and agricultural output. Global COE Hi-Stat Discussion Paper, Series 223, 1–38, Hitotsubashi University (in Japanese).

Takimoto, S. 1927. The complete series of Nobuhiro Sato family, Vol. 2, ed. Nobuhiro Sato family, 19–421. Tokyo: Iwanami Shoten (in Japanese).

Torao, T. 1995. *Engishiki*, Book Series of Japanese History No. 8, ed. The Society of History of Japan, 0–253. Tokyo: Yoshikawa Kobunkan (in Japanese).

Unknown authorship, 1979. *Hyakusho-Denki*. In Series of Agricultural Books no. 16, eds. T. Yamada, J. Iinuma, M. Oka, and S. Morita, 69–116. Tokyo: Nobunkyo (in Japanese).

Yamada, Y. 1970. Fertilizer application for rice cultivation. In *Study on Rice Cultivation in Japan*, ed. The Society of Kyushu Agriculture, 65–79. Tokyo: Ochanomizu Shobo (in Japanese).

Yamane, I. 1981. *Soil Science for Agricultural Land*. Tokyo: Nobunkyo (in Japanese).

17 Soils in Farming—Centric Lessons for Life and Culture in Korea

Rog-Young Kim, Su-Jung Kim, E. Jin Kim, and Jae E. Yang*

CONTENTS

17.1 INTRODUCTION

We live in an information age where we can find all the information whenever and wherever we need it. However, we do not pay any attention to our natural environment to gain information that may be useful in predicting weather or practicing farming. Our ancestors in Korea settled and have farmed since the Neolithic period (from about 12,000 to 6000 years before present [BP]), observing many natural phenomena (Lee 2005). As a result, many farming and weather-related proverbs, which were expressed in brief and familiar phrases, have been passed down orally from generation to generation (JARES 1979; RDA 1989). Our ancestors had communicated their knowledge through proverbs. Most of their knowledge is useful even to this day (GiATC 2012; KATC 2012). The proverbs not only provided us with the scientific principles of agriculture, but also an insight into the philosophy and social life of the people in the agrarian society of Korea.

* Department of Biological Environment, Kangwon National University, Chuncheon 200-701, Republic of Korea; Email: yangjay@kangwon.ac.kr

Owing to the long history of agriculture in the Korean peninsula, agriculture and land were always regarded as the roots of people's lives. A well-known proverb says, "Agriculture forms the basis of national existence"; this was written in a Korean agricultural book *Nongsa jikseol,* which was published in 1429 during the early period of the Korean *Joseon* dynasty (1397–1897). Another well-known proverb says, "Body and land are not two but one;" this has been recorded in a Korean medical book *Hyangyak Gypseongbang*, which was published in 1433 during the same period as *Nongsa jikseol.*

The word "proverb" is called *sokdam* in Korean, literally meaning folk sayings. *Sokdam* was first mentioned in the book *Samguk Yusa*, a collection of folktales, legends, and historical accounts related to the era of the Three Kingdoms of Korea (57 BC–668 AD) that was compiled by the Buddhist monk *Ilyeon* at the end of the thirteenth century (Korean Britannica online 2012). The first collection of *sokdam*, named *Sunohji*, in which about 130 *sokdams* are listed, was compiled by *Manjong Hong* at the beginning of the seventeenth century. It was not until the eighteenth century that the word *sokdam* itself and various *sokdams* were widely used among the people (Korean Britannica online 2012).

Today, about 10,000 of the old Korean *sokdams* related to customs, taboos, beliefs, and recommendations in various areas of society are listed in *The Dictionary of Korean Proverbs* (CPDKP 2001). There are many farming proverbs related to the forecast of harvests, weather prediction, seeding/transplantation, field operations, and agrarian culture. They concern (1) crops such as rice, barley, soybean, sesame, sweet potato, red pepper, and radish, (2) weather prediction, including rainfall, drought, fog, wind, thunder, and snow, (3) plants such as azalea, the chestnut tree, Japanese apricot tree, pine tree, and jujube tree, and (4) animals such as the cow, swallow, magpie, cuckoo, skylark, dragonfly, firefly, snake, silkworm, and grasshopper.

We have grouped a selection of farming proverbs into four categories according to their subject matter: (1) proverbs about soils and soil management, (2) proverbs about season-related farming, (3) proverbs about catastrophes, (4) proverbs about agrarian life and culture. We have concentrated on those relating to soils and soil management, but have included examples of those in the other categories.

17.2 PROVERBS ABOUT SOILS AND SOIL MANAGEMENT

17.2.1 Paddy Soils

1. *A bundle of grass carried early in the morning in summer brings one more bag of rice in autumn.* A bundle of grass could be used as a compost to increase the content of organic matter in the rice fields, ensuring soil fertility and rice yields.
2. *A bowl of rice will be provided free of cost to others, but a basket of compost will not be provided to others.* People will not give away compost because it is necessary for farming and is hard to make.
3. *Three buckets of cattle dung bring three sacks of rice.* Animal manure was so important in the olden days that it could be traded for an equal quantity of rice.
4. *Don't be proud of having a good paddy soil, but be proud of having a good rice seedling.* Even with superb paddy soil, the year's harvest would be spoiled if the rice seedlings are not cultivated properly.
5. *Don't be proud of having a good rice seed, but be proud of having a good soil.* Even with a good breed of rice seed, the year's harvest would be spoiled if the soil is not managed properly.
6. *Paddy fields should be drained at least once even if it has to take place in a dream.* Midsummer drainage prevents the roots from rotting by eliminating harmful substances such as organic acids and H_2S formed by soil reduction and vitalizes the roots by providing oxygen. Also, it promotes healthy growth and development of crops by stimulation of N mineralization and reduction of denitrification and increases their resistance to lodging ("falling down") by strengthening the ground troops (roots). So, it is very important that midsummer drainage is done.

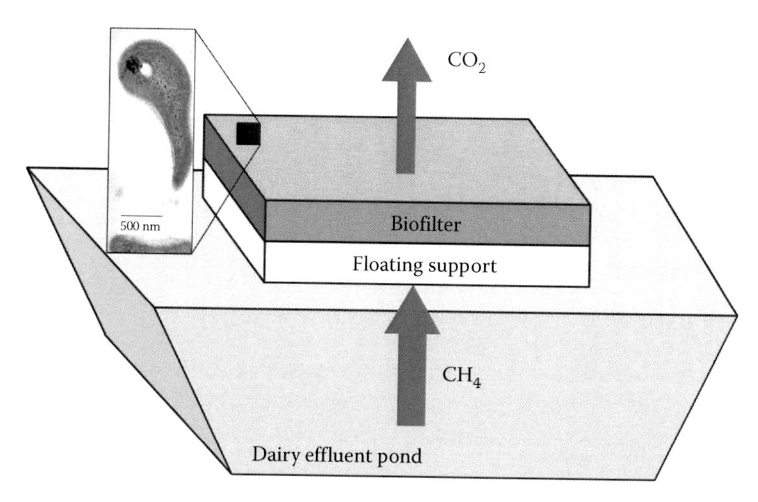

FIGURE 1.5 Diagram showing how an active soil methanotroph population can be used in a cover-biofiliter to remove CH_4 emissions from a dairy effluent pond. (Adapted from Pratt, C. et al. 2013. *Environmental Science & Technology* 47:526–532.). An electron micrograph of one of the methanotrophs (*Methylosinus trichosporium*) is shown on the left. (Courtesy of Dr. Réal Roy and Brent Gowan, Victoria University, BC, Canada.)

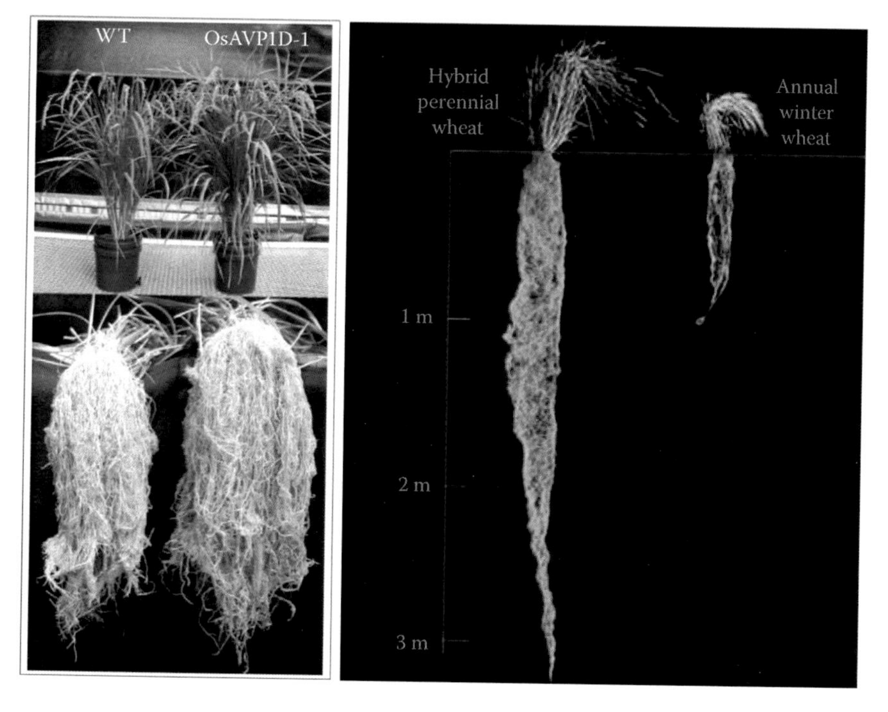

FIGURE 2.2 Root systems of conventional and nutrient-efficient crop plants. (a) (Left) Transgenic rice plants (right-most plant in panel, labeled OsAVP1D-1) have larger root systems that use phosphorus more efficiently than conventional rice cultivars (left-most plant in panel, labeled WT). (Adapted from Yang, H. et al. 2007. *Plant Biotechnology Journal* 5:735–745.) (b) (Right) Root systems of perennial wheat (intermediate wheatgrass) (left-most plant in panel) have much longer and more efficient root systems than conventional annual wheat (right-most plant in panel). (Adapted from The Land Institute. 2013. http://www.landinstitute.org/vnews/display.v/SEC/Press%20Room (accessed February 2013).)

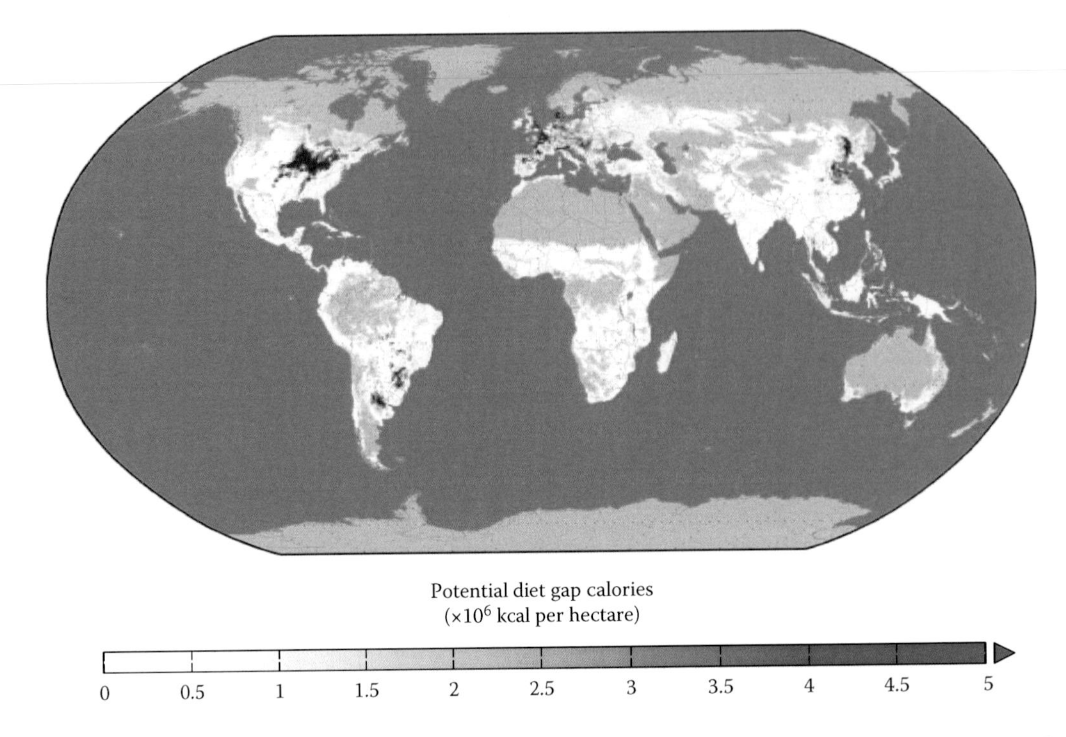

FIGURE 2.3 The global "diet gap" for 16 of the most important staple crops. This analysis shows how much more food energy could be directly available to people if crop production was not allocated first to livestock or biofuels. [Calculated as the difference between the total available calories from crops (if all calories were consumed by people) and the actual, delivered food calories to people (based on allocation of crop energy to food, animal feed, and other uses)]. (Light gray color is area that is not used in this analysis, i.e., land that is not used for the 16 staple crops studied by Foley et al. 2011). (The scale for potential diet gap is applicable to the color version only). (Adapted from Foley, J.A. 2011. *Nature* 478:337–342.)

FIGURE 3.1 Soil erosion by water on cropland in the Eifel region of Germany. The risk of soil loss by water erosion is increasing in many European countries as a consequence of the use of heavy machinery, larger fields, and increasing cultivation of maize to produce biomass. (Image by Nikolaus K. Kuhn.)

FIGURE 3.7 Nutrient-depleted soils in northern Namibia; as a consequence of continued sorghum production, the soils have a strongly reduced productivity of <100 kg per hectare. (Photo by Nikolaus J. Kuhn.)

FIGURE 4.5 Soil museum and exhibition at National Taiwan University: Soil monolith exhibition for students and farmers.

FIGURE 7.1 Ulrike Arnold, Earth Paintings, Bryce Canyon, 2003. (Courtesy of the artist.)

FIGURE 7.3 Elvira Wersche, Sammlung Weltensand. (Courtesy of the artist.)

FIGURE 7.5 Daro Montag, 5 Film Strips from *This Earth*, 2007. (Courtesy of the artist.)

FIGURE 7.7 Koichi Kurita, Soil Library, 2012. (Courtesy of the artist.)

FIGURE 11.7 The ecological infrastructures of the vineyards of Marlborough, New Zealand, also provide valuable cultural services in relation to aesthetics, tourism, recreation, fashion, music, social needs, and cuisine. Each year the Marlborough Wine & Food Festival celebrates the terroir provided by the local viticultural infrastructures.

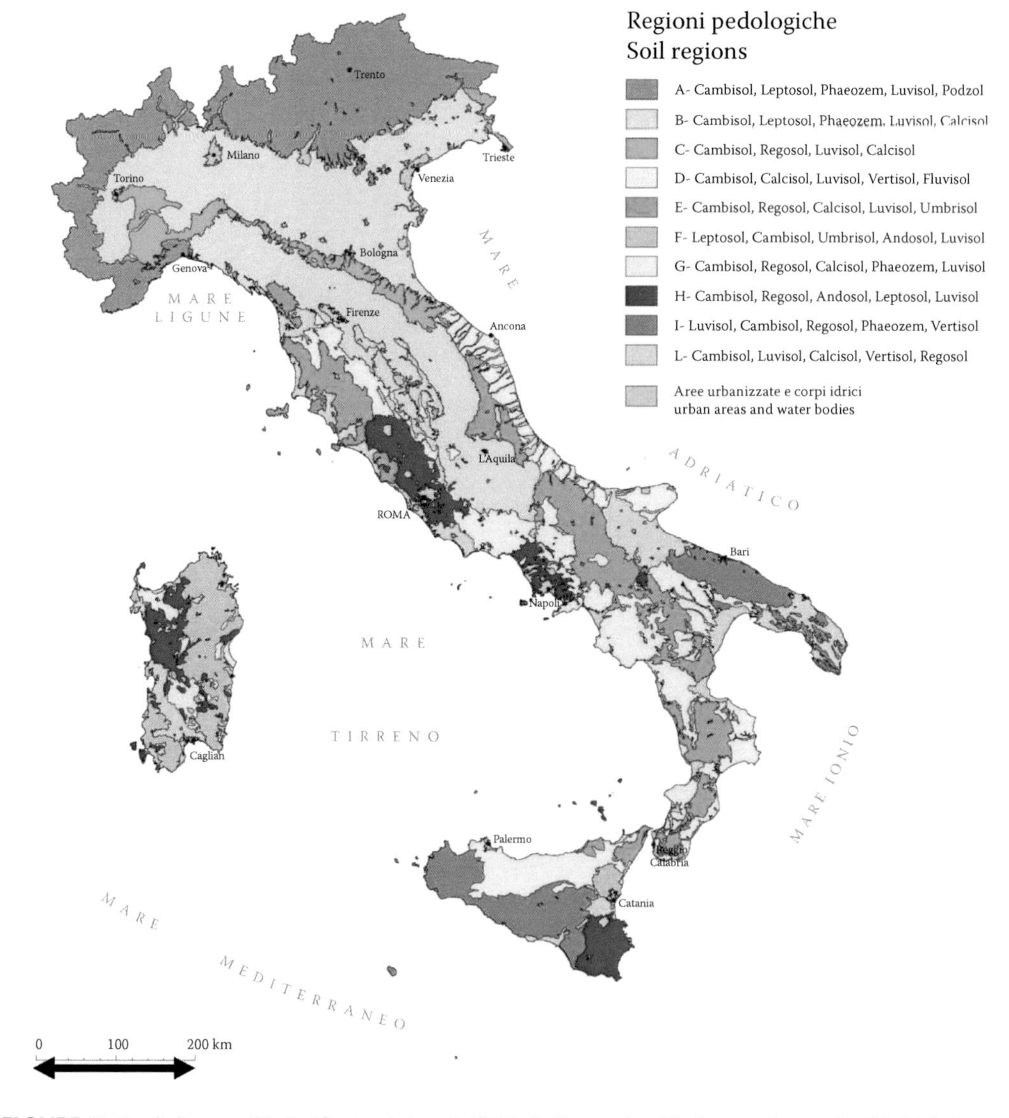

FIGURE 12.2 Soil map of Italy (Costantini et al. 2012). Soils are classified according to the World Reference Base (IUSS Working Group WRB 2006). The combination of different soils inside each area indicates the complexity and the spatial variability of Italian soils.

FIGURE 14.1 View of upper Leipsokouki catchment looking northwest. Photo taken after wheat harvest; fields are covered in light tan wheat stubble. Gray areas are exposed mudstone bedrock. Most trees in view are oaks (*Quercus* sp.). Note narrow valley floor.

FIGURE 17.1 Monthly farm works for February. (The first *Nongga-wolryongdo*; KOCCA 2006.)

FIGURE 17.4 Monthly farm works for May. (The fourth *Nongga-wolryongdo*; KOCCA 2006.)

FIGURE 17.7 Monthly farm works for August. (The seventh *Nongga-wolryongdo*; KOCCA 2006.)

FIGURE 17.10 Monthly farm works for November. (The tenth *Nongga-wolryongdo*; KOCCA 2006.)

FIGURE 19.1 Agricultural research at Moray. (Photo credit: Francisco Mamani-Pati.)

FIGURE 19.2 Terraces are still being constructed today by Bolivian farmers. (Photo credit: Francisco Mamani-Pati.)

FIGURE 19.4 Raised fields or *suka kollus* were first developed in the year 1000 BC Kallutaca, Laja-La Paz. (Photo credit: Francisco Mamani-Pati.)

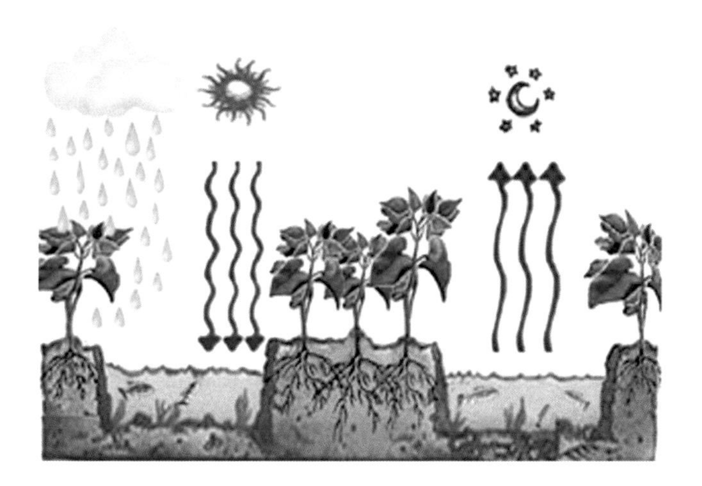

FIGURE 19.5 Diagram of water and heat cycle of *suka kollus*. (Diagram by Francisco Mamani-Pati.)

FIGURE 24.2 SWRT membrane installation equipment showing rolls of polyethylene (PE) film being released from the backs of membrane installation chisels when inserted into the soil to depths of 40–70 cm. Multiple adjacent passes across small and large fields are accurately controlled by satellite-guided global positioning systems (GPS) that are becoming the standard equipment on many farm tractors.

FIGURE 25.1 The author's backyard vegetable garden.

FIGURE 25.5 The author displaying the color pattern in front of the crop.

FIGURE 27.7 Applied soil physics in action: The Forsyth Barr Stadium, Dunedin, New Zealand—the world's first permanently closed-roof stadium pitch supporting a natural turf surface, successfully used for the 2011 Rugby World Cup. (Reproduced by permission of Darcy Schack of JAM Photographics Limited.)

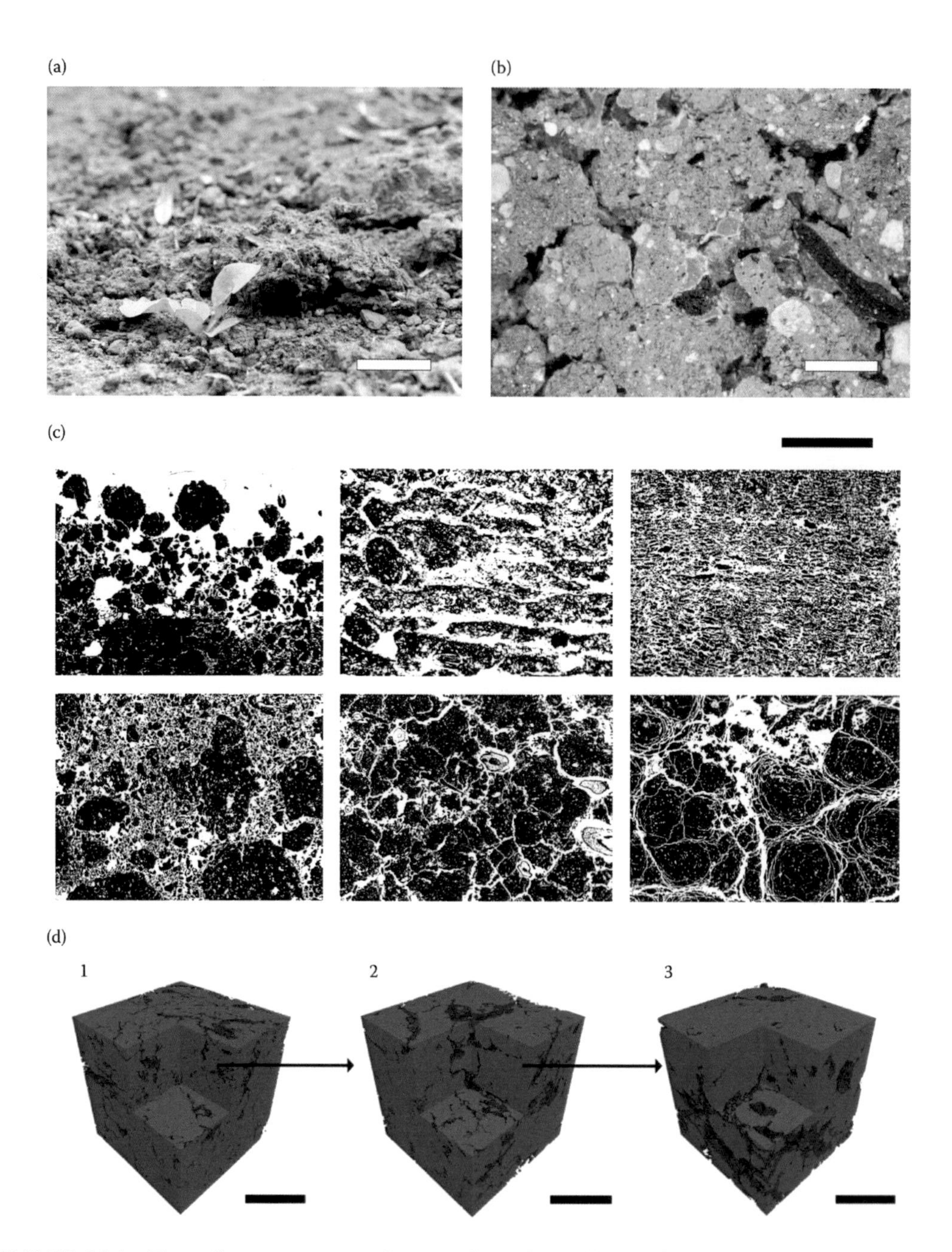

FIGURE 29.1 The soil stage across scales and dimensions. (a) The familiar scale to humans. Scale bar = 10 cm. (b) Soil aggregates and pore network visualized in cut-face of a resin-embedded undisturbed soil. Scale bar = 5 mm. (c) Contrasting pore network morphologies as visualized in two dimensions via thin sections of resin-embedded soils. Scale bar = 10 mm. (d) Three-dimensional reconstruction of soil pore networks at three successive spatial scales derived from x-ray CT scans of a clay loam beneath a grass racecourse. Cube 1: scale bar = 10 mm, pores greater than 50 μm resolved; Cube 2 zoomed-in image of the corner zone of Cube 1, scale bar = 5 mm, pores greater than 25 μm resolved; Cube 3 zoomed-in image of the corner zone of Cube 2, scale bar = 2.5 mm, pores greater than 12.5 μm resolved. ((a), (b) From Karl Ritz; (c) From Young, I.M. and Ritz, K. 2005. *Biological Diversity and Function in Soils*, ed. R.D. Bardgett, M.B. Usher, and D.W. Hopkins, pp. 31–43. Cambridge: Cambridge University Press; adapted from Fitzpatrick, E.A. 1993. *Soil Microscopy and Micromorphology*. Chichester, UK: John Wiley; (d) From Craig Sturrock, CPIB, Hounsfield Facility, University of Nottingham.)

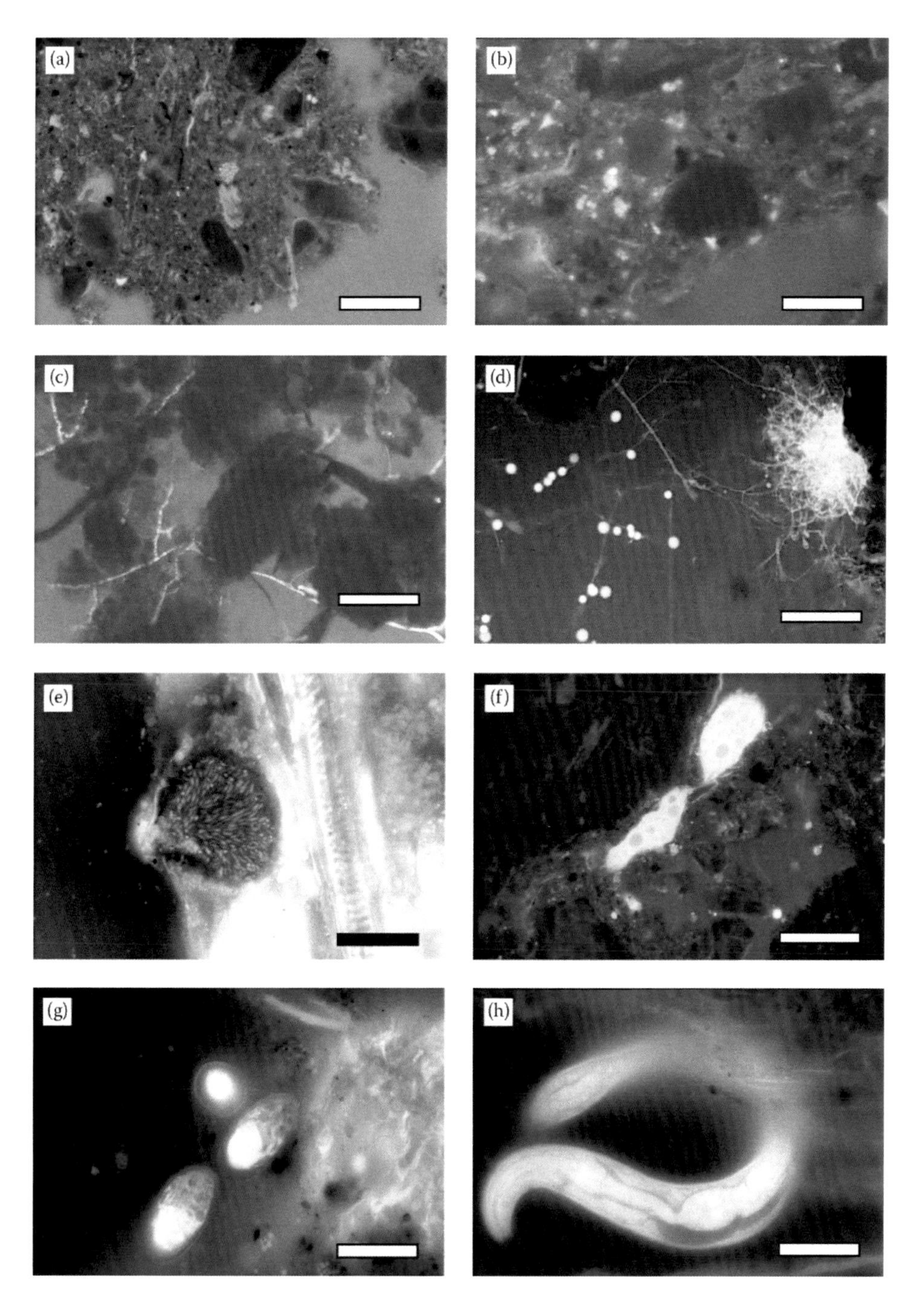

FIGURE 29.2 Soil biota visualized in biological thin sections of undisturbed arable soils. (a) Bacterial micro-colony in topsoil zone of arable soil. (b) Bacterial colonies in earthworm cast. (c) Mycelium of *Rhizoctonia solani* foraging through pore network. (d) Unidentified fungus bridging pore in exploratory mode and proliferating in exploitative mode on localized substrate; the bright spots are spore clusters. (e) Fungal perithecium (flask-shaped fruiting body containing spores) embedded in decomposing root, with opening from which spores are extruded aligned with lumen of pore. (f) Naked amoeba passing through soil pore neck. (g) Testate amoebae; three individuals in vicinity of bacterial colonies located on pore wall surface. (h) Bacterial-feeding nematode. Scale bars (a)–(g) = 20 μm; (h) = 100 μm mm. (From Ritz, K. 2011. *Architecture and Biology of Soils*, pp. 1–12. Wallingford, UK: CABI.)

FIGURE 29.3 Scanning electron micrographs of testate soil amoebae showing silicaceous plate-based construction of the test. Scale bar = 10 μm. (From Wilhelm Foissner, Universität Salzburg.)

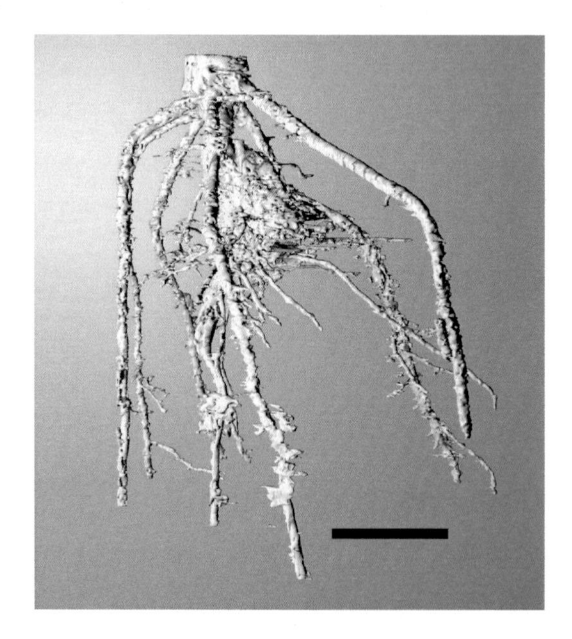

FIGURE 29.6 Computer-assisted tomographic image of soil-grown maize root system. Scale bar = 15 mm. (From Stefan Mairhofer, CPIB, Hounsfield Facility, University of Nottingham.)

7. *Rice paddies can be so dry that the cattle ankle sinks in the paddy fields.* Although water control changes with the level of growth and development of rice, it is still improper to load the fields with too much water or to severely dry them down. Just enough soil water should be kept so that only the cattle ankle sinks in the paddy fields.
8. *If the rice field looks dark, then there will be a lot of rice stalk.* Too much nitrogen causes overgrowth, weakening the crops and making their leaves turn black. Farmers would end up only harvesting rice straw, rather than rice grain, at the end of the year.
9. *Plant stakes deep, but plant rice seedling shallow.* The stake that keeps hold of the cow must be planted firmly and deep so that the cow would not be able to escape, but the seedlings should be planted shallow so that they take root fast, and productive stems develop at the proper time, ensuring a good yield.
10. *Well plowed rice field is productive like a field with additional fertilization.* Plowing the clayey soil in spring and autumn improves soil structure and soil physical properties, and thus enhances crop yields.
11. *Too much manure causes bad harvest.* With too much manure, a rice crop would develop nonbearing tillers and farmers would harvest rice straw instead of rice grains at the end of the year. Also blight and lodging would become more frequent with too much nitrogen fertilizer.

17.2.2 Upland Soils

1. *Don't farm barley crops without application of wood ash.* Wood ash contains potash and phosphate that increase resistance to frost in crops.
2. *Bury a corpse and barley seeds deeply.* Planting barley seeds deeply would help the plants endure cold and drought and prevent them from premature heading in warm winters.
3. *Rich snow in December and January brings hope of a good harvest of winter barley.* Snow keeps soils warmer during the cold winter, providing moisture to the soil. Snow also helps keep the barley seeds from being eaten by bugs.
4. *Winter barley which died from the cold wind comes to life when it snows heavily.* In winter, if the barley faces severe cold before it hardens, it could suffer from frost damage. However, if it is covered with snow, it could conserve heat and moisture and yield a supply of grain.
5. *Rich harvest of barley stem is bad harvest of barley grain.* Too much nitrogen fertilizer could cause overgrowth and lodging. This hinders grain filling and delays harvest maturity, decreasing yields.
6. *Sweet potatoes cultivated in unfarmed soils taste good.* Sequentially cropped, sweet potatoes taste dry and form poor tuberous roots due to lack of micronutrients. So, they taste better when they are cultivated in unfarmed soils with abundant micronutrients.
7. *Soy beans grow from soy bean seeds; red beans grow from red bean seeds.* You reap what you sow. The soil cannot lie.

17.2.3 Soils in General

1. *A high class farmer manages soil, a middle class farmer grows crops, and a small farmer grows weeds.* Even if the farmer cultivates a superb breed of crop in excellent growing conditions, the farmer would not be able to increase yields if the soil is infertile. So, one of the most fundamental principles of agriculture is to improve soil fertility.
2. *Plow the fields deeply. Then, gold will be found.* Plowing the fields deeply helps the crops grow and develop by helping their roots spread far.
3. *Sand rain makes pine tree lush.* Sand rain is yellow dust that blows from China's Gobi Desert and Inner Mongolia toward the Korean Peninsula, usually during the spring season. Sand rain destroys caterpillars by closing their respiratory tract and hence promotes lush pine trees.

4. *Field having some moles shows good soil fertility.* Moles feed on earthworms and cicada larva that live in soil that have rich humus.
5. *Excessive thunder and lightning during a year will bring a bumper harvest.* During a shower, lightning converts the nitrogen in the atmosphere into nitrates that fall to the ground with rainwater, providing nitrogen and stimulating plant growth.

17.3 PROVERBS ABOUT SEASON-RELATED FARMING

To predict seasonal change and natural phenomenon, 24 solar terms in the lunisolar calendar played an important role in agrarian societies in East Asia. The lunisolar calendar was the official calendar in Korea until the Gregorian calendar was officially adopted in 1896. Traditional holidays such as the Korean New Year and *Chuseok* (harvest moon festival) are still celebrated based on the old calendar. It incorporates elements of a lunar calendar with those of solar calendar and shows both the lunation and time of the solar year. Each of the 24 terms of the lunisolar calendar suggests the position of the sun as it travels 15° on the ecliptic longitude (Table 17.1). It falls around the same date every year on the solar calendar with an interval of about 15 days.

17.3.1 Spring

1. *Easterly wind on Ipchun will bring plentiful crops. Ipchun* is the first day of spring and the first solar term of the year, usually falling around February 4 of the solar calendar (Table 17.1). Easterly winds from the Pacific Ocean in the Korean peninsula in spring are warm

TABLE 17.1
The 24 Solar Terms of the Lunisolar Calendar

	Korean Name	Useful Translation	Date on the Solar Calendar	Longitude	
1	*Ipchun*	Start of spring	Feb. 3 or 4	315°	Spring
2	*Usu*	Rain water	Feb. 18 or 19	330°	
3	*Gyeongchip*	Awakening of insects	Mar. 5 or 6	345°	
4	*Chunbun*	March equinox	Mar. 20 or 21	0°	
5	*Cheongmyeong*	Clear and bright	Apr. 4 or 5	15°	
6	*Gogu*	Grain rain	Apr. 20 or 21	30°	
7	*Ipha*	Start of summer	May 5 or 6	45°	Summer
8	*Soman*	Grain full	May 20 or 21	60°	
9	*Mangjong*	Grain in ear	Jun. 5 or 6	75°	
10	*Haji*	June solstice	Jun. 21 or 22	90°	
11	*Soseo*	Minor heat	Jul. 6 or 7	105°	
12	*Daeseo*	Major heat	Jul. 22 or 23	120°	
13	*Ipchu*	Start of autumn	Aug. 7 or 8	135°	Autumn
14	*Cheoseo*	End of heat	Aug. 22 or 23	150°	
15	*Baekro*	White dew, dew curdles	Sep. 7 or 8	165°	
16	*Chubun*	September equinox	Sep. 22 or 23	180°	
17	*Hanlo*	Cold dew	Oct. 8 or 9	195°	
18	*Sanggang*	Frost descent	Oct. 22 or 23	210°	
19	*Ipdong*	Start of winter	Nov. 7 or 8	225°	Winter
20	*Soseol*	Minor snow	Nov. 22 or 23	240°	
21	*Daeseol*	Major snow	Dec. 7 or 8	255°	
22	*Dongji*	December solstice	Dec. 21 or 22	270°	
23	*Sohan*	Minor cold	Jan. 5 or 6	285°	
24	*Daehan*	Major cold	Jan. 20 or 21	300°	

and moisture-laden. People believed that the weather would be warm and that the rain would come, when the easterly wind blew in from the Pacific Ocean.

2. *Westerly wind on Chunbun will bring a lean year of barley. Chunbun*, the March equinox falls on March 21 (Table 17.1). It marks the beginning of the growing season of winter barley. During this season, a warm and moisture-laden easterly wind was expected. If a dry westerly wind blew, people predicted a drought and a bad harvest of barley.
3. *Windy weather on Samjinnal brings bad harvest. Samjinnal* is known as the day when the swallows return from the south and the snakes come out from their winter sleep. It falls on March 3 of the lunisolar calendar and is usually in early April of the solar calendar. If the weather was windy and dry on *Samjinnal*, the seeding should be postponed because of moisture deficiency and drought. People then predicted a bad harvest.
4. *Rain on Gogu brings a hope of bumper harvest. Gogu* means grain rain (rain that helps grain grow) and is the 6th of 24 solar terms, usually falling around April 20 of the solar calendar (Table 17.1). The period around April 20 was the best time to prepare rice seedbeds and to sow seeds of all crops. Therefore, if it rained on *Gogu*, the soil had enough moisture favorable for seed germination. People thought that this would bring a rich harvest (Figures 17.1 through 17.3).

17.3.2 Summer

1. *Sweet potatoes planted until Haji can grow up even just with saliva of cow. Haji* is the 10th of 24 solar terms, and usually falls around June 21 or June 22 of the solar calendar (Table 17.1). Sweet potatoes need temperate climate, abundant sunshine, and warm nights, whereas too much rainfall can damage sweet potatoes. If sweet potatoes were planted up until *Haji*, they grew very well even in the dry season from May to June.

FIGURE 17.1 **(See color insert.)** Monthly farm works for February. (The first *Nongga-wolryongdo*; KOCCA 2006.)

FIGURE 17.2 Monthly farm works for March. (The second *Nongga-wolryongdo*; KOCCA 2006.)

FIGURE 17.3 Monthly farm works for April. (The third *Nongga-wolryongdo*; KOCCA 2006.)

2. *Monsoon rain in July makes even stones grow.* The temperature in July is very high in the Korean Peninsula. The monsoon rain provided enough water to grow crops in this hot season.
3. *If you hear the chirping of cicadas, throw away your rice seedlings.* Cicadas in the Korean Peninsula normally start to chirp in the beginning of July. If cicadas had started to chirp, it was too late to plant rice seedlings (Figures 17.4 through 17.6).

17.3.3 Autumn

1. *Rain on Ipchu brings a good harvest of vegetables. Ipchu* is the 13th of 24 solar terms which falls usually around August 8 of the solar calendar (Table 17.1). *Ipchu* marks the first day of autumn. This season is the time to start sowing seeds for winter vegetables. If it rained during this season, people predicted a good harvest of vegetables.
2. *Rain on Cheoseo reduces the amount of rice in the rice jar. Cheoseo* is the 14th of 24 solar terms that usually falls around August 23 of the solar calendar and means the end of the hot summer (Table 17.1). This is typically the heading time of rice. If it rained on *Cheoseo,* people thought that the rice rotted before it had time to fully ripen (Figures 17.7 through 17.9).

17.3.4 Winter

1. *Bury winter barley seed before Ipdong. Ipdong* is the first day of winter and the 19th of 24 solar terms that falls around November 7 of the solar calendar (Table 17.1). The weather starts getting colder after *Ipdong.* If the sowing time of barley was delayed after *Ipdong,* the germination of barley seeds and the growth of young seedlings could not be ensured due to frost damage.

FIGURE 17.4 (**See color insert.**) Monthly farm works for May. (The fourth *Nongga-wolryongdo*; KOCCA 2006.)

FIGURE 17.5 Monthly farm works for June. (The fifth *Nongga-wolryongdo*; KOCCA 2006.)

FIGURE 17.6 Monthly farm works for July. (The sixth *Nongga-wolryongdo*; KOCCA 2006.)

FIGURE 17.7 **(See color insert.)** Monthly farm works for August. (The seventh *Nongga-wolryongdo*; KOCCA 2006.)

FIGURE 17.8 Monthly farm works for September. (The eighth *Nongga-wolryongdo*; KOCCA 2006.)

FIGURE 17.9 Monthly farm works for October. (The ninth *Nongga-wolryongdo*; KOCCA 2006.)

2. *A halo around the full moon on Daeboreum will bring a good harvest. Daeboreum* means the first full moon; it falls on January 15 of the lunisolar calendar and usually in late February of the solar calendar. A halo around the moon appears when cirrus clouds form because the ice crystals in these clouds refract the moonlight. The cirrus clouds are formed under low atmospheric pressure conditions, so a halo usually foretells rain. Rains in early spring were regarded as a sign of good harvest (Figures 17.10 through 17.12).

17.4 PROVERBS ABOUT CATASTROPHES

17.4.1 Extreme Weather

1. *When the hoe handle falls off, then a drought is coming.* The hoe handle falling off implies that the weather is very dry.
2. *If a cuckoo cries in May, crops will fail.* The cuckoo is a migratory bird that comes to Korea in May. The cuckoo follows warm and dry climates, so a cuckoo's cry means there will be many days that have warm and dry weather, causing drought and bad harvest.
3. *There is an end to drought, but no end to rainy season with regard to their damage.* Even though water is scarce, crops can survive drought with the moisture in the soil. However, in rainy seasons, the crops and seeds are irrevocably damaged by excess moisture. The damage from flood is greater than that of drought, so it is crucial that farmers get prepared with water control projects in rainy seasons.
4. *If magpies build their nests low in a tree, there will be a serious storm.* Magpies begin to build their nest in January. People believed that if a serious storm were approaching, the magpies would nest lower in a tree.

FIGURE 17.10 (**See color insert.**) Monthly farm works for November. (The tenth *Nongga-wolryongdo*; KOCCA 2006.)

FIGURE 17.11 Monthly farm works for December. (The eleventh *Nongga-wolryongdo*; KOCCA 2006.)

FIGURE 17.12 Monthly farm works for January. (The twelfth *Nongga-wolryongdo*; KOCCA 2006.)

5. *Wind-piled snow in mouse hole brings a lean year of winter barley.* Winter crops such as barley can conserve heat under layers of snow. Snow in a mouse hole means strong winds which bring frost and drought damage by blowing away all the snow.

17.5 PROVERBS ABOUT AGRARIAN LIFE AND CULTURE

17.5.1 Family

1. *Frequent spring rain makes the mother-in-law generous.* Frequent rain in dry spring ensured a proper sowing time for all crops due to an adequate soil moisture supply. People became generous because they predicted a good harvest.
2. *Don't visit even your daughter in barley hump.* Barley hump refers to the period in May and June when food stored for the winter ran out and the barley was not ripened yet for harvest. In this time of the year, almost everyone suffered from food shortage, so it was best not to visit anyone, even a close family member.
3. *Don't get related by marriage to the family whose rice leaves are dark green in July.* In July, the rice leaves should be yellow green. Too much fertilization makes their leaves turn black. Farmers would end up harvesting rice straw at the end of the year.

17.5.2 Customs

1. *If the rope swing is dripping with water, there will be a good harvest.* If the rope swing is so wet with rain that it is dripping with water in *Dano* (a festival in Korea that celebrates of the end of sowing season, held in the fifth day of the fifth month in the lunisolar calendar), farmers could transplant rice in the proper time and could expect a bumper year.

2. *Sleeping in the rice field prepared for rain will bring a rich harvest of rice.* When the weather was dry in times when water was necessary, farmers used to stand guard at sluice gates, sleeping in the rice fields and wearing rain gear in order to irrigate the fields as soon as possible. This tells us how important it is to supply water when it is needed.
3. *If hairs of birds fall out, rice grains will be bursting at the seams.* This means that the autumn weather is so clear and warm that the hairs of birds would fall out. With plenty of sunshine, the process of grain filling will be done so successfully that the grains will burst at the seams.

17.6 CONCLUSIONS

In this chapter, we have provided an outline of life and culture in agrarian society in the Korean Peninsula using farming-related proverbs. We learned from these proverbs how precisely our ancestors understood natural phenomena and farming practices. They also knew that one of the most fundamental principles of agriculture is to improve soil fertility, for example, by adding organic matter, deep plowing, and ensuring proper soil moisture. Our ancestors also knew the adverse effect of overfertilization with nitrogen. Many farming proverbs were related to the season and the 24 solar terms. Rice and barley were the main materials of the farming proverbs as they were the staple food in Korea. Rains in the sowing season were as valuable as oil. Most of knowledge and customs of our ancestors is quite scientific and useful even to this day. We can learn our lessons by applying the wisdom of the proverbs and will then understand more about nature and culture in the Korean Peninsula.

REFERENCES

CPDKP (Committee of Publishing the Dictionary of Korean Proverbs). 2001. *Dictionary of Korean Proverbs.* Yeokang Publisher. 4779pp. ISBN: 89-89457-76-9. ISBN13: 978-89-89457-76-3 (in Korean).

GiATC (Gimcheon Agricultural Technology Center). 2012. Farming proverbs. Gimcheon, Republic of Korea. http://www.gca.or.kr/agr_proverb/list.asp (accessed October 2012) (in Korean).

JARES (Jeollanam-do Agricultural Research & Extension Services). 1979. Collection of farming proverbs. Naju, Jeollanam-do, Republic of Korea. 161pp. (in Korean).

KATC (Kyeungsan Agricultural Technology Center). 2012. Farming proverbs. Kyeungsan, Republic of Korea. http://www.gsa.go.kr/agr_proverb/list.php?start=10&mcnt=&mseq=&key=&word=(accessed September 2012). (in Korean).

KOCCA (Korea Creative Contents Agency). 2006. The *Nongga-wolryongdo* (Twelve paintings about the monthly works of the farmhouse of the *Mid-Joseon* Dynasty. http://www.culturecontent.com/search/search.jsp (accessed September 2013).

Korean Britannica online. 2012. Korean proverbs. 2012.12.05 article. http://preview.britannica.co.kr/bol/topic.asp?article_id=b12s2555b001 (accessed September 2012) (in Korean).

Lee, E.W. 2005. Korea's pastimes and customs—A social history. Homa & Sekey Books, Paramus, New Jersey. 264pp.

RDA (Rural Development Administration). 1989. Agricultural proverbs. Suwon, Republic of Korea. 205pp. (in Korean).

18 Terra Preta
The Mysterious Soils of the Amazon

Antoinette M. G. A. WinklerPrins[*]

CONTENTS

18.1 INTRODUCTION

The popular imagery of the Amazon typically conjures up one of two views: rampant environmental destruction on the one hand, and pristine, virgin nature with still-untouched tribes on the other. Neither is a correct representation of the region; the reality lies in between, but the persistence of these imageries hinders a more realistic approach to the region. *Terra preta* (TP) (black earth) soils, the topic of this chapter, challenge both images as they indicate much more human agency in the landscape in the past, a challenge to the persistent pristine myth image, and they represent a way in which soils of the region can be made more sustainable without external inputs, thereby challenging the idea that the only development in the Amazon is the destructive conversion of rainforest into industrial agriculture and predatory forestry. In this chapter, I define and describe TP soils and consider how their existence has contributed to past interpretations of the region, and how knowledge of their formation can contribute to a more sustainable use in the future.

18.2 WHAT ARE *TERRA PRETA* SOILS?

Terra preta, Portuguese for black earth, are part of a continuum of soils often referred to as Amazonian Dark Earths, Archaeologic Dark Earths, or Anthropogenic Dark Earths (ADEs). True or "proper" TP soils are by definition black or a very dark brown, and are at one end of the continuum of ADEs, which extends to the much lighter-colored *terra mulata* (TM) (brown earth) soils. ADEs are anthrosols, soils that exist in the landscape as a consequence of human activity, either

[*] Environmental Studies, Advanced Academic Programs, Johns Hopkins University, 1717 Massachusetts Ave., N.W., Washington D.C. 20036, USA; Email: antoinette@jhu.edu.

through incidental activity such as the accumulation of human waste (anthropic soils), or as the result of deliberate action (anthropogenic soils), most likely associated with long-term sedentary farming, that is, farming that stays in one place, in contrast to shifting cultivation where people and their crops move periodically. The key differential is that TP soils contain ceramic fragments (potshards), whereas TM soils typically do not.

Current thinking is that TP soils are the result of human inhabitation, that is, that they formed as composted material accumulated via incidental human activity (often in debris piles referred to as middens), while TM soils are the result of deliberate manipulation of the soil to improve its quality for agricultural purposes. The degree to which ADEs are anthropogenic or anthropic is debated at length, as this has ramifications for viewing the ways in which people managed their resources in the past, and how they may be able to do so again in the future. However, for the purposes of this chapter, the point is that TP soils are the result of human activity and they represent human agency on the Amazon landscape.

The dark color of this continuum of soils is because they all contain high levels of stable soil organic matter (SOM), on average three times higher than the background or nonanthrosols of the Amazon basin. In addition to their relatively high levels of organic matter, ADEs are high in most other measures of soil fertility such as cation exchange capacity (CEC), phosphorus, nitrogen, calcium, magnesium, and especially carbon. Their pH is much more neutral than the surrounding mostly acidic soils, ranging from 5.2 to 6.4. These chemical attributes result in soils that contain and retain plant-available nutrients. High SOM also improves soil physical properties as the organic matter plays a role in changing the porosity of the background soil matrix, resulting in soils that are better able to retain moisture. This attribute may seem irrelevant for a tropical rainforest location, but the reality is that many areas of the Amazon basin, especially the central Brazilian Amazon, experience a strong dry season that impacts nonirrigated agricultural potential. Also, background nonanthrosols in the region are often either high in clay content resulting in periods of excessive moisture, or exhibit sand-sized aggregates that result in rapid flow-through and poor water retention. Both of these issues are muted by the high organic matter content of ADEs.

What has been mysterious about these soils is their ability to persist in a landscape that common ecological knowledge would dictate they could not. In the generally high humidity and high rainfall environments such as found in the Amazon basin, most nutrients in the soil mineralize and are leached out of the system quickly, and because of this, the vegetation of the region typically absorbs nutrients quickly from the soil once these have been released in the soil through decomposition or other processes. Why then have ADEs, dated to have formed up to 2500 years ago, continue to exist?

Soil chemists studying ADEs have concluded that the unique nature of the carbon in these soils is the key to the stability of the organic matter in ADEs and the key to the mystery of the persistence of ADEs in this landscape. The carbon found in ADEs is aromatic carbon (also known as black or pyrogenic carbon) that is likely a consequence of the incorporation of charcoal into the soil. This particular charcoal is the result not of natural forest fires (there is charcoal found in the Amazon that is the result of fires; these were less frequent in the past, but are now more frequent due to both natural and human-induced disturbance and climate change), but due to slow, cooler burns associated with the use of fire as a management tool in sedentary agriculture (more about this in Section 18.3 below). Soils with this type of carbon are able to retain, even attract nutrients, resulting in more plant-available phosphorus, calcium, and nitrogen, as the aromatic carbon acts as a carrying agent for nutrients. Aromatic carbon is also known to be highly resistant to degradation. This recalcitrant form of SOM is why ADEs persist in a high-leaching environment.

One aspect of ADEs that remain poorly understood, quite mysterious actually, is the microbial complexes associated with them. What is known thus far is that TP soils exhibit greater microbial diversity than background soils, and much of these populations are predominantly fungal. These microbial populations perform an important function in the maintenance of the high fertility of ADEs over time, possibly even contributing to the generation of nonpyrogenic black carbon. They also assist in the improved soil moisture retention of ADEs.

18.3 *TERRA PRETA* FORMATION

The appearance of ADEs in the landscape has generated a wide variety of ideas about their genesis. Although early theories (in the late 19th and early 20th centuries) pointed to likely anthropogenic origins, later theories were convinced of geologic (sediments or volcanic) or ecologic origins. These nonanthropogenic theories of ADE formation that dominated in the middle of the twentieth century reflect the era in which they circulated, an era in which the cultural potential of the Amazon was thought to be quite low and it was inconceivable to believe that human action was responsible for such a valuable resource. The full acceptance of their human origins did not come about until the last few decades, paralleling radically new findings in archaeology and anthropology regarding human activity and cultural achievement in the past (further discussed in Section 18.5 below).

Questions about the differences and/or similarities between ADEs and Histosols (organic soils) persist, but ADEs, though they contain high levels of organic matter, are not Histosols. Histosols are typically located low in the landscape, in a position of accumulation of organic debris and moisture. There are plenty of Histosols in the floodplain areas of the Amazon, but most ADEs are not located in the floodplain and are typically well drained. Most importantly, once Histosols are drained and their organic content exposed to air, the organic matter quickly volatizes and the soil subsides. This is not the case with ADEs.

What is clear today is that ADEs formed *in situ* and from the top, as their mineralogical composition is always the same as the surrounding soil. This helped dismiss the ideas that ADEs were the result of sedimentary processes. Evidence of midden formation around villages helps confirm *in situ* formation. In addition, scientists postulate that agricultural practices in the past included "slash and char," a variant on the well-known slash-and-burn agriculture. Slash and char involves the use of a much cooler, slower burning that is actively managed, through long-term smoldering. Instead of letting biomass burn until ash as is typically done in present-day slash-and-burn agriculture, biomass is charred to the desired charcoal, containing the critical aromatic carbon, which is then used as a soil conditioner and incorporated into the soil over time. Evidence of slash and char comes from observing soil management practices today as practiced by people indigenous to the region. Research with smallholder agriculturalists that I have conducted with Brazilian colleagues has documented the formation of a soil conditioner called *terra queimada* (TQ) (burnt earth). This is created by first sweeping mostly organic refuse into piles where it is left to smolder for an extended period of time, and then the charred residue is applied to desired crops. This process has been observed throughout the basin by anthropologists and geographers. Long-term research with the Kayapó Indians by Susanna Hecht has found a similar process, a practice called "in-field burning," in which small, cool and slow burns are used to reduce weeds and other biomass within fields to produce char, which is then incorporated into the soil.

Interestingly, today's refuse piles contain nonbiomass materials as well, especially plastic, aluminum, and batteries because in places where there is no regular refuse collection (much of the Amazon region), there is nowhere to go with this trash. Clearly, these materials change the chemistry of the char. Locals acknowledge that it may not be good to burn these materials but do not know what else to do with them. In cities, there are increasing efforts to collect trash, even in the informal settlements, so what is burned is more likely to be mostly debris from biomass, but in rural areas, the composition of the char likely contains remains of nonbiomass items. Much further research into the composition of present-day char is needed to fully understand the degree to which TQ may be an analog to what Amerindians were doing in the past.

18.4 WHERE ARE *TERRA PRETA* SOILS LOCATED?

TP soils are patches within a landscape consisting mostly of highly weathered and acid tropical soils known as Latosols and Acrisols (or Oxisols and Ultisols in the USDA's *Soil Taxonomy*). Their individual extent is not large, most patches range in size from 2 to 350 hectares, with the majority being at the smaller end of that range. The Amazon Basin is of continental size, about the size of the lower 48 states of the U.S.A., and consists mostly of Brazilian territory, but also extends into neighboring Bolivia, Peru, Ecuador, Colombia, Venezuela, and the Guianas. It is estimated that ADEs cover approximately 3% of the basin, although observers of these soils note that it is likely that, once all ADEs are identified, their extent may be as high as 10% of the basin, which would be an area the size of France. Because of their generally small individual extent, ADEs rarely appear as individual classes of soil on soil maps of the region, but are inclusions in more spatially extensive soil classes.

The full extent of TPs/ADEs is not known since the Amazon has not been subject to a full field-based systematic soil survey, so it is highly likely that there are many more patches of these soils. The best map we have to date is one that I created with a colleague and is available as an interactive Geographic Information System (GIS). More information about this GIS and associated database is available in the WinklerPrins and Aldrich 2010 publication.

ADEs are mostly located on bluffs along the main-stem Amazon River (known as the Solimões River between the Brazil/Peru border and the city of Manaus) and its many tributaries. Their location there is likely for two reasons: (1) bluffs are easily accessible via what remains the main form of transportation in the Amazon, river transport, and therefore ADEs in those locations are best known to local Indian and Mestizo populations (who typically live on bluffs) and have been reported to scientists who have documented them in such a way that others know about these locations. (2) Amerindian populations before the arrival of Europeans were predominantly bluff dwellers. Bluffs are topographically advantageous in a flat landscape such as the Amazon, and bluff locations enabled access to resources from the rich riverine environment (e.g., water, fish, turtles, and naturally fertile floodplain soils) as well as those of the nonflooded uplands (e.g., fruit, meat from hunting, medicinals, and lumber). The geographer William Denevan has made a compelling case for the dominance of prehistoric bluff settlement in his 1996 paper given the complementary resources of the upland and floodplain environments.

18.4.1 Are These Soils Only in Amazonia?

Anthrosols occur around the world, so these anthropogenic TP soils are not unique in that sense. Wherever people have occupied the land for a long time, they have modified their soils, but with varying impact and degree of intentionality. Prime examples are the anthrosols found in northern Europe, known as *plaggen* soils, which are the result of deliberate composting of stable waste in order to both raise the land and make it more productive. Another place where soils bear the clear signal of human action is in Meso-America where indigenous people enriched soils due to long-term habitation and made changes in the landscape by constructing terraces and creating raised beds of cultivated soils called *chinampas*. Elsewhere in the Americas, we know of raised beds on the Bolivian *altiplano* and terraces in the Andes, which are a means of water control but are also examples of active modification of soils. One can argue that *chinampas* and terraces are examples of landscape modification more than soils modification, but these activities go hand in hand. By creating terraces, people, in effect, manage downslope sedimentation and augment this with additional inputs such as crop waste, manure, and nightsoil. Similarly, terraces dominate densely populated areas of Asia and though often thought of as primarily a means of managing the water needed for rice cultivation, are in effect also forms of soil management. Inhabitants of many nonvolcanic Pacific islands created fertile soils in "wet gardens" by augmenting natural accumulations of organic debris in limestone hollows. This permitted a more productive agriculture to take place in what otherwise would be rather infertile islands. In Australia, old Aboriginal cooking hearths have resulted in more carbon in soils in places where Aboriginals lived for longer periods of time, but these are not quite the ADEs found in the Amazon. In West Africa, there is evidence of deliberate enrichment of patches of land in savannas to create forest "islands" with desired tree species. These are similar to forest islands of anthropogenic origin in the savanna to the south of the Amazon Basin (knows as the *cerrado*), which were created by Amerindians.

Chemically, ADEs are unique. By virtue of their high organic matter content being stable, even recalcitrant, its pyrogenic carbon, and the high phosphorus and other chemical signatures of ADEs, their microbial complexes, and possibly the composition of the ceramics contained in TP soils especially, they are not the same in other anthrosols found around the world. An obvious question then is, why not? Are there no ADE-like soils in places such as the Congo basin or other tropical locales where natural conditions are not unlike those found in the Amazon basin and where people have lived for a long time? The answer is that we do not yet know for sure that ADE-like soils are not found elsewhere; it is a topic that needs much further research. It might be that there has been an Amazonian bias in research, and that there are similar soils elsewhere, but to date these have not been scientifically reported. Recent research is demonstrating that there may be similar soils in West Africa, but perhaps the differences in lifeways and vegetation used in an agricultural complex are big enough that in one place (the Amazon) living and farming resulted in ADEs, and in another (the Congo basin), it did not. Certainly, the inclusion or lack thereof of large herbivores, including domesticated ones (cattle) as existed and were managed in Africa, may play a role, but again, much more research on this matter needs to be undertaken. It could be that under certain conditions, ADEs could form in any number of tropical locations; the key is to figure out what those conditions are, and for that matter to be open to seeing these soils in landscapes where they are perhaps not expected.

18.5 *TERRA PRETA* SOILS AND AMAZONIAN PREHISTORY

Amazonian prehistory has long been represented as one consisting mostly of "shifting cultivators," "nomadic farming," "simple farmers," and the like. Open any lay publication or textbook about the Americas before Columbus and the Amazon is typically represented as relatively sparsely settled by people with relatively low levels of cultural achievement. Some maps simply portray a void and indicate only the "high" civilizations of the Andes (the Inca Empire) and those found in Meso-America (Aztec and Maya). These portrayals do a disservice to what is indicated (often only a fraction of the complexity and variety of cultures that thrived before European arrival), but an even greater disservice to acknowledging that people inhabited and successfully cultivated places such as in the Amazon Basin.

Although Amazonia has never been as densely settled as other river valleys such as the Nile or the Yellow River, it was not a demographic void. Prehistoric Amazonians did not leave temples or structures as did many other prehistoric peoples; their legacy is their inscription of the landscape, now deciphered in the vegetative patterns and ADEs. Charles Mann, in his book *1491*, provides an accessible synthesis and integration of much of the new knowledge about the lifeways of the people in the Americas before the arrival of Europeans, including a map that is much more detailed and inclusive of the varied cultures that inhabited the Western Hemisphere before 1492. This new revisionist research deconstructs the typical tropical forest trope into a more realistic depiction of what life was like in the region before the arrival of the Europeans. This new knowledge confirms that there were likely more people in the basin (5–6 million, as opposed to the long thought about half of that) prior to Columbus, and that their cultures were much more complex than previously thought. Observers of the region, including historical demographers, have long thought this, but neither the archaeological nor ethnographic evidence had been clear on this matter until recent work.

The evidence comes from multiple sources, the existence and clear human signature of ADEs being a leading piece, but this is strongly supported by ethnohistoric, ethnographic, and ethnobotanical evidence. Long-term research with the indigenous peasantry of the Amazon as well as cultural groups such as the Ka'apor, Kuikuru, and Kayapó, Amerindian populations that have a cultural connection with their deep pasts, demonstrates some of the practices that have resulted in a much more anthropogenic Amazon. Severe depopulation at contact (it is estimated that over 90% of the native population of the Amazon perished as a consequence of disease and/or slavery in the colonial era) however makes this type of research difficult, as significant cultural regression took place. We know from linguistic evidence that remaining populations retreated away from the accessible bluffs deep into upland forest and savanna where they were forced to change their agricultural practices from a more sedentary form (managed fallows and fields of the type that resulted in ADEs) to what is often seen today as "traditional" shifting cultivation.

Clear evidence has emerged from the ethnographic and ethnohistoric research by Michael Heckenberger and colleagues (in 2003 and 2007 articles) that prehistoric Amazonia consisted of large villages, on bluffs or near waterways, consisting of 2500–5000 people, which were highly interconnected with complex transportation and communication networks. People were sedentary and practiced long-term agriculture as part of a complex of forest and fallow management. The landscape was domesticated, and consisted of mosaics of active fields of primarily annual crops, regenerating fallows that were managed and augmented with perennials and tree crops, and regrown forests that were enriched with desired species for food, fiber, and medicinal plants. This contrasts with earlier interpretations of prehistoric Amazonian villages that were thought to have consisted of 50–350 people who moved frequently as people "shifted" their place of living to accompany their shifting cultivation.

William Balée, an anthropologist who worked with the Ka'apor, views the Amazon as a cultural forest. He estimates that at least 12% of the Amazon consists of an anthropogenic forest, one where the human imprint is clear, and this is likely an underestimate. Why does he see the forest this way? As mentioned earlier, prehistoric Amazonians did not leave behind temples or pyramids, but instead deeply inscribed the landscape, the forest, with their actions. There was no steel in Amazonia prior to Europeans. So, in order to fell trees and create openings in the forest, people used a combination of stone axes and fire. Cutting tropical hardwood trees with a stone axe is a slow, inefficient, and incredibly laborious job that one preferred not to do frequently (steel axes are about 20 times faster than stone axes). This is why the evidence indicates that there was a great preference for reuse of previously cut forest patches (i.e., secondary growth) as these contained smaller trees and trees that were easier to cut. We also know from ethnographic and ethnohistorical research that Amerindians created and maintained forest trails wherever they traveled, enriching paths with desired trees and other plants, and preferentially selecting or weeding volunteer species. They practiced a form of silviculture and cultivated fallows as orchards. Over time, constant human management of the vegetation that surrounded and connected settlements changed and manipulated natural succession pathways. Even today, in areas of the Amazon with relatively thin populations, past settlements can be deciphered from the vegetative composition. Patches of higher densities of utilized species, including domesticates and semidomesticates, provide the human signature. Scientists have determined that at abandoned TP sites, the species that grow are domesticated plant species. They also found that this sets in motion a positive feedback in that people are attracted to prior sites of inhabitation as the soil has been improved and there is a concentration of useful species, and then by occupying that same landscape the soils and vegetation continue to be enriched.

What has contributed significantly to a revisionist view of the Amazon beyond the clear new scientific evidence is a new way of framing the relationship between humans and the physical environment. The Amazon has long been the place in which a perspective known as environmental determinism was the dominant view of how people interacted with the physical environment. This idea held that cultural potential was limited by the physical environment, and that the Amazon rainforest, a hot and humid place with poor soils, was a challenging environment for humans to develop complex cultures in, and that any evidence there was of sophistication (as expressed, e.g., in quality pottery) was brought there by people from other places (e.g., the more temperate Andes). Accepting deeper history and greater human agency, a new perspective called historical ecology rejects environmental determinism and focuses on the human rather than the natural history of the environment, contextualizing it within historical and cultural traditions. Various scientists have engaged and developed this perspective, key among them are Clark Erickson and William Balée who state that "culture is physically embedded and inscribed in the landscape as nonrandom patterning, often a palimpsest of continuous and discontinuous inhabitation by past and present people" (p. 2 in their 2006 paper).

The historical ecology perspective is part of a suite of theoretical approaches that emerged in the social sciences in the late twentieth century, often referred to as postmodern or poststructuralist approaches. These approaches offer and permit a much more differentiated view of the world, and are open to variation, differentiation, and nuance. Emerging from the rigid modernist era with its emphasis on generalizability, grand theory, and a rejection of traditional knowledge, the postmodern/poststructuralist framing of ideas is not so fixated on modernity and generalization, and is very open to ideas that emerge from the local, and from peoples whose voices were drowned out in the modern era. This has provided an opening for less environmentally deterministic views of the Amazon and has permitted an ability to see the human agency all around, especially in TP soils.

A few other factors have also contributed to revisionist views of the Amazon. The first is that there has now been long and sustained focus on and debate about the region because of attention to the rainforest deforestation. This has brought about more scientific investigations of all sorts in the region, and an intensification of discussion about the region's people and ecosystems. Coupled with this is an increased institutional opening of the region to new archaeological research by both nationals from the relevant countries (Peru, Brazil, Bolivia, etc.) and foreign scientists. From an archaeological perspective, the region has been understudied, in part due to the challenges inherent to the preservation of material culture in tropical lowland environments, and in part, due to structural constraints on funding. The former has benefitted greatly from the development of better methods and new dating techniques for the environmental setting that have emerged in the last several decades and that permit better archaeological investigations to be conducted; the latter has improved as well, and is related to the institutional opening of the region to the broader range of archaeologists mentioned above.

18.6 *TERRA PRETA* SOILS AND THE FUTURE OF THE AMAZON (AND OTHER TROPICAL PLACES)

The Amazon region is a contentious place, seen by most people as filled with the potential of one sort or another. Here again, we see dichotomous views. Developers would like to see an increase in production of agricultural commodities such as soybeans, as well as beef production, and continued logging of tropical hardwoods. Environmentalists would rather see a discontinuation of these activities, and continue to work toward having more land set aside in some form of protection from further development. As with many Amazonian dichotomies, in actuality, both are occurring simultaneously, as the governments that oversee Amazonian territories have responded to both pressures. They are interested in developing the resources of the Amazon in order to advance their economies, and have made it a priority that they, and not environmentalists from the Global North, set the agenda. Soybean production has expanded deep into the basin and soybean varieties are being developed to better tolerate rainforest conditions. The quest to better connect the basin with roadways, including a long sought-after road to the Pacific Ocean continue, as do other infrastructure development efforts. To many people who live and work in the basin, these are seen as positive developments.

These same governments have also set aside land to be protected from development, as there is broad consensus that the Amazon rainforest provides environmental services not just locally but globally as well. Most of the protected areas permit sustainable use by local residents, including vast indigenous reservations, thereby acknowledging that people live throughout the basin and are part of its landscape. They have also worked toward enforcing existing laws regarding required conservation on private property, and have increased true protection of areas that are set-asides. All of these efforts are imperfect, have been drawn-out in execution, and vary in their relative success, but as a whole, governance of the Amazon has improved from all perspectives.

The existence of ADEs has challenged conventional views of the physical environment in the conservation versus development debate, yet in much of the discussion about these matters, knowledge of ADEs has been ignored or brushed aside. Environmentalists, especially those in the Global North, continue to hold on to notions of the existence of a nonhuman nature, a notion that has its origins in the development of the original national parks in the American West. They are struggling with the idea of conserving land that is deeply inscribed (in its soils and vegetation) with prior human activity. But this requires a different way of thinking about the forest. The Amazon rainforest is still worth conserving, but the forest needs to be accepted as a cultural forest, and not envisioned as a "pristine wilderness," since, as a cultural forest, it still maintains extensive biodiversity and provides critical ecosystem services.

Concern has been raised regarding the emergence of revisionist perspectives of Amazonian prehistory—namely, that by acknowledging more human agency in the past, it opens up the Amazon to future unabated development. The people who hold this view are mostly those who imagine the Amazon as a place predominantly of pristine nature with little evidence of human activity, akin to what I discussed above. But there is a scalar tension here. Human activity, a form of disturbance, was in prehistory much milder and smaller in scale, functioning at the town/large village level and environs. It involved the manipulation of fields, fallows and forests, but on the whole, did not remove and replace the landscape with a completely different land cover. Today, large landholders ("largeholders") degrade the forest by extracting many high-value hardwoods from their land, leaving many trees that are incidentally felled to rot. They convert the forest to pasture or soybean, some of these fields are the size of the state of Rhode Island (or more), so the scale of these disturbances is much more extensive and the functioning of these new landscapes is entirely different from what had been. The interest in sustainable agriculture based on knowledge embedded in TP soils is less important to these largeholders who operate in the global commodity political economic arena. What may be of interest to them, however, is that by incorporating elements of knowledge of TP soils and its modern derivative, biochar (discussed further below), they can contribute to global carbon sequestration efforts—currently this is something they can volunteer to do, but may ultimately be something that they are required to do as a consequence of global climate change treaties.

The scientific investigation and documentation of TP soils has uncovered a form of (quite literally!) buried knowledge whose mysterious formation needs to be further uncovered as it offers possible alternative solutions for the sustainable management of soils of inherently low fertility. TP soils have embedded in them a form of carbon—aromatic carbon—that has the potential to contribute to the creation of a more productive and sustainable landscape that can assist farmers in the future, especially smallholder farmers who either do not have access to artificial inputs or who choose not to use them. And much of the tropical lowlands around the world remain occupied by smallholder farmers who would benefit immensely from a more productive agriculture that uses organic inputs generated from waste that is readily available. This gets me to the topic of biochar.

18.6.1 Biochar

At the end of the 2006 World Congress of Soil Science, held in Philadelphia, Pennsylvania, USA, there was a workshop held to discuss the potential of generating TP-like soil conditioners. At the time, research on TPs was very active and there was considerable excitement generated by the potential of the "ancient knowledge" buried in these soils. Inspired by TP, academics and entrepreneurs came together and formed what has come to be known as the International Biochar Initiative (IBI) (http://www.biochar-international.org/). I quote from their website, "This organization is a non-profit organization supporting researchers, commercial entities, policy makers, development agents, farmers and gardeners, and others committed to supporting sustainable biochar production and utilization systems. Sustainable biochar is a powerfully simple tool to fight global warming. This 2,000 year-old practice converts agricultural waste [via pyrolysis] into a soil enhancer that can hold carbon, boost food security, and discourage deforestation. It's one of the few technologies that is relatively inexpensive, widely applicable and quickly scalable."

Although questions of intellectual property have been highlighted by Kawa and Ayuela-Caycedo in their 2008 paper, the IBI and its affiliated people and organizations have moved forward quickly, seizing on the inherent essence of TP soils, SOM stability as a consequence of aromatic carbon. Researchers, entrepreneurs, and farmers are trying in various ways to create biochar, an organic soil conditioner from organic wastes such as those generated in forestry (especially at sawmills), the waste in charcoal production, agricultural by-products (e.g., *bagasse* from sugar cane processing), slaughterhouse waste, manure, and other sources. Subjecting this waste to pyrolysis (slow, cool, low-oxygen burning; i.e., charring) results in biochar. This is essentially what Wim Sombroek, the late Dutch soil scientist who first brought scientific attention to TP soils and who spearheaded a revival in research on the topic in the 1990s, envisioned when he argued for *Terra Preta Nova* (TPN), new TP. Wim wanted to figure out ways of (re)creating (new) TP, and set out an agenda to do so before his untimely passing away. Johannes Lehmann, a soil scientist instrumental in research on TP soils and the IBI, has argued that it is difficult to figure out exactly how TP soils were formed in the past, as there are too many variables (inputs) that are unknown at this time, and may remain unknown. Although there remains substantial academic interest in pursuing just how TP formed in the past, many believe, cautiously, that what is already known about them is what will enable the provision of a more environmentally friendly and sustainable fertilizer based on biochar, something that is of great benefit to many smallholder farmers throughout the world, as this will improve the fertility of their land by using locally available by-products. Much research remains to be done on and with biochar, as there is great variability in its quality (of inputs and degree of pyrolysis, for example). But fundamentally, it is seen as a way of improving farmer production capacity and stemming land degradation throughout the world, even rehabilitating wastelands, as its use as a soil conditioner helps ameliorate soils not just over the short term, but over the long term as well.

Another aspect about biochar that has global implications is its ability to contribute to carbon sequestration. The conversion of waste biomass into carbon via biochar has important implications for improving global carbon sequestration, a process of global importance. The reason for this is that waste biomass is highly degradable and therefore contributes to atmospheric carbon whereas biochar added to the soil captures carbon that is held in the soils and not subject to atmospheric emissions.

18.7 CONCLUSIONS

A 2011 article by Janzen et al. in the flagship journal of the Soil Science Society of America outlined grand challenges for the field of soil science. Several of their themes overlap with the theme of this book—soils in the context of science-informed sustainability—and especially this chapter. The linkage to these identified grand challenges are worth taking a moment to note. The efforts to find ways to (re)create TP soils, or at least use the buried knowledge of existing TP, especially as exemplified by biochar, can help address two of these challenges. The first, "nutrients," questions whether we can preserve and enhance the fertility of soils while exporting ever bigger harvests. The use of TP-inspired biochar as a soil conditioner is a way of enhancing soil fertility in a sustainable way over the long term, and will contribute to more substantial harvests, especially for farmers most challenged to do so, small-scale farmers with limited resources. The second, "recycling waste," questions how we better use soils as biogeochemical reactors, thereby avoiding contaminations and maintaining soil productivity. Again, biochar is a way of both recycling waste and sequestering carbon, processes with significant benefit to the global environment.

And so, in conclusion, TP soils, the mysterious soils of the Amazon, while they continue to remain mysterious in some ways, provide a new way of thinking about the region, and also provide the potential for a more sustainable future. Achieving their full potential does necessitate confronting the political economic reality of the Amazon and of global agriculture in this era of global environmental change. This means accepting conservation of cultural forests as well as working toward development that is inclusive and respects sovereign rights and works with the environment, including ADEs and ADE derivatives such as biochar.

ACKNOWLEDGMENTS

I thank the many colleagues who work on TP soils and with whom I have interacted and collaborated. Earlier versions of parts of this chapter have been presented at colloquia at various colleges and universities. I acknowledge and appreciate the feedback I received at those opportunities as this helped sharpen my thinking about these soils. The writing of this manuscript was supported by the National Science Foundation (USA) while the author worked at the Foundation. Any opinion, finding, conclusions, or recommendation expressed in this material are those of the author and do not necessarily reflect the views of the National Science Foundation (USA).

FUTHER READING

The study of TP soils is highly interdisciplinary, involving geographers, soil scientists, anthropologists, archaeologists, agronomists, and others. Much of the substance of what I discuss in this chapter is based on material published in four edited volumes on TP soils, which are included in the list below and marked with an asterisk (*). There is an increasing amount of research about these soils and their consequences being published in scientific journals, and I have included a few key pieces in the list below, along with a few classic, foundational pieces.

Balée, W. and C.L. Erickson, eds. 2006. *Time and Complexity in the Neotropical Lowlands: Studies in Historical Ecology*. New York: Columbia University Press.

Barrows, C.J. 2012. Biochar: Potential for countering land degradation and for improving agriculture. *Applied Geography* 34: 21–28.

Denevan, W.M. 1992. The pristine myth: The landscape of the Americas in 1492. *Annals of the Association of American Geography* 82(3): 369–385.

Denevan, W.M. 1996. A bluff model of riverine settlement in prehistoric Amazonia. *Annals of the Association of American Geographers* 86(4): 654–681.

Denevan, W.M. 2001. *Cultivated Landscapes of Native Amazonia and the Andes*. Oxford: Oxford University Press.

Fraser, J., W. Teixeira, N. Falcão, W. Woods, J. Lehmann, and A.B. Junqueira. 2011. Anthropogenic soils in the Central Amazon: From categories to a continuum. *Area* 43(3): 264–273.

Glaser, B. 2007. Prehistorically modified soils of Central Amazonia: A model for sustainable agriculture in the 21st century? *Philosophical Transactions of the Royal Society (B)* 362: 187–196.

Glaser, B. and J.J. Birk. 2012. State of the knowledge on the properties and genesis of anthropogenic dark earths in Central Amazonia (*terra preta do índio*). *Geochimica et Cosmochimica Acta* 82: 39–51.

*Glaser, B. and W.I. Woods, eds. 2004. *Amazonian Dark Earths: Explorations in Space and Time*. Berlin: Springer.

Heckenberger, M.J., A. Kuikuro, U.T. Kuikuro, J.C. Russell, M. Schmidt, C. Fausto, and B. Franchetto. 2003. 1492: Pristine forest or cultural parkland? *Science* 301: 1710–1714.

Heckenberger, M.J., J.C. Russell, J.R. Toney, and M.J. Schmidt. 2007. The legacy of cultural landscapes in the Brazilian Amazon: Implications for biodiversity. *Philosophical Transactions of the Royal Society (B)* 362: 197–208.

International Biochar Initiative. http://www.biochar-international.org/ (last accessed January 25, 2013).

Janzen, H.H., P.E. Fixen, A.J. Franzluebbers, J. Hattey, R.C. Izaurralde, Q.M. Ketterings, D.A. Lobb, and W.H. Schlesinger. 2011. Global prospects rooted in soil science. *Soil Science Society of America Journal* 75(1): 1–8.

Kawa, N.C. and A. Ayuela-Caycedo. 2008. Amazonian Dark Earth: A model for sustainable agriculture of the past and present? *The International Journal of Environmental, Cultural, Economic and Social Sustainability* 4(3): 9–16.

Lehmann, J. 2006. Bio-char sequestration in terrestrial ecosystems—A review. *Mitigation and Adaptation Strategies for Global Change* 11: 403–427.

*Lehmann, J., D.C. Kern, B. Glaser, and W.I. Woods eds. 2003. *Amazonian Dark Earths: Origin, Properties, Management*. Dordrecht: Kluwer.

Lentz, D.L., ed. 2000. *Imperfect Balance: Landscape Transformations in the Pre-Columbian Americas*. New York: Columbia University Press.

Mann, C.M. 2005. *1491: New revelations of the Americas before Columbus*. New York: Alfred Knopf.

Sombroek, W.G. 1966. *Amazon Soils: A Reconnaissance of the Soils of the Brazilian Amazon Region*. Wageningen: Centre for Agricultural Publications and Documentation.

*Teixeira, W.G., D.C. Kern, B.E. Madari, H.N. Lima, and W.I. Woods, eds. 2009. *As terras pretas de índio da Amazônia: Sua caracterização e uso deste conhecimento na criação de novas areas*. Manaus: Embrapa Amazônia Ocidental.

WinklerPrins, A.M.G.A., and S.P. Aldrich. 2010. Locating Amazonian dark earths: Creating an interactive GIS of known locations. *Journal of Latin American Geography* 9(3): 33–50.

*Woods, W.I., W.G. Teixeira, J. Lehmann, C. Steiner, A.M.G.A. WinklerPrins, and L. Rebellato, eds. 2009. *Amazonian Dark Earths: Wim Sombroek's Vision*. Dordrecht: Springer.

19 Modern Landscape Management Using Andean Technology Developed by the Inca Empire

Francisco Mamani-Pati, David E. Clay, and Hugh Smeltekop*

CONTENTS

19.1 INTRODUCTION

The Andean region in South America, named for the Quechua word *anti* which means "high crest," is characterized by the diversity of its ecology: desert coasts, tropical landscapes, and dry, cold plateaus with extreme climatic variability that at first glance seems one of the least favorable environments for human life. The Inca Empire or Tawantin-suyu (the name given by the Incas to their state) was a pre-Columbian culture located in this area that flourished between the fourteenth and fifteenth centuries. This empire occupied a vast territory of South America, including over 3 million square kilometers between the Pacific Ocean and the Amazonian jungle. It included the coastal and mountain regions of Ecuador, Peru, western Bolivia, southern Colombia, northern Chile, and northwestern Argentina. The Inca Empire was founded between 1100 and 1200 AD, and the Inca were the inheritors of those that came before them. The success of the Inca Empire resulted from their ability to organize. The Inca ruled until their defeat by the Spaniards in 1533 (Arkush 2006). Agriculture was conducted by farmer/soldiers who were aided by professionals. Work was communal, and slash and burn agriculture was not normally practiced. They domesticated over 20 corn varieties, 240 potato cultivars, sweet potato, squash, bean, peppers, peanuts, and quinoa. They matched crops and cultivars to expected environmental stresses. Potato was planted as high as 4600 meters, while corn was planted as high as 4100 m (Mamani-Pati et al. 2011). The empire was divided into four states: Chincha-suyu located to the north, Colla-suyu to the south, Anti-suyu to the east, and Conti-suyu to the west. The Inca society established three basic laws of the Tawantin-suyu: *Ama Sua* (do not steal), *Ama Llulla* (do not lie), and *Ama Kella* (do not be lazy). The capital of the Inca Empire was the city of Cuzco, in modern Peru.

* Catholic University of Bolivia "San Pablo," UAC-Carmen Pampa, Coroico, La Paz, Bolivia; Email: fjmpati@yahoo.com

Prior to the arrival of the Incan people in the fourteenth century, native people had been living along the Pacific Ocean coastal regions of South America. These people grew squash, corn, and beans. Between 500 and 1000 AD, there was a period of warm temperatures that allowed the lowland and valley inhabitants to inhabit the higher attitudes (Williams 2002; Dillehay and Kolata 2004; Chepstow-Lusty et al. 2009). In the territory of modern Bolivia, a pre-Inca culture began and flourished from around 400 to 900 AD. There is evidence of a prolonged period of drought that lasted from 900 to 1400 AD. This was the likely cause of the decline of food production in the central Andes (Ortloff and Kolata 1993; Williams 2002), which was concurrent with the collapse of the Tiwanaku (1100 AD) and nearby Wari (1000 AD) cultures in that period, and the migration and the founding of new populations in other areas. With the end of the drought, the Incan people extended their reign to the high plateau. The Inca success was most probably related to their ability to use precision farming to maximize agricultural productivity in highly variable climates.

However, Incan people demonstrated over many centuries their ability not only to survive in such circumstances but also to dominate the geographical environment and create a series of flourishing civilizations. In 1438, the Bolivian highlands were incorporated into the Inca Empire (Tawantinsuyu) as part of the Colla-suyu state, and Quechua was imposed as the official language, although the Aymara as an ancient language has been spoken continuously (Somervill 2005). The economy was based on agriculture, and land use was communal. Each family had its land to cultivate and feed itself; larger families received more land.

The use of sustainable management practices was partially responsible for the success of the Inca culture (Morris 1999). It is estimated that the Inca cultivated about 70 crop species. A large number of crops were needed to farm in the various environments of the region that had extreme climatic variability. Major crops included potatoes (*Solanum* spp.), sweet potatoes (*Ipomoea batatas*), corn (*Zea mays*), chili peppers (*Capsicum* spp.), Andean squash (*Cucurbita maxima*), coca (*Erythroxylon coca*), peanuts (*Arachis hypogaea*), Andean lupin (*Lupinus mutabilis*), oca (*Axalis tuberosa*), maca (*Lepidium meyenii*), achira (*Canna adulis*), arracacha (*Arracacia xanthorrhiza*), isano (*Tropaeolum tuberusum*), ulluku (*Ullucus tuberosus*), beans (*Phaseolus* spp.), quinoa (*Chenopodium quinoa*), canihua (*Chenopodium pallidicauli*), and amaranth (*Amaranthus caudatus*) that were used for food and medicinal purposes (Brush 1982; Kelly 1965; Ochoa 1991; Flores et al. 2003). In the highlands of the Andes, over 200 species of potatoes were grown and harvested, which were distinguished by color (red, white, yellow, brown, black, purple, pink, gray, spots, and stripes) and other characteristics. To prevent decomposition, the Inca people learned to dehydrate potatoes, known as *chuño,* which prevented famine in drought seasons in the Inca nation (Ochoa 1991; Somervill 2005).

The most important crops to the Inca people were quinoa, potato, and corn. Quinoa is drought tolerant, resists extreme weather and low soil fertility, and can be grown at high altitudes up to about 4000 m; it was considered a "sacred food" of the Inca Empire (PROINPA Foundation 2004). Many of the crops domesticated by the Inca were widely distributed by the Spanish. In the highland forests coca plants were cultivated; it was considered a "sacred leaf" consumed as tea and in ceremonies by pre-Inca people.

The Inca also domesticated llamas and alpacas for their wool and meat, as well as for helping with transportation and travel. The manure from these animals was used to fertilize crops. Wild vicuñas were captured and their fine hair was used for clothing and blankets. Also, they planted cotton (*Gossypium barbadense*) used for making clothing (Somervill 2005).

The extreme environmental conditions of the Andean highlands make it incredibly difficult to produce a sustainable food supply. To achieve food security, Andean people have adopted many of the concepts behind precision agriculture, that is, agriculture based on the observation of variation within fields and responding to those variations to increase efficiency, production, and/or profitability. These concepts include an acute awareness of the natural environment that is used to adjust planting patterns, select varieties, and modify their microclimate.

19.2 PREHISTORIC LANDSCAPE MANAGEMENT IN THE ANDEAN HIGHLANDS

Precision farming is used by modern farmers to improve nutrient efficiency, energetic efficiency, and long-term sustainability. Although these modern precision farming tools were not available 3000 years ago, the same concepts were implemented by pre-Colombian people (Earls 1998; Mamani-Pati et al. 2011). Rediscovering knowledge learned by these people can help us manage food security issues today.

Using environmental cues to predict climate is not a new concept. For example, in North America, the number of reddish-brown segments of a wooly bear caterpillar (such as *Pyrrharctia isabella*) is commonly misunderstood to be a predictor of the severity of cold of the coming winter (Wagner 2005). Andean people also used environmental cues to make accurate predictions about their environment, and then used those cues to plan their cropping patterns; the evidence for this is seen in modern practices that use a combination of different environmental cues to predict weather patterns. For example, a rainbow around the sun can indicate a particularly dry season; and when wild ducks on Lake Titicaca build their nests higher among the reeds, rainfall is typically above average (Earls 1998). Orlove et al. (2000, p. 68) indicate that "poor visibility of Pleiades in June—caused by an increase in subvisual high cirrus clouds—is indicative of an El Niño year, which is usually linked to reduced rainfall during the growing season several months later."

Knowledge and observation of the seasons is very apparent in the pre-Colombian constructions of Andean people. The Kalasasaya temple at Tiwanaku (Bolivia) is designed to track the solstices and equinoxes, as well as lunar cycles with internal markers (Janusek 2008); many solar observatories existed both during and before the Inca period (Ghezzi and Ruggles 2007). Other representations of astronomical observations are also apparent, such as the *pachaquipu*, a record made up of a complex combination of cords and knots (Laurencich-Minelli and Magli 2009/2010). Such astronomical observations attest to a great awareness of the advance of the seasons, and the agricultural cycles that follow them.

A prime example of this experimentation and its utility for the development of crops is the potato. Over 5000 varieties of the common potato (*Solanum tuberosum*) are known to grow in Bolivia and Peru (Ochoa 1991; Morris 1999; Flores et al. 2003), each suited to a specific soil type, water availability, and temperature scheme. Andean farmers have also cultivated bitter potatoes (most notably *Solanum juzepczuki*, *S. curtilobum*, and *S. ajanhuiri*), edible species tolerant to frost and grown in altitudes of up to 4600 m above sea level. Likewise, the Chenopodiaceae-family plants quinoa and cañahua (*Chenopodium quinoa* and *C. pallidicaule*), have been adapted by Andean people to an enormous range of growing conditions (PROINPA Foundation 2004).

Additionally, there is evidence that Andean farmers made, and make, optimal use of fallow to manage soil fertility (Orlove and Godoy 1986; Pestalozzi 2000; Sarmiento and Bottner 2002), and vary nutrient resources, utilizing more fertilizer where there is more potential for greater yields. For example, there is much evidence that the Incan people use manure from humans, domesticated camelid species, and tree leaves to fertilize their crops. Bird guano was collected from coastal islands and used as a source of phosphorus. This resource was so valuable that Incan custom mandated that anyone who killed a guano-producing bird would be put to death. On land at higher altitudes, human and llama manure was used as fertilizer on crops, including the incorporation of llama manure when sowing maize seed. Fallow periods of two or more years were also used to restore soil fertility when necessary. Llama manure was also used as fuel for fires. These traditions continue in the higher altitudes of Bolivia and Peru, combining modern techniques with ancient knowledge (Carney et al. 1993; Bandy 2005; Mamani-Pati et al. 2011).

Historically, agricultural activities in the Inca period were based on research conducted at local and empire scales, and land was separated into distinct production zones. At the community and individual scales, research was conducted near their homes or in specific zones where production fields were located (Kelly 1965; Earls 1998). Andean farmers observed the behavior of crops under

FIGURE 19.1 **(See color insert.)** Agricultural research at Moray. (Photo credit: Francisco Mamani-Pati.)

different climatic conditions; the resistance of crops to frost, excessive or deficient rainfall; and the behavior of crops in hot and cold years when exposed to different diseases, insects, and fungi. A crop's behavior in a colder year was likely simulated by sowing at a higher altitude and in a hotter one by sowing it lower down. A crop's reaction to high rainfall could be tested by planting it in an area characterized by heavy rainfall in a normal year and vice versa for low rainfall. Simulation and experimentation are believed to have been routine activities for the Andean farmer, as they are today.

Many people believe that the Inca Empire supported an agricultural research center at a site in Moray, Peru (Figure 19.1). The Moray site is a set of circular terraces arranged vertically, each presenting slightly different conditions. Each terrace produces a microclimate where different crops can be tested. Modern day analysis of the site suggests that this terrace system can produce up to 20 different types of microclimates determined by the location of its terraces: terraces found in the lower parts have higher temperatures, and upper terraces exhibit colder temperatures. Each terrace provides different altitudes, relationship to sun exposure, soil temperature, soil, and soil moisture content. Moray is believed to have served as a model for calculating agricultural production capacity not only in the valleys but also from different parts of the Inca Empire (Earls 1998). Many of the soils within the terraces were transported from the valleys as evidenced by analysis of the soils on site, which are characteristic of soils from other places in the kingdom. This site could have been used to assess soil and temperature effects on plant growth and development. As a result of this experimentation, the Inca people were able to adapt to great climate changes. The agricultural research, apparent in Moray, likely gave them the ability to develop cropping systems adapted to these new environments and thus maximizes agricultural productivity (Ortloff and Kolata 1993; Hodell et al. 1995; Binford et al. 1997; Haug et al. 2003; Peterson and Haug 2005).

To take advantage of warmer temperatures, pre-Inca and Inca people constructed agricultural terraces using glacial-fed irrigation and planted trees which reduced erosion and increased soil fertility. Drought and the need to increase agricultural energy efficiency will undoubtedly be the most devastating environmental problems of the next century.

19.3 ANDEAN TECHNOLOGY AND ITS ENVIRONMENTAL IMPACT

19.3.1 Agricultural Terraces or Andenes

In many parts of the high Andes, the high altitude and cold weather severely limited agricultural production. To expand the available land, Incan people used several techniques which allowed farming

FIGURE 19.2 **(See color insert.)** Terraces are still being constructed today by Bolivian farmers. (Photo credit: Francisco Mamani-Pati.)

on hillsides and at high altitudes. The most important method of agriculture of the Incan people was the use of special terraces called *andenes* (Morris 1999; Branch et al. 2007; Chepstow-Lusty et al. 2009). *Andenes* are artificial agricultural terraces built on steep hills, used to increase cropping area on the otherwise inaccessible Andean hillsides. This also allowed for better water use through the creation of interconnected channels, connecting various levels. The lengths of *andenes* varied from 4 to 100 m, and their widths from 2 to 20 m, with some *andenes* as narrow as 1 m on very steep slopes, and as long as 150 m in very flat areas (Figure 19.2). These flat, stepped areas resisted erosion and landslides, maintained soil fertility by avoiding the loss of mineral nutrients from water running across soil surfaces, stored more water, and took advantages of more predictable microclimates.

The Incan people created an effective irrigation system that was partially responsible for their success. Their knowledge of water flow characteristics is noticeable in the architecture of their gravity-driven water management. The Incan people protected natural resources (air, soil, water, biodiversity) by the following methods: (1) using digging sticks rather than plows to seed their crops; (2) implementing effective drainage and irrigation systems; (3) minimizing erosion by building terraces that were constructed at a great expense; and (4) the adoption of nutrient management practices that produced soil that has remained fertile even after centuries of use.

Andenes were created to protect soils from erosion and manage irrigation water with efficiency. *Andenes* create a favorable microclimate and solve drought and frost problems. The *andenes* store heat, raise the humidity, and change the flow of air (Figure 19.3). *Andenes* also store more heat from solar radiation than a flat surface or slope because part of the reflected rays reflect again on the terrace wall. At night, all of the heat radiated by a natural slope is lost in the atmosphere, whereas on *andenes*, the radiated heat is reflected again from the *andenes* walls that then act like a thermoregulator of the microclimate within the terrace.

The Incan people used the *andenes* to increase agricultural efficiency. The current return to this same ancestral knowledge is helping Andean farmers to address climate change.

19.3.2 *Suka Kollus* or Raised Field Agriculture

Andean farmers used *andenes* on steep slopes, and raised fields (*camellones* in Spanish, *waru waru* in Quechua, and *suka kollus* in Aymara) in flat high landscapes. By the year 1000 AD Andean farmers of the high plateaus created different technologies to adapt them to the harsh climate of the region. Around Lake Titicaca they developed *suka kollus,* most likely to manage flooded lands by draining the excess water to raise the temperature of the soil as a defense against frost, and to make

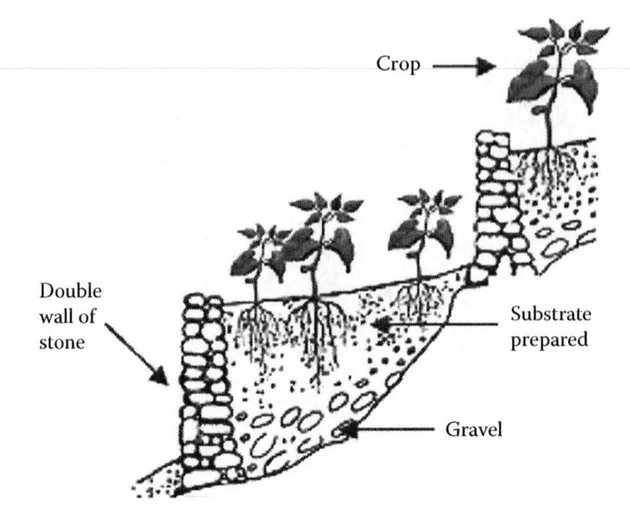

FIGURE 19.3 Diagram of the structure of *andenes*-style terraces. (Diagram by Francisco Mamani-Pati.)

the agroecosystem less vulnerable to drought conditions (Erickson 1988, 1995, 2006; Erickson and Candler 1989; UNEP 1997; Morris 1999; Biesboer 1999; Lhomme and Vacher 2003; Mamani-Pati et al. 2011). These *suka kollus* consisted of beds of soil that varied from 1 to 20 m wide, 10 to 100 m long, and 0.5 to 1 m high, depending on the contours of the land. They were surrounded by channels of water collected from rain, with dimensions of 1–4 m wide and 0.5–1 m in depth. This water management has a dual purpose: to moderate temperature changes and provide irrigation water.

According to Erickson (2000), the agricultural infrastructure of raised fields is the oldest in South America (Figure 19.4). The raised beds were constructed by digging around the bed surfaces to achieve channels for the water. The soil from the channels was distributed over the beds, elevating the original soil surface. Crops were installed on the land beds and surrounding channels were connected to allow a more efficient distribution of water.

The water in the channels between the beds has beneficial effects on the microclimate (Figure 19.5) by regulating the local temperature and absorbing solar energy during the day, which radiates

FIGURE 19.4 **(See color insert.)** Raised fields or *suka kollus* were first developed in the year 1000 BC Kallutaca, Laja-La Paz. (Photo credit: Francisco Mamani-Pati.)

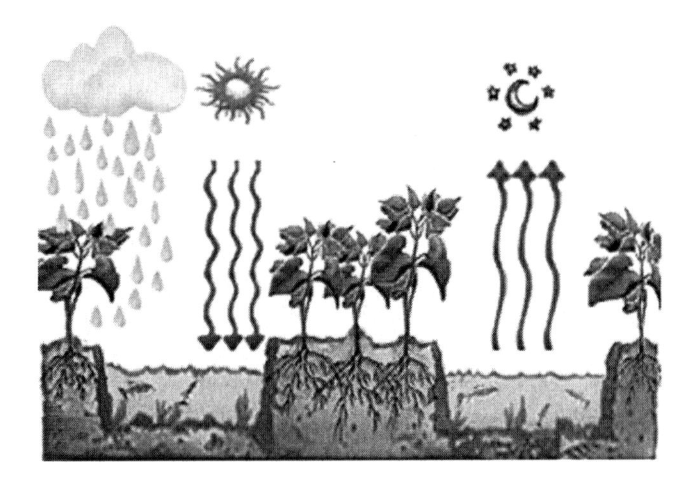

FIGURE 19.5 **(See color insert.)** Diagram of water and heat cycle of *suka kollus*. (Diagram by Francisco Mamani-Pati.)

to the surrounding soil and the air during the night (Kolata and Ortloff 1989; Kolata et al. 1996; Swartley 2000; Bandy 2005; Sanchez de Lozada et al. 2006). Nakajima (2004) indicates that raised fields allow for the concentration and release of heat, thus the maintenance of a stable temperature in beds over night, becoming an agricultural technique suitable to regulate cold and encourage high productivity. Sanchez de Lozada et al. (1998) show that the effect of frost mitigation is essentially the result of processes related to air circulation during the night from the channels to the beds.

The *suka kollu* system has other benefits. Erickson (1992) indicates that the channels accumulate nutrients and form organic sediments that serve to fertilize the crops in the beds. The system results in higher diversity with aquatic plants and animals, some of which, like fish, can have economic benefits. The raised beds serve to drain surplus water, and can be used for irrigation when necessary. The better drainage and use of rainwater instead of groundwater probably also help to manage pH and salinity problems.

Bolivian and Peruvian farmers have been improving their efficiency by combining historical techniques with modern knowledge. Modern-day Bolivians reconstructed some of the ancient *andenes* and *suka kollus* to protect agriculture from the effects of droughts. For example, people in areas where *suka kollus* and *andenes* have been reestablished are improving their ability to produce food by using the continually replenished ground waters. Also, in 1983 Andean farmers restored ancestral irrigation systems to water crops during an intense drought that damaged agricultural production (Erickson 1995, 2000, 2006). Raised fields take advantage of natural phenomena. Andean civilization is highly respectful of the environment, and the Andean worldview incorporates a harmonious relationship with nature.

Lessons from history tell us that investments into agriculture research are critical for providing a sustainable food supply and reducing the potential for societal collapse. As in the past, precision farming is a tool that can help increase energy efficiency and food security.

19.4 CONCLUSION

Although ancient people did not have GIS or GPS, they used precision farming techniques to help them produce a reliable food supply. In the Andes, the Incan people most likely invested in agricultural research and built highly effective systems that produced microclimates required for food security.

The *suka kollus* and *andenes* are ancestral technologies that have been used for over 3000 years for the production of food. The *suka kollus* mitigate the effects of frost through the increased temperature. The removal of excess water is another important function of *suka kollus,* which are located mainly in areas with water tables near the surface. The water in the canal is heated by the sun during the day and this significantly accumulates and concentrates the heat over the *suka kollus* during the night. Both techniques can be used to cultivate Andean plants, highly adapted to environmental constraints, and minimize frost and drought problems.

ACKNOWLEDGMENTS

The authors acknowledge the Catholic University of Bolivia "San Pablo," UAC-Carmen Pampa and South Dakota State University for their support in the development of this chapter.

REFERENCES

Arkush, E. N. 2006. Collapse, conflict, conquest: The transformation of warfare in the Late Prehispanic Andean highlands. In *The Archaeology of Warfare: Prehistories of Raiding and Conquest*, eds. E. N. Arkush and M. W. Allen, 286–335. Gainesville: University Press of Florida.

Bandy, M. S. 2005. Energetic efficiency and political expediency in Titicaca Basin raised field agriculture. *Journal of Anthropological Archaeology* 24: 271–296.

Biesboer, D. D. 1999. Nitrogen fixation in soils and canals of rehabilitated raised-fields of the Bolivian Altiplano. *Biotropica* 31: 255–267.

Binford, M. W., A. L. Kolata, J. W., Janusek, M. T. Seddon, M. Abboti, and J. H. Curtis. 1997. Climate variation and the rise and fall of an Andean civilization. *Quaternary Research* 47: 235–248.

Branch, N. P., R. A. Kemp, B. Silva, F. M. Meddens, A. Williams, A. Kendall, and C. V. Pomacanchari. 2007. Testing the sustainability and sensitivity to climatic change of terrace agricultural systems in the Peruvian Andes: A pilot study. *Journal of Archaeological Science* 34: 1–9.

Brush, S. B. 1982. The natural and human environment of the central Andes. *Mountain Research and Development* 2: 19–38.

Carney, H. J., M. W. Binford, A. L. Kolata, R. R. Marin, and C. R. Goldman. 1993. Nutrient and sediment retention in Andean raised-field agriculture. *Nature* 364: 131–133.

Chepstow-Lusty, A. J., M. R. Frogley, B. S. Bauer et al. 2009. Putting the rise of the Inca Empire within a climatic and land management context. *Climate of the Past* 5: 375–388.

Dillehay, T. D. and A. L. Kolata. 2004. Long-term human response to uncertain environmental conditions in the Andes. *Proceedings of the National Academy of Sciences of the United States of America* 101(12): 4325–4330.

Earls, J. 1998. The character of Inca and Andean agriculture. http://macareo.pucp.edu.pe/~jearls/documentosPDF/theCharacter.PDF (accessed December 3, 2012).

Erickson, C. L. 1995. Archaeological methods for the study of ancient landscapes of the Llanos de Mojos of Bolivia. In *Archaeology in the American Tropics: Current Analytical Methods and Applications*, ed. P. Stahl, 66–95. Cambridge: Cambridge University Press.

Erickson, C. L. 1988. Raised field agriculture in the Lake Titicaca basin: Putting ancient agriculture back to work. *Expedition* 30: 8–16.

Erickson, C. L. 1992. Prehistoric landscape management in the Andean Highlands: Raised field agriculture and its environmental impact. *Population and Environment: A Journal at Interdisciplinary Studies* 13: 285–300.

Erickson, C. L. 2000. An artificial landscape-fishery in the Bolivian Amazon. *Nature* 408: 190–193.

Erickson, C. L. 2006. El valor actual de los camellones de cultivo precolombinos:Experiencias del Perú y Bolivia. In *Agricultura Ancestral. Camellones y Albarradas: Contexto Social, Usos y Retos del Pasado y del Presente*, ed. F. Valdez, 315–339. Quito: Ediciones Abya-Yala.

Erickson, C. L. and K.L. Candler. 1989. Raised fields and sustainable agriculture in the Lake Titicaca basin of Peru. In *Fragile Lands in Latin America: Strategies for Sustainable Development,* ed. J. O. Browder, 230–248. Boulder: West View Press.

Flores, H. E., T. S. Walker, R. L. Guimarães, H. P. Bais, and J. M. Vivanco. 2003. Andean root and tuber crops: Underground rainbows. *Hortscience* 38: 161–167.

Ghezzi, I. and C. Ruggles. 2007. Chankillo: A 2300-year-old solar observatory in coastal Peru. *Science* 315: 1239–1243.

Haug, G. H., D. Gunther, L. C. Peterson et al. 2003. Climate and the collapse of Maya civilization. *Science* 299: 1731–1735.
Hodell, D. A., J. H. Curtis, and M. Brenner. 1995. Possible role of climate in the collapse of the Classic Maya civilization. *Nature* 375: 391–394.
Janusek, J. W. 2008. *Ancient Tiwanaku*. Cambridge: Cambridge University Press.
Kelly, K. 1965. Land-use regions in the central and northern portions of the Inca Empire. *Annals of the Association of American Geographers* 55: 327–338.
Kolata, A. L. and C. Ortloff. 1989. Thermal analysis of Tiahuanaku raised fields in the Lake Titicaca basin of Bolivia. *Journal of Archaeological Science* 16: 233–263.
Kolata, A. L., O. Rivera, J. C. Ramirez, and E. Gemio. 1996. Rehabilitating raised-field agriculture in the southern Lake Titicaca basin of Bolivia. In *Tiwanaku and its Hinterland: Archaeology and Paleoecology of an Andean Civilization. Vol. 1 Agroecology*, ed. A. L. Kolata, 203–239. Washington, DC: Smithsonian Institution Press.
Laurencich-Minelli, L. and G. Magli. 2009/2010. A calendar Quipu of the early 17th century and its relationship with Inca astronomy. *Archaeoastronomy* 22: 1–20.
Lhomme, J. P. and J. J. Vacher. 2003. Frost mitigation in the raised fields of the Andean Altiplano. *Bulletin de l'Institut Français d'Études Andines* 32: 377–399.
Mamani-Pati, F., D. E. Clay, and H. Smeltekop. 2011. Geospatial management of Andean technologies by the Inca Empire. *GIS in Agriculture: Nutrient Management for Improved Energy Efficiency*, eds. D. E. Clay and J. Shanhan, 255–263. Boca Raton, FL: CRC Press.
Morris, A. 1999. The agricultural base of the pre-incan Andean civilization. *Geographical Journal* 65: 286–295.
Nakajima, N. 2004. Los experimentos de suka kollu: Potencialidad y realidad. *Revista Cultural- La Paz* 32: 26–33.
Ochoa, C. 1991. *The Potatoes of South America: Bolivia*. Cambridge: Cambridge University Press.
Ortloff, C. R. and A. L. Kolata. 1993. Climate and collapse: Agro-ecological perspectives on the decline of the Tiwanaku state. *Journal of Archaeological Science* 20: 195–221.
Orlove, B.S and R. Godoy. 1986. Sectoral fallowing systems in the Central Andes. *Journal of Ethnobiology* 6: 169–204.
Orlove, B. S., J. C. H. Chiang, and M. A. Cane. 2000. Forcesting Andean rainfall and crop yield from the influence of El Niño on Pleiades visibility. *Nature* 403: 68–71.
Pestalozzi, H. 2000. Sectoral fallow systems and the management of soil fertility: The rationality of indigenous knowledge in the High Andes of Bolivia. *Mountain Research and Development* 20: 64–71.
Peterson, L. C. and G. H. Haug. 2005. Climate and the collapse of Maya civilization: A series of multi-year droughts helped to doom an ancient culture. *American Scientist* 93: 322–329.
PROINPA Foundation. 2004. *Study on the Social, Environmental and Economic Impacts of Quinoa Promotion in Bolivia*. La Paz: PROINPA.
Sanchez de Lozada, D., P. Baveye, and S. Riha. 1998. Heat and moisture dynamics in raised fields of the Lake Titicaca region (Bolivia). *Agricultural and Forest Meteorology* 92: 251–265.
Sanchez de Lozada, D., P. Baveye, R. F. Lucey, R. Mamani, and W. Fernandez. 2006. Potential limitations for potato yields in raised soil field systems near Lake Titicaca. *Scientia Agricola* 63: 444–452.
Sarmiento, L. and P. Bottner. 2002. Carbon and nitrogen dynamics in two soils with different fallow times in the high tropical Andes: Indications for fertility restoration. *Applied Soil Ecology* 19: 79–89.
Somervill, B. A. 2005. *Great Empires of the Past: Empire of the Inca*. New York: Shoreline Publishing Group LLC Press.
Swartley, L. 2000. *Inventing Indigenous Knowledge: Archaeology, Rural Developmentand the Raised Field Rehabilitation Project in Bolivia*. London: Routledge.
UNEP (United Nations Environment Programme). 1997. *Source Book of Alternative Technologies for Freshwater Augmentation in Latin America and the Caribbean*. Rome: UNEP.http://www.oas.org/dsd/publications/unit/oea59e/begin.htm#Contents (accessed December 3, 2012).
Wagner, D. L. 2005. *Caterpillars of Eastern North America: A Guide to Identification and Natural History*. Princeton: Princeton University Press.
Williams, P. R. 2002. Rethinking disaster-induced collapse in the demise of the Andean Highland States: Wari and Tiwanaku. *World Archaeology* 33: 361–374.

20 Indigenous Māori Values, Perspectives, and Knowledge of Soils in Aotearoa-New Zealand

Māori Use and Knowledge of Soils over Time

Garth Harmsworth and Nick Roskruge*

CONTENTS

20.1 INTRODUCTION

Before Polynesian arrival (~1200 AD), about 85% of Aotearoa-New Zealand (NZ) was covered in indigenous or native forest (Ausseil et al. 2011). At this time, areas not in forest included glacial alpine environments, alpine herbaceous areas, high tussock-lands, and extensive sand country and wetlands. By the mid-eighteenth century, this indigenous forest area had reduced to about 52% of

* Landcare Research NZ Ltd – Manaaki Whenua, Private Bag 11 052, Palmerston North 4442, New Zealand; Email: HarmsworthG@LandcareResearch.co.nz

the total area of NZ (Ausseil et al. 2011), following large-scale burning of forests, much attributed to an early Polynesian hunting era from ~1000 to 1300 AD (e.g., moa and native bird hunters), and additional and repeated clearance of indigenous forest and scrub particularly on lower mountain slopes. Historically, these slopes were converted and harvested by Māori to extensive bracken fern (*Pteridium esculentum*, *Pteridium aquilinium*), along with large tracts of regenerating tea tree or mānuka (*Leptospermum scoparium*) and kānuka (*Kunzea ericoides*). The bracken fern provided starch food crops, such as breads/cakes, while the mixed mānuka/kānuka scrub provided tea, wood for cooking fires, framing for housing, and medicinal extracts. Fires lit by early Polynesians destroyed large areas of indigenous temperate evergreen conifer (podocarp—largely from the *Podocarpus* genus)—broadleaved (hardwoods) forest in the South Island and drier lowlands, particularly in the eastern regions of both the North and South Islands of NZ. However, in many *iwi/hapū* tribal regions of NZ, native forest stayed considerably intact into the nineteenth century before European colonization converted a large proportion of indigenous forest into an extensive pastoral landscape or grassland. The pioneering and development phase between 1860 and 1950 saw rapid and extensive deforestation of indigenous forest through clear-felling and burning, draining of wetlands, with the conversion to farmland and plantation forestry. By the mid- to late-1900s, native forest only covered 23% of NZ's total land area of 26 million km^2 (Ausseil et al. 2011).

Early Māori (from about 1500 to the 1700s) came to rely increasingly on the soil resource, and learnt how to manipulate it to derive maximum benefit. Activity surrounding food production or harvesting was structured with a strong relationship to the gods, stars, moon, and heavens for many crops (Best 1924, 1925; Roskruge 2011, 2012). This became particularly advanced with the advent of the Māori horticulturalist, who developed an acute awareness of different types of soils, and knowledge of how soils could be modified within regional climates to improve crop suitability, yield, growth, condition, survival rate, and how to expand the areas available for cropping (Roskruge 2011, 2012).

From about 1600 AD, Māori were developing a greater sustainability ethic and greater environmental knowledge because of declining natural resource areas and degrading resources. Māori agriculture became prolific from about 1700 and developed extensively in lowland areas in the 1800s with the cultivation of Polynesian and European food plants.

The following sections discuss developments in the uses and knowledge of soils by Māori with time in eras ranging from traditional (Section 20.2), time of first European contact (Section 20.3), and recent times (Section 20.4). These are discussed in the context of Māori beliefs and concepts of soils, the environment, and land that are outlined in our earlier chapter in this book (Harmsworth and Roskruge 2014), which also contains a glossary of relevant Māori terms.

20.2 SOILS AND THEIR USES IN PRE-EUROPEAN TIMES

20.2.1 Traditional Uses for Soils

Traditionally, soils had a wide range of uses. One important use was for organic dyes and these were applied to a range of materials, for example, clothing made from fibers, wood products, instruments, insides of *whare* (houses), and art designs. Many dyes were sourced from clays, muds, organic soils, peats, and plants typically found in wetter areas such as muds or *para* of a swamp. They included Parapara, a soft mud used for dying flax fiber; Paraharaha, a black mud used for dying flax fiber; Pukepoto, dark blue earth used as pigment, found in swamps; Tareha, an ochre; Kokowai, a red ochre; Te uku, a white clay; Uku, a blue clay; and Kura, a red ochre. Paru, a mud dye rich in iron salts, was unfortunately acidic and in later years led to progressive breakdown and damage of cloaks and garments.

Different colors were derived from muds mixed with plant materials and oils, such as

- Black: Hinau (*Elaeocarpus dentatus*) bark soaked in dark-colored mud; Harakeke flax fiber was often dyed black

- Reddish-brown: Tanekaha (*Phyllocladus trichomanoides*) bark, pounded, soaked in water, fiber immersed in the water
- Blue black: Tupakihi (*Coriaria*), Whawhakou (*Eugenia*), Mako (*Aristotelia*)
- Yellow: Raurekau (*Coprosma*)
- Yellow: Tanekaha (*Phyllocladus trichomanoides*) bark and a small plant, Kakariki
- Brownish-yellow: Puriri (*Vitex lucens*) mixed with Tanekaha (*Phyllocladus trichomanoides*)
- Red paint: (shark liver oil with kōkōwai (iron oxide stained earth), kōkōwai—red ochre
- Black: charcoal, soot often mixed with fish oil
- White: halloysite, kaolinite white clays

Clays had many uses. Red ochre, found in clay, was smeared on people's faces and bodies as a sign of chiefly status. It was also used on carved items such as *waka* (canoes) or houses, and even on the bones of the dead. Kōkōwai, one type of red ochre, was rolled into balls, baked in fire or hot ashes, and mixed with shark oil to form dyes. Tākou was another type of red ochre. Red ochre, found in clay, was baked in a fire and mixed with shark oil. Taioma, a white paint, was created by burning and pulverizing the clay, and then mixing it with oil. Pukepoto was a cobalt blue color found in clay. Uku, a white or bluish clay with a soapy consistency and feel, was used for washing. Food was sometimes cooked by enveloping it in clay mixed with water and placing it in a hot fire.

20.2.2 Indigenous Forest and Scrub Plants Cultivated by Māori

From around 1400, Māori increasingly modified soils within forest and scrubland (shrubland) and brought wild plant variants into a kind of cultivation (Best 1925; Buck 1950; Challis 1978; Leach 1984; Park 1995; King 2003; Roskruge 2012). These plants became dominant in early stages of revegetation in cleared and modified sites, providing roots, shoots, leaves, and rhizomes, which, when mixed together, became a major part of the Māori diet. Traditional Māori vegetable foods before the nineteenth century included numerous edible ferns, raupō (*Typha augustifolia*), harakeke—a flax (*Phormium tenax*), and nikau palm (*Rhopalostylis sapida*). Probably the most important food was aruhe, rauaruhe rahurahu, rarahu or bracken fern (*Pteridium esculentum*, *Pteridium aquilinium*), a starchy food that became a staple diet (Shawcross 1967; Challis 1991). The aruhe root was dug out with a stick (a *koo*), harvested, cleaned, dried, soaked, dried again, and beaten or ground to soften it. Areas that produced good aruhe would be burnt in August every 3–5 years to prevent mānuka (*Leptospermum scoparium*) and kānuka (*Kunzea ericoides*) shrubland and other small trees from taking over. The burning, which was said to improve the flavor of the bracken fern roots, was very much controlled and seasonal. Other wild species utilized by Māori included ti kouka—cabbage tree species (*Cordyline terminalis*, *Cordyline australis*), a hardier variety ti pore (*Cordyline stricta*), many podocarp species, the berries from *Coprosma* species, Ngaio (*Myoporum laetum*), kawakawa (*Macropiper excelsum*), fuschia species (e.g., *Fuchsia excorticate*), tutu (*Coriariaceae*), mainly used for juice, hīnau (*Elaeocarpus dentatus*), tawa (*Beilschmiedia tawa*), and karaka (*Corynocarpus laevigatus*). A wide variety of forest and cultivated plants were also used for Māori medicines (Riley 1994).

20.3 SOILS AND THEIR USES AROUND THE TIME OF FIRST EUROPEAN CONTACT

20.3.1 Soils for Horticulture

At the time of the first European contact, Māori already recognized the soil properties and soil types that were most important for the establishment and management of gardens (Best 1925; Leach 1984; Keane 2011a,b). Climate, however, was an extremely important factor for horticulture. New Zealand, with relatively high rainfall, much lower temperatures, shorter seasons, and cold winter seasons than the tropics and subtropics, meant that soils needed to be improved to compensate for the poorer climatic growing conditions.

In conjunction with the required conditions and management of crops, Māori knowledge and Māori taxonomy of soils advanced greatly (Hargreaves 1959, 1963; Macnab 1969; Clarke 1977; Leach 1984; Singleton 1988; Hewitt 1992; Furey 2006; Roskruge 2009, 2011). Māori naming and categorizing of soils was not a systematic taxonomy but apparently intended for the management of root crops. Soil properties and conditions in the temperate climate needed to be improved and many soils were modified accordingly to lift crop success and productivity and extend the range, both geographic/areal and climatic, in which crops could be grown (Clarke 1977; McFadgen 1980; Leach 1984; Singleton 1988; McKinnon et al. 1997; Roskruge 2009, 2011). The soil qualities emphasized most were coarse texture, friable consistence, and fertility (Hewitt 1992). The increasing use of Māori terms for soils and the classification of soils went hand in hand with advances Māori made in horticulture and the planting of crops to increase planting success and yields.

20.3.2 Māori Horticulture

From the 1700s onwards, Māori acquired considerable horticultural and soils knowledge through trial and error (Best 1925; Taylor 1958; Jones 1984, 1986; Leach 1984, 1989; Mitchell and Mitchell 2004b; Horrocks et al. 2008). Agriculture and horticulture were essentially the same thing—subsistence farming of crops, managing natural resources for natural harvests, all in the absence of grazing animals (Keane 2011a,b; Roskruge 2012).

As horticultural knowledge grew, Māori became increasingly aware of the importance of soils, crop selection, crop cultivars, and microclimate, and were able to grow the subtropical sweet potato or kūmara (*Ipomoea batatas*—brought from South America and across to central Polynesia around 700 AD) to near the limit of its climatic tolerance. Some 80 varieties of kūmara were grown. Today's *kūmara*, hardier and more prolific than the original varieties, was introduced by Europeans.

The first crops were mostly tropical in origin and limited in climatic tolerance, and few were suited to the temperate climate of NZ. At first, these plants had a low survival rate. However, new crops became increasingly widespread in small localized gardens in forest clearings, and the staple diet for Māori between ~1600 to the late 1900s regularly included the kūmara (*Ipomoea batatas*), taewa (*Solanum tuberosum*, traditional potato varieties), taro (*Colocasia esculenta*), uwhi or yam (*Dioscorea* sp.), the hue or bottle gourd (*Lagenaria siceraria*, *Lagenaria vulgaris*); these were mainly grown on the soils of the coastal lowlands, floodplains, and low terraces of the North Island and particularly in the north of the South Island in sheltered valleys and plains (Best 1925; Hargreaves 1959; Macnab 1969; Challis 1978; Leach 1984; Molloy 1988; McKinnon et al. 1997; Mitchell and Mitchell 2004b; Furey 2006; Horrocks et al. 2008; WRC 2011; Roskruge 2012).

Māori horticultural practice adhered to elaborate *tikanga* (rules and custom) but was also focused on improving soil drainage and moisture retention, and increasing soil temperature to sustain warmth for crops. Sites for horticulture, therefore, were more sheltered, warmer, and on better-drained, friable soils. Māori seldom used any form of irrigation, so site and landform selection was paramount (Roskruge 2011). Māori also grew crops away from flood-prone areas and avoided foot slopes at the base of steep hills due to higher runoff. Māori cropping and horticulture was commonly located adjacent to and above river and stream terraces, and on raised coastal flats, warm-facing hillsides, and sheltered forest clearings forming suitable microclimates to grow crops. Locations were carefully selected to offer protection from the wind and reduce frost damage risk by providing cold air drainage and constant air movement (Jones 1986; Roskruge 2011). This resulted in slightly higher mean day and mean night temperatures and improved soil moisture retention. Crop rotation was also a common practice and repeated cropping was undertaken for no more than 3 years at most sites (Roskruge 2011).

20.3.2.1 Some Important Horticultural Practices Developed and Applied by Māori

The following practices, to ensure sustainable crop success, were identified by Roskruge (2012):

1. Crop rotation.
2. Controlled burning of fern lands to manage overcrowding and encourage vigorous regrowth and therefore edible fern root production and other regenerating weeds used as greens.
3. Crop storage mechanisms (both storage houses above ground and insulated storage pits below ground).
4. Sophisticated processes developed to transform poisonous or otherwise inedible plants to make them edible, for example, tutu (*Coriaria arborea*) juice had to be strained through finely woven bags to separate it from highly toxic seeds and stems.
5. Managing wild populations of particular plants for use as food, *rongoa* or various utility such as harakeke (flax) para fern or ti.
6. The role of ritual to ensure protection of crops and success. *Tauamata-atua* shrines or stones representing the Ātua Rongo-o-maraeroa were strategically placed and a section of the garden set aside for appeasement of the gods through the first harvest.

20.3.3 Matariki

Traditionally, Māori celebrated a good cropping year and then planned the New Year in the winter months of late May/June/early July, following the harvesting of the year's crops and the planning of new crops for the spring. Māori named this significant time of year Matariki (Tiny Eyes or Mata Ariki—Eyes of God), after the cluster of stars that appears in the pre-dawn sky over NZ to the northeast in late May or early June. Also known as the Pleiades system, Matariki is located at the bottom left of Orion's Belt. The cluster consists of over 500 stars—and usually 6–7 of these stars can be seen clearly to the naked eye. For most *iwi/hapū*, traditional Matariki celebrations, representing the end of one phase and the beginning of another, take place when the first new moon appears after the Matariki have risen and can be seen.

20.4 SOILS AND THEIR USES IN RECENT TIMES

20.4.1 Māori Food Production

Between the eighteenth and early twentieth centuries Māori settlement throughout NZ became increasingly confined to lowland areas, such as river valleys, near river mouths, low terraces, floodplains, coastal terraces, coastal sand country and estuaries, and the edges of lakes. As the Māori population grew, settlements increased, and higher levels of food production and crop yield were needed. By around 1840, the Māori population was estimated at around 100,000–200,000 (King 2003), and Māori agriculture started to increase rapidly in area with a variety of new and old crops (Hargreaves 1959, 1963; Macnab 1969; Leach 1984; Singleton 1988; Roskruge 2011, 2012). Māori increasingly developed knowledge of pests and disease of crops (Roskruge et al. 2012). In the mid-1800s, kūmara was still an important part of the Māori diet, but other European crops (e.g., potatoes, maize, corn, hops) were becoming increasingly popular and economic. Maize was an enjoyed delicacy and by the early 1840s, major potato gardens were evident on many flats in the North and South Islands of NZ. "By 1860 large areas of land in the Waimea, Whakatu, Motueka, Marahau, Riwaka and Golden Bay were beginning to be cultivated with crops like corn and potatoes" (Challis 1978; Mitchell and Mitchell 2004a,b). Farming of animals became a common practice after about 1850 throughout many parts of NZ.

20.4.2 Māori Soil Classification

Soil classification proper began in NZ with the Māori gardeners and horticulturalists (Best 1925; Hewitt 1992; Roskruge 2009) focused on food production. Terminology was often hierarchical, and became progressively more specific and descriptive at the *iwi/hapu* level, denoting specific and local knowledge of soil characteristics and cropping needs at certain localities and places. Māori terms of soil texture and soil types are given in Tables 20.1 and 20.2.

The list of traditional soil names (Table 20.2) was first compiled by Best (1925) and is not comprehensive. Many variations in names occur between regions and iwi/hapū groups. Based on location, and specific descriptions from *iwi/hapu* oral knowledge and literature, Table 20.3 has been compiled showing the "best fit" Māori correlation to the modern NZ soil classification (Hewitt 1998).

20.4.3 Improving Soils for Cropping and Horticulture

Māori realized that to benefit horticulture they had to improve soils in various ways (Best 1925; Challis 1976, 1978; Clarke 1977; McFadgen 1980; Leach 1984, 1989; Singleton 1988; Gumbley

TABLE 20.1
Māori Terms and Descriptors for Soil Texture

Māori Term	Description	Māori Term	Description
Kenepuru	Silt, sandy silt, fresh alluvial deposit	*One*	Sand
One pārakiwai, parakiwai, parahua	Silt	*Onepū, pū*	Sand, loose sand
Oneuku, uku, hāmoamoa	Clay	*Kirikiri, tahoru*	Sand
Keretū	Clay	*One tai*	Sandy soil, coastal/alluvial
Keremātua	Stiff clay	*One hunga*	Marine and estuarine sand, sandy beach, sometimes mixed with mud
One wawata	Lumpy soil	*Poharu, oru, paru (mud for dyeing fiber)*	Mud
Kerepehi, kerepei	Clod of earth or clay loam	*Kaitara, taratara*	Coarse
One matua	Loam	*Pōhatu, kāmaka, kōhatu*	Stone
One haruru	A light but good sandy loam	*Pōhatuhatu*	Stony
Oneone punga	Light spongy soil, lacks substance	*One pākirikiri*	Soil containing gravel
		One kōkopu	Gravel, or very gravelly soil
		Kirikiri, matakirikiri	Gravel, gravelly sand
		Pōhatu	Pebble
		Tokatoka	Rock—hard
		Pōhatuhatu	Rocky
		Toka	Boulder

Source: Adapted from Best, E. 1925 (reprinted 1976). Māori agriculture: The cultivated food plants of the natives of New Zealand: With some account of native methods of agriculture, its ritual and origin myths. *Dominion Museum Bulletin No. 9*. Wellington: A.R Shearer, Government Printer. pp. 42–43; Williams, H.W. 1975. *A Dictionary of the Māori Language*. 7th ed. Wellington, New Zealand: A.R. Shearer, Government Printer. 507p.; Hewitt A.E. 1992. *Australian Journal of Soil Research* 30:843–54; Roskruge, N. 2012. *Tahua-roa: Food for Your Visitors. Korare: Māori Green Vegetables Their History and Tips on Their Use*. Palmerston North: Institute of Natural Resources, Massey University. 110p.

TABLE 20.2
Māori Terms for Soils Types

Māori Term	Description	Māori Term	Description
Oneone	Soil, earth general term	*One, onepū, pū*	Sand, loose sand, sandy soil
One pārakiwai, parahua	Silt	*Onewhero*	Red sand
One mātua	Loam (suitable for all types of horticulture)	*One tea*	White sand
One haruru	A light but good sandy loam	*One tai*	Sandy soil (near beach)
One takataka	Friable soil	*One rua*	Reddish pumiceous sand
One matā	Dark-colored fertile soil, topsoil	*One para huhu, parahua*	Alluvium
Paraumu	Humus	*Kōtae*	Alluvial soil
One paraumu	Dark, fertile, friable soil	*One kopuru*	Soil found in wet situations
One rere	Good draining soil	*Pāra-ki-wai*	Alluvial soils, sediment carried by water, silt
Momona	Good, fertile land	*One rei*	Peat
One punga	Light spongy soil	*Para*	Fragments, sediment, muddy, waste
Kirikiri tuatara	Fertile brown soil	*Reporepo*	Soft mud
One tuatara	Stiff brown soil, fertile, but requiring sand or gravel to be worked in	*Keretū*	Clay
Tuatara wawata	Brown friable, fertile soil suitable for kūmara	*One uku, uku, hāmoamoa*	Clay, unctuous clay, white or bluish color
One nui	Rich soil consisting of clay, sand, and decayed organic material	*Taioma*	White soil, pipeclay
One kōkopu	Gravel, or very gravelly soil	*Kōtore, te uku*	White clay
One pākirikiri, one-kirikiri	Gravelly soil, soil containing gravel	*Kerewhenua*	Yellow clay
One wawata	Lumpy soil	*Keremātua*	Stiff clay
One puia	Volcanic soil	*Matapaia*	A clay which when baked hard was used as a stone for cooking
One tea	White soil from sandy volcanic ash	*One pohatu*	Stony soil
Hunua	Infertile, high country	*One hanahana*	Dark soil mixed with gravel or small stones, apparently a talus soil
Pākihi	Flat land, dried up and infertile	*One kura*	Reddish, poor soil

Source: From Best, E. 1925 (reprinted 1976). Māori agriculture: The cultivated food plants of the natives of New Zealand: With some account of native methods of agriculture, its ritual and origin myths. *Dominion Museum Bulletin No. 9*. Wellington: A.R. Shearer, Government Printer. pp. 42–43; adapted from Hewitt A.E. 1992. *Australian Journal of Soil Research* 30:843–54.

et al. 2004; Roskruge 2011, 2012; WRC 2011). Modified soils are found in both the North and South Islands and have been described as "Artefact fill anthropic soils," "Māori plaggen soils" (Rigg and Bruce 1923; Challis 1976, 1978; McFadgen 1980; Gumbley et al. 2004), or simply "Māori soils" (Challis 1976). Modification was mainly carried out to increase soil temperature, improve texture and drainage, improve workability and tillage of the soils for cropping, and to

TABLE 20.3
Moāri Terms and the Modern New Zealand Soil Classification

Soil Order	Diagnostic Features	Best-Fit Māori Translation (Examples)
Organic soils (O)	Organic material, thick litters	*One rei* (peat soils)
Gley soils (G)	Reductimorphic near surface to depth	*One mākū* (wet), *one aro* (bog), *one kopuru* (soil in wet conditions)
Ultic soils (U)	Strongly weathered, acid, illuvial horizon	*One keretū, one keremātua, one tino uku* (very clayey)
Podzols (Z)	An E horizon and sesquioxidic/organic illuvial horizon	*One pākihi* (naked land), *one hahore* (barren), *one inanga* (white), *one mā i runga paraumu pango* (white on dark humus)
Allophanic soils (L)	Dominated by short-range order minerals, volcanic	*One matua puia* (volcanic), *one rahoto* (scoria)
Pumice soils (M)	Dominated by pumiceous or glassy material	*One tāhoata* (pumiceous)
Melanic Soils (E)	Low color value, pedal A horizon and pedal high-base-status B horizon	*One pākeho* (limestone), *One karā* (basalt)
Semiarid soils (S)	Semiarid soil moisture regime, very weakly weathered and weakly leached	*One maroke* (arid, dry), *one pakapaka* (arid, dry)
Oxidic soils (X)	Strongly weathered, low-activity clay, fine polyhedral structure friable	*One uku, one uku taioma* (clays)
Granular soils (N)	Strongly weathered, mod-active clay, pedal cutanic horizon	*One kaitara* (coarse), *one tōpata kirikiri tini* (granular, many grains)
Pallic soils (P)	Pale colors, weakly weathered high base status, low sesquioxides, high slaking potential high subsoil density	*One kōmā* (soils—pale) *One tea* (soils—white)
Brown soils (B)	Yellow brown B horizon, low base saturation	*One pākākā, one parauri* (soils—brown)
Anthropic soils (A)	Soils substantially modified by man	*One whakarerekē* (soil modified)
Recent soils (R)	Distinct topsoil, absent or very thin b horizon, no fluid subsurface horizon	*One o nā noa nei, one para huhu* (soil recently developed, e.g., floods)
Raw soils (W)	Topsoil absent or with fluid subsurface horizons	*One tino pāpaku* (soils—very shallow, thin), *onepū* (soils—sandy, loose, raw, weak development)

Source: Adapted from Hewitt A.E. 1998. *New Zealand Soil Classification*. Landcare Research Science series 1. 2nd ed. Lincoln, New Zealand: Manaaki Whenua Press. 133p.

follow *tikanga* best practice management. Three main methods of soil improvement were defined by Roskruge 2012:

- Application of wood/ash plant material as a fertilizer
- Placement of stones around crops to increase soil temperature by improving heat retention
- Addition of sand or gravel to improve soil structure by lightening heavy clay soils

Archaeological evidence for horticulture during pre-European times shows numerous examples of Māori-modified soils. These features are recognized by redeposited mixtures of sand and gravel; garden walls, terraces, and mounds; relative levels of organic matter and effects from mulching or burning; garden implements such as the wooden *kō* used for digging, garden pits; and structures for storage (Rigg and Bruce 1923; Best 1925; Macnab 1969; Challis 1976, 1978, 1991; Clarke 1977; McFadgen 1980; Singleton 1988; Bagley 1992; Gumbley et al. 2004; Furey 2006; Horrocks et al.

2008). In the Waikato region alone, Māori modified about 2000 hectares of soil for growing crops, of which kūmara was the most important (Clarke 1977; Singleton 1988; WRC 2011), with kūmara gardens located near rivers and on river terraces. Typical kūmara plantations ranged in size from 0.2 ha to over 20 ha. Early Māori horticulturalists were aware that clayey soils needed to be graveled to improve friability and drainage. The preferred growing medium for kūmara was light, warm, and sandy soil. Where this was not available, Māori horticulturalists added gravel and sand, and, less commonly, charcoal and shells, to the existing soil to improve texture, consistency, and drainage. Many soils were also mulched with burnt-off vegetation that was first spread over the soils, and then worked in. The use of gravel was to build up a coarser layer on the original soil into which kūmara were planted. Large amounts of gravel were quarried for this purpose, and the holes left from this are known as borrow pits (Best 1925; Clarke 1977; McFadgen 1980; Singleton 1988; Gumbley et al. 2004). Soils were specific to certain crops; other crops such as taro preferred wetter or damper soils and were therefore grown near stream banks, swampy areas, and adjacent to the coast (Best 1925; Furey 2006).

20.4.4 Fertilizer

To improve soil condition, various materials were added, including mulch, seaweed, shells, and ash. Wood ash and charcoal from plants such as mānuka/kānuka (*Leptospermum scoparium*, *Kunzea ericoides*) was also used extensively as a fertilizer on soils (Best 1925; Buck 1950; Roskruge 2009, 2011). Cultivated areas were often weeded and cleared, and weeds and brushwood were then burned, and the ash worked back into the soil as fertilizer. In areas where ashes were used as a fertilizer, the old cultivation areas have been found to be richer in phosphoric acid than in adjacent areas (analysis by Cawthron Institute).

The application of manure and urine near any form of food crop, was, and still is, strongly repugnant to Māori, probably stemming from the central belief system and worldview that the Earth Mother, Papa-tū-ā-nuku, provides food, and it would be unwise to offend her by using something "unclean" and "*tapu*" (sacred). The laws of *tapu* (initially made to prevent human sickness) prevented any form of excrement being associated with food. In many cases, this opposition to manure was entrenched through Māori custom and lore, which continued through several generations. Before Europeans settled in NZ, there were few animals capable of producing manure—but even after stock were first introduced, Māori did not wish to use manure for agriculture. Māori agriculture was thus probably unique in that traditionally manure was never used as a fertilizer.

20.4.5 Treating Effluent and Sewage—Land-Based Treatments

As described above, traditional Māori beliefs mean that using manure or human sewage in soils was repugnant and offensive to Māori, and still is (Awatere 2003; Pauling and Ataria 2010); the only soil amendments used, therefore, were the addition of gravel (Singleton 1988) sand, fish waste, seaweed, or wood ash (Pohlen 1957).

Much concern focused on the treatment and disposal of human effluent, especially where it was discharged into water, and the need to protect significant *mahinga kai* and *wāhi tapu* (Pauling and Ataria 2010). For Māori, there is an overwhelming preference for impure water (e.g., mixed water, polluted water, land effluent, treated sewage, industrial waste) to be treated through land and earth first (Awatere 2003), rather than enter directly into natural water ecosystems. This affirms Papa-tū-ā-nuku (earth mother) as the appropriate filter for impure water (e.g., such as through terrestrial and artificial wetlands), and emphasizes the importance of maintaining the integrity of the *mauri* of each water body, such as a lake, river, or stream (Durie 1994). These issues are of ongoing concern to Māori (Pauling and Ataria 2010).

20.5 CONCLUSIONS

This chapter provides insight into Māori knowledge, their use and management of soils, and how this knowledge continually evolved through time. The acquisition of this knowledge, together with its use and application, emerged from an elaborate set of traditional beliefs, values, and concepts (Harmsworth and Roskruge 2014). Before the coming of Europeans, Māori had an extremely detailed appreciation of landscapes and soils, largely based on spatial description and linked to tribal history. Soil classification proper began in NZ with the Māori gardeners and horticulturalists (Hewitt 1992; Roskruge 2012) focused on food production.

There has been renewed interest in and a revitalization of *mātauranga Māori* (Māori knowledge) in NZ, recognition that it is an important dynamic knowledge base that should be used alongside Western science to help broaden our perspective and worldview. New and innovative knowledge is often created at the interface of combined knowledge systems. In NZ and globally, we are faced with an array of challenging and increasingly complex issues, including deleterious impacts to the health of land, soils, and water ecosystems. To help protect and manage our land and soil environments sustainably, and to maintain and enhance human well-being, we need to respect the importance of different knowledge forms and perspectives, such as indigenous knowledge. To achieve intergenerational sustainability and management of our precious natural resources, such as soil, it is essential to include a range of societal, community, and indigenous values and expectations in all aspects of decision making, planning, and policy.

REFERENCES

Ausseil, A.E., J.R. Dymond, and E.S. Weeks, 2011. Provision of natural habitat for biodiversity: Quantifying recent trends in New Zealand. In *Biodiversity Loss in Changing Planet*, eds. O. Grillo and G. Venora. Rijeka, Croatia: InTech. 318p.

Awatere, S. 2003. *Wastewater Technical Review: Tangata Whenua Perspectives*. Prepared for the Land Treatment Collective, Rotorua: Forest Research. 17p.

Bagley, S. 1992. Summary of Ian Barber's findings. Nelson, New Zealand: Department of Conservation. In H. Mitchell, and J. Mitchell 2004, *Te Tau Ihu o Te Waka: A History of Maori of Nelson and Marlborough.* Vol I, 51p.

Best, E. 1924. Māori religion and mythology. *Dominion Museum Bulletin No. 10.* Wellington: Board of Māori Ethnological Research and Dominion Museum.

Best, E. 1925 (reprinted 1976). Māori agriculture: The cultivated food plants of the natives of New Zealand: With some account of native methods of agriculture, its ritual and origin myths. *Dominion Museum Bulletin No. 9.* Wellington: A.R. Shearer, Government Printer. 315p.

Buck, P. (Te Rangi Hiroa) 1950. *The Coming of the Māori.* 2nd edition. Wellington: Māori Purposes Fund Board/Whitcombe & Tombs. 551p.

Challis, A.J. 1976. Physical and chemical examination of a Māori gravel soil near Motueka, New Zealand. *New Zealand Journal of Science* 19: 249–54.

Challis, A.J. 1978. *Motueka: An Archaeological Survey.* Auckland: Longman Paul. 118p.

Challis, A.J. 1991. The Nelson-Marlborough Region: An archaeological synthesis. *New Zealand Journal of Archaeology* 13: 101–42.

Clarke, A. 1977. Māori modified soils of the upper Waikato. *New Zealand Archaeological Association Newsletter* 20(4): 204–22.

Durie, M. 1994. *Whaiora: Māori Health Development.* Auckland: Oxford University Press. 238p.

Furey, L. 2006. *Māori Gardening: An Archaeological Perspective.* Wellington: Department of Conservation/Te Papa Atawhai. 137p.

Gumbley, W., T.F.G. Higham, and D.J. Lowe. 2004. Prehistoric horticultural adaptation of soils in the middle Waikato basin: A review and evidence from S14/201 and S14/185, Hamilton. *New Zealand Journal of Archaeology.* 25: 5–30.

Hargreaves, R.P. 1959. The Māori agriculture of the Auckland province in the mid-nineteenth century. *Journal of the Polynesian Society* 68: 61–79.

Hargreaves, R.P. 1963. Changing Māori agriculture in pre-Waitangi New Zealand. *Journal of the Polynesian Society* 72: 100–17.

Harmsworth, G. and N. Roskruge. 2014. Indigenous Māori values, perspectives, and knowledge of soils in Aotearoa-New Zealand: A. Beliefs, concepts of soils, the environment and land. In *The Soil Underfoot*, eds. G.J. Churchman and E.R. Landa. pp. 111–126, Boca Raton: CRC Press.

Hewitt A.E. 1992. Soil classification in New Zealand: Legacy and lessons. *Australian Journal of Soil Research* 30: 843–54.

Hewitt A.E. 1998. *New Zealand Soil Classification*. Landcare Research Science series 1. 2nd edition. Lincoln, New Zealand: Manaaki Whenua Press. 133p.

Horrocks, M., W.G. Smith, S.L. Nichol, and R. Wallace, 2008. Sediment, soil and plant microfossil analysis of Māori gardens at Anaura Bay, eastern North Island, New Zealand: Comparison with descriptions made in 1769 by Captain Cook's expedition. *Journal of Archaeological Science* 35: 2446–64.

Jones, K.L. 1984. Dune soils and Polynesian gardening near Hokianga, North Head, North Island New Zealand. *World Archaeology* 16(1): 75–88.

Jones, K.L. 1986. Polynesian settlement and horticulture in two river catchments of the Eastern North Island, New Zealand. *New Zealand Journal of Archaeology* 8: 5–32.

Keane, B. 2011a. Oneone—soils—Uses of soils. Te Ara—the Encyclopaedia of New Zealand, URL: http://www.teara.govt.nz/en/oneone-soils/1 (accessed February–September 2012).

Keane, B. 2011b. Oneone—soils—Soil in Māori tradition. Te Ara—the Encyclopaedia of New Zealand, URL: http://www.teara.govt.nz/en/oneone-soils/2 (accessed February–September 2012).

King, M. 2003. *The Penguin History of New Zealand*. Wellington: Penguin Books. 570p.

Leach, H. 1984. *1000 years of Gardening in New Zealand*. Wellington: Reed. 157p.

Leach, H. 1989. Traditional Māori horticulture: Success and failure in Aotearoa. *New Zealand Agricultural Science* 23: 34–35.

Macnab J.W. 1969. Sweet potatoes and Māori terraces in the Wellington area. *Journal of the Polynesian Society* 78(1): 83–111.

McKinnon, M., B. Bradley, and R. Kirkpatrick. 1997. *The New Zealand Historical Atlas: Visualising New Zealand. Ko Papa-tū-ā-nuku e Takoto Nei*. Auckland: David Bateman & Department of Internal Affairs, New Zealand. 290p.

McFadgen, B.G. 1980. Māori plaggen soils in New Zealand, their origin and properties. *Journal of the Royal Society of New Zealand* 10(1): 3–19.

Mitchell, H. and J. Mitchell. 2004a. *Te Tau Ihu o Te Waka: A history of Māori of Nelson and Marlborough: Vol. I. The People and the Land*. Wellington, NZ: Huia Publishers in association with the Wakatu Incorporation.

Mitchell H. and J. Mitchell. 2004b. Māori horticultural skills and their soils: http://www.theprow.org.nz/Māori-horticultural-skills-and-their-soils/ (accessed February–September 2012).

Molloy, L. 1988. *Soils in the New Zealand Landscape: The Living Mantle*. Wellington: Mallinson Rendel. 239p.

Park, G. 1995. *Ngā Uruora: The Groves of Life. Ecology and History in a New Zealand Landscape*. Wellington: Victoria University Press. 376p.

Pauling, C. and J. Ataria. 2010. *Tiaki Para: A Study of Ngai Tahu Values and Issues Regarding Waste*. Landcare Research Series No. 39. Lincoln: Manaaki Whenua Press. 42p.

Pohlen, I.J. 1957. The history of soil science. In *Science in New Zealand*, ed. F. R. Callaghan, 153–63. Wellington: Reed.

Rigg, T. and J. Bruce. 1923. The Māori gravel soil of Waimea West, Nelson, New Zealand. *Journal of the Polynesian Society* 32(126): 85–93. http://www.jps.auckland.ac.nz/document/Volume_32_1923/Volume_32,_No._126/The_Māorigravel_soil_of_Waimea_West,_Nelson,_New_Zealand,_by_T._Rigg_and_J._Bruce,_p_85–93/p1

Riley, M. 1994. *Māori Healing and Herbal: New Zealand Ethnobotanical Sourcebook*. Auckland: Viking Sevenseas. 528p.

Roskruge, N. 2009. Hokia ki te Whenua: Return to the land. Unpublished PhD thesis, Massey University, Palmerston North, New Zealand.

Roskruge, N. 2011. Traditional Māori horticultural and ethnopedological praxis in the New Zealand landscape. *Management of Environmental Quality* 22(2): 200–12.

Roskruge, N. 2012. *Tahua-roa: Food for Your Visitors. Korare: Māori Green Vegetables Their History and Tips on Their Use*. Palmerston North: Institute of Natural Resources, Massey University. 110p.

Roskruge, N., A. Puketapu, and T. McFarlane. 2012. *Nga Porearea Me Nga Maemate O Nga Mara Taewa. Pests and Diseases of Taewa (Māori potato) crops*. Palmerston North: Institute of Natural Resources, Massey University. 72p.

Shawcross, K. 1967. Fern root, and the total scheme for 18th century Māori food production in agricultural areas. *Journal of the Polynesian Society* 76: 330–52.

Singleton P.L. 1988. Cultivation and soil modification by the early Māori in the Waikato. *New Zealand Soil News* 36: 49–57.

Taylor, N.H. 1958. Soil science and prehistory. *New Zealand Science Review* 16(9–10): 71–79.

Waikato Regional Council (WRC) 2011. Māori and soils. http://soilsdev.waikatoregion.govt.nz/Topic-Basics_Of_Soils/Māori_and_Soils/ (accessed February–October 2012).

Williams, H.W. 1975. *A Dictionary of the Māori Language.* 7th edition. Wellington, New Zealand: A.R. Shearer, Government Printer. 507p.

21 Potash, Passion, and a President
Early Twentieth-Century Debates on Soil Fertility in the United States

Edward R. Landa[*]

CONTENTS

21.1 INTRODUCTION

On the face of it, the history of soil science would appear to have only limited appeal. First, it is history, and to some, like the British schoolboy in the 2006 film *The History Boys*, that reduces it to "just one bloody thing after another." And when it comes to an applied, practical science such as soil science, impassioned disputes do not immediately spring into most people's minds, as they might with the histories of fields such as atomic physics, genetics, and climate science. It is, nevertheless, a history of people, ideas, and complex systems, and as such, an arena not immune from the passion and intrigues of controversy. As well, it deals with the ultimate source of our food—the soil.

By the dawn of the twentieth century, many of the key factors of crop nutrition had been established. For example, carbon dioxide from the air was accepted as the major source of carbon in plant biomass, and the ability of legume crops to assimilate nitrogen gas from the atmosphere and via the action of bacteria in root nodules, convert it to a chemical form that was able to meet the physiological needs of plants was already known (Tisdale and Nelson 1966). But an issue that remained in dispute was the proper assessment of nutrient supply from the soil, and the possible role of other root-zone substances upon plant growth. The debate on this was carried out both within the U.S. Department of Agriculture's Bureau of Soils (USDA/BOS) and between USDA bureaus, and between BOS and the outside scientific community. Soil scientists and others who were deeply

[*] Department of Environmental Science and Technology, University of Maryland, College Park, Maryland 20742, USA; Email: erlanda@umd.edu

invested in their research, or their more parochial interests, waged it in both public forums and behind closed doors. Egos, emotions, and political connections would all come to the fore as the battle played out over the course of more than two decades in an era of emerging concerns about the sustainability of natural resources.

21.2 MILTON WHITNEY AND THE BUREAU OF SOILS

The USDA's Division of Agricultural Soils (later BOS) was ruled by Milton Whitney (1860–1927) from its founding in 1894 until his death (Figure 21.1). A Maryland native who studied chemistry for three years under Ira Remsen at Johns Hopkins University, Whitney held state-level soil science positions in Connecticut, North Carolina, South Carolina, and Maryland before his appointment as head of the federal soil survey and research program (Helms 2002; Helms et al. 2002). Whitney proved to be skilled at securing appropriations,[1] and under his leadership, soil survey activities grew at an astonishing pace. The bureau's budget increased from $15,000 in 1896, to $109,000 in 1902, and to $200,000 in 1907. This rapid growth was all the more remarkable given the steadily declining topographic mapping production rate by the U.S. Geological Survey during the 1880–1930 era (Whitney 1907; Gardner 1998; Thompson 1987). Staffing at the BOS grew during his tenure from 10 in 1895, to 120 in 1903, to 218 in 1927 (Weber 1928; Jenny 1961).

Whitney was a good judge and recruiter of scientific talent, and among his early hires as PhD-level physicists at the BOS were Lyman J. Briggs, who later served as a Presidential adviser on the initiation of the Manhattan Project; Edgar Buckingham, who pioneered theories of soil water flow while at BOS; and Noah Dorsey, who went on to develop methods for the measurement of ionizing radiation. But Whitney was far less skilled at managing the workplace in a manner that encouraged retention of these scientists. Each of these researchers had brief careers with the BOS, and then went on to distinguished careers at the National Bureau of Standards (NBS). At least for Briggs, who rose to become the Director of NBS, the cause was clearly Whitney's view of mathematically sophisticated approaches to soil science as inappropriate for the task at hand (Landa and Nimmo

FIGURE 21.1 Portraits of the players: F.H. King (courtesy of the Soil Science Society of America), Eugene W. Hilgard (courtesy of the Bancroft Library, University of California, Berkeley), Harvey W. Wiley, Clifford Richardson (courtesy of the Chemical Heritage Foundation Collections), Milton Whitney, James Wilson (courtesy of the Special Collections Department/Iowa State University), and Theodore Roosevelt.

2003; Nimmo and Landa 2005). For another Whitney hire—Franklin Hiram King, a seasoned agricultural physicist and hydrologist from the University of Wisconsin, who in 1895, authored one of the first college-level soil science textbooks (King 1895; Simonson and McDonald 1994)—the split was traumatic and life changing.

F.H. King (1848–1911) was hired by Whitney in the fall of 1901, moving from the University of Wisconsin to become the chief of a new, Washington, DC-based research unit (the Division of Soil Management) within the BOS at the start of 1902 (Beatty 1991; Tanner and Simonson 1993). King's expertise was in agricultural engineering[2] and soil water movement.[3] This later experience looked to be a good fit for Whitney's view of soils. Despite his formal training in chemistry, Whitney's firmly held beliefs about the capability of soils for crop production focused primarily on their physical properties—namely, texture (the proportions of sand, silt, and clay) and moisture status.

21.3 "OLD SCHOOL": MILTON WHITNEY AND EUGENE W. HILGARD (1890–1903)

In the summer of 1890, the 30-year-old Whitney, then in South Carolina, began correspondence with Eugene W. Hilgard of the University of California and one of America's most prominent soil scientists, seeking review of his manuscript *Soil Investigations*. Whitney put forth the view that crop yields were based upon a soil's texture and its moisture-holding capacity. Fertilizers were judged to exert their effect by changing the physical arrangement of clay particles—a process termed "flocculation." Specific matchings of given textures with specific crops was deemed to be one of the primary goals of soil investigations. His position negated the value of soil chemical analyses:

> Why will not truck *{i.e.,"truck farming," meaning growing vegetable crops on a large scale for shipment to distant markets}*, tobacco, wheat and grass grow equally well on all soils? It is not so much a matter of plant food as of the texture of the soil. (Whitney 1891, 264)

Whitney's report was published as part of the 1891 annual report of the Maryland Agricultural Experiment Station (AES). Whitney's boss, Henry E. Alvord, the president of the Maryland Agricultural College, asked Hilgard, a friend and confidant, to review it for *Agricultural Science*, the journal of the Society for the Promotion of Agricultural Science, for which Hilgard was an associate editor. Alvord was proud of the fact that the soils division headed by Whitney was the first of its kind at any state AES, and he was obviously hoping for an endorsement of the efforts underway in this fledgling group.

Despite a polite exchange of letters with Whitney in 1890, the substance of the report did not sit well with Hilgard, who, in 1892, published a highly critical review in *Agricultural Science*. Whitney responded in a nondeferential tone, restaking his claim to a "distinct advance in agricultural science," and calling the senior scientist Hilgard "a very able exponent of the view of the 'agricultural old school'" (Amundson 2006, 158).

The following year, Charles Dabney, Whitney's mentor from his days in North Carolina, became the Assistant Secretary of Agriculture under President Grover Cleveland. In 1894, Dabney appointed Milton Whitney as chief of the new Division of Agricultural Soils. Whitney assumed the leadership role for soil investigations at the USDA, bringing with him these strongly held beliefs about the limited value of soil chemical analyses in considerations of crop yield (Jenny 1961; Tanner and Simonson 1993; Gardner 1998; Helms et al. 2002; Amundson 2006; Willis 2007).

21.4 BULLETIN NO. 22

In the summer of 1903, Whitney and his aide-de-camp, physical chemist Frank Cameron, published BOS Bulletin no. 22, *The Chemistry of the Soil as Related to Crop Production*, whose data they purported showed:

> ...contrary to opinions which have long been held, that there is no obvious relation between the chemical composition of the soil as determined by the methods of analysis used and the yield of crops, but that the chief factor determining the yield is the physical condition of the soil under suitable climatic conditions. (Whitney and Cameron 1903, 13)

The *de facto* stamp of USDA approval that accompanied this publication outraged Hilgard, who wrote a long rebuttal that was published in *Science* on December 11, 1903.

> Were such statements to emanate from a private laboratory, on a mere personal responsibility, it would be likely to be passed over and allowed to run its course. But when it emanates from the head of the Bureau of Soils in the United States Department of Agriculture, and is expressly and persistently given as the opinion of that bureau, it can not be thus passed over unchallenged. (Hilgard 1903, 755–756)

Hilgard argued, among other points, that the analysis of nutrients in a soil water extract was not indicative of plant-available nutrients in the soil, noting that the role of organic acids secreted by roots and acting as potent agents of nutrient release, was ignored in this approach. In his view, Whitney and Cameron had set up a straw man of no merit—the failure of this water-extraction method did not discredit chemical methods (which Hilgard and most agricultural chemists admitted were not yet developed to the stage of adequately assessing plant available nutrient levels in soils), nor, by default, support the view that physical properties of the soils were the chief determinant of crop yield.

During the next few years, other prominent soil scientists from the United States and Great Britain joined Hilgard in criticism of Bulletin no. 22 (e.g., Russell 1905). But the person most immediately and severely impacted was Whitney's subordinate at the BOS, F.H. King. Just two years earlier, in November 1901, King had left a chair in agricultural physics at the University of Wisconsin (National Cyclopaedia of American Biography 1926) for the seemingly golden opportunity to do soils investigations on a larger scale, and a better-staffed and better-funded level, at an up-and-coming federal research institution. And indeed, that initially was the case. King led a series of multistate experiments that are impressive in their scope and complexity, even by today's standards (King 1905). But things were to change dramatically in 1903–1904:

1. King objected to the interpretation offered in the Bulletin no. 22, and asked that his name be omitted as an author.
2. On January 18, 1904, within 6 months of the release of the Bulletin, Whitney requested King's resignation (Jenny 1961; 97, 101, 137). His separation from the BOS officially occurred on June 30, 1904 (Whitney 1905a).

This above story represents the public telling to-date of the Whitney–King controversy in the soil and plant science literature. Other facts, however, emerge from correspondence housed at the National Archives in College Park, Maryland (http://www.archives.gov/), and the Wisconsin Historical Society in Madison (http://digital.library.wisc.edu/1711.dl/wiarchives.uw-whs-wis000nc). These sources reveal a back-story of openly hostile workplace relations before the firing, and a behind-the-scenes, postfiring struggle that played out in academic and government circles.

21.5 BLIND-SIDED

The period between January 18 and June 30, 1904 must have been an exceptionally tense period in the halls of the BOS. King prepared a series of six data-rich bulletins describing work during the 1902 and 1903 field seasons, submitting the last two to Whitney around June 27, 1904 (King 1904a). On the evening of Friday, July 1, 1904, the day following his official separation from the BOS, and perhaps his last day in Washington, D.C., before returning home to Madison, Wisconsin, King met with Secretary of Agriculture James "Tama Jim" Wilson. He learned that Whitney had, that day, prepared a transmittal letter to the Secretary of Agriculture regarding the six submitted bulletins (King 1904a; Wilson 1904b). Typically, a *pro forma* paragraph of bureaucratic approval, Whitney instead recommended that only three of the papers be published by USDA, and that these should be prefaced with the disclaimer that the conclusions drawn "must be considered as the personal views of the author, and in the main, do not carry the indorsement [*sic*] of this Bureau." As for the other three papers, the BOS Chief deemed them "not sufficiently mature to justify departmental publication of the results so far obtained, and in their present form." Whitney recommended that USDA not publish them, but that King be allowed to submit them to journals if he chose to do so (Whitney 1905a).

The conclusions drawn by King in the three accepted reports, destined to be published in combined form as Bulletin no. 26, did indeed contradict what Whitney and Cameron stated in Bulletin no. 22. The studies carried out by each side relied on the measurement of water-soluble nutrients in soils (Figure 21.2). Massive amounts of data were presented in both bulletins. Statistical testing was not a routine part of the scientific process of this era[4]—data were considered by investigators and then conclusions voiced. Thus, the type of acceptance/rejection criteria that we typically use today to evaluate such large data sets was not operative. Whitney and Cameron tended to average nutrient concentrations across different soils, and then focus on mean concentrations in the soil solution, while downplaying the wide range of values observed (see, e.g., Whitney and Cameron 1903, 37). Whitney and Cameron saw these mean concentrations as representative of both soils broadly classed as either productive or unproductive. In their view, "practically every soil contains all the common rock-forming minerals" (Cameron 1911), and thus, when contacted by soil moisture, these sparingly soluble substances impart to the soil solution the same type and quantities of solutes. They saw no relation between the chemical composition of the soil/water extract ("the soil solution") and the yield of crops.

In contrast, King saw the differences as real—for example, among eight soil types sampled in four states, the total salts (including the essential nutrient ions potassium (potash), calcium, magnesium, nitrate, phosphate, sulfate, and other ions (bicarbonate, chloride, and silica)) recovered from the top 120 cm (4 feet) of the soil, ranged from 2400 kg/ha for the Norfolk sand to 6900 kg/ha for the Janesville loam (King 1905, 65). Furthermore, he tied these nutrient amounts to quantitative crop yield data, presenting graphs of corn and potato yields that closely paralleled the potash and potash + nitrate + phosphate concentrations in the water extract of the respective soils (King 1905 (Figures 7 and 8; 119, 121)). He noted, with caution, not wishing to generalize too far without further investigation, that:

> ... there is a well-marked tendency for larger amounts of water soluble salts to be recovered by the methods adopted from the soils upon which crops have made the largest yields. (King 1905, 123–124)

The line was drawn between the two camps.[5]

21.6 PUBLISHED BY THE AUTHOR

Working from Madison,[6] King privately published the rejected papers in book form (postpaid price: 50 cents) on August 18, 1904, the title page noting that the papers "have been refused Departmental publication by the Chief of the Bureau of Soils" but are "Published by the author, with permission of the Secretary of Agriculture." King's preface to the volume explained the linkage of these three papers to the ones destined for USDA publication:

> The three papers here presented form but portions of a single investigation systematically planned to throw new light upon important problems in soil management, and the full significance of them, as parts of the whole, can only be seen by considering them in connection with the three papers from which they have been severed in that they were not allowed to appear as Departmental publications.

(a)

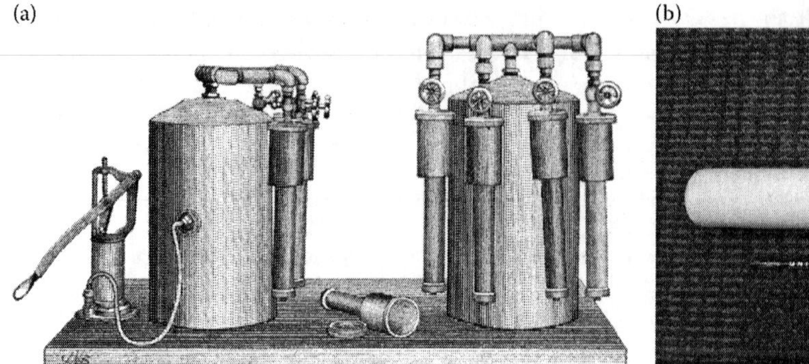

(b)

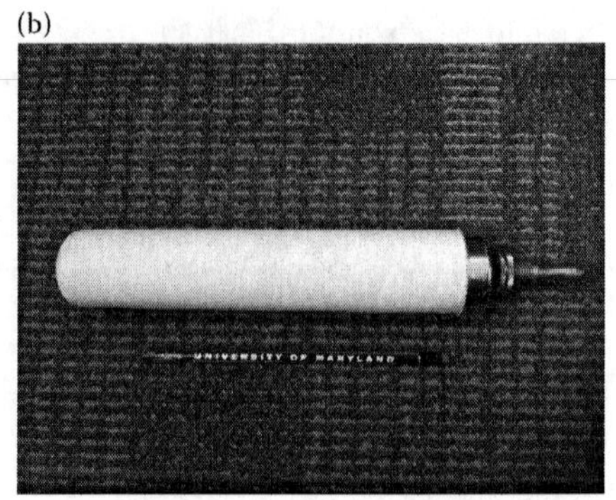

FIGURE 21.2 (a) Device used in the BOS experiments by Whitney, Cameron, King, and coworkers for preparing soil extracts to assess concentrations of nutrients in solution. (From King, F.H. 1905. Investigations of soil management. USDA Bureau of Soils Bulletin no. 26, Figures 2 and 30.) The goal here was to prepare an extract that was free of soil particles. This was a longstanding need in soil chemistry studies; King's pre-BOS studies in Wisconsin used a tea-bag-type initial soil/solution separation using a muslin sack, followed by settling or filtration, presumably using filter paper (King 1899a; King and Whitson 1901, 38). In the latter, BOS studies, a slurry of 100 g of soil and 500 mL of distilled water was mixed for 3 min and then allowed to settle in a Mason jar for 20 min. The supernatant was then decanted into the filtration device shown here. Compressed air was used to force the solution phase through a porous, unglazed porcelain "Pasteur–Chamberland" filter to obtain a clear filtrate for chemical analysis. The hand-powered device, designed by BOS physicist Lyman Briggs, enabled filtration in BOS field laboratories. Briggs had also designed a high-speed centrifuge for the separation of soil water in the laboratory (Landa and Nimmo 2003). This centrifuge was initially tested as the means to obtain actual soil pore water (the true "soil solution"), but the volumes obtainable in reasonable times precluded use of the method on the scale needed for these studies, and the soil/water slurry method was thus adopted as an approximation (Whitney and Cameron 1903). Hardened filter papers of the type used by analytical chemists to separate precipitates were also tried using a vacuum filtration funnel, but clay particle breakthrough was a problem. In the end, the porcelain filters (despite their fragility and need for careful washing and periodic heating to redness with a Bunsen burner to destroy microorganisms plugging the pores) proved to be the best method for the large number of samples handled in the BOS field laboratory (Briggs and Lapham 1902; Whitney 1905c). (b) The Pasteur–Chamberland filter (also called a "Mandler" filter), which found many applications in soil science laboratories during the early-to-mid-twentieth century. In addition to the collection of particle-free filtrates, such filters (also called "porcelain biscuits" (i.e., bisque) (Hilgard 1906), or "filter candles," the latter presumably due to their shape) could be used to concentrate soil clays from settled slurries by application of suction via the metal fitting. It made use of a technology invented in the 1880s by Louis Pasteur's assistant, Charles Chamberland, for the removal of bacteria from water (bacteria are about the same size as clay particles). Filters of this type are still widely used in gravity-fed, household water purification units throughout the world to produce water free from protozoa and bacteria (see, e.g., http://www.consciouswater.ca/gravity-water-filter/). The illustrator (AHB) for the drawing (a) was likely Albertus Hutchinson Baldwin (1865–1944), a classically trained artist with four paintings in the permanent collection of the Metropolitan Museum of Art (Springer and Murphy 2009). Baldwin illustrated other reports for the BOS (see, e.g., cover of *Soil Science Society of America Journal* (2003) 67(3), for a Baldwin drawing of another device designed by Lyman Briggs).

> It is believed that the subjects of the six papers, and the data presented in them, merit adequate discussion, but this was withheld to avoid, as far as possible, antagonizing the published views of the Bureau and the three papers are presented here as they were originally submitted.
>
> In addition to these statements, it is due the writer and his associates in this investigation to say that the data presented have lost very much of fullness and value through changes in plan made in the midst of the investigation but over which we had no control. (King 1904a)

The second paragraph of the preface is a bit ambiguous, but implies that the BOS hierarchy suppressed the discussion of data expected in a research paper. Hilgard clearly portrayed it this way, writing in a letter to *Science* published on November 4, 1904:

> It is to the conclusions deducible from the facts given, then, that we must look for the substance of these papers, and for the possible cause of their falling under condemnation. (Hilgard 1904a, 605)

While King was circumspect in his public writing in the preface as to what specific BOS views were not to be antagonized, Hilgard was direct:

> ... it is easy to see that these results are wholly incompatible with the remarkable utterances of 'Bulletin 22' ... (Hilgard 1904a, 607)

Hilgard apparently ordered a supply of reprints of his *Science* letter for pamphleteering "where they will do the most good" in the campaign against Whitney (Hilgard 1904c).

Whitney seems to have learned about the book around September 21, 1904 from colleagues at Pennsylvania State College (Armsby 1904) and elsewhere.[7] Angered by the preface, Whitney wrote to George Hill, the USDA publications chief, on October 14, 1904, suggesting that Secretary now reverse course and deny USDA publication of all the papers (Whitney 1904c). After consideration of the request by Wilson,[8] Hill wrote to Whitney on November 16, 1904:

> The Secretary instructs me to advise you that in his opinion the objectionable tone of Professor King's preface, above referred to, should not be allowed to make any difference in the conclusion arrived at by you originally, as expressed in your letter of transmittal of July 1st. His conviction is that we ought to go ahead without any reference to Professor King's expressions of dissatisfaction, and that if the papers you selected to recommend for publication originally were worthy of being published by the Department then, they should still be so. (Hill 1904)

Bulletin no. 26 was published by USDA in April 1905.[9]

21.7 THE ROAD TO RESIGNATION

Let us take a step back several years and examine what had led up to this schism. King had negotiated a $4000 per year salary before accepting the 1901 BOS offer—indeed, he was paid more than Whitney (Whitney 1904b). Whitney had indicated that annual research funding of $25,000–$30,000 was likely for King's newly established Division of Soil Climatology (by May of that year renamed the Division of Soil Management) (Whitney 1901b). Soon after his arrival, he was allowed to hire another University of Wisconsin faculty member, Edgar Buckingham, as an assistant (Whitney 1902; Nimmo and Landa 2005), although apparently with some restrictions dictated by Frank Cameron (King 1904b).

King mounted major field efforts with 12–16 junior scientists on his staff during 1902 and 1903 (Rice 1904; Strahorn 1904). Despite the scale of these ongoing research efforts, his budget was cut to $20,000 in November 1902 (King 1903a), and in mid-May 1903, Whitney sent King a stinging letter rejecting a submitted paper on soil fertility as verbose and of little consequence. It went on to accuse him of not yet having developed an organization and research plan, and of being abusive to subordinates. Owing to both the alleged poor performance and a less-than-expected appropriation, King's budget for the 1903 fiscal year was cut to $15,000 (Whitney 1903a).

The correspondence of June 1903 revealed the rift developing between Whitney and King on the interpretation of the soluble salt data that would eventually be published in Bulletins no. 22 and no. 26. Whitney accused King of cherry-picking data from the large sampling in order to make his case, and advised:

> Until these main fundamental differences are found, I do not think it would be wise to publish the minor differences found in your work last year as unless the fact that these are minor differences is very strongly emphasized, it would be misleading, and would be in apparent conflict with what the Bureau will publish to the effect that these chemical differences are not major controlling factors. (Whitney 1903b)

Whitney pushed King to focus his attention on crop yield studies addressing soil moisture, rather than chemical factors.

Faced with a budget shortfall and a large staff of field assistants to whom he was loyal, King—a novice to USDA, but a creative thinker when it came to dealing with its bureaucracy—proposed a novel solution: to pay at least one of the technical staff as a clerk from the general pool of administrative monies. The BOS Chief was not amused nor swayed (Whitney 1903c), and continued to press for the reassignment of at least five of King's research workforce to mapping positions in the soil survey.

Tense exchanges of letters on the staffing for his project and the content of his reports continued. In early August 1903, an exasperated King broke the chain of command and wrote directly to Secretary of Agriculture James Wilson. His 10-page letter, documenting what he saw as a war being waged on him and his staff by Whitney, ended:

> My men look upon the rulings as an effort to kill the Division of Soil Management, and the inference, it must be admitted, is a natural one. (King 1903b)

The appeal apparently fell on deaf ears, and the workplace environment became even more toxic in the coming weeks.

Earlier that summer, at Whitney's request,[10] King reviewed the manuscript version of Bulletin no. 22 and provided comments to the BOS chief. On August 18, 1903, Whitney sent the galley proofs to King, whose name he had added as a coauthor (Whitney 1903d). King's reply of August 20 urged Whitney to await the results of his group's 1903 field season before rushing ahead with publication, and cautioned:

> ... views of such a radical character as those expressed are certain to bring unfavorable criticism unless backed strongly by data which are not only unquestionable in themselves but which are also clearly applicable to the case in hand.
>
> As I have indicated to you several times, I feel that the data you have presented, collected as they have been, will be regarded as having little bearing upon the main propositions, one way or the other, and it was this conviction regarding our own data collected last fall at your request which led me, as you know, to adopt the lines of control which we did in our work this season. Having the convictions here expressed, I feel that it would be a matter of dishonesty on my part to willingly appear as a joint author of the bulletin; and unless there is some other reason more urgent than the mere matter of courtesy to me I shall be obliged to you if you cause my name to be removed from the title page of the bulletin. (King 1903c)

On January 8, 1904, King wrote to Wilson one more time, again pleading his case for intervention by the Secretary to preserve his Division's investigative independence and the jobs of his junior scientists. If no relief could be offered, he indicated that he would resign, noting:

> I came here for the opportunity to work. When this opportunity ceases I am ready to go. (King 1904b)

Wilson, clearly in concert with Whitney, accepted King's resignation on January 18, 1904, with the provisos that King had requested—the use of his laboratory until February 1904, and ample time (until June 30, 1904 was granted, a period of time suggested by Whitney (1904a)), to write up the results for publication (specified by Wilson to a wary King as "submitted to the Chief of the Bureau of Soils") (Wilson 1904a).

21.8 *LET'S HEAR IT FOR THE BOYS*: KING AND HIS ALLIES (1904–1905)

Hilgard and King had been past correspondents, and King obviously knew about Hilgard's December 11, 1903 *Science* letter savaging Bulletin no. 22. King had apparently written to Hilgard at some point circa January 1904 suggesting a petition to Wilson for an investigating committee targeting Whitney. Hilgard seems to have floated the idea among his fellow agricultural experiment station directors. But the timing was bad for King—the directors feared that such action might jeopardize the "Adams Bill" (Hilgard 1904b), a key piece of Congressional legislation, sponsored by Representative Henry C. Adams of Wisconsin, which would provide increased federal research funding to the stations. King's allies in the agricultural sciences, likely led by Hilgard, did however quickly get the word of King's dismissal to Adams, who in addition to his legislative leadership role on this key issue, was the representative from King's home congressional district. These academic allies armed Adams with 25 reprint copies of Hilgard's *Science* paper, and a confidential dialog was then initiated in February 1904 by Adams with King (Adams 1904).

King also began corresponding with his former field assistants. One of these was W.C. Palmer, who had left the BOS to become coowner of a newspaper in North Dakota, *The Minot Optic* (http://history.nd.gov/archives/cities/minot.html). Around June 1904, King apparently got wind of the fact that Whitney had solicited letters from King's staff and associates as to working relations with him. He confirmed this with Palmer, and then began to query people who might have been requested to submit such letters. In the coming months, letters began to arrive in Madison—the early ones were reserved and cautious, but confirmed that written, confidential statements had been requested for the Secretary of Agriculture (Schreiner 1904). King wrote to Secretary Wilson on July 22, 1904; it was brief and to the point:

> I desire very much to know the exact nature of the criticisms which have been submitted to you by men working under me while I was associated with the Bureau of Soils. Will you do me the favor of sending me copies of each of the letters or statements which have been made by the several men who have expressed themselves regarding my relations with them?
>
> I make this request because statements have come to me from a [n]umber [*sic*] of sources which are misleading, untrue, and which are doing me great injustice. (King 1904e)

Wilson replied on July 26, 1904:

> The papers you refer to are confidential reports, made at my request some time subsequent to my acceptance of your resignation. They, therefore, have no bearing whatever upon the severence of your connection with the Department. Only in the event of your having been dismissed from the Department on charges filed would you have a right to copies of the confidential records pertaining to your case. As it is, these records cannot now be given out.
>
> These papers have been held strictly confidential, even inside the Department, and have only been shown, in confidence, to one person, about which you were advised at the time. (Wilson 1904c)

The "one person" was likely Henry Adams, and indeed the Congressman's inquiry to Wilson on King's behalf may well have triggered the solicitation of these statements; indeed King (1904i) asserted this in a December 6, 1904 letter to Hilgard.

The letters were collected from 14 BOS employees by Whitney and transmitted to Wilson. Two of the letters were from November–December 1903 (i.e., before the resignation). These were from a messenger-janitor and a stenographer, both of who wrote to the BOS Chief Clerk A.G. Rice to complain about King's disagreeable personality. The other letters were written between March 30 and April 8, 1904, and came from the scientific staff. They too noted King's irritable personality, authoritarian manner, and treatment of junior colleagues as unskilled laborers rather than as scientific colleagues. One of the staff, J.W. Nelson, wrote of thinking of himself as a "Mechanical Machine," with no opportunity to think for himself (Nelson 1904a).

But on August 2, 1904, King received a letter from W.C. Palmer. Palmer had resigned from the BOS on December 31, 1903 (Anonymous 1904) and moved to Minot. He had remained in correspondence with at least one of King's remaining staff in Washington. That individual—among the group of King's former staff who Palmer referred to as "the boys"—wrote Palmer a confidential letter, which asserted that two letters on the subject of King as a supervising scientist were drafted at home on March 29, 1904 by Frank Cameron. These "models" were then circulated to the BOS staff in Washington for the writing of their own letters. Each letter was then allegedly reviewed and approved by Cameron prior to its signing and submission (Palmer 1904). King began to query "the boys." J.W. Nelson, still employed at the BOS in Washington, was at first scared, writing to King on August 4:

> I can only answer your inquiry by a personal conference and I am not in a position to voice the sentiment of the other men in regard to the matter.
>
> Hoping you will understand the situation . . . (Nelson 1904b)

However, later letters from Nelson and from J.O. Belz[11] to King and his wife were warm and provided gossip about the current inner workings of the BOS, all with an anti-Whitney/Cameron flavor. King (1904i) speculated that inquiries by Congressman Adams to the Secretary of Agriculture on his behalf had triggered the efforts by Whitney and Cameron to secure these letters.

Still in denial of nutrient limitations in soils as a factor in crop yields, the focus of Cameron's and Whitney's research attention in the years following King's departure was on an alternative mechanism—toxic substances in soils and their possible suppression of plant growth, a phenomenon known as "allelopathy." Willis (2007) provides details of this later work at the BOS in a comprehensive history of allelopathy from antiquity to the modern era, and notes:

> Without doubt, the most tumultuous period in the history of allelopathy was that associated with the United States Department of Agriculture (USDA), and its Bureau of Soils, during the first two decades of the twentieth century. (Willis 2007, 209)

Despite the controversy and turmoil in the aftermath of the publication of Bulletin no. 22 and of King's ouster, work in the 1905–1919 era by biochemists at the BOS did produce some valuable information on the nature of soil organic matter.

"The boys," now under the supervision of King's replacement Frank Gardner, were assigned to the toxicity work, which centered upon the lawn at Whitney's home in Takoma Park, Maryland (Schlundt 1908). Writing to King in September 1905 about the new BOS work on the "pernicious principle," Nelson commented:

> It is not my intention to ridicule the work or men in the Bureau; but I write as I have so you will understand actual conditions. I can truthfully say too that there is scarcely a man that does not laugh at the work the Bureau is putting out. We Soil Management men and Mr. Gardner have been trying for a long time to get them to do some practical work. We have been turned down every time but I think they will come to it soon from out side [*sic*] pressure if nothing else. (Nelson 1905)

21.9 SUSPICIOUS MINDS

What King did not know when he came to Washington and the USDA was that he was entering a bureaucracy where all of the bureau chiefs did not know how to play well together. He soon became a player in struggles within the BOS, and as he making his exit, he became involved in the interbureau clashes, in the person of physician and chemist Harvey Washington Wiley (1844–1930). The Pure Food and Drugs Act of 1906, a landmark piece of legislation during the Progressive Era in the United States, was also known as the "Wiley Act," named for the Chief of the USDA Bureau of Chemistry (BOC) from 1883 to 1912. Wiley sought to centralize all chemical work within USDA under the BOC. This got him into turf battles with chiefs in charge of other fiefdoms such as animal industry and forestry, but none more so than his battle with Milton Whitney. As early as 1898, Wiley and Whitney were skirmishing on whose unit should be doing the soils work within the Department. In 1903, Wiley joined forces with Hilgard and others in attacking the technical merits of Bulletin no. 22, but his was an-inside-the-walls-of-USDA assault, with Wiley and the head of the BOC Soil Analysis Laboratory (Charles C. Moore) going to see Secretary Wilson to denounce Bulletin no. 22 as extreme and erratic. Wiley lost that battle, and effective June 30, 1904 (the same day that King left the BOS), Wiley was ordered to cease soil investigations within the BOC (Anderson 1958).

On that last day, he submitted to Wilson a 593-page manuscript summarizing 8 years of soil fertility research conducted by the BOC (Wiley et al. 1904). He intended for this to be published as a BOC Bulletin. The letter of transmittal noted the focus of a portion of the work on "the relative solubility of plant food in soils and its relation to crop production," a descriptor that Wiley clearly chose to link it to the Bulletin no. 22 controversy. It does not appear to have been published by USDA. Wiley served under Wilson during 15 of the 16 years of the Secretary's tenure at the Department of Agriculture, and they clashed on other issues, such as the BOC Chief's autonomy in running the Bureau (Wiley 1929, 279–280). Wiley's assessment of the Secretary's performance and judgment during those 15 years were succinctly stated: "... he had the greatest capacity of any person I ever knew to take the wrong side of public questions" (Wiley 1930, 190–191).

King's correspondence indicates that Wiley met with him just before he left Washington for Madison, Wisconsin in July 1904 (King 1904d). But King seems to have been unaware of the lead-up to this point—the tumultuous events of February–March 1904 involving Clifford Richardson, who Wiley described in his 1930 autobiography as "an associate of sterling character and loyalty" (Wiley 1930, 185).

Clifford Richardson (1856–1932) was a Harvard-trained chemist who served as Wiley's deputy when he became the head of BOC in 1883. He left government service in 1887 and went on to become a leading figure in the asphalt paving testing industry (http://www.asphaltinstitute.org/public/award_winners/ROH/Richardson_Clifford.pdf).

Richardson had close ties to Theodore Roosevelt, having graduated from Harvard in 1877, three years ahead of Roosevelt, and married the sister of the first Mrs. Roosevelt (Alice Hathaway Lee Roosevelt, 1861–1884, who died giving birth to her first child, Alice Roosevelt Longworth) in 1880, remaining close to the future president and his daughter after the death. Early in February 1904, Richardson submitted a brief, typewritten memorandum focused on Bulletin no. 22 to President Roosevelt:

> As an old agricultural chemist and former employee of the Department of Agriculture, my attention has been called to a bulletin, No. 22, of the Bureau of Soils in which there seem to me to have been presented some very remarkable and mischievous conclusions. These conclusions have been vigorously attacked by the leading agricultural chemists of the country and in a most able way by that well known authority on soils, Dr. Hilgard of California.
>
> I think it would be a matter of interest to you to ask the Secretary of Agriculture in regard to the matter and I would very much like to know what his policy is in regard to the Bureau of Soils and what answer his Department will make to the severe criticisms that have been made of its work. (Richardson 1904)

The memo was sent by the President's secretary on February 13, 1904 to Wilson, who then passed it to Whitney. The BOS Chief immediately shifted into crime-scene-investigation mode. On February 17, Whitney (1904a) wrote a four-page reply to the Secretary in which he pointed the finger at Wiley as the likely source of the memo, noting uniquely identifying characteristics of some of the typed letters (i.e., unusually long hyphen, "s" somewhat out of alignment, crossline of "e" slanting slightly upward), and other typing characteristics that were similar to a letter written by Wiley on February 12, 1904.

Wilson confronted Wiley with the charges. Wiley requested a few days to locate Richardson, but clearly feared that Wilson was on a path to dismiss him. After several unanswered letters sent to Richardson's asphalt testing company in Long Island City, New York, Wiley tracked him down on a trip to California. Caught in a trap, he wrote a desperate letter to Richardson on February 25, 1904[12]:

> Great things have been going on since you were here. The memorandum which you prepared in this bureau to give to the President found its way into Whitney's hands and he accuses me of having prepared it in this bureau. I admitted, of course, that it was typewritten in this bureau, but said I had nothing to do with preparing it nor did I know for whom it was being prepared. Whereupon Whitney appeared in my office last Saturday and demanded my resignation, threatening that if I refused to resign to have me dismissed. (Wiley 1904a)

With little subtlety, Wiley gave his recall of events, laying out in his letter to Richardson what he would like to be included in a statement from Richardson to Wilson:

> ... but so far as I can recall I never suggested to you to take the matter to the President in any shape manner nor form...
>
> If I recall the circumstances, in talking with me after the memorandum was prepared, you said that it had been done at the request of the President. If I am correct in this fact, a statement to that effect in your letter to the Secretary would certainly have weight...

He closed the letter saying:

> Your ideas concerning Bulletin 22 were necessarily formed on reading it and especially in reading in connection with it the bulletin of Hopkins {*Cyril G. Hopkins of the University of Illinois, Whitney's most vocal critic*} and the paper of Hilgard. It seems to me that it would be impossible for any informed man to reach any other conclusion after reading the published data above mentioned. (Wiley 1904a)

Richardson apparently read between the lines and wrote what was asked of him in a telegram to the President. Roosevelt then wrote to the Secretary Wilson on March 4, confirming that the memo was written at his request, and indicating "... and of course no charges are to be made against any one about it" (Anderson 1958; 111, 296). Wiley had dodged a bullet. Most people would have breathed a sigh of relief and retreated, but the animosity between him and Whitney was apparently so strong that he was soon drawn back into the fray, now using King as his ticket to the fight.

21.10 A RADICAL PLAN

As noted earlier, Wiley seems to have met with King just prior to his departure from Washington. Responding to a suggestion made by Wiley at that meeting, an obviously anxious and politically inexperienced King wrote from Madison in an obtuse manner, clearly with the view that the less put in writing, the better:

> I have only just reached home and have not yet had the time yet for the conference suggested. Moreover, I feel that there is great danger of doing the wrong thing or doing the right thing in the wrong way, especially for me on account of the relations which have existed and to a certain extent still exist...
>
> Do you feel sure that so radicle [*sic*] a move as the one suggested is the best one to make for the advancement of Agriculture in the U.S. provided it could be successfully made? Do you have any assurance that such a plan would meet with the hearty support of the other Divisions and Bureaus? Do you think wise at this time to let him under stand [*sic*] that you think such a change would be desirable and feasible?
>
> I feel that this is a matter in which haste should be made slowly and with a good deal of discretion. Moreover, I cannot think of doing any thing [*sic*] which can be construed as having a personal end as its aim. (King 1904d)

Who was the conference suggested to be with?—Congressman Adams of his home district seems the likely person (King 1904g). And what was the radical plan? Was it an attack by King on Whitney to be published in *Science*? Was it contact with President Roosevelt through a member of Congress or a trusted adviser? Was it the abolition of the BOS or its "radical reformation" (Wiley 1904b)? Was it a realignment to shift all soils work from the BOS to the state agricultural experiment stations? All of these options were being discussed between King and his allies during the latter half of 1904. On Christmas Eve, 1904, Wiley wrote to King, in a letter marked *Confidential*:

> It has been almost impossible to find out what is going on at the present time. The Agricultural Committee is considering the bill by a sub-committee to which Mr. Adams does not belong. I have had a long talk with him and he is anxious to do the right thing, but does not know exactly how it should be done. He doesn't know the attitude of the sub-committee and will not until the bill is reported to the full committee.
>
> Without my ever mentioning the subject, he stated to me that he thought the soil work should be radically changed and he believed that Mr. King should be put at the head of it. I did not give him any suggestion at all in regard to this matter, either before or after his remark. It shows, however, that he is alive to the situation. (Wiley 1904c)

So was "the radical plan" a reorganization of USDA in which King would dethrone Whitney? The record available beyond this point is unclear. Suffice it to say, this coup did not happen, and King grew despondent over his future in agricultural research. Aside from what seem to be some consulting work in Cuba and Montana (King 1907, 1908b), his employment since leaving Washington appears to have been limited; he wrote and self-published a book on "Ventilation for dwellings, rural schools and stables" (King 1908c), and proposed to the University of Wisconsin Extension Division a plan to develop a set of correspondence courses in agriculture aimed at rural school teachers and their pupils, farmers, and the general public (King 1908a).

On February 21, 1906, in the final week before the passage of the Adams Act that decentralized soils research by its federal appropriations to state experiment stations for their agricultural research activities (Rosenberg 1964), King wrote to Adams looking for the option of a far less radical return to the BOS:

> My suggestion to Professor Henry {*W.A. Henry, Dean of the College of Agriculture at the University of Wisconsin, and a mentor to Adams on agricultural legislation (Rosenberg 1964)*} was to the effect that, as a matter of simple justice to the interests of agricultural investigation, from the view-point [*sic*] of problems of national character, The Department of Agriculture, and especially the Bureau of Soils should now be willing to set aside their jealousies and prejudices and permit me to come back and resume the investigations in Soil Management under conditions of reasonable academic freedom.
>
> So far as my personal feelings are concerned, regarding my relations with the Chief of the Bureau of Soils, there is nothing which would stand in the way of my taking up the work under his direction, and in view of the increased appropriation which the Bureau of Soils is asking, and particularly in view of the very severe criticism, undoubtedly just, regarding the character of results which the Bureau of Soils has thus far published, it seems to me that Congress is justified in directing the expenditure of a portion of that appropriation along lines which seem to it wise. (King 1906)

Nothing came of this reconciliation plan, and Adams died in July 1906.

21.11 CONCLUSION

King died on August 4, 1911. In remarks prepared for reading at the funeral, Cyril Hopkins wrote of him:

> Professor King's work was remarkable. In his mastery of details he had few equals; the thoroughness of his investigation was such that when he completed a piece of work it was finished; and his loyalty to scientific truth was so absolute that he resigned his position as expert for the United States Bureau on Soils—a position in which his salary was higher than that of the Chief of the Bureau—a position in which he was rightly recognized as occupying the foremost place in America as a soil investigator—he resigned that exalted and highly influential and remunerative position because he would not endorse, or even appear to sanction, the drawing of unwarranted (and what proved to be erroneous) conclusions from his own data.

> Agriculture is the basis of all industry and of all industrial prosperity, and the fertility of the soil is the foundation of every form of agriculture. With King's support the erroneous teaching of the Federal Bureau of Soils that soils do not wear out, that fertility of the soil is automatically and permanently maintained, would almost certainly have found general credence, and wide acceptance for a generation, with the resultant untold detriment to this country; King was loyal to science and loyal to truth. (Hopkins 1911)

The soil science building at the University of Wisconsin at Madison was renamed for King in 1934, and largely due to his posthumously published book *Farmers of Forty Centuries* (King 1911), the result of his 1909 trip to Asia (funded from cashing in his life insurance policy; Tanner and Simonson 1993), F.H. King is now seen as visionary in the organic farming and sustainability communities (see Clothier and Kirkham, this volume).

Milton Whitney would stay at the helm of the BOS until his death in 1927.[13] His ability to secure funding for soil survey activities and recruit good talent are a lasting part of his contribution to soil science. Yet sadly, his primary enduring legacy is as a polarizing figure within the soil science establishment. Indeed, his alienation of leaders at the land grant colleges was a strong driver in the 1906 passage of the Adams Act (Jenny 1961, 99–100; Rosenberg 1964).

Whitney's view of soils as an inexhaustible source of plant nutrients, with crop production limited only by *in situ* toxins, never found broad support outside of the walls of the BOS leadership circle, and did not persist beyond his reign there. But an examination of the path to this outcome reveals more general aspects of the nature of large-scale, coordinated science programs emerging in the early twentieth century. Science program leaders—with their individual belief systems, operating styles, and resource allocation—could play a critical role, for better or for worse, in setting research directions. The Whitney versus King/Hilgard/Wiley struggle was not just a fundamental disagreement on data interpretation. It was also a battle of egos, and the exercise of power and domain. The sequence and character of events here is only partially revealed in the publications of the era. The human dimensions emerge from the trove of documents that the players chose to retain, and which archivists have preserved.

ACKNOWLEDGMENTS

I wish to thank Dennis Merkel, Department of Biology, Lake Superior State University, Sault Ste. Marie, Michigan (visiting professor at the University of Maryland-College Park, Department of Environmental Science and Technology (ENST), Spring 2013) for his review of the manuscript. The catalyst for our coming to know each other was ENST Chair William Bowerman. Thank you Bill, for the creating a community-of-interest on the early history of the BOS that was literally just a wall apart.

My thanks to John Swann, historian at the Food and Drug Administration, Rockville, Maryland, for pointing me toward Oscar E. Anderson's *The Health of a Nation*; Samuel Stalcup, University of Oklahoma and former historian at the USDA Natural Resources Conservation Service, for the help in locating the source of the Hugh Hammond Bennett quote; and Ray Weil of the University of Maryland for telling me about the current usage of versions of the Pasteur–Chamberland filter as household water purification devices.

The archive staffs of the Wisconsin Historical Society (Harry Miller), University of Maryland (Anne Turkos), USDA National Agricultural Library (Sara Lee), and the National Archives/College Park provided invaluable and much-appreciated assistance.

NOTES

1. As a 1907 example of Whitney's aggressive pursuit of funding, he confidentially wrote to the editor of a Texas farm and ranch magazine asking him to lobby the Texas congressional delegation to increase funding of the BOS from $200,000 to $500,000, with the *quid pro quo* that such steps would increase soil survey activities in the state (Whitney 1907). The use of interpersonal networks and quasi-public campaigns for funds and legislative authority was apparently common among the bureau chiefs under

Secretary of Agriculture James Wilson as a means of extending the reach, reputation, and impact of the USDA (Carpenter 2001, 226).

2. In combing the Internet for information on F.H. King, one now often sees speculation about his cylindrical-tower storage silo design being the inspiration for Frank Lloyd Wright's design of the Guggenheim Museum in New York City—see for example this 2010 blog post from The Chronicles of Higher Education (http://chronicle.com/blogPost/Happy-Birthday-FH-King-/24605/), or the Wikipedia page on King (http://en.wikipedia.org/wiki/Franklin_Hiram_King. accessed June 14, 2012; flagged with a "who?" for too-vague attribution in January 2012). A query to Neil Levine (Emmet Blakeney Gleason Professor of History of Art and Architecture at Harvard University, and author of *The Architecture of Frank Lloyd Wright*, Princeton University Press, 1997) indicated that he had never heard of any such connection (e-mail to author; June 17, 2012).
3. Despite his main focus as a soil physicist, the kind of data-rich chemical studies that King would initiate at the BOS would have come as no surprise to Whitney. He conducted very similar work during 1899–1902 at the University of Wisconsin (King and Whitson 1901, 1902). These studies of nitrates and other soluble salts showed the greatest yields of corn to be associated with the largest amounts of soluble salts, foreshadowing the findings in King's BOS Bulletin no. 26. The salt concentrations were measured with an electrical conductivity device designed by Briggs and Whitney; they had frequent correspondence with King during 1899–1901, with the BOS providing technical expertise and instruments. King saw the soluble salt measurements as promising, noting in an 1899 letter to Whitney: "I have found your method so convenient a one to use, have been abel [*sic*] to duplicate results so closely from different samples in the field, and it has indicated changes occurring in the soil in so marked a way that I have been very much in hopes that it could be used in some positive way to indicate quantitatively the actual changes which do take place in the amount of soluble salts in a soil under different treatments" (King 1899b).

   King's focus on large data sets was apparent in the University of Wisconsin investigations with Whitson; these 1899–1900 studies included analysis of nitrate in more than 3000 one-foot (30 cm) soil cores (King and Whitson 1901).
4. The statistical testing of data sets to determine if real differences existed between test cases was not common practice in this era. However, much of the practical impetus for the development of such methods would soon come from industries and activities tied to agricultural production and research. For example, the Student *t*-test was developed in 1908 by chemist William S. Gosset for the Guinness brewery in Ireland for quality control in the production of beer (Ziliak 2011), and statistician Ronald Fisher was first employed by noted soil scientist E.J. Russell (author of the classic text *Soil Conditions and Plant Growth)* in 1919 at the Rothamsted Experimental Station in England to retrospectively examine crop yield data collected there in a series of landmark fertilizer-application field experiments begun in 1843 (written communication, Nancy S. Hall, University of Delaware; November 14, 2012)
5. Hilgard summarized his objections to the findings of Bulletin no. 22 in a footnote to his 1906 Soils textbook: "Whitney (Bull. 22, U.S. Bureau of Soils) claims on the basis of a large number of (three-minute) extractions of soils made with distilled water, {1} that these solutions are essentially the same composition in all soils; {2} that all soils contain enough plant-food to produce crops indefinitely; and {3} that the differences in production are wholly due to differences in moisture supply, which he claims is, aside from climate, the only governing factor in plant growth. The tables of analytical results given in Bull. 22 fail to sustain the first contention; the second is pointedly contradicted both by practical experience, and by thousands of years of cumulative culture experiments made by scientific observers; the third fails with the second, except of course in so far as an adequate supply of moisture is known to be an absolute condition of both plant growth, and the utilization of plant-food. It is moreover well known that it is not water alone, but water impregnated more or less with humic and carbonic acids, that is the active solvent surrounding the plant root" (Hilgard 1906, 321–322; {numbers} added above for clarity).

   The last point, critical of Whitney and Cameron's use of distilled water extraction, also cut both ways when it came to Hilgard's ally F.H. King. Hence, the addition of this footnote by Hilgard on the following page of the textbook:

   > The investigations of King (On the Influence of Soil Management upon the Water Soluble Salts in Soils and the Yields of Crops, Madison, 1903 {*sic*; 1904}) show that from some soils at least, a sufficiency of plant-food ingredients for a season's crop may be dissolved by distilled water alone, if the soil be repeatedly leached and dried at 110°. Whether such a supply can be expected under field conditions, remains to be tested. (Hilgard 1906, 323)

   Hilgard's general assessment was that King's water extraction results were worthy of further development to assess the immediate productive capacity of soils, "But that it is not likely to give any definite clew [*sic*] as to the *durability* of such lands" (Hilgard 1906, 333).

6. King was not able to return to his old job as Professor of Agricultural Physics at the University of Wisconsin; that position had been filled in the fall of 1901 by his former assistant, A.R. Whitson (Beatty 1991).
7. Despite separate requests from Whitney and later Cameron, King refused to send either of them copies of the privately published book, noting to Cameron his limited supply and the fact that "... the original mss of these three papers was retained in this office of the Bureau of Soils" (King 1904h). King found Whitney's request for three or four copies to reflect his former boss's "cheek" or impertinent boldness, noting in a October 5, 1904 letter to Hilgard:

> In regard to this request a friend of mine remarked "If you could secure a sufficient area of Professor Whitney's cheek to be used for plating purposes, you could construct a war ship that would defy the world—16-inch Harveyized steel would be nothing compared to it." (King 1904f)

{"Harveyized" refers to the Harvey process, developed in 1890, for producing nickel-bearing, face-hardened steel, largely for use in naval armor.}

Whitney provided King with 100 copies of the original print run of 1000 copies of Bulletin no. 26 (Whitney 1905b). This bulletin seems to have been a popular item and within a month, the supply for official distribution was exhausted. Despite his reluctance to see it published, Whitney immediately requested that an additional 1000 copies to be printed (Arnold 1905).
8. "Tama Jim" Wilson, the Nation's longest-serving cabinet member, served as Secretary of Agriculture from 1897 to 1913 under three presidents (McKinley, T. Roosevelt, and Taft) (Iowa State University 2000). His unprecedented long term in the cabinet no doubt reflected a savvy politician who viewed the public debate emerging with E.W. Hilgard's letter to *Science* of November 4, 1904 as a cautionary signal.
9. The individual papers within the total suite of six papers were designated as B through G on the cover page of the 1904 privately published volume. Papers B (Amounts of plant food readily recoverable from field soils by distilled water), C (Relation of crop yields to amounts of water-soluble plant-food materials recovered from soils), and G (Relations of differences of yield on eight soil types to differences of climatological environment) were those destined to be published by USDA as BOS Bulletin no. 26, and were only noted by title on the cover page. (The letter designation allowed for reference to them in the papers published here, as Bulletin no. 26 was not published until the following year.) The remaining three papers were presented in the book in the order E (Influence of farm yard manure upon the water soluble salts of soils), F (Movement of water-soluble salts in soils), D (Absorption of water soluble salts by different salts). The ordering does seem to have some unspecified logic—going from source of the salts (E), to their lateral and vertical movement (F), and finally to their retention or lack thereof by the solid matrix of the soil (D). However, their rather odd, out-of-alphabetical order designation, self-assigned by King, is not explained, nor is there any hint as to what "A" might have been.

Whitney's stated objection to papers E, F, and D are not detailed in his Letter of Transmittal for Bulletin no. 26, but the three rejected papers are characterized as "preliminary" and "not sufficiently mature to justify departmental publication" (Whitney 1905a, 3). Perhaps the reason for the rejection of papers E, F, and D lies in their direct support of the value of fertilizers in improving crop yields. Whitney had long seen fertilizers as of only indirect importance to plant nutrition, insisting that they exerted their effects by increasing the rate of mineral weathering and improving the structure of the soil (Whitney 1901a), or by neutralizing toxic substances in the soil (Whitney 1909).

This general stance by the BOS against the need to fertilize soils became part of the wider public debate in the coming decade. For example, in an April 1910 speech at the City Club in Chicago, King's ally Cyril Hopkins of the University of Illinois lambasted Secretary of Agriculture James Wilson for supporting Whitney's view on soil fertility. A quote from Whitney's writings in the previous year that got specific attention in the press coverage spelled out his view of the permanency of soil fertility as a national asset:

> The soil is the one indestructible, immutable asset that the Nation possesses. It is the one resource that cannot be exhausted; that can not be used up. (Whitney 1909, 66)

Wilson's counterpunch was to accuse Hopkins of having formed a company involved in the sale of phosphate fertilizers to farmers (The Daily Review 1910).

Whitney's extreme position in this statement did not sit well with some at the BOS. Noted soil conservationist Hugh Hammond Bennett (1881–1960), who began his professional career with the BOS under Whitney in 1903, would later comment.

> When I read that unqualified assertion, I learned that it is possible to put much misinformation into two short sentences. (Bennett 1947, 262)

10. Letters between King and Whitney of September 14–15, 1903 elucidate the timeline of two intersecting events. Bulletin no. 22 was apparently sent up the chain of command at USDA from Whitney to Secretary Wilson, in manuscript form, on July 31, 1903. King's technical review, as a nonauthor colleague, apparently took place in parallel to this bureaucratic review, with him receiving the manuscript that same day. King's later refusal of Whitney's offer to be on the paper as an author occurred on August 20, 1903.

    A short paper, dated July 30, 1903, describing methods used in determining readily water-soluble salts found in soils, and presenting a snapshot of data collected by King's team at their Janesville, Wisconsin field site on May 1, 1903, was submitted by King to the journal *Science*. The paper did not hint at the disagreement that King had with interpretations offered by Whitney and Cameron, but King had not sent the paper to Whitney for review or approval prior to its submission. The BOS Chief only became aware of its existence when it was published in the September 11, 1903 issue of *Science* (King 1903d). On September 14, he wrote to King that besides violating USDA protocols on publication permission, the timing of the paper was offensive, noting the coincidence of the July 30 and July 31 dates:

    > You therefore knew that this publication *[Bulletin no. 22]* was coming out as an official publication of the Department, and having declined to associate your name with this official matter, it was in addition to a violation of discipline, a matter of grave discourtesy for you to have published any associated or related figures until this had appeared, no matter what feelings may have induced you to decline to allow your name to appear on the Department publication. (Whitney 1903e)

    King's contrite reply of September 15 indicated that the paper was prepared and mailed to *Science* several days prior to receipt of Bulletin no. 22 for review on July 31. For a man who we know was meticulous in maintaining correspondence files, he claimed to be unsure of the exact date that he sent the paper, suggesting that July 30 was perhaps the date when it was accepted by the editor or sent to the printer (King 1903e). One might conclude that King was trying to stake a public claim in the arena of water-soluble salts and crop yields prior to the release of the Whitney and Cameron report.
11. J.O. Belz had worked for King as a graduate student in soil physics on the nitrate studies carried out at the University of Wisconsin just prior to King's move to the BOS (see note 3) (King and Whitson 1902, 17).
12. Whitney was able to secure a copy of Wiley's confidential and self-damaging February 25, 1904 letter to Clifford Richardson. In a May 3, 1904 letter to the Secretary of Agriculture, Whitney explains how it arrived in his office via the Dead Letter Office of the Post Office:

    > This correspondence came into my hands in a peculiar way, having been received today from the Dead Letter Office. It seems to have been sent to me as a copy of my letter was on top with the Department of Agriculture, Bureau of Soils, at the top of the sheet, while Dr. Wiley's letter was simply dated from Washington, was not on his letter head paper, and his office title is not given. (Whitney1904d)

    Improbable events do happen, but it seems more likely that there was a mole in Wiley's office.

    The file of materials at the National Archives that included this "Dead Letter Office letter" is an eclectic collection of sensitive materials—for example, the 1904 letters from BOS staff on King and other matters related to King's resignation; materials on the Wiley matter and on BOS employee salaries. While these documents date from 1903 to 1908, they were archived with the 1927–1936 BOS correspondence. It seems likely that these materials were locked away in Whitney's office, only seeing the light of day with that post-mortem clean out by the Chief's office staff.

    While a staple of detective stories and courtroom dramas of the precomputer/printer era of the early- to mid-twentieth century, the examination by criminal investigators of typewritten documents for peculiar, identifying characteristics was still in its infancy in 1904. At least three years prior to the publication of any technical reports on forensic typewriter examination, Sherlock Holmes commented such usage in Sir Arthur Conan Doyle's 1891 short story "A Case of Identity" (Calamai 2008).
13. Following Whitney's death, the BOS was reorganized and merged with parts of Wiley's BOC. Wiley, who left USDA in 1912 to head up laboratories at *Good Housekeeping* magazine, railed against the reorganization in his 1929 book *The History of a Crime against the Food Law*. The remainder of the BOC went on to become the U.S. Food and Drug Administration (FDA). The Harvey W. Wiley Building, headquarters of the FDA Center for Food Safety and Applied Nutrition, sits at the edge of the author's campus in College Park, Maryland. While the 1904 battle with Whitney was damaging to him within the walls of USDA, by the following year, he would emerge as a household name in America for his lead role in the passage of the Pure Food and Drug Act of 1906. For his 20-year campaign in this arena, Harvard University political scientist Daniel Carpenter ranks him above fellow bureau chief, forester and well-known conservationist Gifford Pinchot, as the dominant figure in USDA during the Progressive Era (Carpenter 2001; 2, 442).

REFERENCES

Archive Abbreviations:

NARA, RG 54: National Archives and Records Administration-College Park, Maryland; Record Group 54: Records of the Bureau of Plant Industry, Soils and Agricultural Engineering.

WHS/FHK: Wisconsin Historical Society, Madison, Wisconsin; F.H. King papers; http://digital.library.wisc.edu/1711.dl/wiarchives.uw-whs-wis000nc

Adams, H.C. 1904. Letters to F.H. King; February 26, 1904 and March 22, 1904. WHS/FHK.

Amundson, R. 2006. Philosophical developments in pedology in the United States: Eugene Hilgard and Milton Whitney. In *Footprints in the Soil: People and Ideas in the History of Soil Science*, ed. B.P. Warkentin, 149–165. Amsterdam: Elsevier.

Anderson, O.E. 1958. *The Health of a Nation: Harvey E. Wiley and the Fight for Pure Food*. Chicago, IL: University of Chicago Press..

Anonymous 1904. Status of Prof. King's Division December 31, 1903. NARA, RG 54, General correspondence 1927–1936, box 98, file 35473.

Armsby, H.P. 1904. Letter to Milton Whitney: September 19, 1904. NARA, RG 54, General correspondence 1901–1927, entry 181, box 51, file 5205.

Arnold, J.A. 1905. Letters to Milton Whitney of May 20 & 23, 1905. NARA, RG 54, General correspondence 1901–1927, entry 181, box 51, file 5205.

Beatty, M.T. 1991. *Soil Science at the University of Wisconsin: A History of the Department, 1889–1989*. Department of Soil Science, University of Wisconsin-Madison.

Bennett, H.H. 1947. Development of our national program of soil conservation. *Soil Science* 64: 259–274.

Briggs, L.J. and M.H. Lapham. 1902. Capillary studies and filtration of clay from soil solutions. USDA Bureau of Soils Bulletin no. 19.

Calamai, P. 2008. The real Sherlock Holmes. *Cosmos Magazine*, issue 19, February 2008; available on line at http://www.cosmosmagazine.com/node/1890 (accessed December 28, 2012).

Cameron, F.K. 1911. *The Soil Solution: The Nutrient Medium for Plant Growth*. Easton, PA: The Chemical Publishing Co.

Carpenter, D.P. 2001. *The Forging of Bureaucratic Autonomy: Reputations, Networks, and Policy Innovation in Executive Agencies, 1862–1928*. Princeton: Princeton University Press.

{The} Daily Review (newspaper; Decatur, Illinois). 1910. Stinging criticism and answer by Wilson: Hopkins says Secretary is in error—Latter intimates Hopkins is not disinterested; April 9, 1910; 1.

Gardner, D.R. 1998. The National Cooperative Soil Survey of the United States. Historical Notes no. 7. Natural Resources Conservation Service, USDA, Washington DC (reprint of 1957 Harvard University, Graduate School of Public Administration doctoral thesis).

Helms, D. 2002. Early leaders of the Soil Survey. In *Profiles in the History of the U.S. Soil Survey*, eds. D. Helms, A.B.W. Effland, and P.J. Durana, 19–64. Ames: Iowa State Press.

Helms, D., A.B.W. Effland, and S.E. Phillips. 2002. Founding of USDA's Division of Agricultural Soils: Charles Dabney, Milton Whitney, and the state experiment stations. In *Profiles in the History of the U.S. Soil Survey*, eds. D. Helms. A.B.W. Effland, and P.J. Durana, 1–18. Ames: Iowa State Press.

Hilgard, E.W. 1903. The chemistry of soils as related to crop production. *Science* 18: 755–760.

Hilgard, E.W. 1904a. Soil management. *Science* 20: 605–608.

Hilgard, E.W. 1904b. Letter to F.H. King; February 3, 1904. WHS/FHK.

Hilgard, E.W. 1904c. Letter to F.H. King; September 28, 1904. WHS/FHK.

Hilgard, E.W. 1906. *Soils: Their Formation, Properties, Composition, and Relations to Climate and Plant Growth in the Humid and Arid Regions*. London: Macmillan.

Hill, G.W. 1904. Letter to Milton Whitney; October 14, 1904. NARA, RG 54, entry 181, Bureau of Soils, general correspondence 1907–1927, box 51, folder 5205.

Hopkins, C.G. 1911. Remarks prepared to be read at F.H. King funeral. WHS/FHK.

Iowa State University (Department of Animal Science). 2000. James "Tama Jim" Wilson: 1835–1920. http://www.ans.iastate.edu/history/link/wilson.html (accessed November 13, 2012).

Jenny, H. 1961. E.W. *Hilgard and the Birth of Modern Soil Science*. Pisa: Collana della Rivista "Agrochimica."

King, F.H. 1895. *The Soil: Its Nature, Relations, and Fundamental Principles of Management*. London: Macmillan.

King, F.H. 1899a. Letter to Lyman J. Briggs; September 30, 1899. NARA RG 54, Correspondence received 1894–1901, box 12, folder 2.

King, F.H. 1899b. Letter to Milton Whitney; October 19, 1899. NARA RG 54, Correspondence received 1894–1901, box 12, folder 2.

King, F.H. 1903a. Letter to Secretary (of Agriculture James) Wilson; June 23, 1903. WHS/FHK.

King, F.H. 1903b. Letter to Secretary (of Agriculture) James Wilson; August 5, 1903. WHS/FHK.
King, F.H. 1903c. Letter (draft?) to Milton Whitney; August 20, 1903. WHS/FHK [see copy (final?) in King 1904b].
King, F.H. 1903d. The amounts of readily soluble salts found in soils under field conditions. *Science* 43(454): 343–345.
King, F.H. 1903e. Letter to Milton Whitney; September 15, 1903. WHS/FHK.
King, F.H. 1904a. *Investigations of Soil Management: Being Three of Six Papers on the Influence of Soil Management upon the Water-Soluble Salts in Soils and the Yields of Crops*. Madison, WI: Published by author, 168 p.
King, F.H. 1904b. Letter to Secretary of Agriculture James Wilson; January 8, 1904. WHS/FHK.
King, F.H. 1904d. Letter to H.W. Wiley; July 6, 1904. WHS/FHK.
King, F.H. 1904e. Letter to (Secretary of Agriculture) James Wilson; July 22, 1904. WHS/FHK.
King, F.H. 1904f. Letter to E.W. Hilgard; October 5, 1904. WHS/FHK.
King, F.H. 1904g. Letter to H.W. Wiley; October 31, 1904. WHS/FHK.
King, F.H. 1904h. Letter to F.K. Cameron, November 7, 1904. NARA, RG 54, General correspondence 1901–1927, entry 181, box 51, file 5205.
King, F.H. 1904i. Letter to E.W. Hilgard; December 6, 1904. WHS/FHK.
King, F.H. 1905. Investigations of soil management. USDA Bureau of Soils Bulletin no. 26.
King, F.H. 1906. Letter to H.C. Adams; February 21, 1906. WHS/FHK.
King, F.H. 1907. Letter to C.E. Thorne; May 15, 1907. WHS/FHK.
King, F.H. 1908a. Letter to L.E. Reber; February 22, 1908. WHS/FHK.
King, F.H. 1908b. Letter to C.S. Sheldon; May 29, 1908. WHS/FHK.
King, F.H. 1908c. *Ventilation for Dwellings, Rural Schools and Stables*. Madison, WI: Published by author.
King, F.H. 1911. *Farmers of Forty Centuries; or, Permanent Agriculture in China, Korea, and Japan*. Madison, WI: Mrs. F.H. King.
King, F.H. and A.R. Whitson. 1901. Development and distribution of nitrates and other soluble salts in cultivated soils. University of Wisconsin Agricultural Experiment Station Bulletin no. 85.
King, F.H. and A.R. Whitson. 1902. Development and distribution of nitrates in cultivated soils. University of Wisconsin Agricultural Experiment Station Bulletin no. 93.
Landa, E.R. and J.R. Nimmo. 2003. The life and scientific contributions of Lyman J. Briggs. *Soil Science Society of America Journal* 67: 681–693.
National Cyclopaedia of American Biography. 1926. *Franklin Hiram King*, 19: 292–293. New York: Pames T. White & Co.
Nelson, J.W. 1904a. Letter to Milton Whitney; March 31, 1904. NARA, RG 54, General correspondence 1927–1936, box 98, file 35473.
Nelson, J.W. 1904b. Letter to F.H. King; August 4, 1904. WHS/FHK.
Nelson, J.W. 1905. Letter to F.H. King; September 18, 1905. WHS/FHK.
Nimmo, J.R. and Landa, E.R. 2005. The soil physics contributions of Edgar Buckingham. *Soil Science Society of America Journal* 69: 328–342.
Palmer, W.C. 1904. Letter to F.H. King; July 29, 1904. WHS/FHK.
Rice, A.G. 1904. Letter to Milton Whitney; March 31, 1904. NARA, RG 54, General correspondence 1927–1936, box 98, file 35473.
Richardson, C. 1904. Memorandum to President Theodore Roosevelt (undated); transmitted to Secretary of Agriculture James Wilson by Wm. Loeb Jr. (Secretary to the President) February 13, 1904. NARA, RG 54, General correspondence 1927–1936, box 98, file 35473.
Rosenberg, C.E. 1964. The Adams Act: Politics and the cause of scientific research. *Agricultural History* 38: 3–12.
Russell, E.J. 1905. The recent work of the American Soil Bureau. *Journal of Agricultural Science* 1: 327–346.
Schlundt, H. 1908. Letter to F.H. King; May 17, 1908. WHS/FHK.
Schreiner, O. 1904. Letter to F.H. King; July 12, 1904. WHS/FHK.
Simonson, R.W. and P. McDonald. 1994. Historical soil science literature in the United States. In *The Literature of Soil Science*, ed. P. McDonald, 379–434. Ithaca: Cornell University Press.
Springer, V.G. and K.A. Murphy. 2009. Drawn to the sea: Charles Bradford Hudson (1865–1939), artist, author, army officer, with special notice of his work for the United States Fish Commission and Bureau of Fisheries. Marine Fisheries Review 71 (no. 4); available at http://spo.nmfs.noaa.gov/mfr714/mfr714.html (accessed January 3, 2013).
Strahorn, A.T. 1904. Letter to Milton Whitney; April 3, 1904. NARA, RG 54, General correspondence 1927–1936, box 98, file 35473.

Tanner, C.B. and R.W. Simonson. 1993. Franklin Hiram King—Pioneer scientist. *Soil Science Society of America Journal* 57:286–292; available at http://www.soils.wisc.edu/soils/people_old_newer/emeritus/king/fh_king.htm (accessed November 5, 2012).

Tisdale, S.L. and W.L. Nelson. 1966. *Soil Fertility and Fertilizers*. London: Macmillan.

Thompson, M.M. 1987. *Maps for America*. 3rd edition. Reston, VA: U.S. Geological Survey.

Weber, G.A. 1928. *The Bureau of Chemistry and Soils: Its History, Activities and Organization*. Baltimore: The Johns Hopkins University Press.

Whitney, M. 1891. Report of the Physicist: Soil investigations. In *Fourth Annual Report of the Maryland Agricultural Experiment Station*, pp. 249–296. Maryland: College Park.

Whitney, 1901a. Exhaustion and abandonment of soils. Testimony of Milton Whitney, Chief of Division of Soils before the {United States} Industrial Commission. USDA Report no. 70.

Whitney, M. 1901b. Letters to F.H. King; July 24, 1901 and October 3, 1901. WHS/FHK.

Whitney, M. 1902. Letter to F.H. King; March 24, 1902. WHS/FHK.

Whitney, M. 1903a. Letter to F.H. King; May 13, 1903. WHS/FHK.

Whitney, M. 1903b. Letter to F.H. King; June 12, 1903. WHS/FHK.

Whitney, M. 1903c. Letters to F.H. King; June 17 and 18, 1903. WHS/FHK.

Whitney, M. 1903d. Letter to F.H. King; August 18, 1903. WHS/FHK.

Whitney, M. 1903e. Letter to F.H. King; September 14, 1903. WHS/FHK.

Whitney, M. 1904a. Letter to Secretary of Agriculture; February 17, 1904. NARA, RG 54, General correspondence 1927–1936, box 98, file 35473.

Whitney, M. 1904b. Letter to Secretary of Agriculture; March 31, 1904. NARA, RG 54, General correspondence 1927–1936, box 98, file 35473.

Whitney, M. 1904c. Letter to G.W. Hill (Chief, USDA Division of Publications); March 31, 1904. NARA, RG 54, General correspondence 1901–1927, entry 181, box 51, file 5205.

Whitney, M. 1904d. Letter to Secretary of Agriculture; May 3, 1904. NARA RG 54, General correspondence 1927–1936, box 98, file 35473.

Whitney, M. 1905a. Letter of Transmittal. USDA Bureau of Soils Bulletin no. 26 (F.H. King Investigations in soil management).

Whitney, M. 1905b. Letter to F.H. King, April 19, 1905. NARA, RG 54, General correspondence 1901–1927, entry 181, box 51, file 5205.

Whitney, M. 1905c. Letter to E.O. Fippin, July 11, 1905. NARA, RG 54, PI 66, Entry 216, Reports of soil surveys 1899–1927; file: 1905_ Tomkins Co., New York.

Whitney, M. 1907. Letter to J.H. Connell, December 20, 1907. NARA, RG 54, General correspondence 1927–1936, box 98, file 35473.

Whitney, M. 1909. Soils of the United States. USDA Bureau of Soils Bulletin no. 55.

Whitney, M. and F.K. Cameron. 1903. The chemistry of the soil as related to crop production. USDA Bureau of Soils Bulletin no. 22. Washington DC.

Wiley, H.W. 1904a. Letter to Clifford Richardson; February 25, 1904. NARA, RG 54, General correspondence 1927–1936, box 98, file 35473.

Wiley, H.W. 1904b. Letter to F.H. King; July 18, 1904. WHS/FHK.

Wiley, H.W. 1904c. Letter to F.H. King; December 24, 1904 WHS/FHK.

Wiley, H.W., C.C. Moore and E.E. Ewell. 1904. Comparative fertility and nitrifying power of soils. Bureau of Chemistry draft report (typescript) submitted to the Secretary of Agriculture June 30, 1904; 593 p.; available at http://archive.org/details/harveywashington00wile (accessed March 15, 2013).

Wiley, H.W. 1929. *The History of a Crime against the Food Law; The Amazing Story of the National Food and Drug Law Intended to Protect the Health of the People; Perverted to Protect Adulteration of Foods and Drugs*. Washington, DC: Published by author.

Wiley, H.W. 1930. *An Autobiography*. Indianapolis: Bobbs-Merrill Co.

Willis, R.J. 2007. *The History of Allelopathy (Chapter 10: The USDA Bureau of Soils and Its Influences)*. Dordrecht: Springer.

Wilson, J. 1904a. Letter to F.H. King; January 18, 1904. WHS/FHK.

Wilson, J. 1904b. Letter to F.H. King; July 12, 1904. WHS/FHK.

Wilson, J. 1904c. Letter to F.H. King; July 26, 1904. WHS/FHK.

Ziliak, S.T. 2001. W. S. Gosset and some neglected concepts in experimental statistics: Guinnessometrics II. *Journal of Wine Economics* 6: 252–277.

22 Soil and Salts in Bernard Palissy's (1510–1590) View

Was He the Pioneer of the Mineral Theory of Plant Nutrition?

*Christian Feller** *and Jean-Paul Aeschlimann*

CONTENTS

22.1 INTRODUCTION

How plants obtained their nutrition undoubtedly puzzled farmers and other observers of the countryside in ancient, and also early historical, times. Up to the first half of the nineteenth century the idea largely prevailed that soil humus, that is, the organic component of the soil, was one of the plants' key nutriments and supplied them with part of the carbon they needed, the soil minerals hence being not really fertilizers but only contributing indirectly to their growth. This was indeed Thaer's opinion (Feller et al. 2003), which has been identified as the "humus theory." Based on several previous investigations, Liebig (1840) reversed this conception in showing that the plant nutriments withdrawn from the soil are minerals originating in fact from degraded organic matter or from the soil's mineral constituents and that all of the plants' organic carbon is exclusively derived from air or soil carbon dioxide and not from humus. Liebig's (1840) publication actually produced an abrupt shift of paradigm as regards views on the nutrition of plants. Indeed, this new "mineral theory" of plant nutrition opened up the era of mineral and chemical fertilization (Boulaine 1989, 1992).

Even so, starting in 1840, Bernard Palissy keeps being mentioned as a pioneer of the mineral theory. Who was Bernard Palissy, where does this bold claim originate, and how true is it?

* UMR Eco&Sols IRD, (Ecologie Fonctionnelle & Biogéochimie des Sols & des Agroécosystèmes), SupAgro Cirad-Inra-IRD, 2 Place Viala (Bt. 12), 34060 Montpellier cedex 2, France; Email: christian.feller@ird.fr

Bernard Palissy was the first to advocate the use of the auger for detecting the occurrence of marl in the soil. Since he also clearly formulated the law of restitution to the soil of what has been taken away by the harvests, his many merits in terms of advancing our understanding of the soil remain undisputed.

As a potter, Bernard Palissy is one of the most legendary figures among France's industrial innovators, thanks to his rediscovery of white enamel (Poirier 2008). At the same time, he is also known as an exceptional scientist, owing to his remarkable ability to observe nature; he is considered to be one of the founding figures of modern geology and palaeontology[1] (Cap 1961 and Orcel 1961, in Palissy 1961). Palissy had a keen interest in soils for two main reasons—on the one hand as a key resource of raw materials for manufacturing industries, and on the other with regard to his view on what has been later called the theory of the mineral nutrition of plants. The aim of this contribution is to critically assess these various aspects.

22.2 THE PERSON BERNARD PALISSY

This section is mainly documented after Boudon-Duaner (1999) and Gascar (1980).

We know almost nothing of the life of the very famous Palissy apart from what he wrote himself. He was born in 1510, most probably in the vicinity of Agen (southeast France) and died in 1590 in Paris. He claimed to be of poor extraction, self-taught, without any knowledge of either Greek or Latin. His father was a glass painter and the son made a living as a glassblower. As a young man, he undertook a long journey through France in 1530–1539 during which he studied the various regional cultures and natural features, particularly the soils, rocks, vegetation, and waters.

He married at the age of 29 years and settled in Saintes (eastern France) where he earned his living as a land-surveyor, potter, and stained-glass maker, developing thereby a passion for white enamel. He strove for 15 years to rediscover the industrial secrets of that particular enamel. He forged his own legend, telling how he put all his money into inventing and constructing ovens and furnaces able to reach high temperatures. As there was no money left when he felt his goal was in sight, he went as far as to burning all his furniture and even the floorboards. He finally enjoyed great success, and his beautifully ornamented plates were highly appreciated up to the royal court, creating a genre that has been largely carried on and imitated up to the present day (Amico 1996).

Palissy, however, was not just a craftsman. He was also open-minded and became interested in the Reformation which made its first steps in the region. He was not only one of the first to convert to Calvinism (1545–1546), but also was one of its first preachers and therefore was threatened. Ironically, he was always protected by some of those who fought against his religion. In 1563, he was imprisoned in Bordeaux for heresy (a crime punishable by death penalty) but was saved by Anne de Montmorency, a high royal representative who made him "Inventor of the King's and the Duke's adorned figurines." In the same year, he published a small book titled *The Veritable Recipe* (Palissy 1563), with many discourses about agriculture and discussions on happiness in those days of threat and persecution.

From 1565, following a request from the Queen Mother, Catherine de Médicis, he set up his pottery workshop in the garden of the Tuileries in Paris. Palissy had help from his two sons; his royal figurines sold well and provided a good living. It is worth mentioning that in the course of the 1986 excavations associated with the expansion and modernization of the Musée du Louvre, some extraordinary remains of Palissy's workshop have been discovered in the immediate vicinity to the royal housing.

The role of water in nature attracted much of Palissy's attention. He was curious about both natural caves and man-made grottos. He constructed artificial, ornamented grottos for Montmorency (just north of Paris), and additional ones for other nobles, especially for the Queen in the Tuileries. He decorated these grottos with all sorts of aquatic and terrestrial plants, and animals such as reptiles, shellfish, snails, insects, frogs, fish, and so on, similar to those that were to be found on his famous plates.

Quite apart from the potter, there also was the scientist who observed nature, and dug into the soils to understand their secrets. He had a true calling for research and communication, and in 1575 decided to give public courses in Paris; these were announced by posters which constituted perhaps the first advertising campaign in France. Listeners paid one crown, and the lecturer agreed to give back four crowns if an error in his teaching could be demonstrated, which apparently never happened. From 1575 to 1586, the most famous scientific names, such as Ambroise Paré (who is considered as the father of modern surgery and developed healing techniques for injuries, in particular the clipping of an artery), but also lawyers, artists, architects, physicians, many lords, and members of the royal court attended his courses. Their content most probably made up the bases of his *Admirable Speeches* (Palissy 1580), a treatise on natural history in the form of a conversation between two characters, one called "Théorique" who keeps questioning and referring to the ancients, and the other "Practique" that is, Palissy himself, who discusses his own understanding of the waters, stones, metals, fossils, and salts, as well as of the earth, marl, and agricultural sciences.

Having become now an important figure close to the royal family, the Protestant Palissy had not been worried at all during all those years, not even during the St. Bartholomew's Day massacre (*Massacre de la Saint-Barthélemy*).[2] In 1587, however, he was again imprisoned by the newly formed Catholic League, and sentenced to death. Apparently to protect him, the Catholic King Henri III put him into the Bastille jail where he died in 1590.

22.3 PALISSY AS THE FIRST USER OF THE AUGER IN SOILS

Describing soil profiles is one of the basic activities of the pedologist. To this end, an instrument called an auger has been in use quite extensively from the second half of the nineteenth century. Palissy, however, was the first to provide a detailed description of a soil-sampling auger in his *Admirable Speeches* (Palissy 1880, new edition, 114; Feller et al. 2006). He formulated it dramatically based on questions and answers exchanged by Théorique and Practique (T and P) on the amendment of soils with marl (calcium carbonate-rich mud) to improve crop production.

"T: You treated me with many learned arguments, yet I am not wholly satisfied as to the most expedient way to discover the aforesaid marl.

P: I can advise you as to no more expedient a method than that I should use myself. Should I desire to find any marl in a province where the necessary tool does not exist as yet, I should look for all and any of the augers which the potters, brick-makers, and tile-makers use for their trade, and I should use them to manure a plot in my field to see if the soil had been improved in any way, and then I should seek for one that was long enough, and had a hollow socket at one end, where I would adjust a stick, and across the other end I should make sure it had a handle ... in the shape of an auger, and this done, I should drill and bore away with the full length of the whole handle at all the ditches in my estate, and then I should take the said tool out and examine the concave side to tell which kind of soil had been dug out, and after cleaning the hollow, I should remove the first handle and insert one that was much longer, and I should drill the auger back deeper into the same hole in the soil with this second handle, and thus, having several handles of different lengths, I should determine the nature of the deepest layers; and not only should I dig the ditches of my estate, but I should bore all over the fields, until at length some testimony of the presence of the aforesaid marl were found at the end of my tool, and if it were, I should dig, at the very place it had been found, a sort of well.

T: Indeed. But what if there should be some hard rock beneath the surface? For such is the case in many places the rocks are soft and tender, especially when they are still planted in the ground.

P: In truth that would be tiresome, though in many a place the rocks are soft and tender, especially when they are still planted in the ground."

In this text, the auger is being mentioned as if such a tool was already fairly well known and commonly utilized throughout the country. The instrument had actually been invented for a completely different use, that is, by potters to look for useful clay deposits, and Palissy mainly adopted it for prospecting soils. Additionally, he suggested a new technological development to the auger that consisted in adding several extension pieces to the handle, one after the other, to allow for lower soil layers to be reached and thereby for the soil depth to be adequately measured. Further, he made it clear that the tool had been designed not solely for prospecting soils but more specifically for the detection of marl which he recommended to the farmers as a fertilizer for improving the quality of cultivated fields. It is worth emphasizing that Palissy was excellent as a scientific observer who noted the weathered layer at the top of the rock.

A detailed search into the origins of the term auger (or "*tarière*" which is the French equivalent to "*terebra*," a genus of marine gastropods with a shell shaped like a long, tapering spire; common name: auger snails) in French seems to demonstrate that Palissy was indeed the first to use this tool for prospecting soils. Many other famous scientists of the seventeenth (Olivier de Serres) or eighteenth century (Buffon) also described a similar instrument (also called "gauge" which is the equivalent to the French word "*sonde*") for the investigation of soils (cf. Feller 2007 in Robin et al. 2007).

22.4 PALISSY AND THE SALTS THEORY OF PLANT NUTRITION

During the past centuries most scientists dealing with the concept of plant nutrition have systematically referred either to the theory of Palissy (1777, 1880) and his "salts," or to that of Thaer (1809–1812, 1811–1816, Vol. 2) and his "humus," or to that of Liebig (1840, 1841) and his "minerals." Nowadays most soil science historians tend to believe that Palissy was a brilliant pioneer; Thaer, a disastrous practitioner; and Liebig, the savior.

It is the correctness of this perception of Palissy as a pioneer of the salts theory of plant nutrition that will be examined next in three separate steps.

22.4.1 Grandeau Makes Palissy the Pioneer of the Mineral Theory

In a chapter in his *Chemistry and Physiology as Applied to Agriculture and Silviculture* titled "History of the agricultural doctrines, Liebig's forerunners," Grandeau (1879, p. 33) devoted more than four pages to Palissy in relation to the mineral theory of plants' nutrition. He writes, in particular:

> One can hardly refrain from expressing a deep feeling of admiration for the sagacity of this great character upon reading some of the following fragments excerpted literally from his Treatise on the various salts and of agriculture published in 1563, which could perfectly well have been written by a contemporary scientist.

Palissy's (1777, p. 217) new edition reads indeed as follows:

> The salt lets vegetate and grow every seed. And even if very few persons know the cause for which manure serves the seeds and just bring it for habitude and not for philosophy, the manure that one gives to the fields would be of no use except for the salt that both straw and hay leave therein through rotting. ... the point is quite easy to believe, if you don't believe it, look at the plowman who has brought manure to his field, he will discard it into small piles and a few days later will have distributed all over the field leaving nothing at the very place of the said piles. After such a field has been sown to wheat, however, you will find that the wheat is more beautiful, greener and denser at the places where the piles have been deposited than at any other one, and this happens because the rains that fell on the piles have taken the salt away in passing through and going down into earth, you can thereby recognize that it is not the manure that is the cause of generation but the salt that the seeds have taken from the earth. ... and you will thereby also understand the cause for which all these feces may contribute to the generation of seeds. I say all feces be they from man or from animal. ... if someone sows a field for several years in a row without manuring it the seeds will draw all the salt from the earth for the growth and the earth will hence be deprived of salt and will not be producing any more, this is why one needs to manure it or let it rest for several years. ... I am not speaking about a common salt only, but about vegetative salts.

Grandeau (1879, pp. 33–38) further referred to the first mention of the word "salt" and went on to state:

> Under 'salt' Palissy obviously understood the mineral matters.... Doubt does not seem to be allowed any longer: Palissy had as clear and correct an idea of the necessity of mineral matters for plant nutrition and for him the value of manure resided principally in its mineral contents. ... is not it really extraordinary to discover in a text of the 16th century the foundations of a doctrine (that of Liebig) that appeared very new and in strong contradiction to the current ideas some forty years ago? In writing the history of the theories on the plants' mineral nutrition, I could not but render to the immortal potter the place he deserves and highlight his ideas that are both original and in accordance with what we know today. ... one would in vain look for any sentence remembering Palissy's ideas through the many books on agriculture that have been published between 1580 and 1840.

Grandeau made reference here to some famous contributions of de Saussure (1804) and Davy (1813) for instance, yet surprisingly not to Sprengel (1826 and 1828, cf. Ploeg et al. 1999) at all.

22.4.2 Palissy in Books on the History of Agriculture

A survey of reference works published in this field in chronological order from the nineteenth century for a French audience shows, for instance that:

- Soon after the release of a French translation of Liebig's (1841) publication, Hoefer (1843, p. 91) devoted a full chapter of his *History of the Chemistry from the Earliest up to the Present Times* to Palissy with the following conclusion:

> How much sagacity, judgment and spirit in those few words! Almost three hundred years after Palissy, nowadays experiences have perfectly confirmed these ideas. It is clear that the salts, and in particular the ammonia salts (sulphate, carbonate and hydrochloride) do play the most important role in the action of the fertilizers. One does not recognize here this rigid observer ... but we are still in the 16th century ... apart from some hypotheses as to the ... the generative water considered to be the fifth basic element, there are facts in this treatise that bear testimony of its author's sagacity....

- In his *Short History of Agriculture* Fabre (~1880, Tome 1) made no mention whatsoever of Palissy, but referred explicitly to de Serres.
- Schloesing Fils (~1892) produced a short introduction on plants' nutrition in his *Notions of Agricultural Chemistry* where he stated (relating to Grandeau 1879) that "... Palissy understood the importance of the mineral matters for the nutrition of plants, but he had no follower...."
- Gain (1918, p. 11) gave an account of half a page on Palissy's (1563) *Treatise on the Salts and the Agriculture* and summarized the opinion of the latter as follows:
  - "The origin of the ashes left once plants have burnt is in the soil.
  - To sustain soil fertility one has to return to it what the harvests have removed (law of restitution).
  - The main value of manure resides in its richness in mineral matters that were taken away from the soil by the plant.
  - The feces of both man and animals ought to go back to the cropped soils because they originate from substances that were removed from them at harvest.

    These four axioms still form the bases of our knowledge of the plant's mineral nutrition and justify the use of mineral fertilizers in agriculture. Between Palissy's time and 1840, however, one does not find the slightest reference to the ideas he so precisely formulated in his treatise. Palissy's sagacity hence was three centuries ahead of Liebig's theories."
- André (1924) made a short presentation of the various agricultural doctrines in his *Agricultural Chemistry* and relied on Grandeau (1879, p. 63) to write:

> ... 250 years before Lavoisier, i.e. in 1563 already, Palissy had expressed remarkably correct ideas regarding the fertilizing properties of manures .... His views were not understood neither by his contemporaries nor by his successors for almost 300 years. The notion to restitute to the soil the mineral matter that is being continually exported with the harvests and not the organic matter which originates from the air has been formulated in a very precise way by Palissy. He said that the manure is only useful because of the salt that straw and hay leave through rotting. No doubt that under this term of 'salt' one has to understand this stable residue left over by the plant parts either after complete decomposition or burning....

- In *Growth of Cultivated Plants* Demolon (1946, p. 95) wrote:

> Although Palissy clearly perceived the role of mineral matters in plants' nutrition as of 1563 the question was only precisely defined during the period 1792 (Lavoisier) to 1840 (Liebig).

- Boulaine (1992; cf. also Boulaine and Legros 1998) finally, the most well-known contemporary French historian of agricultural and soil sciences, provided the following comments in his *History of the Pedologists and of the Science of Soils* (Boulaine 1989, p. 29):

> As for the soil, Palissy recorded the "importance of the salts" ... He also said elsewhere "If I knew all the virtues of the salts, I would be able to do marvelous things." Palissy's masterpiece is the *Treatise of various salts and of the agriculture* in which the mineral nature of plants' nutrition in the soil is clearly outlined on the basis of observed experience.

Among the foreign authors some of the references worth emphasizing are:

- Russell (~1940, p. 10) who in a long historical introduction to *Soil Conditions and Plant Growth* mentioned Palissy (1563) and described his presentation as being remarkable. He reproduced a long quotation dealing with manure, the positive effect of burnt straw, the salt stone (i.e., the solid residue left after burning straw or plant piles used at that time by chemists or to complement cattle feed), and more generally salt to conclude:

> For every confirmed hypothesis one will find several that were not and the commencement of the agricultural chemistry must be found later when men had recognized the necessity of experimenting.

- Thanks to his *Agricultural Chemistry* Browne (1944, pp. 29–30) is admittedly one of the most important American historians of agricultural chemistry. He was appreciative of, yet remained perfectly objective with regard to Palissy to whom he devoted a full page just after Paracelsus (whose writings were based on the theory of the four elements: air, fire, earth, and water and who believed in transmutation), indicating in particular:

> He was a younger contemporary of Paracelsus, but to turn from the books of the latter with their accounts of medieval superstitions to the pages of Palissy is like a leap from the Middle Ages into modern times.... He has been called the early founder of agricultural chemistry and the designation is not wholly unmerited when it is considered that Palissy's views upon some phases of the subject anticipated the work of three centuries later.

- Krupenikov is the Russian historian of soil science. He used the following terms regarding Palissy in his *History of Soil Science* (1992, pp. 85–86):

> ... he suggested that plants are fed by the 'salts of the soil' and that the soil is important for them by reason of the fact that it contains salts... Palissy emphasized that plants take different salts from the soil. It is interesting that he also advocated the clearing-burning system of farming as a method of supplying nutrient salts to the soil... Palissy anticipated by almost three centuries the ideas of Liebig who was inclined towards the mineral nutrition of the plants and the necessity of returning to the soil the nutrients taken from it. The salt theory of Palissy was supported 60 years later by G. de Brosse (1621) who confirmed that soil 'without salt is useless for fruit bearing or, more correctly, salt is the father of fertility'.

It transpires quite clearly from all these French and other references that Palissy has actually been considered as a brilliant forerunner of Liebig. The main reason for this appreciation, however, resides in the feeling that Palissy gave to the notion of "salts" the same meaning that has applied to it from the nineteenth century.

22.4.3 What Palissy Really Said and How He Was Understood in His Times

22.4.3.1 Palissy's Conception

The term "salts" occurred very often in all the works of Palissy, who used it with a somewhat surprising definition, as for sugarcane, for instance (Palissy 1880, pp. 29–30):

> ... it's a nodded, hollow grass similar to rye but made like a reed: it is from that grass, however, that the sugar comes from, which is nothing else but the salt... It is true that all salts do not have the same flavor, the same appearance, and the same action, not the same virtue, notwithstanding they are all salts, and I dare say and maintain that there is no plant nor grass species on earth that does not have some form of salt in itself... And what is more, I dare say that if there was no salt in fruits they would not have any flavor, nor virtue, nor odor.

Palissy produced many examples of plants which leave a salt after burning; this is very much a mineral. In other instances it is, however, something different as is the case of the tannins in the oak bark. His conception of the definition and properties of the salts are summarized in the *Admirable Speeches* (Palissy 1880, pp. 299–305) as follows:

> ... let us speak of its virtues that are so great that no human ever knew them perfectly. The salt whitens everything, the salt hardens everything, it conserves everything, gives flavor to everything, it is a cement that binds and cements everything.... The salt rejoices the human beings, it whitens the flesh, provides beauty to the reasonable creatures, keeps alive friendship between male and female because of the vigor it confers to the pudenda: it helps towards the generation, gives voice to the creatures as well as to the metals. The salt allows for several finely ground stones to be combined into a single mass and form glass and all sorts of crockery, with the salt everything may be made almost transparent. The salt lets vegetate and grow all seed.... It is a stable, tangible body ... conserving and generating everything ... wood, plants and minerals. It is an unknown, invisible body like a spirit and nevertheless ... supporting the thing in which it is embedded.

As regards water, Palissy (1880, pp. 266–267) distinguished "common" and "generative" waters:

> The water is the beginning and origin of every natural thing ... the generative water of the human and brutal seed is not common water ... it is the one that germinates all the plants and trees, that supports and maintains their development.

The conception of Palissy is that everything be it plant, animal, or rock partakes of the generative water which is not the common water. This generative water contains salts that are of both organic and mineral origin, and are indeed responsible for all growth phenomena. There is a cycle of the generative water which links soil properties to the growth of plants and to the formation of rocks. According to him rocks and plants are formed from earth thanks to the generative water, which in turn may be returned at some stage to the soil as vegetable or animal ingredients. This generative water is the "quintessence," the so-called fifth element.

22.4.3.2 What Were Salts in the Sixteenth through the Seventeenth Centuries?

For quite some time, the very meaning of the word "salt" remained highly confusing (Viel 1997), as was also the case for the terms "organic" and/or "mineral" with regard to chemical compounds. According to Bram (2005), Lémery (1645–1715) was apparently first to clearly separate the chemicals according to their vegetable, animal, or mineral origin. In the antiquity and up to the late middle ages, any water-soluble material was hence called a "salt," mineral as well as vegetable ones. All "salts" were believed to possess the two properties of marine salt, that is, flavor and solubility (Viel 1997). Hoefer (1866–1869, Tome 2) went even as far as to consider that no chemist before Palissy had ever classified so many substances as salts. The confusion was to last for at least a century after Palissy. One obviously has to consider that alchemy still strongly dominated the science scene of the sixteenth century, and, as a consequence, Palissy's ways of thinking as well. According to Rassenfosse and Gueben (1928) and Bram (2005), the alchemists' philosophy was largely based on Plato's four elements and on Aristotles' four qualities to which (a) the quintessence, and, later on, (b) the three principles (sulfur, mercury, and salt) were added. According to Anonymous (1721, pp. 53–54) the three principles of vegetation were:

> ... the Water, the Salts and the Heat; if the Salts are the soul of the vegetation, Water is needed to dissolve them, and Heat to put them into action .... The Spirit of Life that God distributed everywhere is just a veritable and universal salt....

The whole conception expressing the alchemist view had been particularly well developed by Paracelsus (1493–1541), yet he was using these terms to convey a meaning completely different from modern usage (Rassenfosse and Gueben 1928), that is:

- The sulfur or combustibility (masculine) principle referred to the heat (fire, volatile) and solid (earth, stable);
- The mercury (feminine) principle referred to the humid (water, visible) and to the volatile (air, delicate);
- The salt or quintessence principle represented the ideal mixture of all known principles and allowed for sulfur and mercury to be coherently associated.

Despite his being highly critical toward the alchemists, Palissy was of course wholly pervaded by the current terminology, and thus also employed expressions like "fifth element" for water, differentiated between two sorts of waters and referred to salt as a "principle" (Browne 1944). As Metzger (1969) convincingly demonstrated, it is impossible therefore to equate the modern definition of salt as a purely chemical substance which has nothing in common with any organic compound, with the ancient one of salt principle. The term of salt was only given a more modern sense after the mid-eighteenth century, as in Valmont de Bomare's (1757, pp. 224–239) *Dictionary of Natural History* for instance:

> The natural salts are fossil substances that have the property of dissolving themselves in a more or less large quantity of water, of crystallizing through evaporation...In general chemists differentiate and separate the salts into acidic, alkaline and neutral ... one assumes with some degree of plausibility that the acids are the basis of all the other salts.

Similar definitions were also proposed later by Diderot and d'Alembert (1780, Tome 30, pp. 515–516) and Rozier (1796, Tome 9, pp. 168–169).

22.4.3.3 Palissy in His Followers' Works

A total of 33 direct references to Palissy (cf. Feller 2007) have been recorded from the scientific literature between 1584 and 1776 by Faujas de Saint Fond and Gobet (in Palissy's reedition of 1777), of which 17 may be viewed as neutral, objective accounts on his biography or his publications, that is, neither negative nor particularly laudatory. Except for one somewhat negative statement emanating from a Catholic priest, all the other 15 were hence positive to highly enthusiastic. Whereas most articles up to the mid-seventeenth century dealt with the notions of salt and generative water, the contributions published afterwards concentrated a lot more on the pioneering role of Palissy in the fields of geology, paleontology, and the fossils, and his ideas have been unanimously applauded (Hoefer 1843).

22.5 CONCLUSIONS

To sum up the above investigations one has to give particular attention to the role played by Grandeau (1879) with respect to Palissy. He is certainly recognized as a major specialist and historian of the agricultural sciences (cf. Boulaine and Feller 1985; Boulaine and Legros 1998), yet his reading of Palissy's works were far too selective. He systematically neglected all sentences going into different directions to only keep references that supported his own perception of Palissy considering the salts as being strictly mineral substances that he formulated as:

> "Under 'salt' Palissy obviously understood the mineral matters (. . .)."

Grandeau was hence obviously biased and he manipulated the facts to a large extent with a view to actually project Palissy into a true French forerunner of Liebig with regard to plants' nutrition. He thereby simultaneously exalted himself as the very rediscoverer of a brilliant scientist that had unfortunately been overlooked for centuries, an argument he has developed without even mentioning several just recently published reports that he must have been aware of which contained exactly the same finding (Hoefer 1843, 1866, 1872; Figuier 1870).

As a result there is no reason for considering Palissy any longer as one of the fathers of the mineral nutrition theory of plants, but he should undoubtedly be credited with a clear formulation of the so-called law of restitution. In addition, his many other contributions to our understanding of the soil sciences, such as his pioneering use and modification of the auger, remain undisputed.

NOTES

1. Palissy was a pioneer in modern geology.
   According to Orcel (1961), a famous professor of mineralogy at the "Museum National d'Histoire Naturelle," Paris, France, Palissy was an important pioneer in the following fields:
   - Palaeontology, whereby he explained how fossils found in different locations in the landscape (even at high altitudes) had sedimented naturally in seas or lakes near mountains and not only during flooding periods.
   - Crystallography, whereby he compared the natural formation of crystals in the environment to the crystallization of salt in water.
   - Volcanology, whereby he proposed an explanation for eruptions and earthquakes as being due to the pressurization of water vapor created by a subterranean fire.
   - Water cycle, whereby he tried to understand the origin of mineral water and thermal sources and was first to explain the phenomenon of artesian wells.
2. The St. Bartholomew's Day massacre (*Massacre de la Saint-Barthélemy* in French) occurred in 1572 and was the organized killing of mainly leading Huguenots (French Protestants) by Roman Catholics during the French religious war. The massacre lasted several weeks and, after Paris, affected the whole country. Modern estimates for the death toll vary from 5000 to 30,000 (*Source*: http://www.folger.edu/template.cfm?cid=1624).

REFERENCES

Amico, L.N. 1996. *A la recherche du paradis terrestre, Bernard Palissy et ses continuateurs*. Paris: Flammarion.
André, G. 1924. *Chimie agricole. Chimie végétale I.* 3e édition. Paris: J.B. Baillère.
Anonymous 1721. *La nouvelle maison rustique ou économie générale de tous les biens de campagne*. Tome 2. 3e édition. Paris: C. Prudhomme.
Boudon-Duaner, M. 1999. *Le potier du Roi*. Carrières-sous-Poissy: La Cause.
Boulaine, J. and C. Feller 1985. L. Grandeau (1834–1911) Professeur à l'École Forestière. *Revue Forestière Française* 37(6): 449–455.
Boulaine, J. 1989. *Histoire des pédologues et de la science des sols*. Paris: Inra.
Boulaine, J. 1992. *Histoire de l'agronomie en France*. Paris: Lavoisier.
Boulaine, J. and J.-P. Legros 1998. *D'Olivier de Serres à René Dumont. Portraits d'agronomes*. Paris: TECDOC.
Bram, G. 2005. *Histoire de la Chimie*. http://histoirechimie.free.fr

Browne, C.A. 1944. A source book of agricultural chemistry. In *Chronica Botanica,* ed. F. Verdoorn Vol. 8. MA, USA: Waltham Publisher.

Cap, P.A. 1961. *Œuvres complètes de Bernard Palissy, édition conforme aux textes originaux imprimés du vivant de l'auteur avec des notes et une notice historique par Paul-Antoine Cap.* Nouveau tirage augmenté d'un avant-propos de Jean Orcel. Paris: Librairie Scientifique et Technique A. Blanchard.

Davy, H. 1813. *Elements of Agricultural Chemistry.* 1st edition. London: Longman.

Demolon, A. 1946. *Principes d'Agronomie. Tome 2. Croissance des Végétaux Cultivés.* 3e édition. Paris: Dunod.

de Saussure, T. 1804. *Recherches chimiques sur la végétation.* (Facsimile, 1957. Paris: Gauthiers-Villars). Paris: Nyon.

Diderot, D. and d'Alembert J. 1780. *Encyclopédie ou dictionnaire raisonné des sciences, des arts et des métiers. Tome 30. Article SEL.* Berne, Lausanne: Edition conforme à celle de Pellet, Sociétés Typographiques.

Fabre, L. ca. 1880. *Cours élémentaire d'agriculture pratique appliqué aux contrées méridionales de la France.* 4 vol. Paris: C. Delagrave.

Feller, C., L. Thuries, R. Manlay, P. Robin, and E. Frossard. 2003. "The principles of rational agriculture" by Albrecht Daniel Thaer (1752–1828). An approach of the sustainability of cropping systems at the beginning of the 19th century. *Journal of Plant Nutrition and Soil Science* 166: 687–198.

Feller, C., E. Blanchart, and D.H. Yaalon 2006. Some major scientists (Palissy, Buffon, Thaer, Darwin and Müller) have described soil profiles and developped soil survey techniques before 1883. In *Down to Earth: A Soil Science History,* eds. B.P. Warkentin and D.H. Yaalon, 85–105. Amsterdam: Elsevier.

Feller, C. 2007. Une fausse rupture ou de l'intérêt du retour aux sources en histoire de l'agronomie : l'exemple de la nutrition minérale des plantes et du "genial" Palissy. In *Histoire et Agronomie: entre ruptures et durée*, eds. P. Robin, J.-P. Aeschlimann, and C. Feller, 181–201. Paris: IRD, Coll. Colloques et Séminaires.

Figuier, L. 1870. *Vie des savants illustres. Savants de la Renaissance.* Paris: Hachette.

Gain, E. 1918. *Précis de chimie agricole.* 2e édition. Paris: J.B. Baillère.

Gascar, P. 1980. *Les secrets de Maître Bernard.* Paris: Gallimard.

Grandeau, L. 1879. *Chimie et physiologie appliquée à l'agriculture et à la sylviculture. 1. La nutrition de la plante.* Paris: Berger-Levrault et Cie.

Hoefer, F. 1843. *Histoire de la chimie depuis les temps les plus reculés jusqu'à notre époque.* Tome 2, Section *Palissy*, 72–98. Paris: Hachette.

Hoefer, F. 1866–69. *Histoire de la chimie depuis les temps les plus reculés jusqu'à notre époque.* 2 Tomes. Paris: Didot Frères.

Hoefer, F. 1872. *Histoire de la Physique et de la Chimie depuis les temps les plus reculés jusqu'à notre époque.* Paris: Hachette.

Krupenikov, I.A. 1992. *History of Soil Science. From its Inception to the Present.* New Delhi, Calcutta: Oxonian Press.

Liebig, J. 1840. *Die Chemie in ihrer Andwendung auf Agrikultur und Physiologie.* Braunschweig: Vieweg u.S.

Liebig, J. 1841. *Chimie organique appliquée à la physiologie végétale et à l'agriculture.* Paris: Masson et Cie.

Metzger, H. 1969. *Les doctrines chimiques en France du début du XVIIe à la fin du XVIIIe siècle.* Nouveau tirage (d'après édition de 1922). Paris: Librairie Scientifique et Technique A. Blanchard.

Orcel, J. 1961. *Avant-Propos. In Œuvres complètes de Bernard Palissy, édition conforme aux textes originaux imprimés du vivant de l'auteur avec des notes et une notice historique par Paul-Antoine Cap. Nouveau tirage augmenté d'un avant-propos de Jean Orcel.* Paris: Librairie Scientifique et Technique A. Blanchard.

Palissy, B. 1563. *Recepte véritable par laquelle tous les hommes de France pourront apprendre à multiplier et augmenter leurs thrésors…* La Rochelle: Imprimerie B. Berton.

Palissy, B. 1580. *Discours admirables de la nature des eaux et fontaines, tant naturelles qu'artificielles, des sels et salines, des pierres, des terres, du feu, des émeaux, avec plusieurs autres excellents secrets des choses naturelles; plus un traité de la marne, fort utile et nécessaire à ceux qui se mellent d'agriculture; le tout dressé par dialogues, ès quels sont introduits la théorique et la pratique.* Paris: Martin le jeune.

Palissy, B. 1777. *Œuvres de Bernard Palissy, revues sur les exemplaires de la Bibliothèque du Roi avec des notes de M. Faujas de Saint Fond et Gobet.* Paris: Ruault Libairie.

Palissy, B. 1880. *Les Œuvres de Bernard Palissy publiées d'après les textes originaux avec une Notice historique et bibliographique et une Table analytique par Anatole France.* Paris: Charavay.

Palissy, B. 1961. *Œuvres complètes de Bernard Palissy, édition conforme aux textes originaux imprimés du vivant de l'auteur avec des Notes et une Notice historique par Paul-Antoine Cap. Nouveau tirage augmenté d'un Avant-Propos de Jean Orcel.* Paris: Librairie Scientifique et Technique A. Blanchard.

Ploeg van der, R.R., W. Böhm, and M.B. Kirkham, 1999. On the origin of the theory of mineral nutrition of plants and the law of the minimum. *Soil Science Society America Journal* 63: 1055–1062.

Poirier, J.P. 2008. *Bernard Palissy. Le secret des émaux.* Paris: Pygmalion–Flammarion.

Rassenfosse, A. and G. Gueben. 1928. *Des alchimistes aux briseurs d'atomes*. Paris: G. Douin.
Robin P., J.P. Aeschlimann, and C. Feller (eds.) 2007. *Histoire et Agronomie: entre ruptures et durée*. Paris: IRD, Coll. Colloques et Séminaires.
Rozier, l'Abbé. 1781–1805. *Cours complet d'agriculture théorique, pratique, économique et de médecine rurale et vétérinaire*, 12 vols. (Tome 9, 1796). Paris: Rue et Hôtel Serpente.
Russell, E.J. ca. 1940. *Les conditions du sol et la croissance des plantes*. Translated from the 4th English edition by G. Matisse. Paris: E. Flammarion.
Schloesing Fils, T. (ca. 1892). *Notions de chimie agricole*. Paris: Gauthier-Villars.
Sprengel, C. 1826. Ueber Pflanzenhumus, Humussäure und Humussaure Salze. *Archiv für die Gesammte Naturlehre* 8: 145–220 (cited by Ploeg et al. 1999).
Sprengel, C. 1828. Von den Substanzen der Ackerkrume und des Untergrundes. *Journal für Technische und Ökonomische Chemie* 2: 243–474, and 3: 42–99, 313–352, and 397–421 (cited by Ploeg et al. 1999).
Thaer, A. 1809–1812. *Grundsätze der rationellen Landwirthschaft*, 4 vol. Berlin: Realschuchbuchhandl.
Thaer, A. 1811–1816. *Principes raisonnés d'agriculture*. Translated from German by E.V.B. Crud. 4 vol. Paris: J.J. Prechoud.
Valmont de Bomare 1757. *Dictionnaire raisonné universel d'histoire naturelle*. Nouvelle édition, Tome 8. Paris: Brunet Libraire.
Viel, C. 1997. Histoire chimique du sel et des sels. http://www.tribunes.com/tribune/sel/viel.htm.

Section IV

Technologies and Uses

"While the narrative is deeply rooted that humanity only disturbs, destroys, or even conquers nature, soil scientists are tasked to help humanity become a more sustaining soil-forming agent." Daniel deB Richter and Dan H. Yaalon *Soil Science Society of America Journal* 76:766–778 (2012).

23 Poetry in Motions
The Soil–Excreta Cycle

Rebecca Lines-Kelly[*]

CONTENTS

In nature there is no waste: all the excreta of living things are used as raw materials by others.

—Winblad et al. (2004)

When I was a child, we used to collect sheep dung to fertilize our vegetable garden. With my siblings, I crawled under shearing sheds to gather black pellets piled high from years of frightened sheep waiting their turn for the shears on the boards above. On our hands and knees, we shovelled the hard black peas into bags and dragged them out into the fresh air and sunshine. Back at our house, we poured the pellets onto the vegetable garden and my father dug them into the soil. We knew they helped the vegetables but had no idea how.

Many years later in another town, another state, and a new century, our neighbors invited us over to see the results of a year's deposits in their compost toilet. Intrigued, slightly horrified, ready for nausea, we inspected the wheelbarrow of composted feces and urine. All we could see was a brown crumbly material, similar to used, dried tea leaves. There was no smell, no disgusting mess, nothing to indicate its original source. Our neighbors distributed it around their fruit trees. Within days, the only evidence of its presence was a forest of cherry tomatoes growing from seeds that had survived the composting process. The fruit trees flourished, their leaves glowing with health.

[*] Wollongbar Primary Industries Institute, NSW Department of Primary Industries, Wollongbar NSW 2477 Australia; Email: rebecca.lines-kelly@dpi.nsw.gov.au

My neighbors, I realized, had created their own cycle of nourishment. They nourished their bodies with food. Their excreta nourished the soil. The soil nourished the food they grew and ate. Their bowel motions/movements were part of the poetry of life. Their cycle was not a closed one by any means; they still brought in food from outside, but they were making use of nature's principal fertilizer, urine, and feces. Nature, after all, has designed a very efficient system. Think of animals in the wild: they forage, eat, urinate, defecate, and move on. They leave their wastes on the ground for soil organisms to feed on; the organisms' activities release nutrients for use by plant roots, and build soil structure that allows plant roots to move easily through the soil to obtain nutrients and moisture.

But something has happened to my own circle of nourishment. It is broken at the point where I should be returning my wastes to the soil. I press a button and my urine and feces disappear with water down a pipe and out of my life. In my case, they do not go very far; just to a septic tank near the house that drains into the surrounding soil, so most of the water and nutrients are eventually taken up by nearby plants. The remaining solids, valuable organic matter, are pumped out every couple of years and delivered to a municipal sewerage system so that I do not have the unpleasant experience of sewage overflowing into the garden. I am part of the sewered culture, part of the "flush and forget" culture that allows us to largely ignore our excretions once we have flushed them away. As a result, we have forgotten that they are a vital nutrient resource for our soils. What would it take, I wonder, to reconnect the human excreta–soil and restore nature's cycle of nourishment?

23.1 NUTRIENTS

> All the human and animal manure which the world wastes, restored to the land instead of being cast into the water, would suffice to nourish the world…
>
> **—Victor Hugo (1862)**

Every year, we each excrete, on average, 500 L of urine, and 50 kg of feces (10 kg dry matter), which together contain around 5.7 kg nitrogen, 0.6 kg phosphorus, and 1.2 kg of potassium although the actual amounts vary according to our body weight, the climate we live in, our water intake and our diet, especially protein content. Most of the nutrients are in our urine, which contains 90% of the nitrogen, 50–65% of the phosphorus, and 50–80% of the potassium (Heinonen-Tanski and van Wifk-Sijbesma 2005). Urine from just one person in one year is sufficient to fertilize 300–400 square meters of crop with 50–100 kg nitrogen per hectare (Richert et al. 2010). When you multiply this by the global population, that is a lot of nitrogen fertilizer. Phosphorus is a particularly important excreta nutrient. It is essential for all life, has no substitute in crop growth, and cannot be manufactured or destroyed. Globally, humans consume approximately 3 megatonnes of phosphorus in food every year. Close to 100% of this is excreted, but only an estimated 10% is recirculated back to agriculture or aquaculture. The shortfall is currently made up by mining rock phosphate, but this is a finite quantity, so human excreta represents a valuable phosphorus resource (Cordell et al. 2009) (see also Hall 2014).

Human feces contain fewer nutrients than urine and consist mainly of undigested organic matter, which, when treated for pathogens and applied to soil, can increase soil organic matter content, thus providing food for soil organisms whose activities improve soil structure and water-holding capacity (Winblad et al. 2004).

The nutrient resource of human urine and feces has long been recognized. Ancient Minoans are believed to have irrigated their wastewater on agricultural crops from 3500 BC, as did ancient Greek civilizations a thousand years later (Tzanakakis et al. 2007). China and Vietnam have a long history of composting human excreta for use on their soils (Winblad et al. 2004). In the Middle Ages, there are records of wastewater being used to irrigate crops in Germany and Scotland (Tzanakakis et al. 2007). As urban settlements developed, human excreta was collected at night (hence the term "night soil") and sold to farmers for fertilizer.

"It was a sensible system with much to admire. Nothing was wasted; everything was recycled. The nutrients ingested by humans in food were taken from their cesspits and placed back into land that would grow more food, which would be consumed by more humans, who would in turn produce more useful 'waste.' It was a harmonious recycling loop that also managed to be lucrative. It satisfied the demands of nature and of capitalism" (George 2009). Even today nearly 200 million farmers in China, India, Vietnam, sub-Saharan Africa, and Latin America use human excreta to irrigate and fertilize nearly 49 million acres (20 million hectares) of cropland (Drechsel et al. 2010).

23.2 PATHOGENS

Civilizations over time have perceived defecation as a very lowly necessity.

—Avvannavar and Mani (2008)

We know from history that human excreta has been valued as a soil fertilizer, and is still used as such in some regions, so why is excreta recycling not widespread around the world? The answer lies in the pathogenic contents of feces. While urine is virtually sterile, just 100 mL of feces contain 10^7–10^9 fecal coliform bacteria (Tilley 2008). Feces are the main source for transmission of enteric infectious diseases and parasites. Bacterial pathogens such as *Salmonella*, *Campylobacter*, and enterohemorrhagic *Escherichia coli*, and viruses are significant microbial risks in recycling human excreta. Cultures around the world have evolved practices to avoid fecal contamination. In the Bible, Deuteronomy 23:13 notes "Outside the camp you shall have a place set aside to be used as a latrine. You shall also keep a trowel in your equipment and with it, when you go outside to ease nature, you shall first dig a hole and afterward cover up your excrement." Most cultures are disgusted by the thought, sight, and handling of feces, possibly an evolutionary mechanism to defend the body from pathogens and parasites (Curtis and Biran 2001). "The emotion of disgust is probably common to all humans in all cultures, and serves to help us to avoid those things that were associated with risk of disease in our evolutionary past. Disgust is thus the name we give to the motivation to behave hygienically" (Curtis 2007). Disgust with bodily excretions, and concerns about contamination and disease, were reinforced by child-rearing practices, by religions that mandated strict hygiene practices and, in the 1880s, with the discovery of the microbes themselves, which led "to an almost obsessive fear of disease transmission by pathogen-laden wastewater" (Shuval 2010). But disgust with feces is not absolute. It varies with the type of feces. Animal feces are not generally regarded as disgusting as human feces (Drangert 1998). Decomposed feces such as those in septic tanks evoke less disgust than fresh feces; feces of babies and family members are more acceptable than those of strangers (Curtis and Biran 2001). Disgust and squeamishness are also borne of habit. People used to defecating in the open find the idea of indoor toilets disgusting, while people with indoor toilets find outdoor defecating disgusting. Disgust may be a primal emotion, but the degree of disgust varies according to culture, tradition, and familiarity (Avvannavar and Mani 2008). Disgust with feces does not apply to urine. Many cultures are relaxed about handling and using urine, possibly because it is essentially sterile. Urine has many uses, including therapeutic drink, antiseptic, insecticide, and production of gunpowder, detergent, dye, and fertilizer (Drangert 2004).

From earliest times, communities have used water to protect themselves from fecal pathogens, often developing elaborate sewerage and drainage systems to remove human excreta from residential areas (Tzanakakis 2007). Modern sewerage systems have their origins in nineteenth-century London. As the city's population expanded, so too did the amount of human excreta, especially once the market for night soil collapsed when enterprising businessmen began importing guano (bird excreta) for fertilizer. Unwanted excreta literally flowed in streets and into streams contaminating drinking water and causing thousands of deaths from cholera. The appalling stench from the Thames River, virtually an open sewer, was the catalyst in 1858 for sewering London, and using water to flush excreta down the pipes and away from the city (Black and Fawcett 2008), a practice decried by Victor Hugo (1862) as leading to "the land impoverished, and the water tainted. Hunger arising from the furrow and disease from the stream." Around this time, the term "sanitation" began to be used to describe the management of human feces and urine.

23.3 WATER-BASED SANITATION

> Frankly, how much sense does it make to clean water to drinking-water standards and then use a good portion of that precious resource to flush waste down an expensive pipe system—a pipe system that even the richest nations can barely afford to rehabilitate?
>
> **—Frank Rijsberman (2011)**

Water-based sanitation uses water to flush excreta along pipes and into treatment works to kill off pathogens before discharge onto land or into waterways. It has become the standard for excreta management in industrialized and developed countries, but is facing a number of challenges. Concerns about water shortages, wasteful water use, environmental pollution, and rising power costs of urban sewerage systems are leading to innovations such as water recycling, artificial wetlands that treat sewage and provide bird habitat, use of hydropower and methane from feces decomposition to power sewerage plants, and biosolids recovery for use in agriculture (Sydney Water 2012).

The term "biosolids" refers to sewage sludge that is treated according to environmental regulations for use as surface-applied soil fertilizer. The word was created in 1991 by the U.S. wastewater industry to distinguish treated sewage sludge from raw sludge, to facilitate public acceptance for its application to land (Lu et al. 2012). Land application of biosolids is becoming a preferred government option because it recycles nutrients, improves soil fertility and condition, prevents water pollution, and avoids environmental impacts of landfill. Its phosphorus value is also increasing as mineral resources decline (Clarke and Smith 2011). However, use of biosolids often faces community resistance due to odor and perceived health risks. Complete sterilization of fecal pathogens is difficult to achieve (Sidhu and Toze 2009), and the chemicals in industrial wastewater that also passes through sewerage treatment works have the potential to move into the food chain. These chemicals include heavy metals such as cadmium and zinc; organic chemicals such as perfluorochemicals and polychlorinated alkanes that have been detected in human blood and environmental samples and can accumulate in plants and grazing animals (Clarke and Smith 2011); and nanoparticles and prions (Torri et al. 2012). These concerns have led to a ban on biosolids use in some countries, and strict regulations on biosolids use in other countries. There is also increasing interest in using biosolids to restore degraded soils rather than use them on agricultural soils (Torri et al. 2012).

Water-based sanitation removes excreta efficiently and safely from households, but is costly for governments to build and maintain. The use of potable water for flushing away excreta does not make sense. As mentioned earlier, we each produce, on average, 50 kg of feces and 500 L of urine annually. A normal flush toilet uses an additional 15,000 L of drinking water per person per year, while water from kitchens and bathrooms (graywater) adds an additional 35,000 L per person per year depending on the location (Stockholm Environment Institute 2006). This means the 50 kg of feces contaminates the relatively harmless urine, the pure water used for flushing, and the graywater (Winblad et al. 2004). While some countries have the capacity to treat this water before it is recycled or discharged, only a fraction of the world's sewage and drainage water is treated before being discharged into waterways (Stenström et al. 2011). More than one billion people, 15% of the global population, have no sanitation facilities at all; in 11 countries, most people still practice open defecation (UN 2012).

These concerns have prompted an international search for sustainable sanitation options that do not require large amounts of water. Leading the search has been the ecological sanitation movement, initiated in 1993, which along with conserving water, aims to return nutrients to the soil. The movement was given new impetus after 2000 with the setting of the UN Millennium Development Goal of a 50% drop by 2015 in the number of people without proper sanitation (UN 2012). In 2001, a Singaporean entrepreneur established the World Toilet Organisation (2012) along with World Toilet Day on 19 November each year to improve sanitation conditions for people globally through "powerful advocacy, inventive technology, education and building marketplace opportunities locally." In 2002, Finland established the Global Dry Toilet Association of Finland (2012) to protect Finland's and the world's waters, and to promote the implementation of the natural nutritional cycle. In 2008, the International Year of Sanitation was declared. In 2011, the Bill and Melinda Gates Foundation (2012) initiated its Reinvent the Toilet Challenge to create toilets that, among other things, recover valuable resources such as water, energy, and nutrients.

23.4 ECOLOGICAL SANITATION

> Ecological sanitation makes use of human excreta and turns it into something useful and valuable with minimum pollution of the environment and in a way which poses no threat to human health.
>
> **—Peter Morgan (2012)**

The aim of ecological sanitation is to reconnect the nutrient loop that links food, humans, excreta, and soil and thus restore nature's cycle of nourishment. There are four main principles: separation at source of urine, feces, and graywater; containment of each product; sanitization and treatment; and recycling of nutrients, humus, and water back to soil and agricultural systems (Stockholm Environment Institute 2006).

23.4.1 Separation at Source

Urine is essentially sterile when it leaves the body, so can be more easily managed for fertilizer use if uncontaminated by fecal pathogens. Separation also largely eliminates odor and fly breeding, and makes storage, transport, and sanitization of both products easier. By not mixing excreta with water, only a comparatively small volume of urine and feces needs to be managed, and dry feces can be transported easily for secondary and tertiary treatment (Winblad 2002). Keeping urine and feces apart is a long accepted concept in parts of China, Japan, and in other parts of the world where simple toilets with urine diversion have been in use for centuries (Winblad et al. 2004).

23.4.2 Containment

Urine needs to be stored in closed containers to avoid ammonia emissions. Storage can be at any scale, from jerry cans and small tanks at the backyard scale, to rubber bladders and slurry tanks as occurs in apartment blocks in Sweden. There is interest in applying and incorporating urine in the soil during the dry season, followed by normal cultivation of the nutrient-rich soil in the cropping season. Research is needed into the fate of the nutrients stored in the dry soil, because nitrogen losses could be substantial and phosphorus could become less available to plants (Richert et al. 2010). Feces need to be stored initially for 6–12 months in a secure space, usually below where they are deposited, and covered with soil, ash, or organic matter to reduce odor and the risk of attracting flies.

23.4.3 Sanitization and Treatment

Ecological sanitation relies on environmental variables such as storage time, temperature, dryness, pH, ultraviolet radiation, and competing natural soil organisms to kill off pathogens (Winblad et al. 2004). Using urine safely in agricultural soils requires it to be collected separately from feces and water, and stored for weeks to months to kill off any pathogens associated with fecal contamination. There is also some concern that pharmaceuticals in urine represent a health hazard, but using urine in soils is likely to be safer than mixing it with water in conventional wastewater management and a high concentration of soil microorganisms are adapted to degrade organic molecules (Richert et al. 2010).

The general principles for ecological sanitation of feces are to minimize moisture by diverting urine; add ash or lime to raise the pH and create unfavorable conditions for pathogens; and store feces for 6–12 months to dehydrate, which kills off many of the pathogens. In a well-managed system, pathogen counts should be considerably reduced after this initial treatment, with the feces dry and light, odorless, easily handled without producing dust, and resembling soil. The feces can then be transported offsite for further treatment to make them safe enough to add to the soil. This secondary processing can include high-temperature composting, chemical addition of urea, longer storage times, or even incineration for completely pathogen-free nutrients (EcoSanRes 2008a). There is increasing interest in pyrolyzing feces to produce a sterile, nutrient-rich source of carbon known as biochar for incorporation into soil (Terra Preta Sanitation Initiative 2013).

23.4.4 Recycling

Urine can be applied pure or diluted to soil and needs to be incorporated as quickly as possible to minimize air contact. Urine nutrients are best utilized if the urine is applied prior to sowing, and during the growing period. Urine salt levels are an issue as they have the potential to increase soil salinity, so monitoring is needed to ensure urine fertilization that is adapted to the particular climate, soil condition, and crop requirements (Richert et al. 2010).

Sanitized feces should be applied before planting or sowing, as the high phosphorus content is beneficial for root formation of young plants. They need to be thoroughly mixed in and covered by soil before cultivation starts. Fecal matter should be within reach of the plant roots but should not be the only growing medium (EcoSanRes 2008b). A period of at least 1 month between application and harvest is recommended both for urine and for treated feces to further reduce the risk of pathogens.

23.5 SOIL-BASED SANITATION SYSTEMS

A toilet can be transformative.

—Nature (2012)

Ecological sanitation's principles of separation, containment, sanitization, and recycling underpin a multitude of toilet designs developed to return excreta to the soil. The diversity of designs reflects both the adaptations that have evolved to meet community needs and aspirations, and emerging technological innovations. Early designs developed in the 1990s drew on the dehydrating and composting toilets that had been used in rural areas and small communities around the world for millennia.

23.5.1 Dehydrating Systems

Dehydrating systems separate urine and feces, and store the feces below the toilet for 6–12 months to decompose and dehydrate naturally with the use of ventilation, additional dry material, and ash, lime, or urea to increase pH. The fecal material is then removed and, ideally, subjected to a secondary treatment of further storage, alkaline treatment, high-temperature composting, or carbonization/incineration (Winblad et al. 2004). The Vietnamese above-ground double-vault toilet is an example of a functional dehydrating system. One vault collects feces and ash, and drains urine into a jar. When the vault is two-thirds full, the contents are leveled with a stick and the vault filled with dried powdered earth and tightly sealed with lime mortar or clay. The second vault then comes into use and when it is ready to seal months later, the first vault is opened at the base and the odorless contents removed for fertilizer. The Vietnamese design has been adapted successfully for local conditions in urban settlements in Sweden, China, Mexico, Guatemala, India, and Sri Lanka (Winblad et al. 2004).

23.5.2 Composting Systems

Composting toilets for use in weekend houses were introduced in Sweden more than 50 years ago. Since then, a wide variety of models have come on the market and they are now used in different parts of the world, including North America and Australia. Feces, or feces and urine, are deposited in a processing chamber along with organic household and garden refuse and bulking agents (straw, peat moss, wood shavings, twigs, etc.). Temperature, airflow, moisture, carbon materials, and other factors are controlled to varying degrees to promote optimal conditions for decomposition. After 6–8 months, the partly decomposed material can be moved to a garden compost or an eco-station for secondary processing through high temperatures (Winblad et al. 2004).

Composting toilets that combine urine and feces are being installed in many buildings around the world, including Australia (Davison and Walker 2003). The U.S. practitioner Joseph Jenkins describes them as "humanure" systems that require both feces and urine to ensure there is enough moisture and nitrogen required for thermophilic composting. They also require substantial amounts of primary carbon cover material such as sawdust, peat moss, rice hulls, grain chaff, or paper products for the toilet, and secondary cover materials for the pile such as woodchips. The high dependency on carbon sources means it is not an option in areas where there is little spare organic matter available (Jenkins 2009).

23.5.3 Soil Composting Systems

In this system, feces, or feces and urine, are deposited in a processing chamber together with a liberal amount of soil. Most pathogenic bacteria are destroyed within 3–4 months as a result of competition with soil-based organisms and unfavorable environmental conditions. The material is then removed and can be subjected to any of the secondary treatments. Twelve months' composting in shallow pits is recommended before application to gardens to kill off remaining pathogens (Winblad et al. 2004).

The Arborloo is one soil composting system where urine and feces are deposited into the pit and covered with soil and materials such as wood ash and leaves after each deposit. The Arborloo has a portable slab and structure, which is removed and placed over a new pit as required. When a pit is nearly full, the slab and structure are removed, the pit is topped up with additional soil and a tree is planted. Every 6–12 months, a new pit is added and a new tree is planted, thereby gradually creating an orchard or woodlot (Morgan 2011).

The *Fossa Alterna* works on a similar principle and consists of two shallow pits dug at the same time and close to one another. One is used for about a year until nearly full, at which point the slab and structure are removed and placed atop the second pit; the first pit is then topped up with additional soil. When the second pit is nearly full, the first pit is opened and the contents removed and mixed with low-nutrient topsoils for use in crop-growing areas. The toilet slab and structure are then placed back on the empty first pit and the second pit is closed (Winblad et al. 2004).

23.5.4 Biogas Systems

Some toilet designs use human excreta to make biogas: urine and feces are digested anaerobically to produce gas that can be used for cooking, lighting, and heating. The solid effluent that remains can be safely used on soils. Household biogas systems using animal and human excreta are an important part of China's sustainable energy program (Chen et al. 2010). Municipal sewage treatment plants around the world are producing biogas from the waste to defray their energy costs (Zhou et al. 2013).

23.6 URBAN ISSUES

> A key challenge is how ecological sanitation can be cost-effectively scaled-up in an increasingly urbanized and economically constrained world.
>
> **—Haq and Cambridge (2012)**

Toilet designs that recycle human excreta back to the soil have been successfully introduced in many rural areas around the world where there is space for their construction, a small number of users (often a family), and agricultural land for application of treated excreta. These conditions for success mean they are unlikely to be successful in urban areas, where space is limited, and populations large. Urban urine and feces create nutrient "hot-spots" that need to be recycled back to agricultural soils to prevent soil nutrient decline. A recent review (Otterpohl and Buzie 2011) identified three main options for urban excreta in developed countries: high-tech systems based mainly on vacuum technology, membrane bioreactors, and biogas plants; urine-diverting flush systems; and dry sanitation systems. Depending on design, the high-tech systems create a variety of end products, including biogas, liquid fertilizer, and recycled water for toilet flushing. The urine-diversion systems collect and store urine for later use on farms. Most waterless sanitation systems are unsuitable for large-scale use because of the difficulty of sterilizing all pathogens but, at Hamburg University of Technology, Otterpohl and Buzie have developed a prototype *terra preta* sanitation system that diverts urine and uses lactic acid bacteria to kill off pathogens by anaerobic digestion. The acid is mixed with biochar, a form of charcoal produced specifically for use in soil and also known as *terra preta*. The biochar absorbs odor and binds nutrients. The treated feces are then composted using earthworms. The compost can be incorporated into soils for nonfood crops after a few months, and in food crop soils after 2–3 years. In 2012, Hamburg University of Technology and the World Toilet Organisation joined forces to offer prizes for the best toilet designs to incorporate the *terra preta* technology (Terra Preta Sanitation Initiative 2013).

Introducing ecological sanitation principles into urban areas with little existing sanitation infrastructure is difficult. Urine-diversion dry toilets are a low-cost option for urban slums, but depend on land for construction and a market for the end product. Anaerobic codigestion of excreta and organic solid waste is also feasible for decentralized treatment of sludge and solid waste to produce biogas, provided the digestate is sanitized for safe reuse or disposal, and there is skilled labor for process control (Katukizaa et al. 2012). Rijsberman and Zwane (2012) suggest three solutions worthy of large-scale investment: community-led total sanitation, which emphasizes behavior change and community responsibility; sanitation as a business, combining technical innovation and fee for use; and reinvented toilets to recycle waste at the household scale.

Examples of innovative urban sanitation systems are emerging around the world. For instance, in India, the humanitarian organization Sulabh International (2012) has built 200 user-pay public toilet complexes in recent decades to collect excreta and produce biogas for power. Material remaining after biogas digestion can be used for fertilizer. The toilets are clean, safe to use, and have become community centers. In Kenya, the sanitation group Sanergy (2012) is building a network of low-cost sanitation centers in slums, distributing them through franchising to local entrepreneurs, collecting the waste produced, and processing it into electricity and fertilizer. Individuals pay to use these urine-diverting toilets, which are kept clean and safe. The urine and feces are deposited in air-tight containers, which are collected daily in handcarts for central processing to create biogas, generate electricity, and produce fertilizer for sale to farmers.

The field of urban sanitation and nutrient recovery is becoming an exciting field for researchers. In 2011, the Bill and Melinda Gates Foundation issued a challenge to universities to design toilets that capture and process human excreta without piped water, sewer or electrical connections to create useful resources, such as energy and water, at an affordable price. Winners of the challenge were announced at a "Reinvent the Toilet" fair in August 2012, and included a solar-powered toilet that generates hydrogen and electricity; a toilet that produces biological charcoal, minerals, and clean water; and a toilet that sanitizes feces and urine and recovers resources and clean water. Other winners included a toilet that uses a hand-operated vacuum pump to dehydrate excreta and transform it into fuel/fertilizer; a self-contained toilet system that disinfects the liquid waste and turns the solid waste into fuel or electricity; and a toilet that uses a solar dish and concentrator to disinfect liquid–solid waste and produce biochar. The winners have all received funding to develop their designs beyond the concept stage (Bill and Melinda Gates Foundation 2012).

23.7 CULTURAL CONSIDERATIONS

> When a new sanitation system is to be introduced into a new area, the religious, cultural and spiritual values in the local context must be considered.
>
> **—Malin Falkenmark (1998)**

These innovative technologies would amaze our ancestors, but their success will largely depend on our readiness to stop thinking of our urine and feces as a problem to be disposed of and realize their value as nutrients and organic matter in a world of declining resources. While there seems to be no limit to human ingenuity in finding ways to recycle our excreta, successful recycling depends on our willingness to change our attitudes. The biggest lesson of 20 years of ecological sanitation development is that recycling our excreta back to the soil will take not just a new toilet design, but a whole new way of thinking about our urine and feces. Any changes to sanitation systems, particularly systems that require active management, need to be introduced in full consultation with intended users, and benefit the users.

As an example of what can go wrong, apartment households in an 2006 urban sanitation project in Mongolia insisted, 3 years later, on a return to flush toilets. The innovative project comprised urine-diversion dry toilets in 832 apartments, graywater treatment, an eco-station for waste composting and agricultural reuse of the products. A review of the project in December 2009 (Jurga 2009) identified a combination of causes for the project failure, including lack of testing and demonstration beforehand, the toilet itself that was inconvenient to use and socially hard to accept, persistent odor due to design errors and construction faults, and fear of long-term maintenance costs. While residents had a positive attitude to excreta recycling in general, neither they nor nearby farmers understood the value of urine on its own. As a result, the residents did not understand the importance of urine-diversion toilets, and the farmers did not want the urine (Gao 2011).

On the other hand, a 1999 pilot in Mozambique won the 2008 Goldman Environmental Prize for Africa, and the toilets are now being built throughout the country. The permanent *Fossa Alterna* twin pits offer a viable system in small backyards, compared with the pit latrines that eventually fill and have to be relocated elsewhere in the yard. The addition of soil and ash to cover the excreta means the pits do not omit odor, attract flies, or harbor malarial mosquitoes. The use of ash encourages hand washing, which has greatly decreased transmission of pathogens. The shallow pits do not generally threaten quality of groundwater used for family water supply, and the excreta compost provides valuable fertilizer that has increased garden yields, encouraged its use in crop fields, and generated surplus crops for sale. The main operating issues have been establishing the right quantity of ash and soil to completely cover the feces, and holding enough dry soil to control odors at the height of the wet season. The Mozambique government now recognizes and actively promotes the sanitation system nationally (Quayle 2012).

The Mozambique and Mongolian experiences highlight the importance of consultation, education, and training in communities when any new sanitation system is to be introduced. A quick search of the ecological sanitation literature reveals many case studies similar to those cited above, and all of them emphasize the importance of working with communities to deliver sanitation systems that work. People's attitudes to excreta management are culturally ingrained, and concerns about odor, disease, peer pressure, and cultural taboos must be acknowledged and dealt with, along with training in safe handling, and the vital role of excreta nutrients in soil fertility and food production. No single toilet design will suit all situations or all cultures, and the most successful designs are developed in collaboration with users, to ensure their suitability for the site and the community. "To achieve the direct effects of containment and reduction of pathogenic organisms the system should be technically appropriate, economically viable, socially acceptable, and institutionally manageable which are factors that all affect the health outcomes" (Stenström et al. 2011).

23.8 CHALLENGES

> To mainstream the development of nutrient reuse, concerted efforts are needed in the policy arena of national and local governments, in particular within the sectors of health, environment and agriculture.
>
> **—Rosemarin et al. (2008)**

It is exciting to see the range and scale of sanitation innovations emerging in response to urgent global concerns about human health, limited water supplies, declining soil fertility, and threats to food security, but restoring nature's cycle of nourishment faces a number of challenges. Changing cultural attitudes, and finding systems that work for urban areas are two of the biggest issues as outlined above. As well, urine and feces have no legal status in most countries (Stockholm Environment Institute 2006), which presents significant implementation challenges for governments. For instance, urine recycling to recover phosphorus is hampered by the lack of an institutional or organizational home in Australia because our "flush and forget" systems make recycling a peripheral concern to policymakers (Cordell et al. 2009). Ironically, developing countries with little sanitation infrastructure in place have greater potential to implement nutrient recovery because innovative technologies can be integrated with sanitation developments. Regions with extensive water-based sanitation will need to retrofit existing sanitation technology and transform people's "flush and forget" behaviors (Mihelcic et al. 2011). Life cycle analysis suggests that while ecological sanitation systems are a promising alternative to small-scale wastewater treatment, larger-scale conventional wastewater treatment plants are not likely to be replaced in the short term (Benetto et al. 2009).

23.9 CONCLUSION

Like everyone else on Earth, I produce urine and feces every day. But until recently I gave them little thought, due to personal distaste and the ability to flush them out of my life. It took a visit to my friends' composting toilet to kindle my interest in excreta recycling, an inherently fascinating topic, and one that is now gaining political traction given its capacity to simultaneously alleviate so many of our global problems: pollution, disease, water shortages, soil nutrient decline, and food security. But we need to do more—and soon! We need to talk to each other about our excretions, about their nutrient value and the immense resource they represent. We need to find out what is happening with human urine and feces in our local area, and share the amazing ideas and the extraordinary sanitation innovations that are happening around the world. We must understand the costs to human and environmental health of doing "business as usual." It is time to realize that our cultural aversion to discussing or dealing directly with our excreta is actually causing disease, increasing pollution and reducing soil fertility. Our excretions are not a problem, they are a key component of the soil fertility cycle, as novelist Victor Hugo (1862) so poetically recognized in *Les Miserables**:

> Those heaps of filth at the gate-posts, those tumbrils of mud which jolt through the street by night, those terrible casks of the street department, those fetid drippings of subterranean mire, which the pavements hide from you,—do you know what they are? They are the meadow in flower, the green grass, wild thyme, thyme and sage, they are game, they are cattle, they are the satisfied bellows of great oxen in the evening, they are perfumed hay, they are golden wheat, they are the bread on your table, they are the warm blood in your veins, they are health, they are joy, they are life.

REFERENCES

Avvannavar, S.M. and M. Mani. 2008. A conceptual model of people's approach to sanitation. *Science of the Total Environment* 390: 1–12.

* In his novel *Les Miserables*, Victor Hugo passionately decries the sewers that channel Parisians' valuable excreta out to sea and away from the soil. The anti-sewer chapter 'The land impoverished by the sea' is the first in the second book *The intestine of the Leviathan* of Volume 5.

Benetto, E., D. Nguyen, T. Lohmann, B. Schmitt, and P. Schosseler. 2009. Life cycle assessment of ecological sanitation system for small-scale wastewater treatment. *Science of the Total Environment* 407: 1506–1516.

Bill and Melinda Gates Foundation. 2012. Water, sanitation and hygiene: Reinvent the toilet challenge. Fact sheet. Global development program November 2012. http://www.gatesfoundation.org/watersanitation-hygiene/Documents/wsh-reinvent-the-toilet-challenge.pdf (accessed 11 March 2013).

Black, M. and B. Fawcett. 2008. *The Last Taboo*. London: Earthscan and UNICEF.

Chen, Y., G. Yang, S. Sweeney, and Y. Feng. 2010. Household biogas use in rural China. *Renewable and Sustainable Energy Reviews* 14: 545–549.

Clarke, B.O. and S.R. Smith. 2011. Review of "emerging" organic contaminants in biosolids and assessment of international research priorities for the agricultural use of biosolids. *Environment International* 37: 226–247.

Cordell, D., J.O. Drangert, and S. White. 2009. The story of phosphorus. *Global Environmental Change* 19: 292–305.

Curtis, V. and A. Biran. 2001. Dirt, disgust and disease. *Perspectives in Biology and Medicine* 44: 17–31.

Curtis, V. 2007. Dirt, disgust and disease: A natural history of hygiene. *Journal of Epidemiology and Community Health* 61: 660–664.

Davison, L. and S. Walker. 2003. A study of owner-built composting toilets in Lismore, NSW. In R.A. Patterson and M.J. Jones (eds.), *On-Site '03: Proceedings of On-Site '03 Conference: Future Directions for On-Site Systems, Best Management Practice*, University of New England, Armidale, NSW, 30 September–2 October, Lanfax Laboratories, Armidale, NSW, pp. 115–122.

Drangert, J.O. 1998. Fighting the urine blindness to provide more sanitation options. *Water South Africa* 24: 157–164.

Drangert, J.O. 2004. Norms and attitudes to ecological sanitation and other sanitation systems. EcoSanRes report 2004/5. Stockholm Environment Institute.

Drechsel, P., C.A. Scott, L. Raschid-Sally, M. Redwood, and A. Bahri (eds.). 2010. *Irrigation and Health: Assessing and Mitigating Risk in Low-Income Countries*. London: Earthscan.

EcoSanRes. 2008a. Guidelines for the safe use of urine and feces in ecological sanitation systems. EcoSanRes fact sheet 5 May 2008. Stockholm Environment Institute.

EcoSanRes. 2008b. Guidelines on the use of urine and feces in crop production. EcoSanRes fact sheet 6 May 2008. Stockholm Environment Institute.

Falkenmark, M. 1998. Wilful neglect of water: Pollution—A major barrier to overcome. Stockholm. International Water Institute Waterfront, Stockholm, Sweden. Cited in Stenström, T.A., R. Seidu, N. Ekane, and C. Zurbrügg 2011. *Microbial Exposure and Health Assessments in Sanitation Technologies and Systems*. EcoSanRes Series, 2011-1, Stockholm Environment Institute.

Gao, S.W. 2011. Ecological sanitation in urban China: A case study of the Dongsheng project on applying ecological sanitation in multi-storey buildings. Student thesis, Linköping University, Norway.

George, R. 2009. *The Big Necessity: Adventures in the World of Human Waste*. London: Portobello Books.

Global Dry Toilet Association of Finland 2012. Website. http://www.drytoilet.org/intro.html (accessed 3 February 2011)

Hall, S. 2014. Soils and the future of food: Challenges and opportunities for feeding 9 billion people. In *The Soil Underfoot,* ed. G.J. Churchman and E.R. Landa. pp. 17–35, Boca Raton: CRC Press.

Haq, G. and H. Cambridge. 2012. Exploiting the co-benefits of ecological sanitation. *Current Opinion in Environmental Sustainability* 4: 431–435.

Heinonen-Tanski, H. and C. van Wifk-Sijbesma 2005. Human excreta for plant production. *Bioresource Technology* 96: 403–411.

Hugo V. 1862. The land impoverished by the sea, Chapter 1, Book 2, Volume 5 in *Les Miserables*. http://www.online-literature.com/victor_hugo/les_miserables/323/ (accessed 3 February 2013).

Jenkins, J. 2009. Humanure sanitation. *Proceedings, Dry Toilet Conference 2009*, Finland. http://www.drytoilet.org/dt2009/full.html (accessed 3 February 2013).

Jurga, I. 2009. *Erdos Eco-Town Project: Lessons Learned and Ways Forward*. Report of Workshop 7–8 December 2009, Beijing. Stockholm Environment Institute (accessed 3 February 2013).

Katukizaa, A.Y., M. Ronteltapa, C.B. Niwagabab, J.W.A. Foppenc, F. Kansiimed, and P.N.L. Lensa. 2012. Sustainable sanitation technology options for urban slums. *Biotechnology Advances* 30: 964–978.

Lu, Q., Z.L. He, and P.J. Stoffella. 2012. Land application of biosolids in the USA: A review. *Applied and Environmental Soil Science* Article ID 201462, 11 pages.

Mihelcic, J.R., L.M. Fry, and R. Shaw R. 2011. Global potential of phosphorus recovery from human urine and feces. *Chemosphere,* 84(6) 832–839.

Morgan, P. 2011. *Trees as Recyclers of Nutrients Present in Human Excreta*. Stockholm Environmental Institute.

Morgan, P. 2012. Ecological sanitation. Aquamor website. http://aquamor.tripod.com/page2.html (accessed 3 February 2013).

Nature. 2012. The bottom line. Ecodesign column. *Nature* 486: 186–189.

Otterpohl, R. and C. Buzie. 2011. Wastewater: Reuse-oriented wastewater systems—Low- and high-tech approaches for urban areas. In *Waste: A Handbook for Management* eds. T.M. Letcher and Vallero, D.A., 127–136. Boston: Elsevier.

Quayle, T. 2012. *Innovative Marketing: Removing the Barriers to Ecological Sanitation in Lichinga, Niassa Province, Mozambique*. ACCESSanitation, Freiburg, Germany.

Richert, A., R. Gensch, H. Jönsson, T.A. Stenström, and L. Dagerskog. 2010. *Practical Guidance on the Use of Urine in Crop Production.* Ecological Sanitation Res Series, 2010-1. Stockholm Environment Institute.

Rijsberman, F. 2011. Why we need to reinvent the toilet. October 6, 2011. http://www.gatesfoundation.org/speeches-commentary/Pages/frank-rijsberman-2011-water-and-health-conference.aspx (accessed 3 February 2013).

Rijsberman, F. and A.P. Zwane. 2012. Sanitation and water challenge paper. Copenhagen Consensus 2012: Solving the World's Challenges. http://www.copenhagenconsensus.com/sites/default/files/water%2Band%2BSanitation.pdf (accessed 6 December 2013).

Rosemarin, A., N. Ekane, I. Caldwell, E. Kvarnstrom, J. McConvite, C. Ruben, and M. Fodge. 2008. *Pathways for Sustainable Sanitation—Achieving the Millennium Development Goals.* EcoSanRes Program, Stockholm Environment Institute.

Sanergy. 2012. Building sustainable sanitation in urban slums. http://saner.gy/ourapproach/ (accessed 3 February 2013).

Shuval, H. 2010. Foreword. In *Irrigation and Health: Assessing and Mitigating Risk in Low-Income Countries,* eds. P. Drechsel, C.A. Scott, L. Raschid-Sally, M. Redwood, and A. Bahri, xv–xvii. London: Earthscan.

Sidhu, J.P.S. and S.G. Toze. 2009. Human pathogens and their indicators in biosolids: A literature review. *Environment International* 35: 187–201.

Stenström, T.A., R. Seidu, N. Ekane, and C. Zurbrügg. 2011. Microbial exposure and health assessments in sanitation technologies and systems. EcoSanRes Series, 2011-1, Stockholm Environment Institute.

Stockholm Environment Institute 2006. EcoSanRes Phase 2. 2006-2010 Project, Project Document Final Draft (3rd) February 22, 2006. Stockholm Environment Institute.

Sulabh International. 2012. Biogas technology. http://sulabhinternational.org/?q=content/biogas-technology (accessed 3 February 2013).

Sydney Water. 2012. Our systems and operations. http://www.sydneywater.com.au/OurSystemsandOperations/ (accessed 3 February 2013).

Terra Preta Sanitation Initiative. 2012. The 2012 WTO International Toilet Design Award. http://www.terra-preta-sanitation.net/cms/index.php?id=19 (accessed 3 February 2013).

Tilley, E., C. Lüthi, A. Morel, C. Zurbrügg, and R. Schertenleib. 2008. *Compendium of Sanitation Systems and Technologies.* Eawag. Dübendorf, Switzerland. http://www.eawag.ch/forschung/sandec/publikationen/sesp/dl/compendium_high.pdf (accessed 3 February 2013).

Torri, S.I., R.S. Correa, G. Renella, A. Vadecantos, A. Vadecantos, and L. Perelomov. 2012. Biosolids soil application: Why a new special on an old issue? *Applied and Environmental Soil Science* Article ID 265783, 3 pages, 2012.

Tzanakakis, V.E., N.V. Paranychianaki, and N. Angelakis N. 2007. Soil as a wastewater treatment system: Historical development. *Water Science & Technology: Water Supply* 7: 1 67–75.

United Nations. 2012. *The Millennium Development Goals Report 2012.* United Nations, New York.

Winblad, U. 2002. *Final Report SanREd 1992–2001.* Sida Stockholm.

Winblad, U., M. Simpson-Hébert, P. Calvert, P. Morgan, A. Rosemarin, R. Sawyer, and J. Xiao. 2004. *Ecological Sanitation: Revised and Enlarged Edition.* Stockholm Environment Institute. http://www.ecosanres.org/pdf_files/Ecological_Sanitation_2004.pdf (accessed 3 February 2013).

World Toilet Organisation. 2012. Website. http://worldtoilet.org/wto/ (accessed 3 February 2013).

Zhou Y., D.Q. Zhang, M.T. Le, A.N. Puah, and W.J. Ng. 2013. Energy utilization in sewage treatment—A review with comparisons. *Journal of Water and Climate Change* 4: 11–10.

24 Global Potential for a New Subsurface Water Retention Technology

Converting Marginal Soil into Sustainable Plant Production

Alvin J. M. Smucker and Bruno Basso*

CONTENTS

24.1 INTRODUCTION

Substantial areas of soils underfoot that require increased water-holding capacities are encountered in agricultural fields, athletic fields, golf courses, parks, home lawns, and gardens, as well as many natural ecosystems. There is a growing fundamental global imperative to convert some of these marginal droughty soils into productive agricultural lands by increasing their longer-term internal water-holding capacities. These conversions of highly permeable coarse-textured soils will contribute to food and biomass production needs associated with growing global populations and renewable energy resources. In addition, these conversions will sequester more carbon, improve soil quality, and reduce groundwater contamination. Continuous cover cropping of these improved soils will also increase water infiltration and reduce erosion of surface soil into freshwater lakes, streams, and rivers.

The processes of plugging a plethora of continuously connected macropores within sandy soils have been occurring since God established sandy soils. Accumulations of thin clay-enriched layers, located at 45–65 cm depths, cause the sandy soils located above these layers to hold more water. Consequently, a few sandy soils are able to retain adequate soil water and nutrient contents, enabling them to produce more grain and biomass. Thousands of years ago, farmers in Northern Africa and Iran established additional water retention systems in irrigated soils (Stein, 1998). These attempts included burying porous clay containers below the root zones of cultivated crops. This slow leakage provided additional water for prolonged periods of excessive evaporative transpiration. Some more

* Michigan State University, East Lansing, Michigan, USA; Email: smucker@msu.edu

industrious farmers removed surface soils and installed layers of various organic and inorganic materials to reduce soil water losses from plant root zones. More recently, longer-lasting asphalt and polymer films have been placed at various depths below plant root zones in continuous layers that doubled or even saturated soil water-holding capacities with significant increases in sustainable production of grain, fruits, vegetables, and cellulosic biomass for conversion into liquid biofuels (Smucker et al., 2014a).

As these new technologies enhance natural fine-textured layers in the soils by disrupting draining macropores at soil depths below the soil surface, our living standards will improve. These accelerated attempts to improve soil water retention at rates greater than natural clay processes have led to the establishment and testing of newly designed concave polymer membranes that gave birth to the new subsurface water retention technology (SWRT) (Smucker et al., 2014a). These newly developed U-shaped engineered polymer membranes (Figure 24.1) must have functional integrities for long periods of time lasting for at least 40 years with modeled projections of up to 300 years when buried beneath the soil surface. We now have commercialized new and innovative SWRT membrane installation machinery (Figure 24.2) that installs long-term water-saving membranes, which more accurately control the retention of optimal quantities of soil water for both irrigated and nonirrigated agriculture, even if there are major changes in local and regional climates.

The goal of this new SWRT water-saving membrane conversion of marginal soils into sustainable agricultural production lands is to incorporate an environmentally safe polymer technology into soils in a manner that blurs the distinction between them and natural clay layers at strategic depths within highly permeable sands. Sustainable plant production on these modified sandy soils will continue to improve as more water, nutrients, and soil organic carbon are retained in plant root zones for prolonged periods of time.

Additionally, SWRT water-saving membranes are designed to overflow during excessive rainfall events, yet they intercept and retain up to nearly 100% of reduced irrigation water volumes needed for maximum plant production. It is anticipated that these new SWRTs will generate additional innovations that increase water and fertilizer use efficiencies for maximum plant growth, while reducing deep leaching of plant nutrients, pesticides, heavy metals, salts, and other toxic substances on large and smallholder farms located in arid and semiarid regions.

Water, the world's most finite natural resource, ensures economic, environmental, political, and social stability. Large cities struggle daily with water quality and related issues while populations

FIGURE 24.1 Excavated SWRT membrane having an aspect (width-to-depth) ratio of 2:1. These membranes retained the majority of roots inside and above each membrane. As Figure 24.3 clearly conveys, some roots grow down and around these membrane configurations.

FIGURE 24.2 **(See color insert.)** SWRT membrane installation equipment showing rolls of polyethylene (PE) film being released from the backs of membrane installation chisels when inserted into the soil to depths of 40–70 cm. Multiple adjacent passes across small and large fields are accurately controlled by satellite-guided global positioning systems (GPS) that are becoming the standard equipment on many farm tractors.

increase. Efficient water use is being addressed by industry, urban centers, and agriculture. Soil scientists and engineers at Michigan State University have developed and commercialized a patented membrane installation implement (Figure 24.2) that inserts strategically spaced water-saving membranes into highly permeable sand soils in a manner that conserves at least 2.25 million liters (592 thousand gallons) of irrigation water per hectare annually. Properly positioned SWRT water-retaining membranes, designed to retain up to 100% irrigation water, are positioned to drain excess rainwater and permit extended root growth beyond depths of these SWRT membranes, are installed in two overlapping layers across fields (Figure 24.3). SWRT membranes are designed to double soil

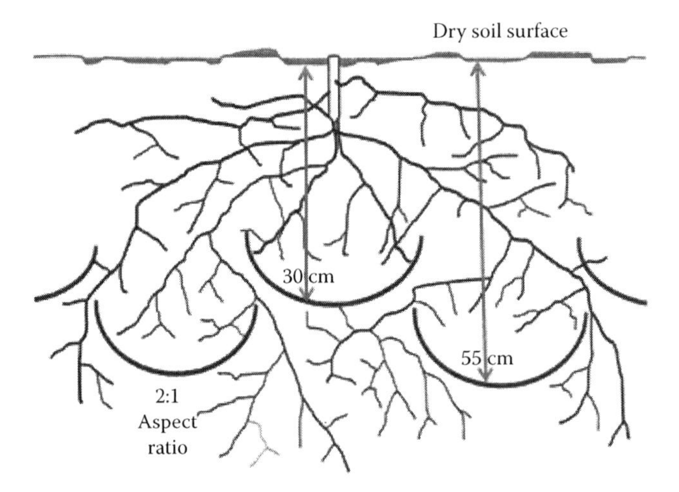

FIGURE 24.3 Diagrammatic locations of strategically positioned layers of LLDPE water-saving membranes having 0.0381–0.0762 mm (1.5–3.0 mil) thickness with individual configurations designed to intercept up to 98% of all vertical soil water flow infiltrating the root zone directly below the soil surface. SWRT membrane spacings permit excess soil water to drain and provide space for aggressive root growth beyond membrane depths. Crops may be planted at any angle to the directions of SWRT membrane installations.

water-holding capacity in plant root zones (Smucker et al., 2014b) in a manner that maximizes water use efficiencies essential for the dramatic expansion of food, fiber, and cellulosic biomass production needed by the rapidly expanding global populations. It is anticipated that these water savings will also diminish the growing competition for water among regional and economic sectors.

Soil water deficits and associated plant water stresses comprise the greatest abiotic hindrance to sustainable plant growth. Essential supplemental irrigation without water conservation practices such as SWRT is expensive, time consuming, and requires additional training. Regional and national surface water reservoirs and irrigation canals are costly, politically vulnerable, and promote the production of mosquito vectors of malaria, elephantitis, and sleeping sickness, and tsetse flies that transmit trypanosomosis in cattle. Expansion of surface and subsurface drip irrigation in arid regions of the world has the potential for substantially increasing agricultural production in every part of the globe. Selected combinations of rapidly expanding supplemental irrigation associated with digital technologies promote precision applications of water and nutrients to plants. Therefore, when SWRT membranes are added to sandy soils, specific quantities of water and nutrients can be added to the root zones of plants and retained in a manner that emulates pharmaceutical prescription management of biological needs. This brings specific control of plant growth and production to the local level, avoiding associated regional competition and control. Cellular phone communication by farmers enables them to produce more crops food with less irrigation and fertilization. SWRT membranes have enabled production of more food with less irrigation and fertilization and have tremendous potential for establishing food value chains that improve nutritional and economical livelihoods of smallholder and larger farmers while improving their soils and landscapes in a changing world. Small-scale irrigation technologies and precision agricultural management practices from water harvesting and collection (see Section 24.1.1) to storage within the root zones of plants will help optimize water usage for each farmer. Precise applications of irrigation water, where yields are increased up to 400%, will dramatically improve food security, nutrition, and incomes.

24.1.1 Modeling SWRT Applications

Harvesting water where it falls has been the greatest limitation to crop production and will remain as such through the next decades of changing weather conditions. Its efficiency is an essential requirement for a sustainable global food security. Subsurface soil water retention technologies, installed within plant root zones, comprise a self-regulating type of technology that improves the production of food and cellulosic biomass and increases water use efficiencies by retaining more plant- available water and nutrients. System Approach to Land Use Sustainability (SALUS) models (Basso et al., 2007) were used to predict field crop production responses to the new SWRT membrane enhancement of crop growth by protecting plant health from the negative influences of heterogeneous soil types and changing climate conditions. Basso's SALUS model predicted production increases of 283% for nonirrigated maize and 68% for nonirrigated wheat grown on sand soils equipped with water-retaining membranes (Figure 24.4).

SWRT membrane conversions of marginal highly permeable soils must become a major contributor to feeding the nine billion people expected to inhabit our planet by 2050 for both large and smallholder farms. This new technology, coupled with precision water and nutrient management, has the potential to transform agriculture, reduce poverty, and improve nutrition among the rural poor globally. Overcoming short-term and long-term water deficits for agricultural plants is a primary step forward to maximize newly developed hybrids, associated with best management and protection of harvested produce. Although estimates of food insecurity vary, Barrett (2010) concludes that feeding these many people requires more than incremental changes. Fedoroff et al. (2010) outline how scientists and engineers can make a big difference at every step from field to fork and we propose it will take a trilogy of new technologies to produce more crop grain and biomass per drop of water as defined by Ash et al. (2010).

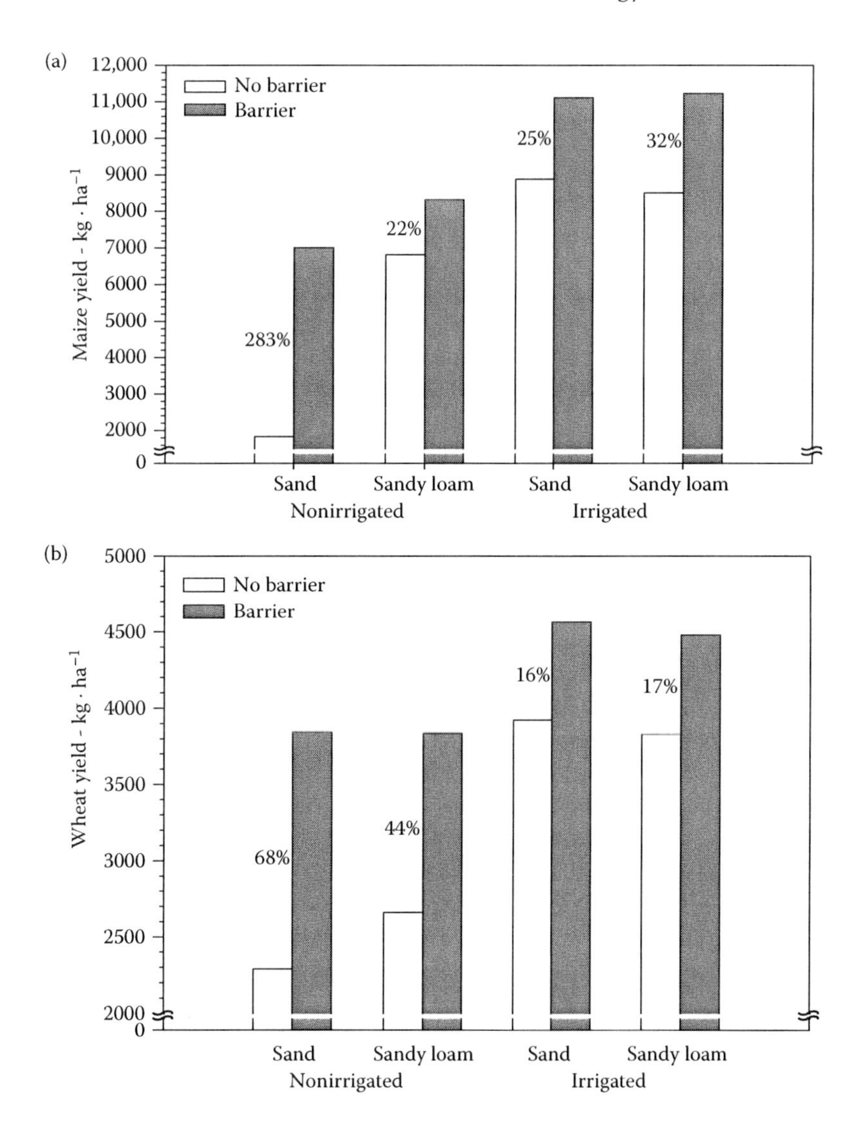

FIGURE 24.4 SALUS models predict grain yield increases of 283% by maize grown on sand soils (a) coupled with 68% increases in wheat yields (b) when SWRT water-saving membranes are properly placed below the root zones of maize (a) and wheat (b) grown without supplemental irrigation, in Brisbane, Australia. These production gains generate, at 2012 prices, additional US$945 for maize and US$327 for wheat. When supplemental irrigation is added to SWRT membrane-improved sand soils, both maize and wheat profits increase even more. Although smaller yield increases are predicted for SWRT-membrane-improved finer-textured sandy loam soils, profits continue to increase.

24.2 SWRT APPLICATIONS

24.2.1 Lysimeter Studies

Large boxes of sand (Smucker et al., 2014b) were used to identify the most ideal depths and arrangements of these SWRT membranes. Initial comparisons of water permeability within soil horizons demonstrated how water-impermeable membranes, properly designed to retain and redistribute water within plant root zones, simulate natural clay layers. Soil water permeability losses by drainage in these fine sands without water-saving membranes were 8200 liters per square meter per day

TABLE 24.1
Aboveground and Belowground Plant Growth of Maize Plant Growth in Sand Containing Water-Saving Membranes

Treatments	Plant Biomass (g)	Roots/100 cm^2	Shoot:Root Mass Ratio
Control, no membrane	57.3	44	1.3
SWRT membranes	208.5	70	3.0

Note: Membranes below the root zone increased biomass by 364% with concomitant 230% increases in shoot-to-root ratio.

(L/m^2/d). Thin clay layers in sandy soils reduced permeability to <120 L/m^2/d. We calculated that natural clay horizons, located at 35–50 cm depths, should reduce water losses while augmenting plant production as described by Yang et al. (2012).

To further test additional concepts of SWRT membrane improvements of soil water-holding capacities, we designed concave engineered linear low-density polyester (LLDPE) membranes that doubled soil water volumes in plant root zones and maintained adequate plant-available water in sandy soils for periods up to 4 days (Smucker et al., 2014a). Furthermore, we identified the best aspect ratios (Figure 24.3), which describe the width-to-depth ratio within each SWRT membrane; these should be 2:1 for maximizing water storage by sandy soils located in humid regions. We assume that drier and more arid environments will require smaller aspect ratios that enable greater storage and redistribution capacities of soil solutions. Lateral distances between each long trough of SWRT membranes installed across a field, allowed for root bypass of these subsurface soil water reservoirs (Figure 24.3). Subsoil water retention membranes increased shoot-to-root ratios by 230%, producing 340% greater aboveground cellulosic biomass than lysimeter control sands without water-retaining membranes (Table 24.1). Observing relationships between the SWRT membrane positions and the water-holding capacity of these water storage membranes led us to conclude that two depths of SWRT membrane positions (Figure 24.3) were required to maximize water retention and the homogeneous rate of redistributed stored soil water as it is wicked upward into the root zone. This uniform presentation of additional soil water to plant roots improved plant water uptake and reduced the shoot-to-root ratios of plants (Smucker et al., 2014b). All these soil water data were compared to yield data for specific plants grown over barriers installed in sands at different depths and volumetric configurations to identify the best membrane geometries required for maximum improvements in soil water-holding capacities by sandy soils improved by SWRT membranes installed in fields.

Plant parameter responses to these varied water retention membranes in the lysimeters with and without water-saving membranes are essential for identifying specific depths and spatial distributions. Maize plant height increased nearly 180% when growing on sand equipped with SWRT water-retaining membranes (Figure 24.5). This accelerated growth rate resulted in cellulosic biomass yields by maize grain and stalks by up to 300% beyond controls. Similar accelerated total biomass production was modeled by the soil–atmosphere–plant SALUS model (Basso et al., 2007).

24.2.2 Field Studies

Real-world reliability of the SWRT water-saving membranes has been thoroughly tested in replicated field trials at two locations in Michigan. Initial construction and field testing of a machine designed to install the SWRT LLDPE membranes required multiple years of development. A two- and four-chisel SWRT membrane installation implement (Figure 24.2) required considerable human intervention; currently, we are developing more streamlined commercial models requiring less human input. SWRT membranes (Figure 24.1) were installed at depths regulated by the capillary rise or wicking of water above the SWRT membranes. Membrane depths for the irrigated

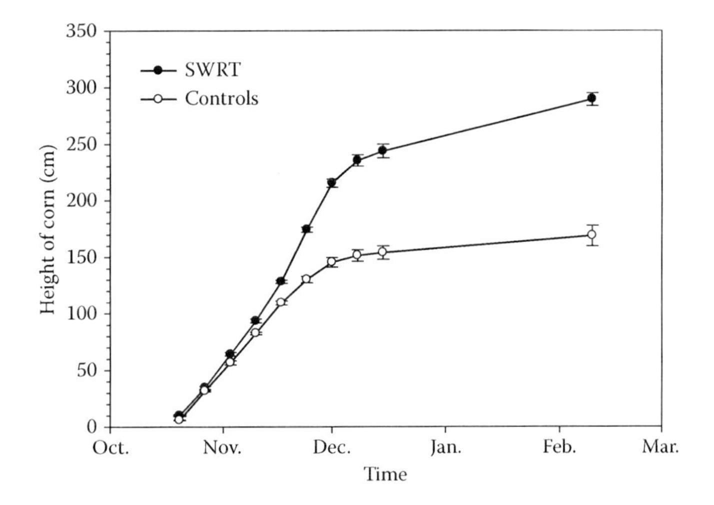

FIGURE 24.5 Increased plant height by SWRT water retention membrane installations below the plant root zones of irrigated maize planted at 20 cm row spacing and 10 cm spacing within rows.

maize grown on coarser soil were 51 and 28 cm. SWRT membrane depths for the irrigated pepper and cucumber on finer sand soils were 56 and 36 cm.

Two row spacings of conventional 76 cm and narrow 38 cm row spacings of maize were grown on sandy soils on the campus farms of Michigan State University. Irrigated green pepper and cucumber grown on the sand soil, improved by SWRT membranes increased production by 128% and 144%, respectively, at the research and extension horticultural farm in SW Michigan. SWRT membrane-improved and irrigated sandy soils also increased grain yields of maize, planted at 38 cm row spacing, by 174%. Maize giving poor yields on sandy soils without SWRT membranes experienced severe plant wilting during much of the early growth and later mature stages of flowering and grain-filling periods during the 2012 droughty year. The very droughty year at the Michigan State University research farm reduced nonirrigated maize yields of control plants shown growing in the three rows along the left side of Figure 24.6 to just 13% of the nearly

FIGURE 24.6 Three rows of narrow row maize without irrigation (on left) served as external border rows to irrigated narrow row maize (on right) planted across SWRT membrane.

TABLE 24.2
SWRT Membranes Increased Irrigated Maize Plant Biomass Grown on a Medium Conover Sand Soil by 138% for 76 cm Row Spacing and 193% for 38 cm Row Spacing

Treatments	Maize Planted at 76 cm Row Spacing (kg/ha)	Maize Planted at 38 cm Row Spacing (kg/ha)
Irrigated control, no membranes	18,719	18,126
Irrigated SWRT membranes	25,741	34,998
SWRT membrane increase	138%	193%

35,000 kg/ha of total plant biomass produced by irrigated narrow row maize growing on SWRT membranes.

Soil depths and spatial configurations of contoured LLDPE membranes can be more precisely positioned below plant root zones when combinations of capillary rise, soil water retention, and hydraulic conductivity are accurately modeled. Depths, aspect ratios, and spacing of soil water retention membranes are constantly being investigated and further modeled to obtain the best fit for each soil type, climate type, and the majority of plants in the crop rotation management system. Water losses to soil depths below SWRT membranes that varied between 3% and 5% indicate there are many additional opportunities for discovering other advantages this new SWRT has for saving water and nutrients in the root zone of most plants. These tremendous reductions in soil solution losses in sandy soils have expanded our research into identifying how SWRT membranes and associated soil biogeochemical mechanisms can be coordinated to reduce or even eliminate the deep leaching of pesticides, endocrine-destructive compounds, and other toxins that contaminate groundwater supplies. Therefore, we believe the wicking of these stored soil solutions upward within plant root zones of the soil, greatly increased plant growth (Figure 24.6) and increased plant yields (Table 24.2).

Returns on investment of installing SWRT membranes for grain crops and vegetables range from 4 to 1 seasons of production. These very short returns on the investments, required for installing SWRT water- and nutrient-saving membranes, which may function well beyond 200 years, will attract most owners of sandy soils to adopt this new technology with the goal of producing much higher grains and plant biomass for producing renewable biofuels.

24.3 SUMMARY

SWRT water- and nutrient-saving membranes can bring marginal lands into profitable agricultural production resulting in

- More efficient use of natural soil and water resources
- Doubling of soil water-holding capacities in plant root zones
- Retention and uniform distribution of water and nutrients in plant root zones

SWRT offers long-term solutions for all sandy soils, including the transformations of arid and desert sands into substantial production regions on planet Earth in a manner that changes lives and landscapes. We now have, with this new SWRT, an innovative and long-term technology that controls optimal quantities of one of the world's most finite critical resources: soil water. We anticipate these new SWRTs will generate additional innovations that increase water use efficiency for maximum plant growth while reducing deep leaching of plant nutrients, toxins, and salts on farms and gardens located in humid, semiarid, and the most arid deserts. The brief time periods ranging from one harvest year for horticultural crops to approximately 4 years for maize crops will expand the application of SWRT-improved sand soils beyond agriculture and into sports turf, reforestation, and other water- and land- saving applications requiring water conservation technologies. As human populations increase and water conservation is more highly mandated, the commercialization potential for SWRT-induced water- and nutrient-saving opportunities will become highly attractive to small and large holders of agricultural lands and associated agricultural industries in all countries.

ACKNOWLEDGMENTS

The project was supported in part by the USDA NRCS Michigan CIG, Michigan Initiative for Innovation and Entrepreneurship, The John Deere Corporation, and Michigan State University AgBioResearch.

REFERENCES

Ash, C., B.R. Jasny, D.A. Malakoff, and A.M. Sugden. 2010. How to get better yields, soil fertility, more crop per drop, better seeds, pest free. *Science* 327: 808–809.

Barrett, C.B. 2010. Measuring food insecurity. *Science* 327: 825–828.

Basso, B., M. Bertocco, L. Sartor, and E.C. Martin. 2007. Analyzing the effects of climate variability on spatial pattern of yield in a maize–wheat–soybean rotation. *European Journal of Agronomy* 26: 82–91.

Fedoroff, N.V., D.S. Battista, R.N. Beachey et al. 2010. Radically rethinking agriculture for the 21st century. *Science* 327: 833–834.

Smucker, A.J.M., Y. Kavdir, and W. Zhang. 2014a. Root zone soil water retention technology: A historic review and modern potential. *Soil Science Society of America Journal* (in review).

Smucker, A.J.M., A.K. Guber, B. Basso, and Y. Kavdir. 2014b. Optimization of soil water content in the root zone. *Soil Science Society of America Journal* (in review).

Stein, T.M. 1998. Development and evaluation of design criteria for pitcher irrigation systems. *Beiheft No. 66, Selbstverlag des Verbandes der Tropenlandwirte. Witzenhausen e.V., Witzenhausen.* ISBN3-88122-971-X.

Yang, Z., A.J.M. Smucker, G. Jiang, and X. Ma. 2012. Influence of the membranes on water retention in saturated homogeneous sand columns. *International Symposium on Water Resource and Environmental Protection (ISWREP)*, X'ian City, China, 1590–1593. 978-1-61284-340 7/111©2012IEEE.

25 Double Loop Learning in a Garden

*Richard Stirzaker**

CONTENTS

I have been a passionate vegetable gardener from my youth. My first garden was in Cape Town, South Africa, where the soils were sandy and the long summers devoid of rain. This was a tough place to learn the trade: the soil water was quickly depleted, the nutrients easily leached, and my harvest often disappointing. Much later, I spent a gap year between school and university on farms in the hinterland of the north coast of New South Wales, Australia. This was dairy and potato country on rich red volcanic soils. Never had I seen such fertile soils effortlessly producing such luxuriant plant growth.

I had already decided to study agricultural science at university. As I paged through the faculty handbook, I became convinced that this degree held the answer to all my soil questions: What made a soil fertile? How much water could it hold? Was there any way to manage a sandy soil to give the kind of yields I saw on those volcanic soils? I completed my four-year degree and, for the most part, thoroughly enjoyed it. At university, I was replacing learning by trial and error in a garden with the bedrock of theory. In my childhood garden, I had worked out that the tomatoes in sandy soil needed water every day in midsummer; trying to stretch it to the second day inevitably led to wilting. Now, theory allowed me to calculate the correct irrigation interval for any crop on any soil for any type of weather.

My first job was to run trials on solar-powered irrigation. These were the early days of drip irrigation and the aim was to test out the most effective combination of solar panels, pumps, and irrigation equipment on a commercial scale. I had suddenly become a very big vegetable gardener, and since this was a high-profile project on the university experimental station, it was no place for trial and error. I was armed with four years of theory, and I was going to put it to good use. But I was soon to find out that theory and practice do not fit together as easily as I had expected.

I have now been doing research on soil, water, salt, and nutrients for almost 30 years. I have had a few more vegetable gardens, all on different soil types with their peculiar challenges. The current garden has spread to encompass every sunlit part of our 900 square meter block in suburban Canberra and includes many fruit trees and vines among the vegetables (Figure 25.1). The garage has been converted into a glasshouse and laboratory. The garden is a maze of pipes and taps, with wires popping out of the ground and snaking their way to data loggers. I have spent countless hours in this garden, filled many notebooks with measurements and observations and used it as a place to contemplate the global business of turning water into food.

Colleagues often ask me how I find the time to run such a garden. The answer is that I have made it a priority because the kind of learning I do in my garden is so different from that which I experience from my formal research or reading the literature. Experimentation in the garden is a bit like playing, with continual probing, wondering, and reflecting. It is where the double loop learning

* 3 Mulga Street O'Connor, Canberra Australia; Email: richard.stirzaker@csiro.au

FIGURE 25.1 **(See color insert.)** The author's backyard vegetable garden.

takes place, which I will explain in more detail later in this chapter. A number of ideas have sprung up out of the garden, and have grown into research projects and even commercial products.

The first of these ideas became the subject of my PhD. I had taken up a job as a research assistant on the university research farm. Living for the first time in the country, I started a garden far bigger than I could manage. I planted the unused portion with various cover crops, one of which had a spectacular effect. It was a type of winter-growing annual clover that died naturally in the early summer. There are many varieties of this clover, each matched to a different length of rainy season, with day length providing the trigger to flower, followed by seed set and death before the hot summer. My short-season clover had completed its life cycle in late spring and collapsed to form a thick *in situ* mulch. This made a great impression on me, especially as things back at my real job were going badly. I had spent weeks on the tractor preparing the ultimate seedbed for planting a hectare of tomatoes. Then the rains had come and turned the beautiful seedbed I had created into a soupy, slippery sludge. I returned home dejected, only to find that the same soil, experiencing the same rain under my *in situ* mulch, was in the most beautiful condition. This was the start of several years of work on no-till vegetable production (Stirzaker et al. 1993).

Several other garden-inspired ideas became formal research projects, but the overlap has been greatest in the area of irrigation: how to get the highest yields from the least possible amount of water. This has been a passion bordering on obsession since the day I took my first job on solar powered irrigation. In describing this passion, I immediately face a problem because it is near impossible to talk about water in the soil without using a lot of technical terms. All too frequently, I read popular articles on water that are so simplistic that they are more likely to lead to error than enlightenment. A major problem is that no two soils are the same. There is no guarantee that what worked so well for crop A in location A will work for crop B or location B. On the other hand, I am acquainted with many professional soil scientists who are not very proficient gardeners or irrigators. Just knowing the theory is not good enough either.

A couple of years ago I had a wonderful opportunity to write a book about how water is turned into food. My brief was to write for a lay audience and the book starts in my own backyard vegetable garden, which acted as a lens through which to explore the major themes of how the world feeds itself (Stirzaker 2010). I introduced the difficult concepts of soil physics by using the analogy of a sponge in the bath. Soil is a bit like a sponge, in that it is composed of solid material and holes, and the holes contain air or water. When we immerse the sponge in the bath, all the pores fill with water. When we lift the sponge out of the bath, water starts to drain from the bottom—quickly at first and then slowing down to regular drips. After a while the drips stop, but the sponge still holds a lot of water. If we give the sponge a mild squeeze, more water will come out. If we squeeze harder we will extract even more water, until we reach the point that we get no further water no matter how hard we squeeze.

I will now put the more technical terms to this analogy. When we lift the sponge out of the bath it is "saturated." Soils that have impeded drainage stay saturated for a long time and the plants die because the roots cannot get oxygen. Our sponge, however, drains freely and soon reaches a point where the dripping stops. We call this the "full point" because the sponge, or soil, is holding as much water as it can against the downwards pull of gravity. Squeezing of the sponge is analogous to a plant sucking water out of the soil, because a plant must exert some force to extract the water away from the soil pores and particles. The water removed by the mild squeeze is called "freely available water" and after the mild squeeze the sponge is at the "refill point." That means all the easily extractable water has been sucked out by the crop and it is time to irrigate or refill the soil profile. By the time we have completed the hard squeeze, the soil is at "permanent wilting point"—no matter how hard the plant sucks, it cannot get any more water out of the soil, so it will wilt and eventually die. Plants will continue to use water between the refill point and permanent wilting point, but the stress of needing to suck so hard means that crop yields will be well below potential. Irrigators need to know when they are approaching the refill point and also to know when to turn the water off at the full point.

Now it is time to apply these simple concepts to the problem I faced at my first job on solar-powered irrigation. My hectare of lettuce and tomatoes were out in the baking sun and I had to get the high yields attained by the best farmers. The volume of water I could pump was limited by the size of my solar panels, which meant that there was no water to waste by sloppy irrigation practice. I dredged up all the notes from my soil science classes to make sure I was on top of the theory and the practical tools that could help me. I decided that the tool I most needed was a tensiometer, as this instrument measures how hard the plants must suck to extract water from the soil. The tensiometer measures this suction force in pressure units such as the Pascal, but for our purposes, it simpler to measure suction as the equivalent height we would need to suck water from a cup using a very long straw.

My soil science lab workbook told me that a soil that was at the full point was analogous to sucking through a straw 3.3 m long. The average adult would need a good set of lungs to suck water through a straw this long, but this is just about as good as it gets for a plant. After all, the full point represents the time water has just stopped dripping from the sponge, before we apply the mild squeeze. The force required to suck water up to a height of 3.3 m is equivalent to the suction force the roots must exert to pull the water out of the soil pores after gravity has done its work.

By the time the soil reaches permanent wilting point, it would be like sucking through a straw 150 m long. Physics students know that even if we attached a suction pump to the straw, it is impossible to lift water more than 10 m high. So how does a tree 100 m tall get water to the top? In reality trees and plants do not even use their own energy to remove water from the soil, but rely on the sun's energy to draw water up the trunk or stem. As water is evaporated from the leaves at the top of the tree, a "pull" is created on the water in small tubes called xylem vessels, which go all the way down the stem and into the roots. Water molecules have very strong cohesion forces when in a narrow vessel, so it is a bit like pulling up a chain. The transpiration at the top of the tree pulls water upwards against gravity and away from the suction force of the soil.

Back on the farm, my instruction booklet that came with the tensiometers told me to irrigate when the soil dried to around 3 m of suction and definitely before 5 m. Now I had a problem. My university lab book told me that the full point was equivalent to 3.3 m of suction. How could I irrigate when the soil was wetter than full? Surely, this water would just drain through the soil like the water dripping from the bottom of the sponge. Fortunately, my lab book also gave me a second very practical definition for the full point as follows "saturate the soil and then measure after it has been allowed to drain for 48 h." This is close to the analogy of allowing the saturated sponge to drain after lifting it from the bath. I did this experiment among the lettuce and tomato crops and my tensiometers told me that the full point by this definition was somewhere between 0.5 and 0.7 m of suction. I decided to align myself with the practical definition and tensiometer measurements rather than the theory in my lab workbook.

It is not unusual for theory and practice to be in conflict, as theories are continually modified and upgraded in the face of new evidence. After all, the literature shows that there is no single definition for full point that makes sense for all soil types. What was worrying for me was that many of the soil professionals I came into contact with lived easily with the dissonance. It was most acute in those selling the new water monitoring devices that started to come onto the market during the 1990s. They confidently drew full points and refill points on graphs and the soil monitoring device showed you every hour of the day where you were between these limits. According to them, the irrigation problem had finally been solved.

Full and refill points are useful concepts, especially for sprinkler and furrow irrigation. During the first 24 h after irrigation, water moves down into the soil profile faster than transpiration can remove it. After 48 h or so, this downwards movement of water tends to be small relative to upwards movement via evapotranspiration, hence the idea of the full point, that is, the soil is holding back all the water it can against the downward pull of gravity. In the case of movable sprinklers, the aim is usually to move the sprinklers as infrequently as possible, thus the need to know the driest the soil can become without stressing the crop. Drip irrigation changed all that. Irrigation intervals could be daily, even hourly. The soil would not fluctuate between the full point and the refill point, but rather infiltration, plant uptake, and drainage from the root zone could all occur simultaneously. The gradient from saturation to permanent wilting point occurs not over time, but with distance from the drip emitter.

The proponents of the soil-water monitoring equipment confidently showed how their fields were expertly managed, with the frequent irrigation keeping the water content trace almost horizontal, oscillating around the full point on the graph. I explained that it was possible to have a big sponge with water going into the top at 10 drops per minute and coming out of the bottom at 10 drops per minute. The sponge would be at the full point, but it would be an appalling irrigation outcome. Few were convinced.

This experience, and many others like it, piqued my interest in how we learn new things. In particular, I was drawn to the idea of single and double loop learning (Argyris and Schön 1978). By learning, I do not just mean accumulating a few new facts as may occur, for instance, through reading this story. Genuine learning changes our understanding of how things work and enables us to respond differently to existing and new challenges. This kind of learning is difficult. Experts are sometimes accused of "trained incapacity," meaning the more they know about one thing, the harder it is to change their mind in response to new information. If I can articulate the jump from single to double loop learning, I know I have genuinely grasped a new and deeper understanding.

Put simply, single loop learning occurs when we act out of a commonly accepted framework. We set full points and refill points and try to stay between them. The framework works quite well for sprinkler irrigation so we apply it to drip irrigation. Even though irrigation is so frequent that the definition of full point no longer applies, the framework appears to work. According to the sponge analogy, we may be dripping water in the top as fast as it drips out the bottom, but over-irrigation is a small price many farmers are prepared to pay to avoid the disaster of under-irrigation and loss of yield. We may fine-tune our actions by deploying more sensors or obtaining more accurate sensors or adjusting full and refill points. However, in the absence of any convincing feedback showing that water is being wasted, the commonly accepted framework persists.

Double loop learning breaks out of the existing framework by explicitly trying to identify and challenge the underlying assumptions of the framework (Figure 25.2). In my case, I had to challenge the idea of full points as a useful concept for drip irrigation. It seemed that the real issue for drip

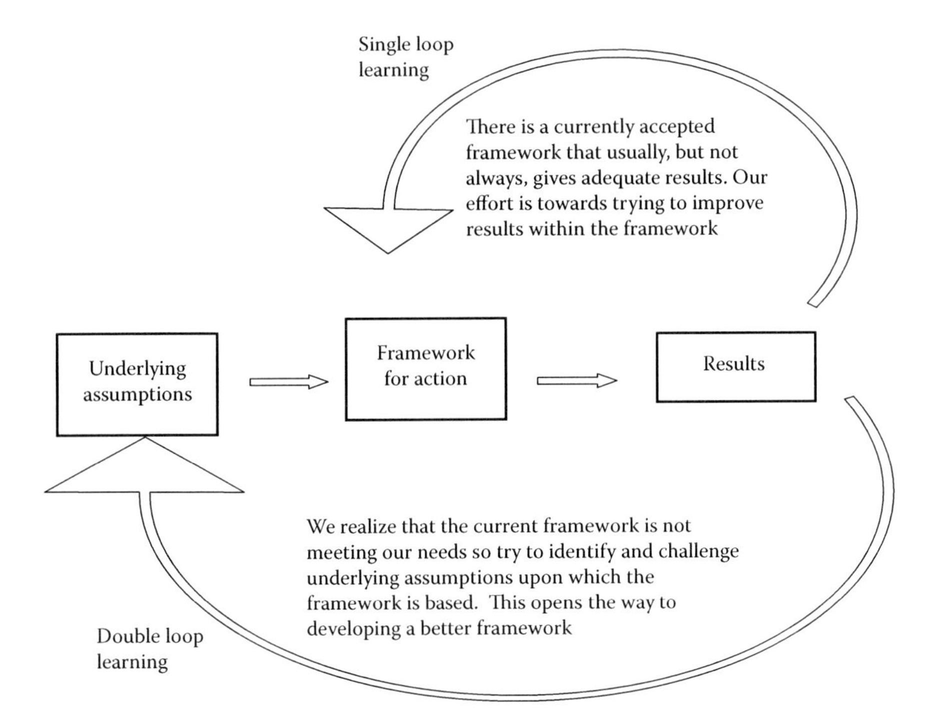

FIGURE 25.2 I was taught a framework for how to irrigate at university, which did not always give the result I wanted. Instead of trying to improve the existing framework, double loop learning invites us to understand and critique the assumptions upon which the framework was constructed.

irrigation was not when to turn the water on (i.e., before the refill point), but when to turn it off. The irrigation interval could be as short as we wanted, provided that it was stopped before water drained from the root zone. A more useful framework could be viewing the soil as a series of "tipping buckets." The top bucket, or soil layer, fills first and then overflows or tips to fill the bucket below and so on. If we break the root zone into just two layers or buckets, and we know the precise time when the top bucket tips into the lower bucket, we could shut the irrigation off shortly after the top bucket tips and minimize drainage from the root zone. This realization resulted in the development and commercialization of a device that could detect wetting fronts moving through the soil during irrigation (see www.fullstop.com.au).

The wetting front detector is composed of a specially shaped funnel with an indicator visible above the soil surface (Figure 25.3a). The funnel is buried in the root zone and water from rain or irrigation percolates through the soil and is intercepted by the funnel. As the water moves down into the funnel, the soil becomes wetter as the cross-sectional area decreases. The funnel shape has been designed so that the soil at its base quickly reaches saturation. It is important to realize that even during irrigation the soil is not saturated, and so the funnel is required to converge the downward flow of water to one point. Once saturation has occurred at the base of the funnel, liquid water flows through a filter into a small reservoir and activates a float. The float trips a magnetically latched indicator, visible to the irrigator (Stirzaker 2003). The detectors are deployed in pairs, with one in the top "bucket" and one in the lower "bucket" of the soil (Figure 25.3b).

This was another project that took its first steps in the vegetable garden. The prototype wetting front detectors were buried in a tomato bed surrounded by very accurate probes for measuring soil water. They performed so reliably that we quickly moved on to more formal experiments. We let simple wetting front detectors manage the irrigation of turf grass by automatically shutting off an irrigation system when a wetting front detector buried at 15 cm depth collected a water sample. The result was as good as we could get with the most elaborate and expensive piece of soil-water

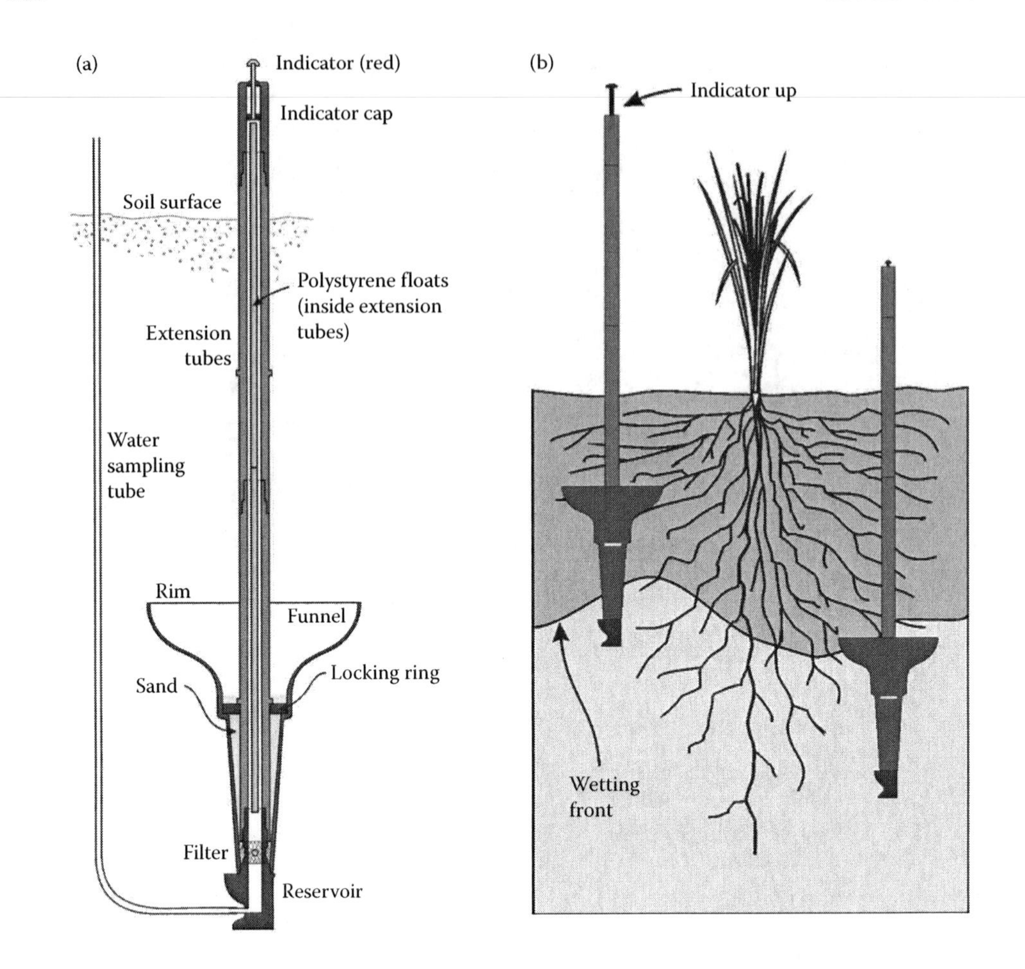

FIGURE 25.3 The components of a wetting front detector (a) and the way it is deployed in the field (b).

monitoring equipment on the market. The wetting front detector won prizes, was commercialized, and for a while I too thought the scheduling problem had been solved.

The detector did sell by the thousands and I followed it to South America where there was an unusually keen interest. After giving a seminar in Peru, a citrus farmer came up to me and said he had driven four hours to hear my talk. I asked what he learned and he said "Not much, I have got 50 detectors on my farm and they have transformed my business: I just came to say thank you." But the glory was short-lived. We visited another farm where the detectors simply did not work. You could pour as much water as you liked on top of the wetting front detector, but the funnel never collected the 20 mL of water necessary to activate the float. It turned out that water accumulated within the funnel as we might expect due to the convergence, but the wetter soil in the funnel sets up a upwards gradient that caused water to be removed from the funnel in the same way that a strip of blotting paper can suck water up and over the edge of a glass of water. Some of these Peruvian soils could suck water up and out of the funnel as fast as it was collected.

The types of soils that foiled the detector are mostly structure-less fine sand or silty soils. I am yet to come across anything quite as bad as what I faced in Peru, but there are some soils in which wetting fronts are harder to detect than others. When we add this to the fact that wetting fronts get more diffused as they move deeper into the soils, there is a problem. There is a certain depth below which the detector will not respond to draining water because the funnel cannot quite bring the soil water status up to saturation point inside the funnel to activate the float. The simple recipe—start the irrigation at a set interval and shut it off when the detector at a set depth responds—did not exist.

It depended on type of soil, width of wetting pattern, and the water content at which the wetting front moved during irrigation.

The early years of commercialization of the wetting front detector were perplexing. There were reports of excellent results from the field, but other studies were ambivalent. Some reported that the detectors responded all the time—others hardly ever. Of course, the former may have been irrigating too much and the latter not enough—it all depends on the placement depths and frequency of irrigation. This was confounded by soils that had very narrow wetting patterns (hence a frequent detector response) and those with wide wetting patterns (an infrequent response), attributes that seemed uncorrelated with soil texture. The point was the detectors were not the foolproof solution I had once dared to hope. You could learn a lot from them, but you had to persevere.

I have at least 50 wetting front detectors in my garden, most with tensiometers beside them. Many times the response of the detector was contrary to my expectations, but the tensiometer showed me my expectations were wrong. Over the years they have taught me two enduring lessons. First, it is very easy to over-irrigate at the start of the season and under-irrigate in the middle. In my attempts to nurture the seemingly fragile young tomato plants, I applied too much water that washed some of the nutrients out of the soil. Then, as the plants entered the exponential growth phase and the heat of summer kicked in, the plants were not getting enough water, despite my best attempts. Water stress sets in around flowering time and early fruit set, which as any tomato grower knows, is disastrous for attaining good yields.

The second lesson I learned in the garden was the vast difference between sprinkler and drip irrigation. I applied too little water too often by sprinkler, so just wetted the surface layers from where the water quickly evaporated. By contrast I applied too much water at one time by drippers and pushed water below the root zone in the immediate zone of the dripper. Most of all, the detectors helped me to be consistent while the crops grew and the seasons changed. No sense in activating a detector at 30 cm depth every second day when the tomatoes are young and then hardly seeing the float pop up when they are flowering.

A short time later, my next journey from single to double loop learning began. I had purchased a pocket electrical conductivity (EC) meter and was measuring the salt content of the water samples collected by the wetting front detectors. I found out that the detectors that were placed deeper in the soil, and rarely collected a water sample, tended to give quite high salt readings. In fact, the salt measured from detectors that were rarely activated could be 10 times greater than those that responded regularly. I had been taught that salt was the arch enemy of irrigation, so elevated readings were not welcomed. The manuals on irrigation and drainage provided calculations for how much extra water needed to be applied to flush the salt out of the root zone. In my experience, most irrigators had excessive leaching as a by-product of dubious irrigation management. There was no need to plan for it.

Around this time I met one of the few irrigation consultants who had made a thriving business out of advising farmers. He insisted that he had refined his irrigation monitoring systems so well that not a drop of water was wasted below the root zone. This, of course, is nonsense because the plants would have been killed by salt. All irrigation water contains at least some salt. Good quality irrigation water has less than 1% of the salt carried by sea water, but irrigation is a salt concentrating business. Small amounts of salt are added to the soil each time we irrigate, but plant roots exclude almost all the salt at their root surface during transpiration. The result is that salt builds up slowly but steadily in the root zone with every irrigation event. Eventually salt becomes very damaging to the plants, unless it is leached during periods of over-irrigation or sustained rainfall.

The framework that defined my single loop learning was that salt was bad and the less salt the better. My double loop learning began when I realized salt build up was inevitable and it provided me with a vital new piece of information. If the salt did *not* build up, then I was probably using too much water and leaching nutrients too. Moreover, the salt would be leached to the water table, and as so often happens in irrigation areas, the water tables would rise and carry the salt back up toward the surface. I started to use salt readings in the detectors to help cut back on irrigation. I tried to ensure there was no leaching until the salt levels in the deeper detector reached a threshold above which the crops could be damaged.

I soon graduated to measuring nitrate in the water sample captured by the detector. At first I just used the test strips that turn shades of purple in response to nitrate. These strips have fairly good resolution at relatively low nitrate levels but become more difficult to read in the high range. Then I found a little pocket reader that measures the strips by reflectometry. The test strips were cheap and the measurements quick and surprisingly accurate. I took my nitrate reader everywhere with me, dipping it into wetting front detector samples on every farm visit I could. Very soon my concerns that irrigation water was not managed very well were compounded by these nitrate measurements. In horticultural soils, the nitrate levels were often 10 times or more above what the plants needed.

Like many others in my field, I have lamented the low usage of soil-water monitoring equipment by irrigators. But almost no one measures the nitrate that moves around with the water. I cajoled quite a number of colleagues around the world into purchasing the nitrate readers. Then, one day, I ordered five more pocket nitrate readers for a new project I was starting in Africa. The supplier told me they had been taken off the market. I got onto the European headquarters of the well-known manufacturer and they confirmed this—adding wryly that there was not much call for measuring nitrogen in agriculture.

The irrigation scheduling problem had suddenly got a lot harder. Not only did we have to keep the water status within limits, we had to get the salt out of the root zone without the nitrate going with it. This set off a new frenzy of activity in my garden and my notebook started to fill quickly. I had now become preoccupied with irrigating in such a way as to *not* activate the deeper wetting front detectors, so I could be reasonably sure the nitrate was not being leached. Figure 25.4 shows my attempt at simultaneously managing water and nitrate to a sweet corn crop in the garden. Nitrate is plotted against the left-hand axis using wetting front detectors buried at 20 and 40 cm. Soil water suction was monitored between the detectors at 30 cm depth, shown on the right-hand axis. The soil has a reasonably heavy clay layer at 50 cm depth containing few roots; hence the monitoring focuses on the shallower layers.

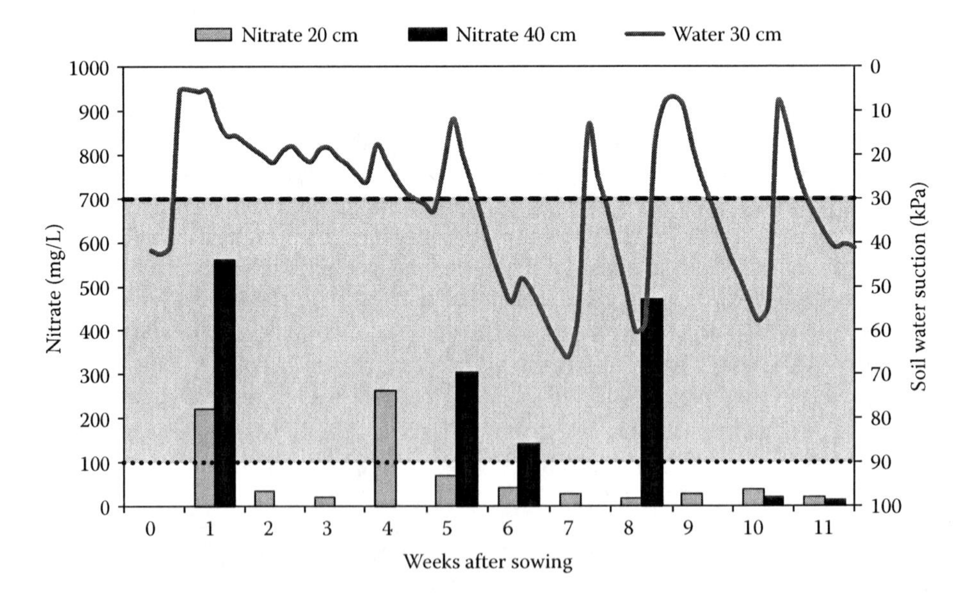

FIGURE 25.4 Managing water and nitrate together for a sweet corn crop in the garden. If the water trace drops below the dashed line, the soil is getting too dry. When the nitrate rises above the dotted line, the soil is vulnerable to leaching. The aim is to keep both graphs out of the gray zone.

My aim was to keep the nitrate between 50 and 100 mg/L and the soil water suction between 10 and 30 kPa (1 and 3 m), ranges that I judged would keep the plants growing at their maximum rate. Irrigation in week two revealed that the nitrate at both depths was way above my target, solely due to mineralization of compost added to the previous crop. I decided on small irrigations for weeks two and three, so as not to activate the 40 cm detector and leach nitrate. Week four was also a small irrigation, but I decided to add a little fertilizer to the water because the nitrate levels at the 20 cm depth were getting low. By this time the soil had dried past my 30 kPa suction limit (right-hand axis) so the sweet corn got a larger irrigation in week five. This activated the deep detector and I found there was still way too much nitrate at that depth.

By week six, the plants were getting big and a large irrigation did not halt the drying of the soil. Nevertheless, I let the soil dry below my suction threshold during week seven in an attempt to force the roots to go in search of the nitrate-laden water deeper in the soil. Then rain intervened, and again a few days later in week eight, activating the deeper detector and showing there was still a lot of nitrate down there. By week 11, I finally had the soil wet and the nitrate low, hoping that most of this nitrate was in the crop itself. The final yield was very respectable on one-fifth of the recommended fertilizer amount, but I am still not sure how much nitrate was leached. The example in Figure 25.4 illustrates the folly of optimizing for one input. Optimizing for water can compromise nitrogen use efficiency and vice versa, and salt adds another layer of complexity (see www.thescientistsgarden.com).

This idea that scheduling irrigation water could be informed by solutes was also taking root among some commercial farmers. An avocado farmer on sandy soils in Western Australia complained that he had used most of the available soil-water monitoring gear, but none fully met his requirements, particularly as he was using slightly salty water and growing a salt-sensitive crop. Then he also worked out how to make use of salt measurements. He irrigated daily according to the tools he was used to and tracked the salt in his wetting front detectors. When the salt was on a rising trend and approaching the threshold salt level, he added 10% more water to his original calculation. When the salt was falling, he did the reverse.

A group of wine grape growers near the mouth of the Murray River, South Australia, had a more serious version of this problem. They also have partially saline water, but they need to under-irrigate in their quest to produce quality wine. Under these conditions, the salt builds up quickly within the root zone. The farmers had been persuaded to purchase soil-water monitoring equipment, and many had done so. However, we found out that water stress due to excess salt was far greater than the stress due to lack of water, which is what the farmers were measuring. These farmers also worked out a way to schedule irrigation by salt monitoring, rather than by water content. First, they applied more water less often, so that the salt would be diluted in a larger volume of soil. Second, they changed the time of leaching from the end of summer to spring. By the end of the long dry summers, the soils were bone dry and leaching irrigations were ineffective. By delaying leaching till the soil profile had been wetted by the winter rains, they could remove much more salt with less irrigation water.

I had now become convinced that it was impossible to irrigate well without regular monitoring of solutes in the root zone. But equally it had dawned on me that wetting front detectors were not enough. They did not give any information until after I irrigated—that is, until the wetting front did/did not pass a certain depth. I really needed to know the water content, or preferably the soil water suction on a more regular basis. A large number of such instruments have come on the market over the last two decades. These instruments measure the amount of water at each depth in the soil at any interval you care to specify and plot the data on a graph. At first I found it amazing to watch the soil wetting at different depths and then see the root extraction patterns in real time. I was so enamored with these water content graphs that I used to study them every day. Today, sitting in the corner of my office is a box full of CDs containing many years worth of such measurements from experiments all over the world. What happened? Ultimately, the problem was data overload.

The problem of data overload sparked off the third bout of double loop learning. I had always believed that more data from more accurate sensors would lead to better decisions. Now, I found that was not the case. A farmer has to get many things roughly right to stay in business and soil water is just one of them. So I embarked on a journey to see what would be the least amount of information I could collect that would make me a better irrigator. I started with a sensor called a "Watermark" that measures soil suction. Suction (i.e., the amount of squeeze on the sponge) is much more useful for farmers than content (how much water is in the soil) because the threshold values for action are similar across all soil types.

I started by installing the Watermark sensors in my garden and read them manually via a handheld reader each day. Then I graduated to a logger with hourly measurements. As always with a new toy, I downloaded every day or so, then weekly, and over time the intervals got longer and longer as other jobs got in the way. Then a third-party company developed a logger for the Watermark that did not need downloading at all (see www.mkhansen.com/). The logging interval was locked at 8 hours, which at first looked to be a backwards step because it gave me far less information. The logger displayed a graph on a small screen showing the last 5 weeks of 8-hourly readings. I could stand in the garden and scroll through six sensors in six different garden beds. I could see the crop at the very same time I was looking at the last 5 weeks of soil water history it had experienced.

This arrangement really worked for me, because it linked up two knowledge domains. The farmer in me looked at the color of the leaves, size and arrangement of leaflets, and aspects of the overall crop vigor that no instrument can measure. The soil scientist in me wanted to see what sequence of soil moisture conditions had produced the crop in front of me, and the effects of skipping or doubling up on an irrigation event. Soon, another ingenious way of displaying the Watermark arrived, called a GDot (see www.mea.com.au). This is a column of seven luminescent yellow dots housed in a clear casing that can be seen from 30 m away. These dots flip over to black as the soil dries. When the full complement of seven dots is on display, the soil is very wet. When the last dot disappears the soil is dry. The other five dots fill in the range.

I could see the GDot from my bedroom window or wandering down a path, so the moisture status of a bed was always in full view. I interacted with the GDot and wetting front detectors in a kind of learning game. I skipped watering for a while to see how long it would take to get to three dots, and then guessed the watering run time that would be needed to get it back to seven dots. Irrigation had become a contest of skills, as I simultaneously tracked water, salt, and nitrate to see how good I could get.

The GDot confirmed my inclinations concerning the need for accuracy. In a monitoring sense, we do not need the exact measurement. Just knowing the number of dots is enough. Then I toyed with the idea that even seven dots were too many, especially as exhaustive tests in my lab showed me that the Watermark was often one or two dots away from the real value measured by my accurate tensiometers. I decided I could get away with three moisture bands reading green (good), yellow/orange (transition), and red (danger). The scientist in me agonized over such a coarse scale. But very soon I realized that it would be much more useful to have three "states" and have these sensors at three depths rather than seven states at one depth. At three depths, there are 27 different combinations of green, orange, and red, and that is more than enough to guide the irrigation decision. Moreover, it turns out that it is easier to make a sensor with two switch points (green to orange and orange to red) than to try to make something that is reasonably accurate across the entire range.

Every spare moment was spent in my lab fine-tuning the design of an inexpensive but reliable sensor and the results from a corn crop in my garden can be seen in Figure 25.5. The *y*-axis is depth, the *x*-axis is time and the colors leap out as the third dimension. Each pattern with depth has a meaning, for example, Day 64 Green–Green–Green = profile full and potentially draining; Day 78 Red–Orange–Green = top down drying; and Day 92 Red–Red–Orange = profile almost dried out. The pattern over time illustrated how my irrigation strategy was playing out.

I called the new sensor a "traffic light" because the meaning of green, orange/yellow, and red is obvious. I have put green, orange, and red thresholds on the salt and nitrate measurements as well. The traffic light and wetting front detector now sit side by side in the garden beds, watching over the movement of water, salt, and nitrate. But the stakes are a lot higher for world agriculture than the games I play in my garden. Estimates are that between 1.5 and 2.5 million hectares of land are

(a)

Day of season	1	8	15	22	29	36	43	50	57	64	71	78	85	92	99	106	113
20 cm																	
40 cm																	
60 cm																	

(b)

FIGURE 25.5 (**See color insert.**) (a) The traffic light signal showing weekly soil tension measurements over three depths for a sweet corn crop in the garden (the chart above shows the change in water status over the season; Green = gray; Orange = hatched; Red = black). (b) The author displaying the color pattern in front of the crop.

seriously degraded each year as a consequence of salinity and waterlogging, of which the root cause is poor irrigation planning and practice. Over the last 50 years when food production doubled, nitrogen fertilizer use increased by seven times. Probably only half the nitrogen fertilizer added to soil is taken up by plants, the rest polluting water and air.

From the day I left university, I searched for the tool that would make me a precise irrigator and I found it did not exist. Then I spent a decade developing a tool I thought could do the job. Eventually, I realized that there was no such tool. The best I could do was to use a combination of theory and tools to learn my way to being a better manager of water, nitrate, and salt in the root zone. The learning journey itself has not been easy. First, I had to "unlearn" some of the concepts I had been taught. Full points were no longer useful for drip irrigation, salt monitoring held vital clues to the way I was irrigating and more data did not always lead to better decisions. Right now, I feel like I know what I am doing, and for the moment, my double loop learning in the garden is complete. But no doubt I will find again that the new frameworks are not always getting things as good as they could be. I will have to go back and revisit the assumptions behind the way I think about these problems. The double loop learning will soon begin again.

REFERENCES

Argyris, C. and Schön, D. 1978. *Organizational Learning: A Theory of Action Perspective*. Reading: Addison Wesley.

Stirzaker R.J. 2003. When to turn the water off: Scheduling micro-irrigation with a wetting front detector. *Irrigation Science* 22: 177–185.

Stirzaker R.J. 2010. *Out of the Scientist's Garden: A story of water and food*. Collingwood: CSIRO Publishing.

Stirzaker R.J., Sutton B.G., and Collis-George, N. 1993. Soil management for irrigated vegetable production. I. The growth of processing tomatoes following soil preparation by cultivation, zero-tillage and an *in situ*-grown mulch. *Australian Journal of Agricultural Research* 44: 817–829.

26 Valuing the Soil
Connecting Land, People, and Nature in Scotland

John E. Gordon, Patricia M. C. Bruneau, and Vanessa Brazier*

CONTENTS

> *That* is The Land out there, under the sleet, churned and pelted there in the dark, the long rigs upturning their clayey faces to the spear-onset of the sleet. That is The Land, a dim vision this night of laggard fences and long stretching rigs. And the voice of it—the true and unforgettable voice—you can hear even such a night as this as the dark comes down, the immemorial plaint of the peewit, flying lost. *That* is The Land—though not quite all. Those folk in the byre whose lantern light is a glimmer through the sleet as they muck and bend and tend the kye, and milk the milk into tin pails, in curling froth—they are The Land in as great a measure. (Lewis Grassic Gibbon 1934, p. 293).

26.1 INTRODUCTION

Soil is inherently part of our natural and cultural heritage, linking the geosphere, biosphere, and human culture. It is the foundation of human life and there are ancient connections between people, land, and soil. In Scotland, the acidic rocks[1] and the cool, wet, Atlantic climate has meant that many soils are poorly drained, inhibiting the breakdown of plant remains and resulting in soils that are highly organic,[2] but with low natural fertility. Peats, gleys, and podzols are the dominant soil types covering over 67% of the country. Only 25% of the land is currently cultivated, with an additional 45% used for rough grazing. Forest accounts for 18% of the land cover (Dobbie et al. 2011). The most iconic soil of Scotland is peat, and the largest extent of blanket bog[3] in Europe occurs in the northern Highlands. However, centuries of land management have also created areas of high productivity, particularly in the south and east. Altogether, soils provide the backbone of Scotland's rural economy, as well as underpinning nationally and internationally valued biodiversity and spectacular and varied landscapes.

* School of Geography and Geosciences, University of St. Andrews, St. Andrews, Fife KY16 9AL, Scotland, UK; Email: jeg4@st-andrews.ac.uk

Over many millennia, people as well as nature have been an integral part of the land in Scotland, exemplified in the opening quotation from the Scottish author, Lewis Grassic Gibbon, writing in the 1930s. Gibbon captured the lives of the ordinary people living and working on the "long, stiff slopes of dour clay"—the red, stony soils developed on the Old Red Sandstone glacial till of the Mearns area in north east Scotland. He presented the harsh realism of life on the land in the early twentieth century as experienced by the local population, in contrast with the picturesque and sublime landscapes portrayed by visitors in the late eighteenth and early nineteenth centuries, or the sentimental, romanticized view of the landscape promoted by Sir Walter Scott and favored by many Victorian visitors to Scotland (Andrews 1987; Gold and Gold 1995). Indeed, the nineteenth century saw the "reinvention" of the Scottish landscape, with the expansion of Highland shooting estates and the popularization of a romantic view of Scottish history (Devine 1994), but this had not changed the hardship experienced by many who worked the land. Nevertheless, even within this more realistic portrait, there was recognition of an elemental beauty in the land, as well as a sense of continuity with the past and the eternal nature of the land compared with human existence:

> ...nothing endured at all, nothing but the land she passed across, tossed and turned and perpetually changed below the hands of the crofter folk since the oldest of them had set the Standing Stones by the loch of Blawearie and climbed there on their holy days and saw their terraced crops ride brave in the wind and the sun. (Gibbon 1932, p. 97)

In more recent times, scientific understanding of the extent and quality of soils has improved greatly with the development of the Soil Survey of Scotland in the mid-twentieth century (Soil Survey of Scotland 1984). Anyone can now access information on the physical and chemical properties of their soil and discover its capability to support agriculture and forestry. But with a more scientific and pragmatic approach, the connection between people and the land has generally become much looser, and although soil is both a natural and a human-made resource, it has tended to be undervalued in both nature conservation and cultural heritage conservation. This leads to questions about how we value the soil and how soil values have changed over time. In this chapter, we explore these questions in the context of four case studies in Scotland that may be considered as "cultural soilscapes" (Wells 2006): Flanders Moss, Muir of Dinnet, Assynt and the North West Highlands Geopark, and the machair of the Western Isles, and offer a Scottish perspective on how the natural and cultural heritage values of soil have changed over time. The examples selected reflect wider patterns and changes in different parts of Scotland.

26.2 FROM WASTELAND TO NATURE RESERVE AND CARBON STORE—FLANDERS MOSS (STIRLINGSHIRE)

Flanders Moss (Figure 26.1) represents the remnants of an extensive area of lowland raised bog developed on the "carselands," or raised estuarine mudflats, in the Forth Valley west of Stirling. The area contains a remarkable archive of past environmental changes and human impacts on the landscape (Smith et al. 2010). It also reveals dramatic changes in land use and values placed on the land over time. Toward the end of the last Ice Age (about 14,500 years ago), as the glaciers retreated, the sea flooded westwards along the Forth Valley, resulting in the extensive accumulation of estuarine deposits. And so began a see-sawing of submergence and emergence from the sea that shaped the landscape and soils, bequeathing a rich resource for people to utilize. As glacioisostatic uplift outpaced the rising sea level, much of the area emerged during the early part of the Holocene, and peat developed across the surface of the poorly drained estuarine deposits. Later, around 9000 years ago, rising sea level for a time exceeded the uplift of the land, and the area once more became part of an extended Forth estuary, with much of the peat becoming buried by estuarine mudflats. By about 7000 years ago, relative sea level was again falling, and the surface of the carse became covered in oak woodland on the drier areas and extensive raised bog elsewhere. Mesolithic and Neolithic

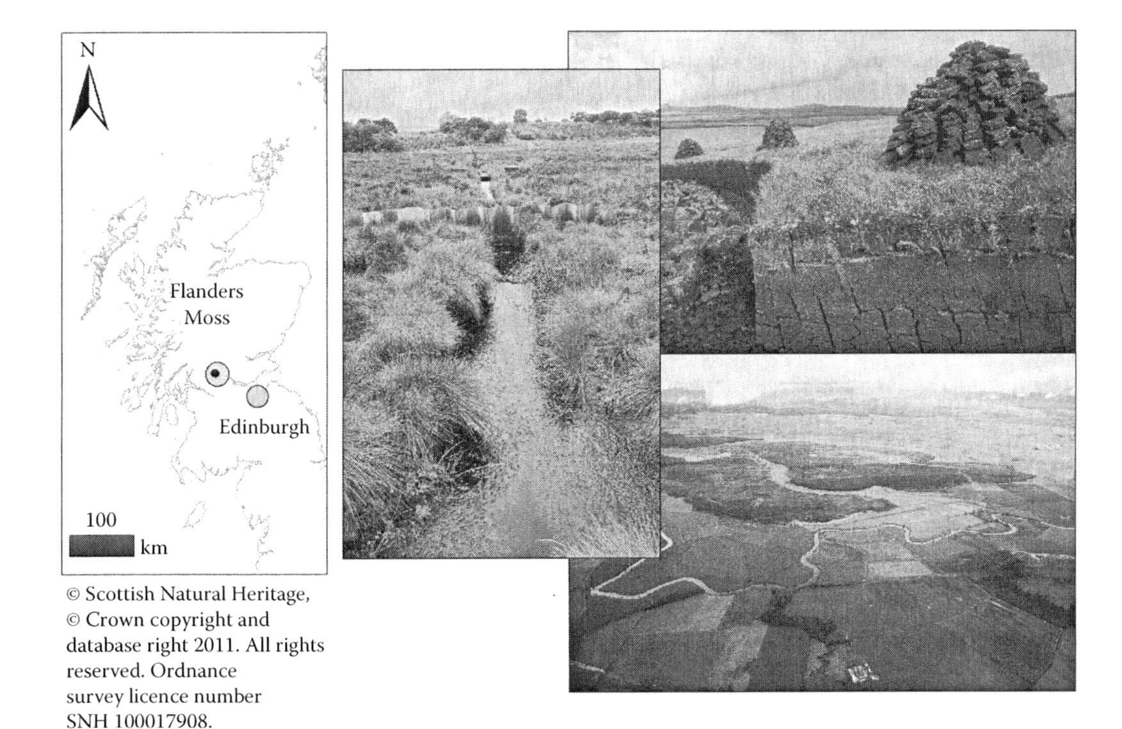

FIGURE 26.1 From wasteland to nature reserve and carbon store—Flanders Moss (Stirlingshire). From top, clockwise: traditional peat cutting in Scotland © L. Gill/SNH; aerial view of Flanders Moss raised bog surrounded by rich agricultural land © P&A Macdonald/SNH; peatland restoration using blocked drainage ditches © L. Gill/SNH.

people made use of the natural resources of the mudflats and creeks, including beds of mussels and oysters, adapting to the changing coastline (Smith et al. 2010). Bone tools found alongside some of the numerous skeletons of stranded whales and seals preserved in the carse deposits suggest that this marine resource was also exploited.

From the Middle Ages until before the agricultural improvements of the 1700s, bogs and other wetlands were regularly managed as productive areas (e.g., for reeds, fishing, and hunting) and as a resource for domestic fuel. During the period of agricultural improvements from the eighteenth century onwards, the bogs were regarded as areas of wasteland that bred disease and pestilence, and large areas of the surface peat were cleared to uncover the fertile carse deposits beneath. One of the leading agricultural improvers was the judge and philosopher, Henry Home, Lord Kames. He leased out land on the Blairdrummond Estate near Flanders Moss on condition that tenants, the so-called "moss lairds," cleared the peat (Rackwitz 2007). The peat, up to 6 m thick in places, was stripped by hand and floated down the River Forth. The practice was finally stopped in 1865 by an Act of Parliament because of the problems it created for salmon and oyster fisheries downstream.

Today, Flanders Moss is the largest remaining area of lowland raised bog in Europe and has a wetland ecosystem that is highly valued for nature conservation. It has been protected for decades under a range of national and international legislation for its unique geological and biological diversity. In recent years, this and other peatlands in Scotland have also become recognized as important stores of soil carbon (over 2700 Mt C in peat soils in Scotland, Dobbie et al. 2011). This additional value of peatland protection and restoration for carbon sequestration is now part of the climate change mitigation agenda. Meanwhile, the surviving peatland at Flanders Moss provides a well-used educational resource and visitor attraction close to the densely populated parts of the Central

Belt of Scotland. It brings people into contact with the beauty and intricacy of peatland landscapes and ecology, as well as revealing their extraordinary archives from the past and their value for carbon storage.

26.3 HIGHLAND TRANSFORMATION—MUIR OF DINNET[4] (ROYAL DEESIDE)

The Dee valley is perhaps best known for its associations with the British Royal Family, following the royal visits to the Highlands in the mid-nineteenth century and the acquisition of the Balmoral Estate by Prince Albert for Queen Victoria in 1852. However, the area has had a long history of human presence and landscape modification dating back to the Mesolithic. The retreating glaciers at the end of the last Ice Age left behind a range of glacial and glacifluvial landforms that support a diversity of soils and habitats associated with the intricate juxtaposition of ridges and mounds of well-drained sands and gravels with wetter enclosed hollows, and large lochs and extensive river floodplains (Figure 26.2). Palynological records from Loch Kinord and Loch Davan reveal that a succession of tree species spread into the area, with birch and Scots pine forest established by around 8000 years ago. The local occurrence of base-rich parent material (limestones) alongside more common acidic materials (derived from granite and schists) accounts for the presence of more versatile magnesian and calcareous soils in the area. Alluvial soils, brown earths, and podzols are common along the valley bottoms and lower slopes, whereas wetter areas and the higher slopes have wet and peaty soils. These soils enabled the establishment of a diversity of habitats and species in a landscape with natural resources, including woodland animals, edible plants and berries, and fish that supported Mesolithic hunter-gatherers. The area around Cromar, Dinnet, and Crathes is rich in archaeological remains, including Mesolithic settlement sites and later Neolithic monuments (standing stones, stone circles, and cairns), Bronze Age cairns, and Iron Age field systems

FIGURE 26.2 The Highland transformation—Muir of Dinnet (Royal Deeside). From top, clockwise: abandoned tenant cottage © L. Gill/SNH; native woodland © L. Gill/SNH; Loch Davan shore © L. Gill/SNH.

as documented in the Royal Commission on the Ancient and Historical Monuments of Scotland online database. Although the impact of the Mesolithic hunter gatherers was probably limited, the transition to farming communities and the establishment of the earliest Neolithic settlement around 6000 years ago near Crathes led to a reduction in woodland cover. At Braeroddach Loch, in the Howe of Cromar, analysis of pollen and sediments has revealed the changing pattern of land use through time (Edwards and Rowntree 1980), indicating phases of woodland clearance and soil erosion during the Neolithic and later during the Bronze and Iron Ages. The onset of pastoral farming and forest clearance after about 6000 years ago was accompanied by landscape disturbance and enhanced soil erosion and a threefold increase in sediment deposition in the loch. This accelerated around 4000 years ago when the start of cereal cultivation was accompanied by substantial woodland reduction. Later, during the historical period of the last 400 years or so, liming and drainage were necessary to support more intensive agriculture. The reasons for subsequent abandonment of such areas probably involve a complex interplay of climate shifts, soil deterioration, and social and economic factors.

The Roy Military Survey (1747–55), which covered the whole of mainland Scotland, depicts an open, forested landscape around Loch Kinord, with woodland resources in this part of Scotland having been protected by sixteenth-century legislation (Outram-Leman 2013). The commercialization of woodland in the eighteenth century with the planting of new tree species, and the adoption of forest management practices from the lowlands led to a change in social structure as well as overexploitation of the woodland resources. The expense of transporting timber has been seen as one of the limitations for exploitation of woodland resources. Until the arrival of the railway in Deeside in the mid-nineteenth century, with its consequent boost to forestry activities, timber was floated down the river. This sometimes had considerable impact on adjacent farmland (through enhancing flooding) and damage to infrastructure. [The 1813 Bridges (Scotland) Act held merchants accountable for any damages incurred during timber-floating operations.] In the late nineteenth century, the impact of overexploitation of woodland was such that, in 1866, the government removed the duty on foreign timber, and as a consequence, the value of Scottish forests plummeted. With timber resources no longer economically viable, the land gained new value as cleared heather moorland, to support game species such as grouse and red deer, and sheep grazing for wool and tallow. The near-treeless landscape of Muir of Dinnet was maintained by regular burning of the vegetation. But in the 1940s, the birch woodland quickly returned during the Second World War, because too few people were available to keep the land clear of trees by muirburn.[5] Soon after, the area was acquired for nature conservation.

Today, the mosaic of landforms and soils of the Muir of Dinnet area provide multiple benefits for nature conservation, cultural heritage, recreation, and education. They are now valued for the nationally and internationally important nature conservation interests and the habitats and species that they support, as well as their history of human occupancy. Lochs Davan and Kinord occupy glacial kettleholes and form the centerpiece of the Muir of Dinnet National Nature Reserve and its associated visitor interpretation center and walking trails. The lochs are designated as conservation areas under national and international legislation for their internationally important numbers of wintering wildfowl and geese. The Muir of Dinnet is also of European importance for its dry heaths, raised bogs, mires, lochs, and otter population. Birch and pine woodland regeneration are also occurring naturally.

26.4 REVALUING THE LANDSCAPE—ASSYNT AND THE NORTH WEST HIGHLANDS GEOPARK

The unique landscapes of Assynt and the North West Highlands Geopark (Figure 26.3) are closely related to the underlying geology and the effects of geological and geomorphological processes over long timescales, involving large-scale crustal dislocations, weathering and glacial erosion. Remarkable isolated, bare, rocky mountains of Torridonian sandstone rise steeply above an

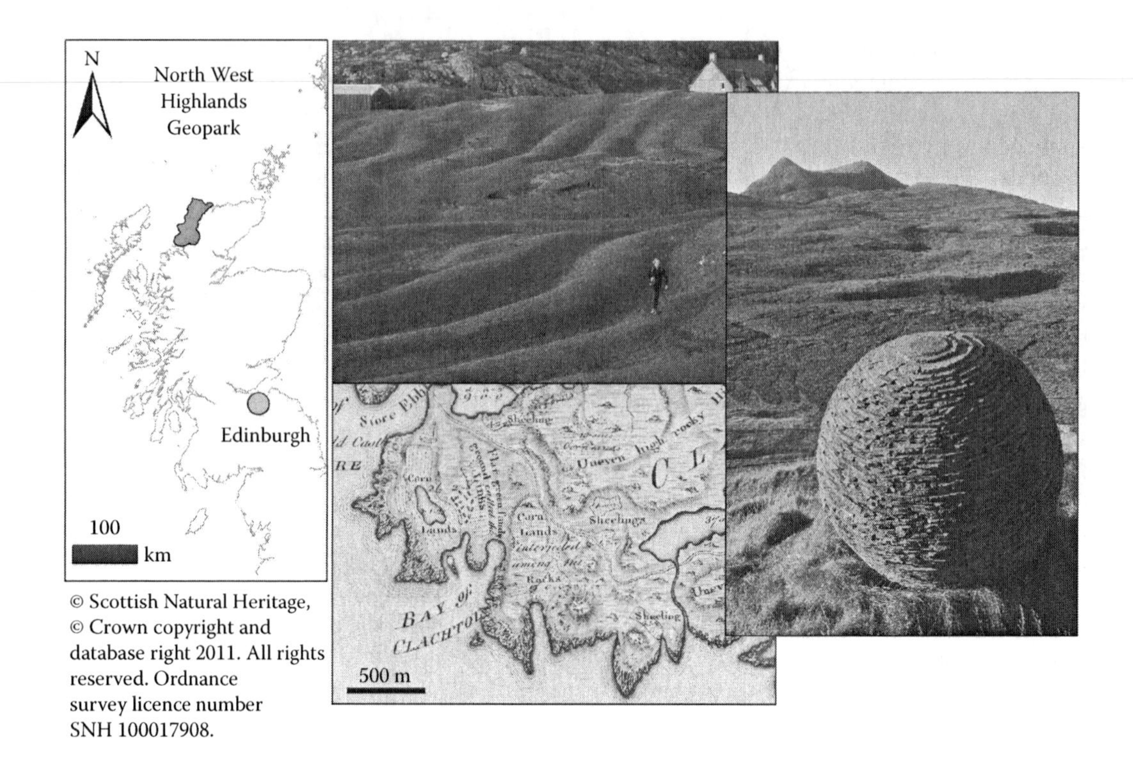

FIGURE 26.3 Revaluing the landscape—Assynt and the North West Highlands Geopark. From top, clockwise: Clach Toll lazy beds © V. Brazier/SNH; rock sculpture, *Globe* by Joe Smith 2001, at Knockan Crag National Nature Reserve © L. Gill/SNH; extract from the 1774 survey of Assynt by John Holme © Sutherland Estates, provided by the National Library of Scotland and reproduced with permission from Lord Strathnaver.

ice-scoured basement of Precambrian Lewisian gneisses. To the east, a series of low-angle thrusts and related folds were generated by the westward movement of older rocks over the top of younger ones by at least 70 km during the Caledonian Orogeny around 430 million years ago. Together, these features form a landscape of international importance for geological studies. Weathering and glacial erosion have emphasized the lines of geological weakness, giving a strong "grain" to the landscape, seen, for example, in the orientation of the main sea lochs and the fjord-like coast, and in the patterns of small lochs and low ice-scoured hills on the Lewisian rocks. Soils have formed on a wide range of drift materials, mostly coarse-grained and acidic. In general, leaching of acidic parent material will give rise to podzolic soils. Here, however, the development of indurated subsoil horizons and the shallow depth of soil have impeded drainage and led to extensive, shallow, surface-water gleys in association with peat within the hollows and channels of the rocky terrain. The surface accumulation of acidic organic matter has favored widespread peat development and typical moorland plant communities. Palynological studies of loch sediments have revealed the postglacial development of birch, birch-hazel, and pine woodland (Pennington et al. 1972). The decline of the pine woodland and expansion of blanket peat formation approximately 5000 years ago probably reflects climatic deterioration to wetter conditions. Locally, calcareous parent materials have produced more fertile, base-rich soils (e.g., in the Durness and Inchnadamph areas) and on the coastal machair. The range of habitats present reveals many close links between geology and biodiversity, with a complex of wet heath, blanket mire, and open water bodies on the ice-scoured Lewisian gneiss plateaux, grass and moss heaths on the drier Torridonian sandstone mountains, and calcareous plant communities on the Durness Limestone.

The human element in this ancient landscape is a relatively recent one, predominantly in the last 5000 years, but the impact on landscape character has been significant, especially in terms of patterns of land use and settlement. Even in this remote corner of North West Scotland, the push to implement the agricultural improvements of the eighteenth-century Enlightenment, combined with far-reaching social changes in the Highlands following the Jacobite rebellion[6] of 1745–1746, provided an environment where land was re-evaluated. The 1774 survey of Assynt by John Holme (National Library of Scotland, not dated) is one the most detailed private surveys of that period, and provides a snapshot of how resourcefully the meager land along the coast was being used at that time for raising livestock (shielings) and farming (corn lands). The accuracy of Holme's map is extraordinary, and features evident in the landscape today, such as the lazy beds,[7] can be matched with the detail on the map (Figure 26.3). Remarkably, the soil surface horizons in the semi-natural humic gleys recorded on the current soil map are significantly more acidic (on average 1.5 pH units lower) and more organic than their cultivated counterparts. Such differences in pH and carbon content between cultivated and semi-natural soils of the same series are well documented in Scotland. It has been reported that some of the more fertile Scottish "brown earth soils" are in fact cultivated podzols that have lost their characteristic surface horizons; these were often recognized as such and mapped as podzols (W. Towers, pers. comm.). Holme's survey, prepared for the owners of the Sutherland Estate, was an economic inventory—an evaluation of the worth of the land. By this time, land had become a commodity for a landowner, rather than a traditional homeland of people under the patronage of their clan chief. Economic forces elsewhere in Britain and Europe influenced how landowners viewed the potential for profit from their land. For the tenants, the threat and reality of eviction, rising rents, and meager returns from this harsh, rocky landscape led many to leave the land, through emigration overseas or by migrating to the fast-growing industrial cities in Central Scotland. In many places, the land was effectively cleared of people and their livestock to enable extensive sheep farming, with the displaced communities concentrating for a time in townships along the western coastal margins. Overcrowding, poor weather, and crop failure led to famines. Later, as elsewhere, the rise in the fashion for sporting estates during the nineteenth century again changed the value of the land, and the sheep were replaced by red deer in the uplands. Large areas of the Highlands became private land, controlled by a few landowners, with crofters[8] confined to making a living from their crofts and from the sea. In this way, much of the land of the far north west of Scotland became empty of its people, and more often visited and valued by those from elsewhere for sport and recreation.

But now the tide is turning in this part of the Highlands. Since the early 1990s, there have been a series of community buyouts in Assynt: in 1993, the Assynt Crofters Trust made history with the first such buyout in Scotland when they acquired North Assynt Estate; Culag Community Woodland Trust bought Little Assynt Estate in 2000; and in 2005, the Assynt Foundation bought Glencanisp and Drumrunie Estates. Since the passing of the 2004 Land Reform (Scotland) Act, a further 16 community buyouts have been achieved elsewhere in the Highlands and Islands. Although community purchase is not a panacea (Warren 2009), these landmark shifts in land ownership have helped to change the way land is valued. Once again the land is a resource for the use of the people who live in the area, rather than an investment or leisure resource for absent landowners. As a consequence of these communities seeking to diversify the economic potential of the land, nature conservation and tourism have gained value. There is now greater emphasis on developing a more environmentally and socially aware approach to land management to reverse centuries of environmental degradation. Although hunting will still continue as part of a more balanced approach to sustainable rural development, deer numbers have been reduced to encourage woodland restoration for the benefit of biodiversity and recreation. In 2011, a group of landowners launched a long-term project—the "Coigach and Assynt Living Landscape" (http://coigach-assynt.org/). This aims to benefit the local people, land, and economy by restoring the health of the whole ecosystem and creating rural employment.

New opportunities have also arisen through the adoption in 2004 of Scotland's first European Geopark, in the North West Highlands (http://www.northwest-highlands-geopark.org.uk/). Community involvement, with support of local government, enabled the area to gain this accolade for its internationally important geological heritage and to help promote sustainable tourism (e.g., through the innovative visitor facility at Knockan Crag National Nature Reserve, including rock art and sculptures). Assynt has long been a great attraction not only for geologists and geotourists, but also for the more general visitor, and has provided a source of inspiration for writers and artists. The poet Norman MacCaig, for example, was fascinated by this landscape and drew inspiration from it. In *A Man in Assynt* (MacCaig 1990, 225), he decried private ownership and the forced clearances of the area:

> Who possesses this landscape?—
> The man who bought it or
> I who am possessed by it?

26.5 LIVING ON THE EDGE—THE MACHAIR OF NA H-EILEANAN SIAR (THE WESTERN ISLES)

The windswept Atlantic coast of the Western Isles, Na h-Eileanan Siar, is home to over half the world's machair, particularly along the west coast from North Uist to Barra (Figure 26.4). "Machair" is Gaelic for "low-lying plain" (Love 2003) and is only found in the remote western fringes of Scotland and Ireland. Machair originated from calcareous sands that have been blown inland from beaches and dunes, giving rise to calcareous, sandy, and nutrient-rich machair soils (Ritchie et al. 2001). The wider machair system supports several different habitats—beach, sand dunes, species-rich grassland plain, wetland, loch, and blackland (mixture of sand and peat) (Angus 2001). The dunes form a natural coastal defense for the flatter machair grasslands, wetlands, and peatlands that lie behind them. In winter, storms bring in a great tangle of kelp and sea weeds that protect the dunes from erosion, and deliver nutrients into the otherwise barren sands. With a shell content

FIGURE 26.4 Living at the edge—the Machair of Na h-Eileanan Siar (The Western Isles). From top left, clockwise: shallow plowing on machair croft © L. Gill/SNH; aerial view of machair on South Uist © P&A Macdonald/SNH; oyster catcher on machair grassland © L. Gill/SNH.

of over 80%, these sands have helped to make fertile the western seaboard of the islands, where otherwise acidic, nutrient-poor, and poorly drained soils formed on Lewisian gneisses and there is widespread development of peat.

People, too, have been important partners in shaping the machair, with a human presence in the Western Isles from the Mesolithic onwards (Gregory et al. 2005). The occurrence of charcoal in the sedimentary record possibly indicates early human involvement in the development of the machair through burning of the vegetation and facilitating movement of wind-blown sands (Edwards et al. 2005). Later human activity has been integral to the evolution of the machair landscape as indicated by the extensive archaeological remains and evidence of woodland clearance and soil erosion (Edwards et al. 2000), while traditional low-intensity agriculture and land use have played a key part in the development of the key characteristics of the machair since Neolithic times (Boyd and Boyd 1990; Angus 2001). The most fertile areas have traditionally been used for the rotational cultivation of crops for animal winter fodder and domestic use, producing a patchwork of strips of arable and fallow ground that provide a variety of habitats for invertebrates and breeding birds. Seasonal grazing of cattle has also been an important part of the land use. Over many centuries and many generations, this form of crofting land use has given rise to a remarkably rich and internationally important coastal habitat, with diverse, colorful flowering plants supporting many invertebrates, small mammals, and an equally diverse and rare birdlife (Angus 1994).

The close connections between soil, landscape, nature, and people have continued from prehistoric to modern times, but sustaining human populations on the exposed Atlantic coast has required resourcefulness and sustainable use of a precious and vulnerable resource. However, the machair system is now at risk from climate change and sea-level rise, and also from some aspects of modern farming. For example, there is increased risk of coastal flooding through storm surges eroding and breaching the protective dunes, and causing salt water incursion into the flat arable area. Some modern farming practices can also inadvertently damage the machair, such as deeper plowing, which can expose the soil to greater risk of wind erosion. It is clear that traditional land use practices have greatly benefited the biodiversity and conservation value of the machair, and for 4 years from 2010, the "Machair Life" initiative is working to encourage and support the retention of sensitive and sustainable farming practices (http://www.machairlife.org.uk/).

The outstanding natural heritage value of the machair today reflects a unique combination of natural processes and human land use. In turn, the machair delivers many valued ecosystem services, including agriculture, nature conservation, natural coastal defense, recreation, and aesthetic values. The challenge is to maintain sustainable management in the future under climate change and pressure of economic and social drivers for less traditional and more intensive forms of land management. Successful nature conservation, as well as economic benefits from tourism, depends on close understanding between the crofters who work the land and those seeking to protect its biodiversity and geodiversity, supported by appropriate funding incentives.

26.6 CHANGING VALUES OF LAND AND SOIL—AND THE RISE OF CONSERVATION

The examples above show how people have developed adaptive farming and land management practices over extensive periods of trial and error and so enabling continuity of local settlements. But many of those communities living at the limits of environmental systems are unable to adjust to abrupt change brought by economic and social factors and exceptional climatic events. This creates cycles of land abandonment and regeneration (Dodgshon 2005; Ross 2008) that have shaped the diversity of the present landscape, soils, and biodiversity, with many of our iconic landscapes (e.g., machair) having only survived through the continuation of more traditional forms of land management.

Although the soil and the land are apparently timeless on a human lifetime scale, human perceptions and values have changed over time, reflecting different social and economic needs as well as environmental change. In turn, this has led to significant landscape changes as illustrated in the above examples. People have derived multiple benefits from the use of soil, including agriculture, nature conservation, coastal defense, and recreation and aesthetic values. The food production needs of early Neolithic farmers were met from the extensive use of the most easily worked land. Essentially, people were working within environmental limits, but with little regard to landscape impacts, particularly in terms of woodland clearance and soil erosion, and generally there were increasing human impacts on soils from approximately 6000 years ago (Davidson and Carter 2003).

The Enlightenment and Agricultural Revolution of the eighteenth century brought widespread changes in the way agricultural land was managed and worked in attempts to overcome the natural limitations of soil fertility (Smout 2000). In Scotland, following the 1745–1746 Jacobite uprising, estates formerly owned by Jacobite supporters were forfeited to the Crown, and the Commissioners for the Annexed Estates introduced the first land surveys and records of agricultural changes. Ownership of these forfeited lands was eventually restored toward the end of the eighteenth century, but by then other landowning families had followed the lead for improvement, like the Sutherland Estate described above. The willingness to introduce more "enlightened" and profitable forms of land management was met initially with great enthusiasm but with varied success, impeded by inherently low soil fertility and a series of harsh climatic events, crop failures, and famines. One of the most dramatic changes to the landscape was the enclosure of large areas of farmland earmarked by landowners for improvement, and originally common land shared by tenants for grazing and access to other resources (e.g., fuel).

The rise in sheep farming in the nineteenth century to meet the demand for wool during the Napoleonic Wars provided further dramatic changes to the Highland landscape and society. This also coincided with improvements in transport networks opening up new markets for export and access to soil fertilization (liming of acidic soils and importation of guano), that in turn boosted arable production. In lowland areas like the Forth Valley, wholesale stripping of peat was encouraged to improve the land, while in the Highlands, the far more serious social consequences of these land management changes resulted in many people forced to leave their homes to settle on the harsh west coast, where the choice they faced was to eke out a living, emigrate overseas, or head for the new industrial centers further south. With the advent of industrialization and growing urban populations in the nineteenth century, came a drive to maximize agricultural yields through mechanization of agriculture and the application of fertilizers, a process that intensified in the post-World War II period. Scotland's population became concentrated in the industrialized Central Belt, and highly mechanized farming became focused on the lowlands of the east coast. The cultural dislocation between people and the land that supported them had set in.

In the Highlands, land was now owned and controlled by a few, and valued as a commodity not only for commercial gain, but also as a mark of social standing. The popularity of the Highland sporting estates and their grand Scottish Baronial-style castles was enhanced by the enthusiasm of Queen Victoria and Prince Albert. Industrialists bought into the Highland idyll, purchasing estates for status and pleasure, as much as economic gain. However, it was not only the affluent who sought out Scotland for their enjoyment (Durie 2003). With improvements in transport and affluence, commercial operators, including Thomas Cook, began to organize excursions in Scotland in the mid-nineteenth century, initially with itineraries that included spectacular natural landmarks and the landscapes associated with Scott's novels and poetry and following the itineraries of earlier travelers and literary figures (Gold and Gold 1995).

Two World Wars in the twentieth century brought an abrupt change in the fortunes of the sporting estates. With the workforce needed elsewhere, many labor-intensive practices (such as the muirburn at Dinnet) were abandoned. People did not return in great numbers to work on these estates, and for some, their landholding assets were no longer economically viable. The postwar period saw new economic and social changes with the expansion of forestry plantations, further growth of tourism and recreation, and the popularization and growth of the conservation movement. Today, Scotland's farm businesses still contribute to economic growth (£700 million annually), but have become overdependent on subsidies from the Common Agricultural Policy and national grants schemes; however, without them, there is a risk of land being left unproductive and derelict (Scottish Government 2012).

There is a long history of environmental awareness among the people in the Highlands that has a strong cultural dimension (Hunter 1995). However, with increasing urbanization, the relationship between people and the land generally has diminished, leading to calls for a cultural renewal (McIntosh 2001; White 2003). Perhaps the real hope for reconnecting people and soil is greatest in those marginal environments like the machair of the Western Isles and in Assynt, where sustainable land use and the economic value of scenery, environment, and wildlife are important for rural development and conservation alike.

In the mid-twentieth century, increased evidence of environmental impacts from the exploitation of the land led to a growing demand for stronger legislative controls over access to the countryside and the development of a national conservation strategy to ensure protection of important sites of conservation value and iconic species shown to be in decline or being overexploited (Vincent 2006). UK nature conservation legislation through the National Parks and Access to the Countryside Act 1949, and the subsequent Wildlife and Countryside Act 1981, emphasized the protection of special sites for valued habitats, species, and geological and physiographical features. Subsequent legislation and policy have enforced the protection of valued habitats and species, but the definition of a specific legal framework for the protection of soil in the United Kingdom has until now lagged behind the development of other aspects of nature and cultural conservation.

Soil and conservation values have also changed over time in response to changing environmental, economic, and social drivers. There has long been a need for a balance between food production and protection from erosion, although not always recognized in practice. But soil conservation is much more than protection from erosion. It is vital for sustainable use of resources and sustainable development. More recently, the need for soil protection has been driven by wider international and national agendas on sustainable development, biodiversity, and climate change. The European Union Thematic Strategy for Soil Protection published in 2006 provided a springboard for increasing awareness of soil protection issues in Europe. It described in detail the roles and threats to soils and provided a rationale to review current soil degradation trends and future challenges to soil protection in European countries. In Scotland, this led to the adoption of a Scottish Soil Framework (Scottish Government 2009), which recognizes a much wider range of essential functions and services supported by soil, including non-agricultural uses and wider benefits derived from healthy soil—quality of life, landscape, and cultural values. There is now a strong emphasis on the importance of soils in ecosystem services and valuation of natural capital. Soils provide multiple functions and deliver essential ecosystem services, including habitat support, biogeochemical and hydrological cycling and prevention of erosion (UK National Ecosystem Assessment 2011). Although not a statutory instrument, the Scottish Soil Framework promotes across government agencies and the wider land user community a common vision and actions for sustainable management and protection of soils under a changing climate (Dobbie et al. 2011). Climate change and carbon management are now a major driver for soil conservation in Scotland (Dobbie et al. 2011; Scottish Government 2011), since the majority of the UK terrestrial carbon stock is contained in Scottish peaty soils (Chapman et al. 2009). The loss of 5–7% of the UK's peatland would equate to the total annual UK human greenhouse gas emissions (Worrall et al. 2011).

26.7 CONCLUDING REMARKS

Robert Burns (1759–1796), Scotland's national poet, was a poet of the soil. Through his agricultural roots and from his time as a tenant farmer, he had a deep connection with the land and the people it supported, clearly expressed in "The Cotter's Saturday Night" (Burns 1786, 136):

> O SCOTIA! my dear, my native soil!
> For whom my warmest wish to Heaven is sent!

For much of human history, "few things have mattered more to human communities than their relations with soil," but in the past century or so, "nothing has mattered more for soils than their relations with human communities" (McNeil and Winiwarter 2004, p. 1627). Urban populations have lost their connections with the soil and awareness of its value and sensitivity, and human impacts have accelerated. Yet, as the ecosystem approach demonstrates, we depend upon soils in so many ways. Writing about the growth of "green consciousness" in the Scottish Highlands, Smout (1993) noted two significant progressions—from a romantic view of nature to a scientific one focused on protected sites, and from the latter to a broader vision of sustainable management of the whole environment. Crucially, the way ahead must encompass both the scientific and the cultural dimensions in an integrated manner.

Learning from the past can help to promote the sustainable management and protection of soils, land, and biodiversity within the limits of economic, social, and environment needs. We now understand better what soil does for us and the multiple goods and services that depend on healthy and functioning soil. But at the same time, there must be recognition that the soil is a fragile and non-renewable resource that is now facing both ongoing and new pressures and threats. Adapting to, and mitigating the impacts of climate change will prove challenging, and how we balance the environmental, social, and economic costs of our actions will be our legacy. Rediscovering our cultural connections with the soil will be an essential part of integrating our understanding of the Earth with its sustainable management.

NOTES

1. Scotland's geology is dominated by "acidic" igneous and metamorphic rocks (granite, gneiss, and several kinds of schist) that are generally resistant to weathering. These "acidic" (or more correctly, felsic) rocks are rich in silica (>65% by weight) and composed mainly of felsic minerals (e.g., quartz, muscovite, orthoclase, and the sodium-rich plagioclase feldspars), leading to acidic, nutrient-poor soils. "Basic" (mafic) rocks are poor in silica (<55% by weight), but contain significant amounts of mafic or ferromagnesian minerals (e.g., olivine, pyroxene, amphibole, and biotite). They include basalt, dolerite, and gabbro, which tend to weather more easily and to form more basic and fertile soils.
2. The total carbon content in the soil is estimated to be equivalent to 186 years of Scotland's current annual greenhouse gas emissions.
3. Blanket bog is a large area of peatland covering the whole landscape and common in areas of high rainfall and low temperature.
4. Muir is the Scots word for moorland.
5. Muirburn is the managed burning of moorland to improve habitats for game birds (grouse).
6. The Jacobite uprisings (so named from Jacobus, the Latin form of James) were a series of rebellions during the late seventeenth and early eighteenth centuries seeking initially to reinstate James VII of Scotland and II of England after he was deposed by Parliament in 1669, and later to regain the throne for his descendants of the exiled House of Stuart. The 1745–1746 uprising, led by Charles Edward Stuart ("Bonnie Prince Charlie"), was comprehensively defeated by the British Government forces at the Battle of Culloden, near Inverness, in 1746. Subsequently, the British Government implemented measures to ensure that another uprising could not take place.
7. Lazy beds are a form of cultivation formerly extensive in the Highlands and Islands of Scotland. Turf and soil dug from the furrows was heaped into a series of parallel, flat-topped ridges, often fertilized with seaweed, and planted with potatoes and grain.
8. Crofting is a traditional system of land tenure in the Highlands and Islands of Scotland. A croft is a small landholding that was usually a tenancy with inherited rights to work the land, but many are now owner-occupied. Crofters also have grazing rights in common grazing land shared with other crofts, which they manage together.

REFERENCES

Andrews, M. 1987. *The Search for the Picturesque: Landscape, Aesthetics and Tourism in Britain, 1760–1800*. Aldershot: Scolar Press.

Angus, S. 1994. The conservation importance of machair systems of the Scottish Islands, with particular reference to the Outer Hebrides. In *The Islands of Scotland. A Living Marine Heritage*, eds. J.M. Baxter and M.B. Usher, 95–120. Edinburgh: HMSO.

Angus, S. 2001. *Moor and Machair. The Outer Hebrides*. Cambridge: White Horse Press.

Boyd, J.M. and I.L. Boyd. 1990. *The Hebrides: A Natural History*. London: Collins.

Burns, R. 1786. *Poems, Chiefly in the Scottish Dialect*. Kilmarnock: printed by John Wilson.

Chapman, S.J., J. Bell, D. Donnelly, and A. Lilly. 2009. Carbon stocks in Scottish peatlands. *Soil Use and Management* 25: 105–112.

Davidson, D.A. and S.P. Carter. 2003. Soils and their evolution. In *Scotland after the Ice Age. Environment, Archaeology and History. 8000BC–AD1000*, eds. K.J. Edwards and I.B.M. Ralston, 45–62. Edinburgh: Edinburgh University Press.

Devine, T.M. 1994. *The Transformation of Rural Scotland: Social Change and the Agrarian Economy, 1660–1815*. Edinburgh: Edinburgh University Press.

Dobbie, K.E., P.M.C. Bruneau, and W. Towers (eds). 2011. *The State of Scotland's Soil*. Edinburgh: Natural Scotland. http://www.sepa.org.uk/land/land_publications.aspx (accessed 2 January, 2013).

Dodgshon, R.A. 2005. The Little Ice Age in the Scottish highlands and islands: Documenting its human impact. *Scottish Geographical Journal* 121: 321–337.

Durie, A.J. 2003. *Scotland for the Holidays. A History of Tourism in Scotland, 1780–1939*. East Linton: Tuckwell Press.

Edwards, K.J. and K.M. Rowntree. 1980. Radiocarbon and palaeoenvironmental evidence for changing rates of erosion at a Flandrian stage site in Scotland. In *Timescales in Geomorphology*, eds. R.A. Cullingford, D.A. Davidson, and J. Lewin, 207–223. Chichester: Wiley.

Edwards, K.J., Y. Mulder, T.A. Lomax, G. Whittington, and K.R. Hirons. 2000. Human-environment interactions in prehistoric landscapes: The example of the Outer Hebrides. In *Landscape the Richest Historical Record*, ed. D. Hooke, 13–32. Amesbury: Society for Landscape Studies, Supplementary Series, 1.

Edwards, K.J., G. Whittington, and W. Ritchie. 2005. The possible early role of humans in the early stages of machair evolution: Palaeoenvironmental investigations in the Outer Hebrides, Scotland. *Journal of Archaeological Science* 32: 435–449.

Gibbon, L.G. 1932. *Sunset Song*. London: Jarrolds.

Gibbon, L.G. 1934. The Land. In *Scottish Scene or the Intelligent Man's Guide to Albyn*, eds. L.G. Gibbon and H. MacDiarmid, 292–306. London: Hutchinson.

Gold, J.R. and M.M. Gold. 1995. *Imagining Scotland. Tradition, Representation and Promotion in Scottish Tourism since 1750*. Aldershot: Scolar Press.

Gregory, R.A., E.M. Murphy, M.J. Church, K.J. Edwards, E.B. Guttmann, and D.A. Simpson. 2005. Archaeological evidence for the first Mesolithic occupation of the Western Isles of Scotland. *The Holocene* 15: 944–950.

Hunter, J. 1995. *On the Other Side of Sorrow. Nature and People in the Scottish Highlands*. Edinburgh & London: Mainstream Publishing.

Love, J.A. 2003. *Machair: Scotland's Living Landscapes*. Battleby: Scottish Natural Heritage.

MacCaig, N. 1990. *Collected Poems*. London: Chatto & Windus.

McIntosh, A. 2001. *Soil and Soul. People versus Corporate Power*. London: Aurum Press.

McNeil, J.R. and V. Winiwarter. 2004. Breaking the sod: Humankind, history and soil. *Science* 304: 1627–1629.

National Library of Scotland. Not dated. John Holme's survey of Assynt in 1774. http://maps.nls.uk/estates/assynt/further_information.html (accessed 2 January, 2013).

Outram-Leman, S. 2013. Ecosystem service provision in the Cairngorms National Park: Case study of past and future management of geodiversity and biodiversity. *Scottish Natural Heritage Commissioned Report No. 554*. Edinburgh: Scottish Natural Heritage.

Pennington, W., E.Y. Haworth, A.P. Bonny, and J.P. Lishman. 1972. Lake sediments in northern Scotland. *Philosophical Transactions of the Royal Society of London* B264: 191–294.

Rackwitz, M. 2007. *Travels to Terra Incognita. The Scottish Highlands and Hebrides in Early Modern Travellers' Accounts c. 1600 to 1800*. Munster: Waxman Verlag.

Ritchie, W., G. Whittington, and K.J. Edwards. 2001. Holocene changes in the physiography and vegetation of the Atlantic littoral of the Uists, Outer Hebrides, Scotland. *Transactions of the Royal Society of Edinburgh: Earth and Environmental Science* 92: 121–136.

Ross, A. 2008. Literature review of the history of grassland management in Scotland. *Scottish Natural Heritage Commissioned Report No. 313*. Edinburgh: Scottish Natural Heritage.

Scottish Government. 2009. *The Scottish Soil Framework*. Edinburgh: Scottish Government. http://www.scotland.gov.uk/Publications/2008/06/27092711/0 (accessed 2 January, 2013).

Scottish Government. 2011. *Getting the Best from Our Land. A Land Use Strategy for Scotland*. Edinburgh: Scottish Government. http://www.scotland.gov.uk/Publications/2011/03/17091927/0 (accessed 2 January, 2013).

Scottish Government. 2012. *CAP Reform Consultation Analysis Report and Consultation Responses*. Edinburgh: Scottish Government. http://www.scotland.gov.uk/Publications/2012/10/3617/0 (accessed 2 January, 2013).

Smith, D.E., M.H. Davies, C.L. Brooks, T.M. Mighall, S. Dawson, B.R. Rea, J.T. Jordan, and L.K. Holloway. 2010. Holocene relative sea levels and related prehistoric activity in the Forth lowland, Scotland, United Kingdom. *Quaternary Science Reviews* 29: 2382–2410.

Smout, T.C. 1993. The Highlands and the roots of green consciousness, 1750–1990. *Scottish Natural Heritage Occasional Paper No. 1*. Edinburgh: Scottish Natural Heritage.

Smout, T.C. 2000. *Nature Contested. Environmental History in Scotland and Northern England since 1600*. Edinburgh: Edinburgh University Press.

Soil Survey of Scotland. 1984. *Organisation and Methods. Handbook 8*. Aberdeen: Macaulay Institute for Soil Research. http://www.scotland.gov.uk/Topics/Business-Industry/Energy/Energy-sources/19185/17852-1/CSavings/Handbook8 (accessed 2 January, 2013).

UK National Ecosystem Assessment. 2011. *The UK National Ecosystem Assessment: Synthesis of the Key Findings*. Cambridge: UNEP-WCMC. http://uknea.unep-wcmc.org/Resources/tabid/82/Default.aspx (accessed 2 January, 2013).

Vincent, M. 2006. Ideas for a UK nature conservation framework. *Conservation Challenge No. 1*. Peterborough: Joint Nature Conservation Committee.

Warren, C.R. 2009. *Managing Scotland's Environment*. 2nd edition. Edinburgh: Edinburgh University Press.

Wells, E.C. 2006. Cultural soilscapes. In *Function of Soils for Human Societies and the Environment*, eds. E. Frossard, W.E.H. Blum, and B.E. Warkentin. London: Geological Society, Special Publications 266, 125–132.

White, K. 2003. *Geopoetics: Place, Culture, World*. Glasgow: Alba Editions.

Worrall, F., P. Chapman, J. Holden, C. Evans, R. Artz, P. Smith, and R. Grayson. 2011. A review of current evidence on carbon fluxes and greenhouse gas emissions from UK peatland. *JNCC Report No. 442*. Peterborough: Joint Nature Conservation Committee.

27 Sports Surface Design

The Purposeful Manipulation of Soils

Richard Gibbs[*]

CONTENTS

27.1 INTRODUCTION

Most people would recognize the value of soil as a basis for sustaining human life. However, soil is also an integral component of the social and recreational fabric of human activity. Whether as individuals playing a sport on natural turf, or simply enjoying watching sports on television or from a seat in a grandstand, every day of the year sees applied soil science in action around the world across thousands of sports and athletic fields, stadiums, lawn bowls clubs, golf courses, racecourses, tennis clubs, cricket grounds, baseball fields, and other recreational areas. Yet, few would have a good appreciation of just how different and challenging the management of soil for sporting use is compared with the management of soil for conventional agricultural or horticultural crop production. This chapter aims to widen that appreciation by examining certain aspects of soil science that

[*] The Sports Turf Research Institute, Bingley, West Yorkshire, BD16 1AU, United Kingdom; Email: richard.gibbs@stri.co.uk

are exploited in order to provide a range of contrasting natural turf sports surfaces able to satisfy the playing quality demands of different sports.

27.2 SOIL MANAGEMENT APPROACH FOR CROP PRODUCTION

Sustainable agricultural or horticultural crop production systems rely heavily on good soil structure along with the biological cycling of carbon as the foundation for generating the ideal soil properties for supporting crop growth. A well-structured soil generated through a complex process of reactions, associated mainly with the clay fraction of soil and its organic matter content, results in particles of sand, silt, and clay forming a system of water-stable aggregates. The micropore spaces within these aggregates are important for retaining soil moisture that is used to supply the growing crop. In between the aggregates are the larger macropore spaces, required for movement of excess water through the profile (i.e., drainage), soil aeration, and unimpeded root growth.

Although particle aggregation is a natural process of soil structure formation, the end result is fragile. For centuries, those managing land for crop production have tried to enhance the process of particle aggregation using soil husbandry techniques developed to achieve favorable soil physical conditions for crop growth. Such techniques include cultivating the land to create soil porosity and a suitable seedbed, and maintaining a high level of soil biological activity through the return of organic material, for example, by spreading and incorporating crop residues or animal manures.

The greater the soil's clay content, the more challenging it may be to manage the soil. However, it is generally expected that agricultural or horticultural crop production systems require soil husbandry techniques to be tailored to the texture of soil available and the growing conditions present, with it being impractical and uneconomic to change soil texture and manipulate environmental growing conditions. It is only when the financial return on a crop is particularly lucrative that it makes sense to take more control of these factors (e.g., in the glasshouse crop production industry).

27.3 USING NATIVE SOILS FOR SPORTING USE

27.3.1 Importance of Soil Physical Properties

Like agricultural or horticultural crop production systems, the extent to which a native soil can be used to support natural turf for sports use also depends on the soil type present and the local climate. However, here the similarity ends. The natural formation of soil structure through the process of particle aggregation to generate ideal soil physical properties is simply too slow and too fragile for sustaining most sports turf soils: sporting activities generally destroy soil structure and damage the turfgrass plant, resulting in loss of turfgrass cover and deterioration in soil physical properties, either deliberately through preparation (e.g., rolling the surface to generate the required playing quality characteristics), or through subsequent use (e.g., playing games on a saturated surface in winter) (Figure 27.1). This loss of soil structure and ground cover is caused not only because sporting activities generally occur over a much wider range of soil moisture conditions than found with agricultural or horticultural traffic, but also because of the sheer intensity of play on some sports surfaces. When one considers that there are thousands of foot imprints into the soil per player per game of football, it is not surprising that a soil that might have behaved as a well-drained agricultural soil may turn rapidly into a grassless quagmire when used as a football field in winter (Adams and Gibbs 1994).

Construction activities also cause a loss of soil structure. Bulk cut-and-fill earthmoving and soil stockpiling activities during construction, common on large sites such as golf courses or multifield sports complexes, result in loss of natural soil horizons and mixing of soil materials from different areas and soil profile depths across a site.

Having some flexibility to use the land in response to changing soil and weather conditions, and the application of traditional principles of soil management are of little help in managing soil

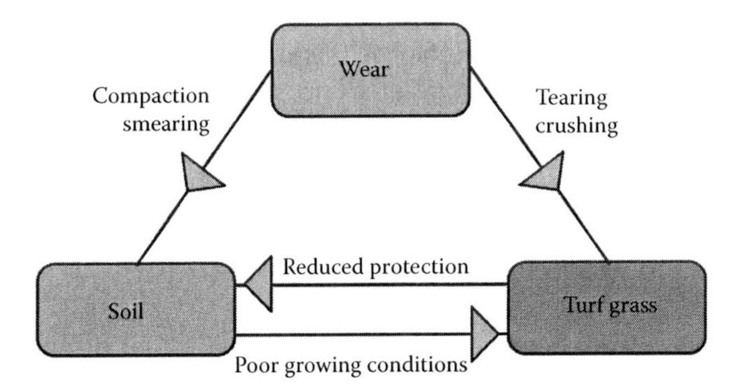

FIGURE 27.1 Summary of the effects of wear on soil and turfgrass.

physical properties of sports turf soils. Conventional cultivation techniques that disrupt surface levels or invert the soil profile are not appropriate for relieving surface compaction, particularly when one considers that specified construction tolerances for some sports surfaces can be as precise as 5 mm of the design level. Scheduled events have to be fulfilled at the arranged times on dates planned months or even years beforehand. Games are then played irrespective of the weather (e.g., under heavy rainfall or in the middle of frozen winter conditions), so there is always the potential for maximum soil structural damage to occur. Moreover, in terms of renovation at the end of a playing season, a great deal of time and effort is actually spent *removing* organic matter off the sports surface during renovation, not adding it. This is because the relatively biologically inert nature of sports turf soils and lack of "inversion" cultivation such as plowing allow organic matter to accumulate as an organic horizon (the thatch layer). The thatch-covered soil is eventually unable to confer the required playing quality characteristics for the sport in question. In fact, organic matter in colloidal form is particularly good at blocking macropores in intensively used and compacted sports turf soils. This contrasts with most natural, horticultural, and agricultural soils, where, almost without exception, the effects of organic matter are beneficial (Adams and Gibbs 1994).

Another important aspect of soil management for crop production is that emphasis is usually placed on optimizing dry matter yields in order to achieve the best financial return from a crop. However, "yield" in terms of grass dry matter production per se is of little relevance in sports turf systems because the reliability of a sports turf soil is actually a reflection of the tolerance to and recovery from wear of the intended turfgrass species under a series of natural and player-induced stresses. If anything, the definition of "yield" in sports turf systems is a function of the number of games, events, or tournaments held on a sports turf surface, or in a pure business model sense, the ticket sales and sponsorship takings at a stadium gate or a golf club.

Thus, the adaptation and management of soils for sporting use introduces constraints and requirements that are not normally experienced under conventional agricultural or horticultural land use. For this reason, soil media and complete profiles tend to be designed, engineered, and constructed for certain sports, particularly at the professional level, in order to provide the best conditions for turfgrass growth along with the ideal playing quality characteristics.

27.3.2 Requirements of Sports Turf Soils

The main requirements of a soil for sporting use are not only that it provides a reliable growing medium from an agronomic perspective, but also that the soil allows a particular quality of play for a given intensity and level of use and maintenance. Design requirements vary significantly depending on the sports surface. Those surfaces that have to be regularly used in wet weather conditions and/or in challenging climates such as the tropics or the desert (i.e., the majority of sports surfaces around the world) require guaranteed drainage rates and/or custom-made rootzones with specified soil physical properties such as predictable moisture retention.

However, it is only in sand and loamy sand soils with a particle size distribution of *at least* 70% sand in the medium-fine particle size range (0.125–0.50 mm) that sufficient macroporosity can be maintained in the absence of particle aggregation to allow rapid transmission of water and a satisfactory state of aeration (Stewart 1994). Since the majority of native soils around the world have silt (0.002–0.05 mm) plus clay (less than 0.002 mm) contents in excess of 30%, adequate natural drainage and aeration in these soil types is ambitious at best when they are used to construct sports surfaces requiring a high quality of play under intensive use and potentially wet soil conditions.

In contrast, summer sports that require dry, hard surfaces to confer adequate ball rebound (e.g., cricket pitches and tennis courts), use heavier-textured (high clay content) soils with high binding strength and soil cohesiveness, so that the surfaces do not disintegrate under play. The cricket pitch is the ultimate test bed of destruction, as will be seen at the end of this chapter. Turfgrasses used for cricket are expected to grow in heavily compacted clays where the soil structure has been deliberately destroyed as part of the preparation of the surface. Cricket is a particular, although fascinating, "sports surface quirk" from a soil science perspective, and many would view the preparation of a cricket pitch to be more closely aligned to the principles of road building than to turf management. At the other end of the soil texture continuum are structureless, sand-dominant rootzones, used as the principal component of thousands of scientifically designed sports surfaces all over the world.

In summary, regardless of the sport, major upgrading of sports turf soils has been achieved by gaining an understanding of, and then control over, their soil physical characteristics. This contrasts with an increase in the value of agricultural and horticultural soils, which has come about largely through better understanding of nutrient management and fertilizer input (Adams and Gibbs 1994). It is just unfortunate that many conventional land management techniques cannot be used on natural turf sports surfaces—hence the challenge in their design, construction, and maintenance.

27.4 DESIGNING DRAINAGE SYSTEMS AND PROFILES FOR SPORTS SURFACES

It is only since the 1960s that significant improvements and innovations have been made in the design of natural turf sports surfaces requiring a predictable quality of play. For the majority of sports surfaces, the main improvement has come about by recognizing that surface compaction is usually the cause of drainage problems in the soil profile, and that a different approach is required to ensure rapid removal of surface water and delivery to an underlying drainage system.

Prior to the 1960s, the only option for improving the drainage performance of a sports surface might have been the installation of a few rudimentary agricultural clay pipe drains, probably into a surface that had been leveled and graded using road-building machinery. Such approaches were often associated with a widespread lack of understanding by soil engineers of the effect of using civil engineering practices on soil compaction when building sports surfaces, resulting in a well-documented catalog of failures.

However, by the end of the 1980s, innovative design and construction techniques based principally around the use of precisely graded drainage gravel and sand-dominant rootzone materials resulted in a significant improvement in the quality of play and an increase in the amount of use various sports surfaces could sustain, as well as a reduction in the number of games postponed or canceled. This was because sand-dominant rootzones did not have complex particle aggregation as in native soils and therefore their physical properties could be more closely predicted than for most native soils. This technology was backed up by well-documented research in the United Kingdom, Europe, and the United States on the selection of suitable materials and methods for the construction and maintenance of natural turf sports surfaces (e.g., Adams 1986, Baker 1990, Hummel 1993, Skirde 1974, Stewart 1994, van Wijk 1980, Waddington 1992).

There are in fact only two principal design options for addressing the physical limitations of sports turf soils so that a calculated improvement in drainage can be achieved:

1. Install a "by-pass" drainage system
2. Modify the soil texture by incorporating sand or by using an imported sand-dominant rootzone

27.4.1 By-Pass Drainage Systems

A by-pass drainage system works on the principle that if a network of continuous macropores or fissures is artificially introduced into the existing soil profile, then preferential channels for water flow will be created. The best example of this technique in an agricultural sense is "mole plowing," where a bullet-shaped instrument attached to the base of a cutting blade is dragged through the soil profile to create temporary drainage channels that are linked to the surface using the fissures created by the blade. This technique has been modified extensively in the sports turf industry to a more permanent by-pass system known as slit drainage. Using mole plow technology was a logical approach for developing what is essentially the sports turf equivalent of agricultural mole plow systems, the difference being that because the profile would not be regularly inverted by cultivation at the end of a cropping season, the mole drain could be made more permanent by backfilling it with permeable material (i.e., uniformly graded gravel and/or sand).

Slit drains are narrow (typically 50 mm wide and 250–300 mm deep), permeable channels of material that are cut into the sports surface at close centers (typically 0.6–1 m apart) to act as a by-pass system through the soil profile. They have proven to be particularly versatile since being developed in the late 1960s in the United Kingdom. The predominant use of these systems around the world is on municipal sports and athletic fields and in provincial, community stadiums and sports clubs. However, increasingly, landing areas on golf fairways, approaches to golf greens, racetracks, and college or university grounds are also being slit drained.

A slit drainage system works by allowing a direct route for water intercepted at the surface to be channeled into an underlying pipe drainage system, thereby by-passing the bulk soil (Figure 27.2). What makes a slit drainage system so attractive is that it offers improved interception of surface water as an excellent alternative to full-scale reconstruction or amendment of the existing soil profile, plus the system can be installed progressively with careful planning as funds become available.

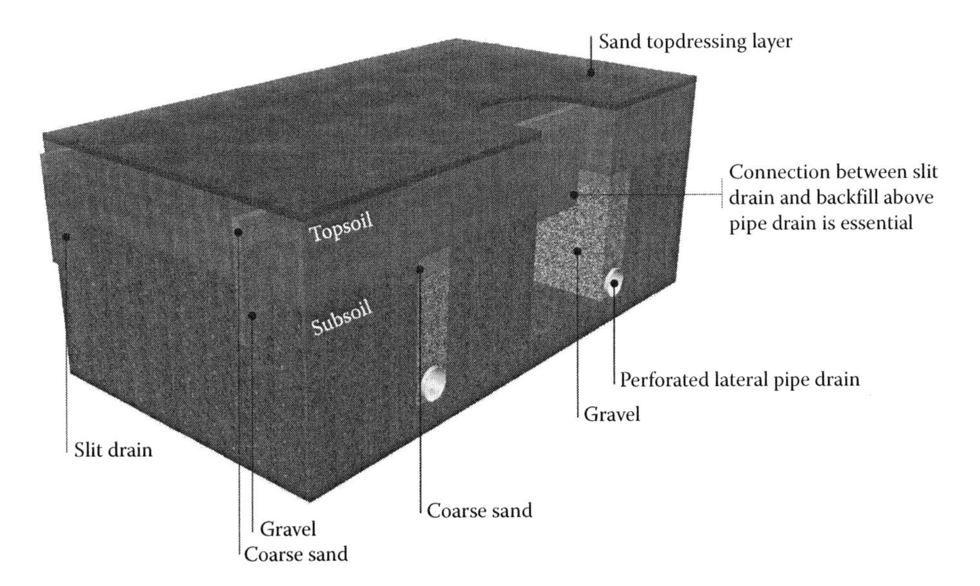

FIGURE 27.2 Conventional pipe/slit drainage system showing excavated slits filled with sand and gravel and topdressed with a shallow layer of sand.

There are many names used to describe slit drainage—a reflection of the way sports turf contractors, scientists, and consultants have dabbled with the system's commercialization, versatility, and flexibility over the years. Sand or gravel banding, sand slitting, sand link system, sand or surface grooving, sand placement, and sand injection are all names used to describe various forms of slit drainage. Although this range of terminology suggests otherwise, there are only two ways in which slit drains can be installed, these being injected or excavated.

27.4.1.1 Installing Slit Drains

With excavated slit drains, existing soil is removed, usually on a conveyor belt, and the slit drain is backfilled in a separate operation with permeable material, usually sand-over-gravel. The combination of gravel and sand in excavated slit trenches has become the most commonly used backfill of an excavated slit drain worldwide.

Injected slit drains are installed by machines with cutting blades that force the soil apart. No soil is removed and backfilling is carried out in a single pass using a hopper feeding behind the cutting blades. Injected slit drains generally use only a single backfill material such as fine gravel (typically termed "gravel banding") or sand (typically termed "sand banding" or "sand grooving"), and are narrower than excavated slit drains (e.g., 25 mm vs. 50 mm width) and spaced closer together (e.g., 0.2–0.5 m apart vs. 0.6–2 m apart).

Because there is no trench excavation with injected slit drains, sports surfaces can be drained for significantly less surface disruption and less cost than with conventional excavated slit drains. However, sand-only injected slit drains are not particularly efficient at removing surface water quickly unless they are connected either to a naturally permeable subsoil or to a lower tier of close-spaced, gravel-filled slits or pipe drains. The reason for this is that the lateral conductivity of the sand used is insufficient to be able to transmit water sideways fast enough over distances of more than approximately 3 m.

27.4.1.2 Protecting Slit Drains

Slit drained sports surfaces are almost universally designed to be deliberately capped with a layer of approved "sports turf" sand (Figure 27.2). The sand layer is used to prevent contamination by soil of the underlying slit drains, and to improve the playing characteristics and drainage of the surface. The final depth of sand applied depends on budget, sand type, sports use, level of play (e.g., recreational vs. professional), and to an extent turfgrass type but typically it varies from as little as 10 mm to over 100 mm. With a deeper sand carpet layer, the carpet itself acts as a homogenous drainage layer in its own right, thereby offering a greater and more uniform drainage and playing quality performance.

27.4.2 Modification or Replacement of Soil Texture

Modification or replacement of soil texture is used when the drainage design expectation is for the *entire* sports surface to intercept and remove rainfall, as opposed to relying on a relatively small percentage of the surface area to do the work as with slit drain systems. Typically, this approach will be used where the sports surface involved is relatively small (e.g., golf or bowling greens), where larger construction budgets are available (e.g., professional sports clubs or stadium venues) or where the sports facility is located on an industrial or urban site where there is no native soil profile present.

27.4.2.1 Modification of Soil Texture

Modification of the soil texture involves ameliorating an existing soil of known texture with sufficient sand of known particle size distribution to turn the existing soil into a sand-dominant rootzone with at least 70% sand in the medium-fine particle size range (see Section 27.3.2). This process can be carried out *in situ* by spreading a specified quantity of sand over an existing soil and cultivating to a precise depth (e.g., 150 mm), but it is a process that requires a great degree of skill and experience. A typical specification set for the UK conditions by Baker (1985) is for the final mix to contain less than 20% fines (particles less than 0.125 mm in diameter), less than 10% silt and clay (particles less than 0.050 mm in diameter), and less than 5% clay (particles less than 0.002 mm in diameter), with proportionally more sand being added if a faster design drainage rate is required.

A properly designed lateral pipe drainage system similar to that shown in Figure 27.2 should be installed directly beneath the sand-ameliorated soil layer to complete the system.

27.4.2.2 Replacement of Soil Texture

A more common alternative to modification of soil texture by *in situ* sand amelioration is to create a sand-dominant rootzone using off-site mixing, where the rootzone material is brought in as part of a complete reconstruction of the profile.

The use of sand-dominant materials as part of a complete reconstruction of the profile was first quantified scientifically using research carried out by soil scientists in the United States in the 1950s. This work culminated in the publication in 1960 of the first practical specifications for golf green construction by the United States Golf Association (USGA) Green Section (Hummel 1993). The basic design is composed of three layers: a gravel drainage layer, overlain by an intermediate or choker layer, overlain by the sand-dominant rootzone (Figure 27.3). The intermediate layer prevents migration of the rootzone into the underlying gravel layer but can be eliminated from the design if certain bridging and particle size criteria of the rootzone and gravel layer are met. The purpose of the gravel layer is to provide a capillary break at the base of the rootzone and in doing so, create a perched water table that allows better control of water in the profile.

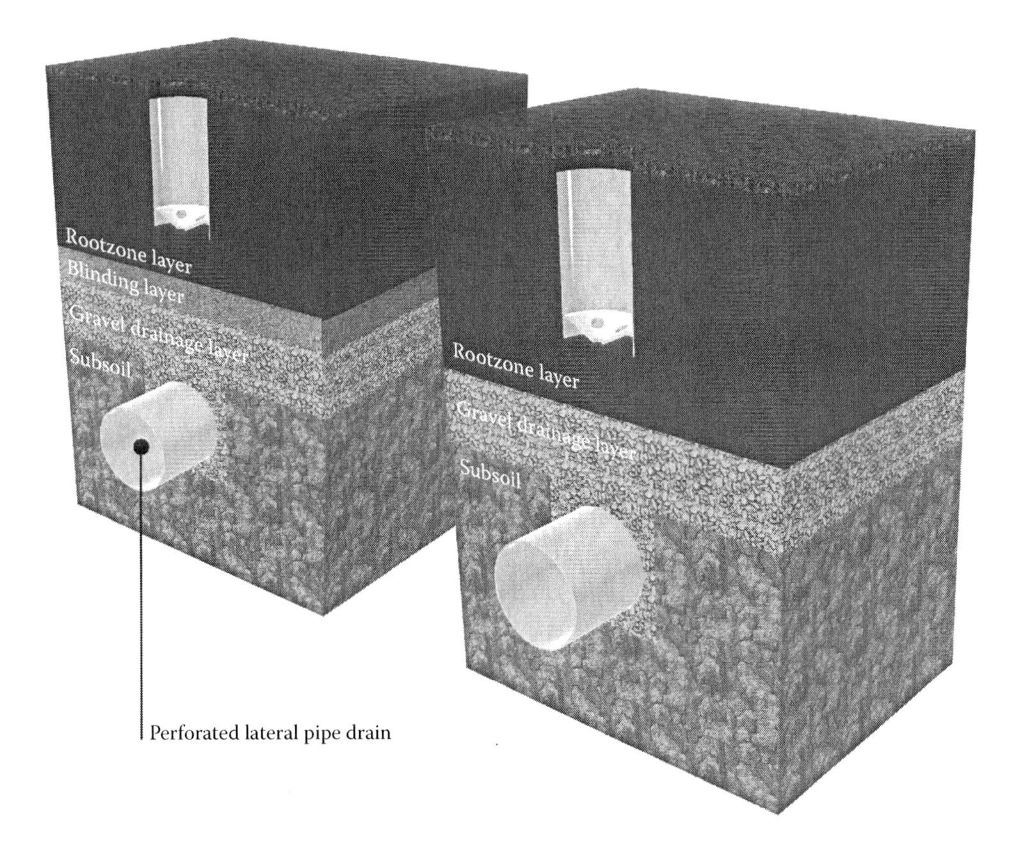

FIGURE 27.3 The USGA Green Section golf green perched water table construction options (on the left with an intermediate blinding layer; on the right without the blinding layer).

Since 1960, research has continued to focus on specifying the ideal make-up of a free-draining growing medium based principally on measurements of water retention/release, saturated hydraulic conductivity (measures the ease of movement of water through the medium when it is saturated), particle size distribution, and organic matter content/nutrient holding characteristics of the growing medium. Now, four revisions later, the "USGA" perched water table sand-dominant rootzone is one of the most utilized design guidelines for golf green construction worldwide. Although not endorsed by the USGA, the guidelines (or variations of the guidelines) have also been applied to other intensively used natural turf sports surfaces requiring play under adverse weather conditions or in challenging climates (e.g., sports or athletic fields in stadiums, and racecourses).

The 1960s and the 1970s saw further developments in the use of sand-dominant rootzones that offered alternatives to the USGA design, principally avoidance of the underlying gravel drainage layer and in many cases the decision to use a "pure" sand rootzone as opposed to an amended or ameliorated sand rootzone. Some construction systems also used an impermeable plastic membrane underneath the rootzone so that the water table could be artificially raised or lowered (e.g., Moesch 1975). All these systems required a sound understanding of the physical properties of sand and application of soil physics for their design and field performance.

There have also been attempts at combining slit drain systems with complete profile reconstructions (e.g., Sport England 2011). A practical example of this design is where the existing native soil is harvested and then placed as a 200 mm layer over an installed gravel drainage layer. This is followed by ameliorating the surface 50 mm or so of the native soil with sand followed by the installation of slit drains to connect the surface to the underlying gravel drainage layer (Figure 27.4). This design has several advantages, including that it is easier to manage than a complete sand-dominant rootzone in terms of soil moisture and nutrient requirements, but at the same time the construction still provides a fast design drainage rate. Also, the gravel drainage layer can be used to attenuate drainage water (i.e., to slow down the rate of flow of drainage

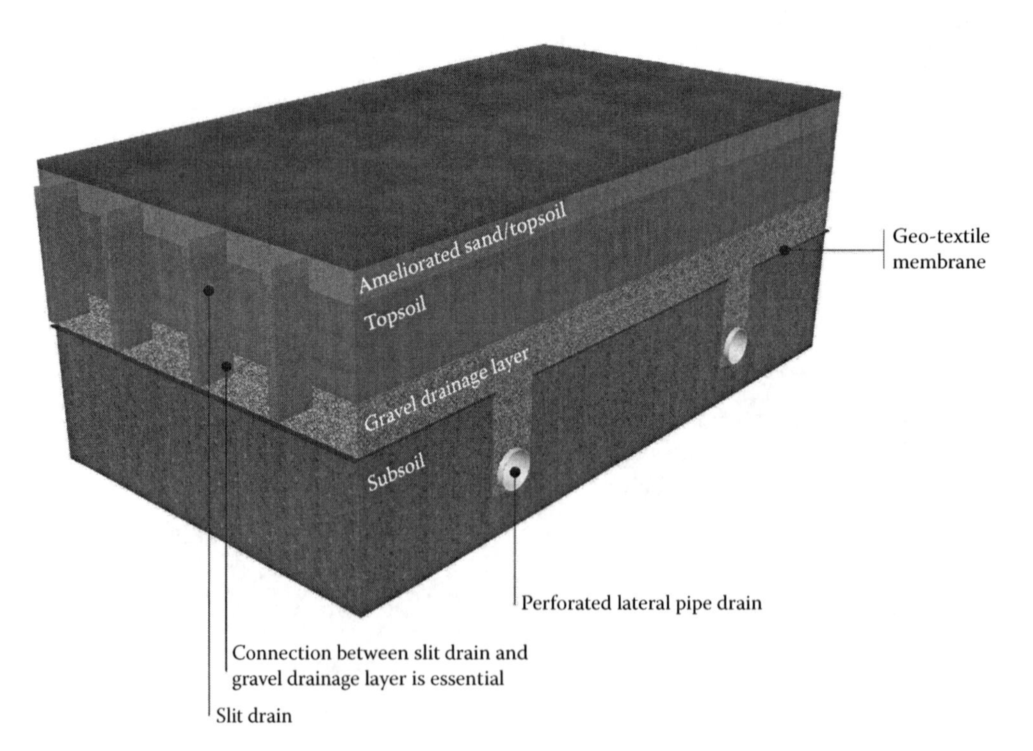

FIGURE 27.4 Pipe and slit drainage system on a gravel drainage layer and with a shallow sand-ameliorated topsoil layer.

water to prevent downstream flooding and erosion, with consequent increase in the duration of flow), an increasingly common requirement for sports field design in built-up urban areas where there are often severe restrictions on discharge of drainage water to a city storm water outlet. For example, the pore space (air-filled voids) in a typical gravel drainage layer under a full-size football field has the capacity to hold around 200 m^3 (cubic meters) of drainage water. This water can be released in a controlled manner to the storm water outlet by using an outflow pipe sized to cope with the permissible discharge. In this way, peak discharge from surrounding hard standing areas (e.g., roads, car parks, and paths) has the opportunity to pass though the storm water system first.

It is the fine tuning of the all the above designs that has been the focus of sports turf rootzone research over the last 30–40 years and that has led to the publication of further scientifically based standards and guidelines for constructing sand-dominant rootzones (e.g., Baker 1988, 2005, Davis et al. 1990, DIN 18-035 1991, FIFA 2011a, USGA 2004).

27.4.2.3 Sand Selection Criteria and Modification for Design

The main requirement for rootzone sands is that they must offer minimum potential for interpacking of sand particles. They therefore need to be uniform and precisely graded. For this reason, the sports turf industry frequently uses grading curves to assess sand quality and for prediction of a sand's physical characteristics (Figure 27.5).

It is generally accepted that rootzone sands need to be predominantly in the 0.125–1.00 mm particle size range, ideally with a near absence of calcium carbonate and soft materials such as pumice (Adams 1986, Baker 2006, Davis et al. 1990, Stewart 1994, USGA 2004). Within this range, different sports surfaces have historically used slightly different sand particle size specifications depending on the extent of surface traction (grip) required by players for the sport: for example, golf greens have a minimal requirement for player traction and have traditionally been built using less stable, amended medium-coarse sands (0.25–1.00 mm). In contrast, sand-dominant sports or athletic fields tend to be constructed using the more stable medium-fine sand range (0.125–0.50 mm). Even this distinction is only a generalization: warm-season (tropical) turfgrasses are better suited to stabilizing medium and medium–coarse sands compared with cool-season (temperate) turfgrasses because of the former's aggressive stoloniferous (overground runners) and rhizomatous (underground runners) growth habit. So, in climates where both warm- and cool-season turfgrasses are viable for

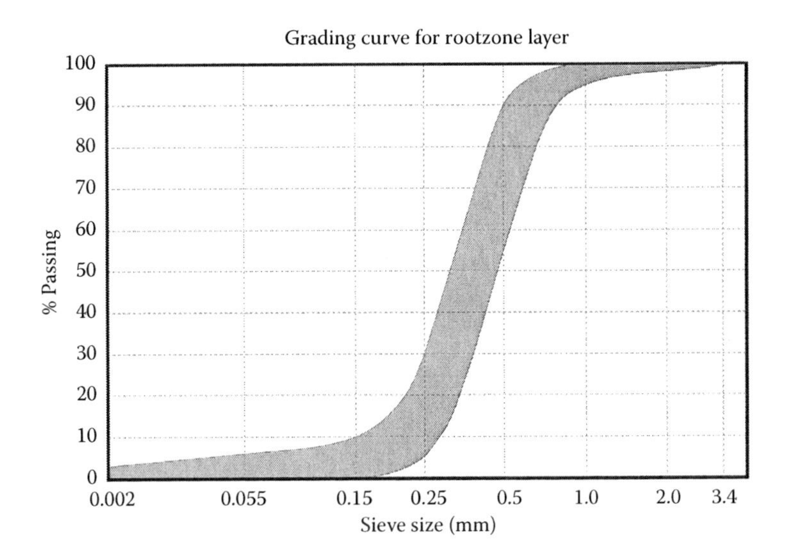

FIGURE 27.5 Typical grading curve for a sports turf rootzone sand.

sports or athletic field construction (i.e., "transition" zones), the range of rootzone sands that can be specified is much broader compared with climates where only cool-season turfgrasses are viable (i.e., temperate zones).

Although a carefully specified sand-dominant rootzone can confer the advantage of free-draining pore space, it will also create difficulties that are less apparent in native soils, these being a potential lack of physical stability, poor nutrient retention, and a very narrow range of soil moisture tension over which available water is released (Adams 2008). Moreover, the requirement for adequate surface stability, a minimum saturated hydraulic conductivity, adequate balance of air- and water-filled porosity, and practical rootzone depths in a constructed profile (typically around 300 mm) places tight limits on the particle size distribution of sand that can be used (Adams and Gibbs 1994). However, it is not just about ensuring appropriate soil physical properties for a rootzone construction; the end product must still be appropriate for healthy turfgrass growth and meet the specific requirements of the sports for which it is intended (Stewart 1994). For these reasons, the type and quality of sand, choice of amendment material (organic or inorganic), mixing ratio of sand to any amendment, and mixing ratio of sand when used to ameliorate a native topsoil are the main factors that are usually altered in a rootzone design in order to optimize the growing environment for the turfgrass plant and its subsequent management.

Specifying a rootzone sand approved for sports turf use is one thing; sourcing an approved material is quite another story. In countries with a well-established market for sports turf construction with access to local sports turf research and consultancy organizations, bespoke materials are readily available. However, in countries without these facilities, it is usually a challenge to find the ideal rootzone sand. The first port of call is usually the local building industry but even if the right material can be found, the reliability and consistency of supply may soon become the next headache. Even in parts of the world where one might immediately think there would be an abundance of the right type of sand, such as the deserts of the Middle East, one may have to trudge over many sand dunes with pocket sieves in hand to find that elusive particle size distribution suitable for sports turf use!

27.5 UNIQUE CHALLENGES OF STADIUM SPORTS SURFACE DESIGN

Despite the various improvements made in the use of sand for designing and constructing sports surfaces in stadiums and sports or athletic fields in the early 1980s, the innovative designs and better construction materials did not necessarily translate into increases in "carrying capacity" (hours of sports use per week) that were predicted by turfgrass researchers at that time. The reason was that even with the best soil physical properties, turfgrasses still had a very finite biological limit of wear, and once worn out, serious problems of loss of surface stability were usually the result. This, when coupled with problems of poor moisture and nutrient retention associated with sand-dominant systems, has resulted in a diverse range of products being developed to help solve the problems encountered.

Two other factors have occurred in the last 25 years or so, which have resulted in a quantum leap in expectation of turf performance: first, the desire by stadium architects to build enclosed (or near enclosed) stadiums that provide improved spectator comfort and facilities. However, the goal to achieve optimum seating capacity and spectator comfort is often at the expense of the turf, with low light levels and cooler soil temperatures in seasonally shaded parts of the stadium; second, because such stadiums are so expensive to build and maintain, there has been a need to diversify and extend usage to increase revenue. Many stadiums around the world now have to operate as fully multiuse venues for economic survival, including having to change the playing surface from one use to another, sometimes overnight.

For example, in Australia and New Zealand, cricket and traditional winter sports (Rugby Union, Rugby League, and Australian Rules Football) often have to be played on the same surface within days of each other. Gone are the days of clear-cut playing seasons and a six-week renovation period between the two types of sport. Cricket is a ball sport played in summer on a sun-baked rolled clay surface (the pitch) with minimal turfgrass cover, whereas the traditional winter sports are essentially a kicking and throwing ball sport requiring a well-drained surface with good traction characteristics and a full cover of turfgrass. There are probably no two types of sport that could be so different from a soils and turfgrass perspective.

The greater challenge is that those multi-use events that include non-sporting events have even worse conflicting surface requirements, particularly when it comes to hosting non-sporting events such as concerts or grand opening ceremonies for which the turfgrass surface has no practical benefit. So, while the need to pay particular attention to television, spectator, and corporate requirements on the one hand underpin a successful stadium, on the other hand, this need creates a situation where the soil physical properties of a high-quality sand-dominant rootzone construction alone are unable to provide a satisfactory sports surface.

To meet the above multi-use challenges, a whole range of innovative construction systems for sports and athletic fields in stadiums now exists, including:

1. Sand-amended rootzones with specialist turf reinforcing products
2. Transportable turf systems
3. Turf management "enhancements" such as injected air systems, vacuum-assisted drainage, undersoil heating/cooling, climate control, and turf protection systems

All of the above systems can be installed and managed using job-specific, highly sophisticated machinery and equipment.

27.5.1 Specialist Turf Reinforcing Products

Turf reinforcing products attempt to combine the playing quality benefits of natural turfgrass with the practical strengthening and engineering advantages of artificial materials. Reviews and guidelines are given by Baker (1997) and FIFA (2011b). Reinforcement materials have been shown to offer a range of potential benefits, including:

1. Increasing load-bearing and shear strength as well as reducing soil compaction
2. Protecting the growing point (crown) of a turfgrass plant and thereby improving grass cover retention and reducing the extent of surface damage
3. Maintaining surface stability once the natural turf has weakened or has been worn away

Turf reinforcing products can be grouped into three broad categories:

1. Intact fabrics or artificial turf carpet placed into or a little below the surface, filled with rootzone material and in which natural turfgrass is grown. This type of system lends itself to being used as part of a big roll turf system for immediate establishment of a playable surface or rapid repair of damaged areas on an existing surface (see next section).
2. Individual strands of artificial turf fibers 200 mm long that are injected vertically into the profile to a depth of 180 mm at very close centers (20 mm) leaving 20 mm of the artificial turf fibers remaining above the surface like blades of grass. This type of system is commonly known as a "hybrid" surface stabilizer and it is particularly good at maintaining surface smoothness and appearance of turfgrass once the natural turfgrass has worn away.
3. Randomly orientated elastic material or plastic (e.g., polypropylene) fibers or mesh elements that are incorporated into the rootzone layer usually before it is laid but sometimes *in situ*. These systems are profile stabilizers and are particularly useful for increasing the load-bearing strength of the rootzone.

The sand rootzone into which the above reinforcing products are installed is frequently further amended with soil and/or organic materials such as peat moss and other organic conditioners. A recent trend has been to use inorganic amendments such as zeolite, a naturally occurring alumino-silicate mineral. This material comes close to the ultimate "designer sand" because it can be graded to sand-sized particles and is highly nutrient- and water-retentive on account of its negative charge and microporous structure.

There are a number of key points to consider when deciding whether or not to include a reinforcing product into a sand-dominant rootzone. They include:

- The cost of the system in relation to the perceived benefits from the product (e.g., will more games be possible with less damage caused to the surface?)
- The specific features of each product in relation to the intended use of the playing surface (e.g., can areas of damaged turf be repaired using conventional turfing procedures?)
- The intended management/maintenance requirements of the playing surface (e.g., what maintenance machinery can be used without damaging the reinforcing product?)
- The species of turfgrass grown (e.g., are some turf reinforcing products better suited to certain turfgrass species than others?)

Unfortunately, there is relatively limited independent, scientific information comparing the different reinforcing products and their management, but they are widely used in sports and athletic fields in stadiums around the world with varying levels of success.

27.5.2 Transportable Turf Systems

The ability to replace large sections of worn or damaged turf in a matter of hours has come about through the development of big roll turf technology (typically 10–20 m lengths of turf up to 2.4 m in width, with turf rolls containing up to 50 mm of rootzone depth) and palletized or modular turf systems (typically of dimensions 1.2 by 1.2 m or 2.4 by 2.4 m with up to 200 mm of rootzone depth). Both systems allow rapid repair and if thick enough, virtually instant play on the repaired surface. In addition, modular turf systems allow sports surfaces to be set up from scratch on hard floor areas that would normally be used for other activities such as concerts, boxing matches, or basketball. Modular turf systems have been used not only at a stadium level where as many as 8000 modules of turf are required to set up one sports or athletic field, but also for establishing natural turf tennis courts and racecourse crossings.

Big roll and modular turf systems are sometimes combined with turf reinforcing products to improve versatility. A key development in the success of big roll turf systems has been the development of machinery that is able to harvest and lay the turf rolls without damage, as well as machinery that can remove accurately up to 50 mm of damaged turf in one pass off the field, leaving only the minimum of releveling required prior to laying new turf.

There are two further examples of innovation in transportable turf systems. One is where an entire sports or athletic field can slide in and out of the stadium on rails, an activity that can take several hours to complete. This is the ultimate transportable turf system. A sliding rail system avoids all the environmental challenges caused by modern enclosed or semi-enclosed stadiums because the turf can be grown outside in its portable rootzone when not in use. However, cost and limited outside space would be the main reason why this type of system is used in only a handful of stadiums around the world to date.

New Zealand and Australia have their own unique type of transportable turf system. These are the portable or "drop-in" cricket pitches that allow the sports of cricket and rugby (or other types of football) to take place in the same season on the same ground. The engineering that has gone into these systems requires a transporter capable of lifting a 25 m long, 3 m wide, and 200 mm deep strip of clay-based soil and underlying drainage layer weighing in the region of 50 tonnes out of the ground without disrupting surface levels or playing quality.

27.5.3 Enhancing the Growing Environment

There are many stadium examples around the world where the ideal sand-dominant rootzone, even when combined with a robust turf reinforcing system, is still unable to provide a satisfactory sports surface, usually because of extreme climate, shade, and multiuse requirements. A wide range of sports surface "enhancements" exists to help maximize the potential of the innovations mentioned so far. The purpose of the enhancements is to allow some control of the stadium environment both above and below the surface, and second some control of the extent of turf damage caused by nonsporting events such as concerts.

Enhancements include powerful pumping and air distribution systems designed to blow conditioned air into the rootzone, which can decrease or increase soil temperature to extend the growing season or control turfgrass dormancy (Figure 27.6). These same systems can also be switched to suction mode so that rapid removal of excess rainfall off the surface can be achieved in a matter of minutes, for example, if there is a major thunderstorm shortly before a game commencing. In colder climates, protection against the effects of frost and snow can be achieved using undersoil heating systems based on water- or glycol-filled pipes or occasionally electrical wires to ensure games can take place during the winter period.

Flat or inflated rain covers and insulating growth covers are also available for entire sports and athletic fields as well as for smaller sports surfaces such as cricket pitches and baseball diamonds. Types of surface covers to control turf damage caused by nonsporting events include rigid pallet-type systems, rollaway flooring systems, and semirigid geotextiles. The ability of turf to survive

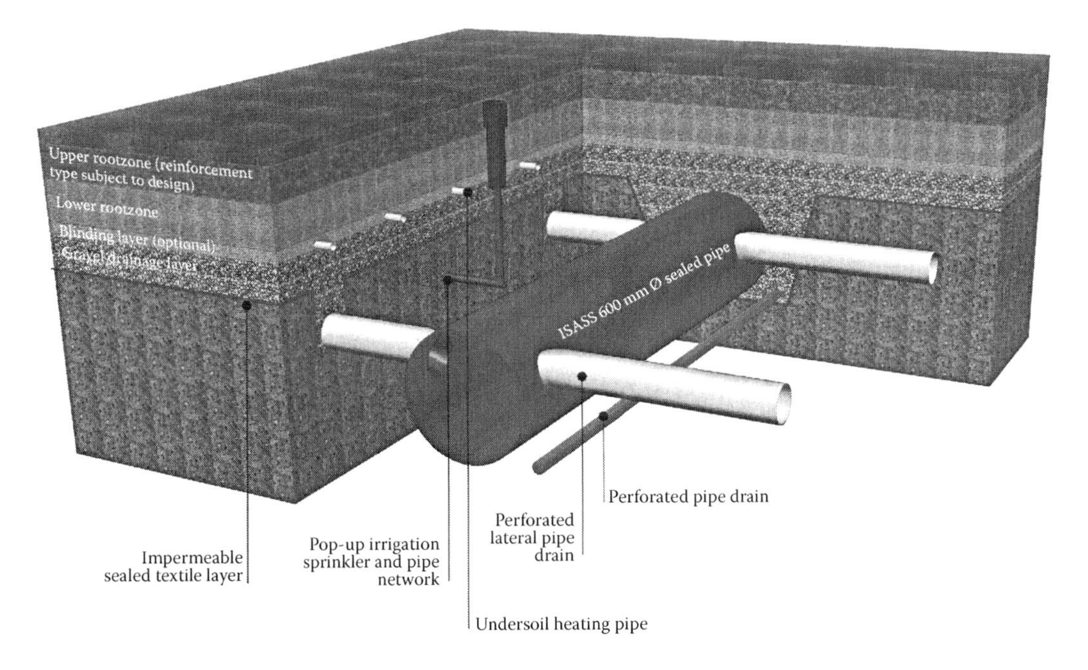

FIGURE 27.6 Layout of a perched water table construction profile used for a stadium field. The multi-layered system has optional turf reinforcement at the surface, an automatic irrigation system, undersoil heating system, optional blinding layer, and an "ISASS" system (*in situ* air sparging system) for vacuum-assisted drainage, aeration/gaseous exchange within the rootzone, and for facilitating cooling or heating of the rootzone.

under turf protection products depends on a range of factors, including the season and species of turf, the condition of the turf, the type of product used, and the amount of time the cover is in place.

One of the more recent developments in the enhancement area has been the use of artificial lighting rigs. These rigs supply light in the photosynthetically active wavelength of 400–700 nm (nanometers) typically producing an average photosynthetic photon flux density of 200–300 μmol/m^2/s (micro-moles of photons per square meter, per second where one mole of photons is 6.0222×10^{23} photons). Over a 10-h period, these rigs can supply turfgrass with over 10 mol/m^2 of photosynthetically active radiation, sufficient to promote active growth of temperate (cool season) turfgrasses in heavily shaded stadiums and allow acceptable recovery from wear during the winter period.

The largest rigs currently have a lighting footprint of between 360 and 590 m^2 (square meters). Typically, between three and nine rigs will be required to provide sufficient light coverage on a sports or athletic field in a stadium depending on the extent of heavy shade. The main advantage of this technology is that a field may not have to be returfed several times a year in order to maintain quality throughout the winter playing season and for many stadiums, the use of artificial lighting to keep turfgrass growing has become more cost effective than returfing. There is no doubt that lighting technology will develop significantly over the next few years so that this method of growth manipulation in the turf industry becomes more energy efficient and cost effective. The most likely developments over the next few years will probably involve greater use of light-emitting diode (LED) bulb technology as an alternative to conventional high-pressure sodium light bulb technology.

The use of artificial lighting technology for the turf industry was originally adapted from the greenhouse crops industry, with their use gaining widespread acceptance in stadiums around the world since about 2006. The decision to install lighting rigs is usually based on a detailed light modeling exercise (a "Hemiview" analysis) of shaded areas within a stadium, or in a situation where a new grandstand is proposed in an existing stadium and it is important to know in advance how the new structure is likely to affect turfgrass growth.

Light modeling for turfgrass growth can also be used at the stadium design stage (i.e., before the actual stadium is built), but it is not common for stadium architects to tailor their design exclusively to allow optimum turfgrass growing conditions. For example, those stadiums with solid (retractable) roofs offer particularly challenging turfgrass growing conditions and almost without exception, they suffer from severe turfgrass and rootzone management problems. Conventional roofed stadiums are not designed to maximize sunlight interception; many have been designed from an engineering, as opposed to an agronomic, perspective where the relationship between stadium architecture and its effect on light reaching the sports surface has not been a principal focus of the design. A notable exception to this approach has been the Forsyth Barr Stadium in Dunedin, New Zealand, where the design team was faced with the challenge of creating the world's first stadium with a permanent natural turfgrass field growing under a permanently enclosed roof (Gibbs 2011).

This southern hemisphere stadium is quite different for two reasons. First, design features were specifically considered for maximizing sunlight interception since the beginning of the project (e.g., the height of the northern stand, the orientation of the field); second, the turfgrass and rootzone requirements of the stadium were thoroughly researched prior to the stadium being built, as opposed to being investigated retrospectively. The outcome of the design was a roof constructed out of ETFE (ethylene tetrafluoroethylene), a transparent polymer originally developed for the space industry weighing just 1% of the weight of glass and able to bear 400 times its own weight. For the first time, a stadium was built completely around turfgrass and rootzone requirements rather than the other way round, along with the ability to eradicate undesirable elements of the environment such as frost, snow, and excessive rainfall (Figure 27.7). A sand-dominant rootzone was still used for the stadium field because this design allowed predictable soil physical properties to be achieved, and in particular, more precise management of soil moisture content through the two irrigation systems installed (subsurface and overhead). This design was considered essential for disease control

FIGURE 27.7 **(See color insert.)** Applied soil physics in action: The Forsyth Barr Stadium, Dunedin, New Zealand—the world's first permanently closed-roof stadium pitch supporting a natural turf surface, successfully used for the 2011 Rugby World Cup. (Reproduced by permission of Darcy Schack of JAM Photographics Limited.)

and optimizing surface moisture conditions for the wide range of sport and nonsport activities hosted by this stadium.

The innovative use of ETFE and the success of the turfgrass growth in this stadium in a temperate climate have set a new benchmark for North American and Northern European venues for the viability of sustaining turf under a permanently enclosed structure. Whether a similar approach can be used for turfgrass growth in much hotter climates (i.e., where spectators and players need to be protected from the extremes of heat, humidity, and tropical rainfall) remains to be seen.

27.6 CRICKET PITCHES: THE LAST WORD IN SPORTS SURFACE DESIGN

It would be remiss to finish this chapter without highlighting the application of soil science to the summer sport of cricket. The quality of a cricket game is judged principally by the nature of the pace and bounce characteristics of a ball as it is bowled and strikes the hard, tightly bound surface of the pitch located in the center of an oval or circular outfield of closely mown turf. For the cricket pitch, itself a thin rectangular strip of intensively prepared soil a little over 20 m in length, pace and bounce characteristics have to take priority over design drainage rates required for so many other sports surfaces. The approach to sports surface design and preparation for cricket is therefore completely different from the examples given in Sections 27.4 and 27.5 of this chapter.

Interestingly, the development of a science-based approach to cricket pitch design, construction, and management is similar in terms of timescale (though not in quantity) to the development of golf green design. That is to say, it has only been from the 1960s that research programs were set up to examine the ideal soil properties required for the sport.

In simplistic terms, the clay content of a soil is responsible for conferring soil binding characteristics, but strong binding is of course achieved only when the soil is sufficiently dry and compact. A general requirement worldwide is that the soil moisture content of a cricket pitch soil should be significantly below field capacity when ready for play. In the 1960s, a simple, practical test was developed for cricket grounds curators to measure soil binding strength. The test, referred to as the Adams Stewart Soil Binding ("ASSB") test, involved making small balls of moist clay called "motties" where all the aggregation had been broken down by hand. The motties were dried for several days and the force in kilograms required to break the motties, when placed between two plates on a set of bathroom scales, was measured. On cricket soils in the United Kingdom, every 2 kg (kilograms) of breaking strength corresponded to approximately 1% of clay, a relationship that worked well for clay contents between 20% and 40% (Stewart 1994).

However, although the ASSB motty test has shown that there is a good general relationship between clay content and soil binding strength, clay mineralogy and soil organic matter content are also important in affecting this relationship. Soils containing kaolinite (a nonswelling clay common in the United Kingdom) as the dominant clay will have a lower soil binding strength than other soils of a similar clay percentage but which contain smectite, a swelling clay uncommon in the United Kingdom but common in Australian and South African cricket soils. Smectites offer substantial shrinkage and high mechanical strength and hardness when dry, properties that lead to faster-paced pitches compared with the United Kingdom. Nevertheless, despite variations in clay mineralogy on a world scale, for UK soils, clay content is still a reasonable guide for predicting soil binding strength, provided organic matter is taken into account (Adams and Gibbs 1994).

Despite the above understanding, predicting pace and bounce accurately is an arduous task for players and more particularly for grounds curators who are required to prepare cricket pitches that offer a fair contest between batsmen and bowlers. In the past, grounds curators had to rely on anecdotal evidence and experience to predict the rebound characteristics of cricket pitches, as relevant scientific information was scarce. Nowadays, cricket pitch preparation models are available, particularly in New Zealand and the United Kingdom, where target values of soil moisture content and bulk density can be set to help grounds curators monitor the quality of their preparation. This approach has been warranted because both moisture content and bulk density are the two properties of soil most readily modified by grounds curators during pitch preparation.

There is no doubt that information about the moisture content, bulk density, and organic matter content of cricket pitch soils has helped grounds curators contribute to a more scientific approach to cricket pitch preparation and construction, which in turn has contributed toward an improvement in the quality of pitches worldwide. As research work continues to unravel the finer detail of the interaction between the cricket pitch surface, soil type, and ball impact, it is possible that we may see "designer" cricket soils being produced in the future to cater for different cricketing needs. Of course, in reality, cricket enthusiasts will know that even when a cricket pitch is prepared in exactly the same way as on a previous occasion, it can still play dramatically differently. Thus, the ball rebound characteristics of a cricket pitch can never really be known until the game has actually started and even then the characteristics will change during the course of a game as the surface continues to dry out and the turfgrass starts to senesce. Clearly, therefore, it should be recognized that every prepared cricket pitch has its own unique playing characteristics and science can only assist grounds curators so far in their quest to construct and prepare better cricket pitches. Rather than viewing pitch variation as a problem as it would be with other sports, many would argue that variation should be viewed as an added attraction of the game of cricket, which helps to prevent it from becoming too stereotyped and predictable!

27.7 CONCLUSION

For many readers, this chapter might be their first awareness of a "branch" of soil science they never knew existed. Indeed, it never ceases to surprise me how many people still do not believe that it is possible to grow grass successfully in sand ("Where is the soil?" they ask, as if comparing the stadium field to their garden vegetable patch). Unfortunately, the sports turf industry is one of those "behind-the-scenes" industries—a good surface rarely gets praised by the media so the general public are generally unaware of the science and technology behind the preparation of a sports surface in a heavily shaded stadium or golf course green. It has to take a fairly global event such as the Olympics, a FIFA World Cup, the Augusta Masters Tournament, or the Wimbledon Championships for the turf to get an agronomic mention in the media spotlight. In contrast, a poor surface or a poor result from a game is quick to hit the news, with the grounds curator or superintendent often being blamed for mismanagement and the surface often being blamed for the result. It can be a very high-pressured business at the professional sporting level with tight deadlines, high expectations, and no second chance to get things right on the day of the match or tournament.

In concluding, it is worth considering whether there are any developments learnt from experience with sports turf management that could contribute to the intensification of plant production to meet increasing global food requirements. Intensive agricultural and horticultural crop production certainly share some common features with sports turf management such as the requirement for growth of a monoculture and an increasing requirement to manage plant growth more predictably. There are some common principles that apply across both industries. However, the purposeful manipulation of soils for sports turf use is probably unnecessarily extreme and costly when it comes to providing the optimum growing medium for agricultural and horticultural crop production—after all, crop production generally does not involve deliberate destruction of plant material and soil physical properties by humans playing sport!

And what does the far future hold for sport turf soils? Well, if the chapter in this book by Haff is to be believed, in a few centuries' time, we could see "smart soils" being created, where soil-grain-size computers are used to create bespoke sports turf rootzones. These rootzones, when used in conjunction with sunlight-powered artificial grass, could then be programmed to create the optimum playing conditions and modified surfaces for different sports at the touch of a button. The only problem is that humans may have evolved to be incapable of taking exercise and playing sport!

ACKNOWLEDGMENTS

I thank the Sports Turf Research Institute (STRI) for permission to reproduce the diagrams used in this chapter, Dr. Stephen Baker for chapter review, and Darcy Schack of JAM Photographics Limited for permission to reproduce photos of the Forsyth Barr Stadium.

REFERENCES

Adams, W.A. 1986. Practical aspects of sportsfield drainage. *Soil Use and Management* 2(2): 51–54.

Adams, W.A. 2008. An overview of organic and inorganic amendments for sand rootzones with reference to their properties and potential to enhance performance. In *Proceedings of the 2nd International Conference on Turfgrass Management and Science for Sports Fields*, eds. J.C. Steir, L. Han, and D. Li, Acta Horticulturae 783: 105–113.

Adams, W.A. and R.J. Gibbs. 1994. *Natural Turf for Sport and Amenity: Science and Practice.* Wallingford: CAB International.

Baker, S.W. 1985. Topsoil quality: Relation to the performance of sand-soil mixes. In *Proceedings of the 5th International Turfgrass Research Conference*, ed. F. Lemaire, 401–409. Paris: Institut National de la Recherche Agronomique.

Baker, S.W. 1988. Construction techniques for winter games pitches. In *Science and Football, Proceedings of the 1st World Congress of Science and Football*, eds. T. Reilly, A. Lees, K. Davids, and W.J. Murphy, 399–405. London: E & FN Spon.

Baker, S.W. 1990. *Sands for Sports Turf Construction and Maintenance.* The Sports Turf Research Institute, Bingley.

Baker, S.W. 1997. The reinforcement of turfgrass areas using plastics and other synthetic materials: A review. *International Turfgrass Society Research Journal* 8: 3–13.

Baker, S.W. 2005. *STRI Guidelines Golf Green Construction in the United Kingdom.* Bingley: The Sports Turf Research Institute.

Baker, S.W. 2006. *Rootzones, Sands and Top Dressing Materials for Sports Turf.* Bingley: The Sports Turf Research Institute.

Davis, W.B., Paul, J.L., and Bowman, D. 1990. *The Sand Putting Green: Construction and Management.* Cooperative Extension Publication 21448. California: Division of Agriculture and Natural Resources, University of California.

DIN 18-035 1991. *DIN 18-035 Sportpläze Part 4 Rasenflächen.* Beuth-Verlag GmbH, Berlin: Deutsches Institut für Normung..

FIFA 2011a. *Football Stadiums Technical Recommendations and Requirements.* 5th edition. Zurich: Fédération Internationale de Football Association.

FIFA 2011b. *Natural Grass Pitch Reinforcements FIFA Guidance Notes.* Zurich: Fédération Internationale de Football Association.

Gibbs, R.J. 2011. Forsyth Barr Stadium—The Undercover Story. *Australian Turfgrass Management Journal* 13.5: 6–12.

Haff, P.K. This volume. The far future of soil. In *The Soil Underfoot*, eds. G.J. Churchman and E.R. Landa. Boca Raton: CRC Press.

Hummel, N.W. 1993. Rationale for the revisions of the USGA green construction specifications. *USGA Green Section Record* March/April 1993: 7–21.

Moesch, R. 1975. Be-und Entwässerung von Rasenflächen nach dem Cell System. *Rasen-Turf-Gazon* 3: 83–85.

Skirde, W. 1974. Soil modification for athletic fields. In *Proceedings of the 2nd International Turfgrass Research Conference*, ed. E.C. Roberts, 261–269. Madison: American Society of Agronomy.

Sport England 2011. *Natural Turf for Sport Design Guidance Note.* Revision 2. London: Sport England.

Stewart, V.I. 1994. *Sports Turf: Science, Construction and Maintenance.* London: E & FN Spon.

USGA Green Section Staff 2004. *USGA Recommendations for a Method of Putting Green Construction.* http://www.usga.org/course—care/articles/construction/greens/USGA-Recommendations-For-A-Method-Of-Putting-Green-Construction(2)/

van Wijk, A.L.M. 1980. *A Soil Technological Study on Effectuating and Maintaining Adequate Playing Conditions of Grass Sports Fields.* Agric. Res. Rep. 903. Wageningen: Centre for Agricultural Publishing and Documentation.

Waddington, D.V. 1992. Soils, soil mixtures, and soil amendments. In *Turfgrass ASA Monograph No. 32*, eds. D.V. Waddington, R.N. Carrow, and R.C. Shearman, 331–383. Madison: American Society of Agronomy.

Section V

Future Strategies

"Enormous opportunities will be generated by the framing of future soil science research needs in the context of contributing to an ecosystems approach that can inform policy and protect the vital functions of soil that support human well-being, the Earth's life support systems, and the diversity of life on this planet." D.A. Robinson et al. *Vadose Zone Journal* 11, p. vzj2011.0051, doi:10.2136/vzj2011.0051(2012).

28 Soil Biophysics
The Challenges

Iain M. Young and John W. Crawford*

CONTENTS

28.1 INTRODUCTION

In a soil profile, soil structure is not unconnected from soil chemistry, a nitrogen compound is not necessarily separate from a root or organic matter, an active microbe is not separate from water, nor is soil made up of only structural units <2 mm in size. It is the connectivity and interdependence of soil that enhances and sustains the many functions of soil, not the isolated parts of soil. Once we achieve an interdisciplinary approach to soil, the chances of new breakthroughs and management solutions will have a far greater likelihood of being reached. The biophysics of soil covers a multitude of disciplines and focuses on the soil architecture and soil water that governs the rate of all soil processes, from the smallest microbe wandering through a profile, to the largest volume of water hitting a profile.

28.2 CHALLENGE 1. SECURING SOIL

How much are our soils worth? Costanza et al. (1997) put a very conservative value of $25 trillion a year on the value of the planet's soils. This value takes into account the functions of soil in terms of a buffer against pollution, a habitat for the vast numbers and biodiversity of biota living in soil, and the value of soil in acting as a chemical reactor.

Soil is exquisitely complex. A dirty, opaque material with more life-forms in a handful of soil than the total number of humans that ever existed, and more biodiversity; dependent on clay type, a surface area of between 20 and 200 m^2 in a sugar cube volume of soil and more biochemical pathways shooting up than any other known material, and a unique biomaterial that contains a hierarchy of pores that allows soil and water to coexist over an impressive range of energy potentials. All this means that soil is able to provide over 97% of all food requirements of humans, in terms of calories, and sustains over 6 billion people on a planet where fertile soil covers only a thin layer on its surface.

Soil is the most complex biomaterial on the planet. It has to be. Without the complexity in structure, life-forms, and chemistry, the food-producing function of soil would not sustain our population.

* School of Environmental & Rural Science, University of New England, Armidale, NSW, 2351, Australia; Email: iyoung4@une.edu.au

And yet, it is the simple things that often defeat us. Dirt has no currency in Western society and has little impact on politicians. It comes under the journalists MEGO category—My Eyes Glaze Over. Bar a few impressive dust storms, we care little of our soil. We do not relate what we eat in our home, buy in our supermarkets, or drink from our Starbucks to the soil. And yet, without soil, we become thirsty, hungry, and we die. Without soil, we become as Mars, with no water, no atmosphere, and only relics of life, with at best distant star gazers trying to figure out the life that could have been.

So, before we examine what we need in terms of new seeds, new chemicals to add to the soil, and new technology platforms that need development, we need to urgently look at legal frameworks that protect our soil asset. So, our first challenge outwith any discipline, any agricultural framework, or any plant species, is to call on governments to implement legal strategies to secure and build our fertile soil reserves. Civilizations have vanished in the past due to their failure to protect their soil resource: the great Mayan cities are now tourist attractions; during the Roman Empire, the second most influential city after Rome, Antioch, choked to death on eroded soil; the Dust Storms in the Great Plains of America blew over Washington DC, denuding plains that should never have been touched by intensive agriculture and the erosion of which were accelerated by promises from Federal government of large land packages on the "inexhaustible resource" of the deep soil of the Great Plains (Egan 2006).

If we ask the question, how much fertile soil do we have left, at current erosion rates, the answer is frightening. Montgomery (2007) suggests that we have on average a 1% net loss of fertile agricultural topsoil each year from erosion. Assuming that we have a depth of 150 mm, that means that we lose half the topsoil by 2080, and we are down to 20 mm by 2200 (Figure 28.1). So, in 100 years, we are well on the way to losing most of our fertile topsoil, and that is without factoring in other pressures: less water, decarbonization of soil, less inorganic inputs, urbanization, and less energy. Also, considering that in the past 40 years, we have already lost 30% of our fertile land; this is a frightening prospect for the planet. Increase the erosion rate to 5% and we lose all fertile soil by 2080. All soil. Gone during the life of our children.

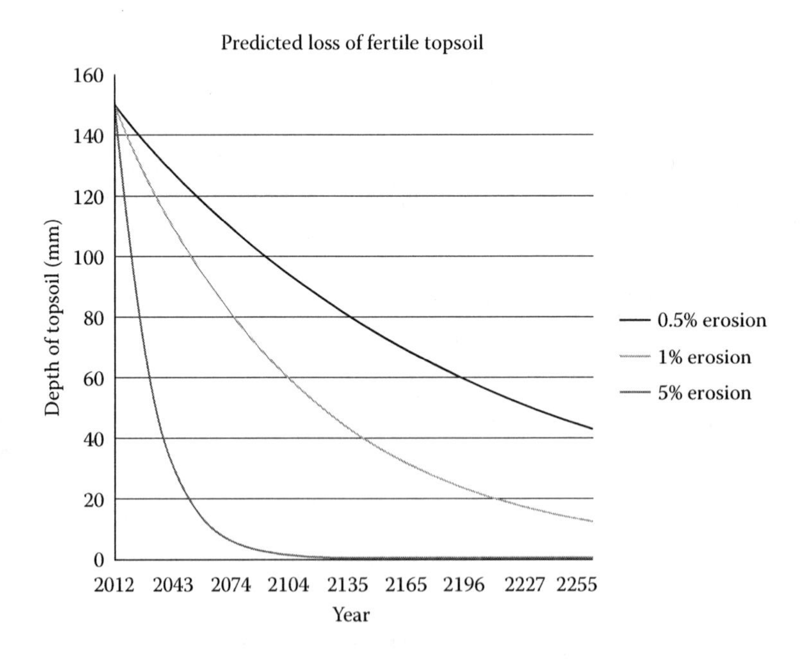

FIGURE 28.1 Predicted topsoil loss from agricultural soil at various net erosion losses per year assuming a starting topsoil of 150 mm.

Today, we have one thing that our forefathers had not, and that is the knowledge of what not to do and what to do to secure our soil resources. As Jared Diamond stated in his book *Collapse* (Diamond 2005), we are now able to choose to succeed or fail, and without far-reaching legislative agreements across the world, we are not convinced that we will choose correctly. The lessons of the past were hard learned but have they taught us anything? These words incidentally are being written as the Mars Rover "Curiosity" is remotely driven across the soil of the Red Planet, firing lasers into soil and rock to learn about possible past life on Mars. Ironic in a sense is that more money is being spent on this singular soil exploration (in excess of $1.5 billion) than has perhaps ever been spent on soils research on the home planet. We may look back in the near future and wonder why.

28.3 CHALLENGE 2. GETTING TO THE ROOT OF THE PROBLEM

Plant breeding in the grains industry has had tremendous success, relying on high-input agriculture with a large reserve of inorganic nutrients available to add to the soil, the production of which relies on cheap and easily accessible fossil fuel. However, such resources are diminishing. Crops grown on land, irrigated, or well favored with rainfall, now approach 80% of the best predicted yields, indicating there is little scope for further significant improvement (World Bank 2008). This has led Passioura and Angus (2010) to suggest that productivity improvements will have to come from less favorable rain-fed environments to meet the increasing world food demand to sustain the additional world population of 2.3 billion people by 2050.

However, investigating roots, the "Hidden Half" of the plant, to improve crop yields, has been virtually ignored, and looking underground where the plant's resource capture and indeed mass is greatest, presents unrivaled opportunities to future breeding programs. Focusing on the top growth of plants has been the main route for plant breeders, for some very good reasons. First, the focus on above-ground plant traits has been highly successful. Second, it is incredibly difficult, expensive, and time consuming to look at roots in soil. The main root architecture work has been done in simple soil constructs that allow us to see roots against a glass front or in hydroponic/soilless systems. However, in the former, we fail to see most of the roots, and such technologies have hardly advanced since the first serious attempt in East Malling Research Station in the United Kingdom in the early 1960s. In the latter, we ignore the environment that we want them to grow in—soil—that is unwise as it is the interactions of roots in soil that are the key to our understanding of plant development (Gewin 2010).

Peering through a glass darkly at roots, or growing them in artificial growth medium does not reflect the growth of roots as they occur in field conditions. As discussed by Lynch (1995) and Hodge (2004), plant roots respond significantly to their soil environment, making the development of a three-dimensional (3D) analysis of the root phenotype growing in soil, potentially a groundbreaking tool in assessing the competitive advantage of different plants in different environments. To be of use, such technology has to be accurate, quick, and cheap, with the ability to achieve high throughput of samples.

A number of technologies are available. One, magnetic resonance imaging, is discounted, as it does not easily work in heterogeneous soil. Another, neutron imaging, has great potential, but is expensive with very limited accessibility to a few research labs. However, x-ray tomography, being lab based, is accessible and has the potential for resolving both soil structure and roots, with great potential for high-throughput analysis. Flavel et al. (2012) provide details on each technology and show the results on experiments linked to nutrient uptake. Essentially, tomography fires x-rays into soil and materials of different electron density attenuate these x-rays. So, soil and air are relatively easy to see, while water-filled pores with roots make it difficult to decouple the roots from water as they have a similar electron density.

In the next 5 years, with suitable funding, we will see a significant increase in work being done in nondestructively analyzing the architecture of roots within heterogeneous soil, with real applications to growers within very short timescales. Taking technology from the medical industry into

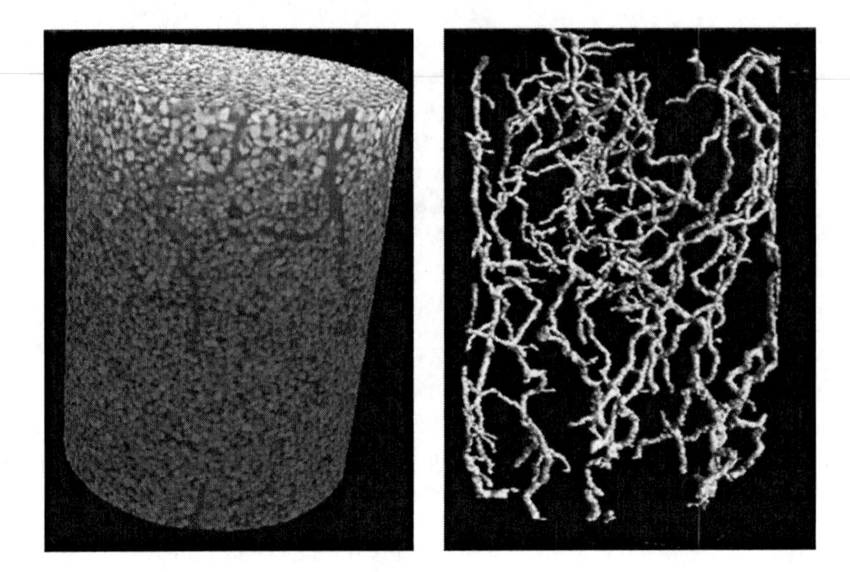

FIGURE 28.2 Microcomputed tomography scanning to segmentation of roots from soil. Reconstructed core in three dimensions: gray curvilinear shapes represent plant roots and reconstructed section of soil with roots. (Adapted from Flavel, R.J. 2012. *Journal of Experimental Botany* 63: 2503–2511.)

the soil labs, we are now able to observe and measure roots growing in heterogeneous soils. Using high-resolution x-rays, we can monitor the impact of roots on soil and vice versa. This emerging technology has the potential to revolutionize the investigation of root–soil interactions, having an immediate impact on plant-breeding programs and on the complex task of understanding such interactions at scales where roots operate: meters to microns. With scan times of as quick as 4 min, with a very high level of environmental control (e.g., light, water regimes, and temperature), we can now observe roots growing in heterogeneous soils, and have the nondestructive measure of root parameters over space and time. Figure 28.2 shows the latest results (Flavel et al. 2012).

One question we can ask ourselves now is what do we know already about root traits that we would screen for? Root architecture is one area that is obviously important in terms of resource capture and one where soil structure plays a significant role in modulating (Lynch 1995). Root elongation rate where fast growth can access water at depth quickly, thus securing available water in times of water restrictions, is another. The recovery of roots say after meeting compact layers in soil, or predators is yet another. The list is already long. That is not to say that such technology does not need to get better. It does. But we already know in what areas we must improve and all are within our grasp: ability to scan larger samples, better and more automated segmentation of roots from soil, faster scans, and improved algorithms for the reconstruction of the 3D volume.

28.4 CHALLENGE 3. DESIGNER SOILS

Soil biodiversity is a growing theme with proposed links to "healthy" soils, whatever that means. However, much of what has been developed is driven by technologies, largely molecular, which now enable us to count parts of the microbial community that were previously unmeasurable. Much less has been invested in trying to find out what we actually need to measure. In particular, it is not what is *there* that is important, but what it *does*. The soil microbial phenotype gives rise to soil function, but is a complex mix of the soil genetic complement and its microenvironmental context in soil. We may have a highly diverse population of soil microbes in a field. So what? How does it relate to the sustainable productivity of that field? Is it anything other than a form of redundancy of functions in soil, or simply a tool that allows scientists to develop more papers looking at the biology of numbers and differences? In one of the most comprehensive, yet least-quoted, analysis of soil biodiversity, a telling statement highlights the lack of application of biodiversity measures to productivity "...no consistent relationships between soil species diversity and soil functions have been found to date" (Turbé et al. 2010, p. 12). Aside from the work on many soil pathogens, this statement holds true.

When we take a step back and ask how much of the microscale detail do we actually need to know to understand the productivity, resilience, and efficiency of soil, we see a much less complicated picture. Work on the biophysics of the soil environment, examining the interaction of the physical habitat and soil microbes, reveals some direct and consistent links with important functions (Crawford et al. 2012; Young et al. 2012). We are on the verge of being able to identify and manipulate specific microbial processes to *design* soils with specific characteristics that may enhance water-holding capacity of soil, increase soil stability, and improve soil structure.

Designing soils will be a new discipline for soil scientists and plant scientists, and one rooted in fundamental science and applied outputs. We now know that surfactants produced from roots allow the whole plant to access water and phosphorus at no extra energetic cost (Read et al. 2003); fungi play a huge role in stabilizing soil and altering gas diffusion in soil systems (Crawford et al. 2012), as well as producing hydrophobic substances that help stabilize soil by lowering water entry into slaking soils (Young et al. 2012). An array of bacteria found in soil produce surfactants that, like roots, can modulate water holding in soils (Fechtner et al. 2011) and exude calcite-binding agents that strengthen soil and through the deposition of inorganic carbon, sequester carbon (DeJong et al. 2006).

We now have the connection between specific microbes and specific functions. The challenge is to find more and take that development from the lab to the field and quantify what an extra millimeter of water means to agricultural production. Where this has often failed in the past is due to our inability to reproduce in the lab, the soil conditions that are present in the field. The key here is the understanding that all soils naturally exhibit high spatial and temporal variability. Interestingly, one area where microbial biodiversity has been constantly linked to is the physical diversity of the habitats in soil (Young et al. 2009). In very much the same way humans do, soil microbes like to live in areas of good resources, low predators, and where relevant, a good selection of mates.

28.5 CHALLENGE 4. A CALL FOR A NEW MODEL ARMY

We know that soil is highly complex. However, is it just too complicated to model, or are there a few key things we need to know about soil that will help us to adequately predict how soil functions?

How do we model soil as an integrated system? Can we not simply use calibrated regression-type "models" that give us statistical trends, in the hope that simple characteristics always behave the same independently and toward each other? There are limited cases where such models operate at a particular time and place. In general, however, they do not provide the necessary level of predictability that is required to manage the soil as a productive dynamic system.

One way forward would be to investigate the use of system modeling approaches. Such approaches use mathematical models to deal with complexity and predict the consequences of large networks of interactions over space and time. These have proved superior to human intuition, especially in situations where we are trying to predict intervention outcomes that are not immediate in a context that is changing. That is, when we are working on almost anything to do with the environment and food production.

The overwhelming response to the idea of applying systems approaches to the soil system is that it is just too complicated. However, where these approaches have been applied in other contexts where very large numbers of components interact (e.g., cancer (Faratian et al. 2009)), it transpires that the details do not matter when the system that is being modeled is resilient to random perturbation (Barabasi 2009). In other words, nature is not fine-tuned, and as such, most of the details of individual interactions do not impact on the resultant behavior of the system precisely because it is complex, and that multiple feedbacks conspire to stabilize the system to perturbation at the component scale. This kind of behavior, or "emergent parsimony," where interactions conspire to make only a small subset of factors matter, is difficult, if not impossible, to predict intuitively, and so, without the application of modeling approaches, progress is unlikely.

These approaches are being used with considerable success in the life sciences where systems biology is starting to make sense of the reams of molecular data, and more importantly, is pointing to new kinds of data and parsimonious descriptions of otherwise overwhelmingly complex systems. At the community level, progress has been made in describing ecological communities and searching for the link between biodiversity and function (Hooper et al. 2005), and to understand socioeconomic systems, and particularly the endogenous instabilities that lead to crises (Farmer and Foley 2009).

Crawford et al. (2012) show the results of a model for the genesis of pore-scale structure in soil by extending the conceptual model of Tisdall and Oades (1982) to include a feedback between microbially generated structure and the quality of the microenvironment in soil. Even this simple model indicates the profound difference in the nature of the dynamics of the system simply by linking microbial activity, oxygen flow, structure, and particle packing. Despite the complexity, however, the results suggest that the details of the interactions and the microbial community do not matter for the emergent behavior.

To progress, we need much closer collaboration between theoreticians and experimentalists to collaboratively decide what needs to be measured and how that measurement should be made. We must avoid the temptation to be seduced by technology that swamps us with data, and might distract us from more worthwhile pursuits.

REFERENCES

Barabasi, A. 2009. Scale-free networks: A decade and beyond. *Science* 325: 412–413.

Costanza, R., R. d'Arge, R. de Groot, S. Farberk, M. Grasso, B. Hannon, K. Limburg et al. 1997. The value of the world's ecosystem services and natural capital. *Nature* 387: 253–260.

Crawford, J.W., L. Deacon, D. Grinev, J.A. Harris, K. Ritz, B.K. Singh, and I.M. Young. 2012. Microbial diversity affects self-organization of the soil microbe system with consequences for function. *Journal of the Royal Society Interface* 9: 1302–1310.

Curtis, T. 2006. Microbial ecologists: It's time to "go large". *Nature Reviews Microbiology* 4: 488.

DeJong, J.T., M.B. Fritzges, and K. Nüsslein. 2006. Microbially induced cementation to control sand response to undrained shear. *Journal of Geotechnical and Geoenvironmental Engineering*, 132: 1381–1392.

Diamond, J. 2005. *Collapse: How Societies Choose to Fail or Succeed* 592pp. New York: Viking.

Egan, T. 2006. *The Worst Hard Time. The Untold Story of Those Who Survived the Great American Dust Bowl*, 340pp. New York: Mariner Books.

Faratian, D., R.G. Clyde, J.W. Crawford, and D.J. Harrison. 2009. Systems pathology—Taking molecular pathology into a new dimension. *Nature Reviews Clinical Oncology* 6: 455–464.

Farmer, J.D. and D. Foley. 2009. The economy needs agent-based modelling. *Nature* 460: 685–686.

Fechtner, J., A. Koza, P. Dello Sterpaio, S.M. Hapca, and A.J. Spiers. 2011. Surfactants expressed by soil pseudomonads alter local soil–water distribution, suggesting a hydrological role for these compounds. *FEMS Microbiology Ecology* 78: 50–58.

Flavel, R.J., C.N. Guppy, M. Watt, A. McNeill, and I.M. Young. 2012. Non-destructive quantification of cereal roots in soil using high-resolution x-ray tomography. *Journal of Experimental Botany* 63: 2503–2511.

Gewin, V. 2010. An underground revolution. *Nature* 466: 552–553.

Hodge, A. 2004. The plastic plant: Root responses to heterogeneous supplies of nutrients. *New Phytologist* 162: 9–24.

Hooper, D.U., F.S. Chapin, J.J. Ewel, A. Hector, P. Inchausti, S. Lavorel, J.H. Lawton et al. 2005. Effects of biodiversity on ecosystem functioning: A consensus of current knowledge. *Ecological Monographs* 75: 3–35.

Lynch, J. 1995. Root architecture and plant productivity. *Plant Physiology* 109: 7–13.

Montgomery, D.R. 2007. *Dirt: The Erosion of Civilizations*. Berkeley: University of California Press.

Passioura, J.B. and J.F. Angus. 2010. Improving productivity of crops in water-limited environments. *Advances in Agronomy* 106: 37–75.

Read, D.B., A.G. Bengough, P.J. Gregory, J.W. Crawford, D. Robinson, C.M. Scrimgeour, I.M. Young, K. Zhang, and Z. Zhang. 2003. Plant roots release phospholipid surfactants that modify the physical and chemical properties of soil. *New Phytologist* 157: 315–326.

Tisdall, J.M. and J.M. Oades. 1982. Organic-matter and water-stable aggregates in soils. *Journal of Soil Science* 33: 141–163.

Turbé, A., A. De Toni, P. Benito, P. Lavelle, N. Ruiz, W.H. Van der Putten, E. Labouze, and S. Mudgal. 2010. Soil biodiversity: Functions, threats and tools for policy makers. Bio Intelligence Service, IRD, and NIOO, Report for European Commission (DG Environment).

World Bank. 2008. *Agriculture for Development. World Development Report.* Washington, DC: The World Bank.

Young, I.M., J.W. Crawford, N. Nunan, W. Otten, and A. Spiers. 2009. Microbial distribution in soils: Physics and scaling. *Advances in Agronomy* 100: 81–121.

Young, I.M., D.F. Feeney, A.G. O'Donnell, and K.W.T. Goulding. 2012. Fungi in century old managed soils could hold key to the development of soil water repellency. *Soil Biology and Biochemistry* 45: 125–127.

29 Life in Earth
A Truly Epic Production

*Karl Ritz**

CONTENTS

29.1 INTRODUCTION

Soil is alive. Very alive. The total mass of living material in a typical arable soil is about 5 tonnes per hectare—in grasslands and forests it can be 20 times this, such that the biomass belowground always equals and sometimes exceeds that aboveground. In a handful of such arable soil, there will be 10 billion bacteria, scores of kilometers of fungal hyphae, tens of thousands of protozoa, thousands of nematodes, and hundreds of worms, insects, mites, and other fauna. Life in soil is diverse. Very diverse. The genetic diversity of the soil biota *always* exceeds that found aboveground and

* National Soil Resources Institute, School of Applied Sciences, Cranfield University, Cranfield MK43 0AL, UK; Email: k.ritz@cranfield.ac.uk

by orders of magnitude—in our handful of soil, there will typically be 10 thousand "species" of bacteria and hundreds of species of the other forms. Most of this life and biodiversity is invisible to humans for two reasons. First, the soil biomass is predominantly microbial in scale, by definition too small to be discerned by the unaided eye. Second, because soil is opaque to visible light, which human eyes have evolved to be sensitive to, it is not possible to look into soil in the same manner that one generally perceives the atmosphere and water bodies. Out of sight and, generally, out of mind. And yet the soil biota are fundamentally important to soil function, since they are essentially the biological engine of the earth, driving and governing many of the key processes that underpin the functioning of terrestrial ecosystems.

By using microscopes and specialized visualization techniques, including x-ray computer-assisted tomography (CAT) scanners, we can visualize the minute and otherwise opaque soil system to understand how soils are spatially organized and the denizens therein. And when we do this, it reveals an astonishing stage. Using the same microscopes, complemented by a suite of biochemical and molecular genetic analyses, a remarkable cast of players is apparent. The scripts that these troupes unwittingly act out are equally remarkable, and the story of how this is manifest is one of a truly epic scale, which would keep any film director unimaginably occupied. This chapter aims to provide a preview of this story by providing glimpses of some of the main features of life below-ground, and how it leads to effective soil function, playing loosely with a filmic analogy.[1]

29.2 THE SET

The iconic 007 Sound Stage at Pinewood studios in England is one of the largest available in the industry at some 5400 m^2. This high surface area is also manifest in our typical handful of arable soil.[2] However, the staging area in soils is manifest as a network of pores of extraordinary character, being a three-dimensional (3-D) labyrinth of great structural heterogeneity across many orders of magnitude, typically from nanometers to centimeters. Tortuous tunnels, the walls of which are also porous, are apparent in all directions whatever scale is considered, or size you are (Figure 29.1). Mineral soils are comprised of a diverse range of inorganic components of different fundamental sizes from millimeters to nanometers, categorized as sand, silt, and clay, plus an equally diverse range of organic constituents of which living organisms typically comprise but a few percent by mass. The remaining non-living organic matter is entirely derived from living material at some point in its history (with the minute exception of any obscure anthropogenically derived inputs). These basic units aggregate via a range of mechanisms, predominantly chemical at the smallest spatial scale, biological at larger scales, and physical across all scales, such that the aggregations are always porous to some extent. Such aggregates are in contact and connect to form larger units across a hierarchy of scales, leading to the pore networks, which, however tortuous, are thus always connected at some scale. The soil pore network is ultimately one contiguous, albeit highly convoluted, surface. These properties of the pore network lead to some remarkable properties, including the ability to simultaneously hold both water and air in close proximity, regulation of the rate at which gases, liquids, solutes, particulates, and organisms move through the soil matrix, and the provision of billions of niches for microbes to reside. This "inner space" is then the arena where the soil biota act out their respective roles. And like the features of any set, the properties of this inner space govern where the actors can be placed and their ability to move, communicate, interact, and function.

29.3 THE CAST

One of the consequences of the huge biodiversity belowground is that there are no dominant headlining individuals among the soil biota, but an absolutely vast cast of diverse players that range in size from μm^3 to km^3.

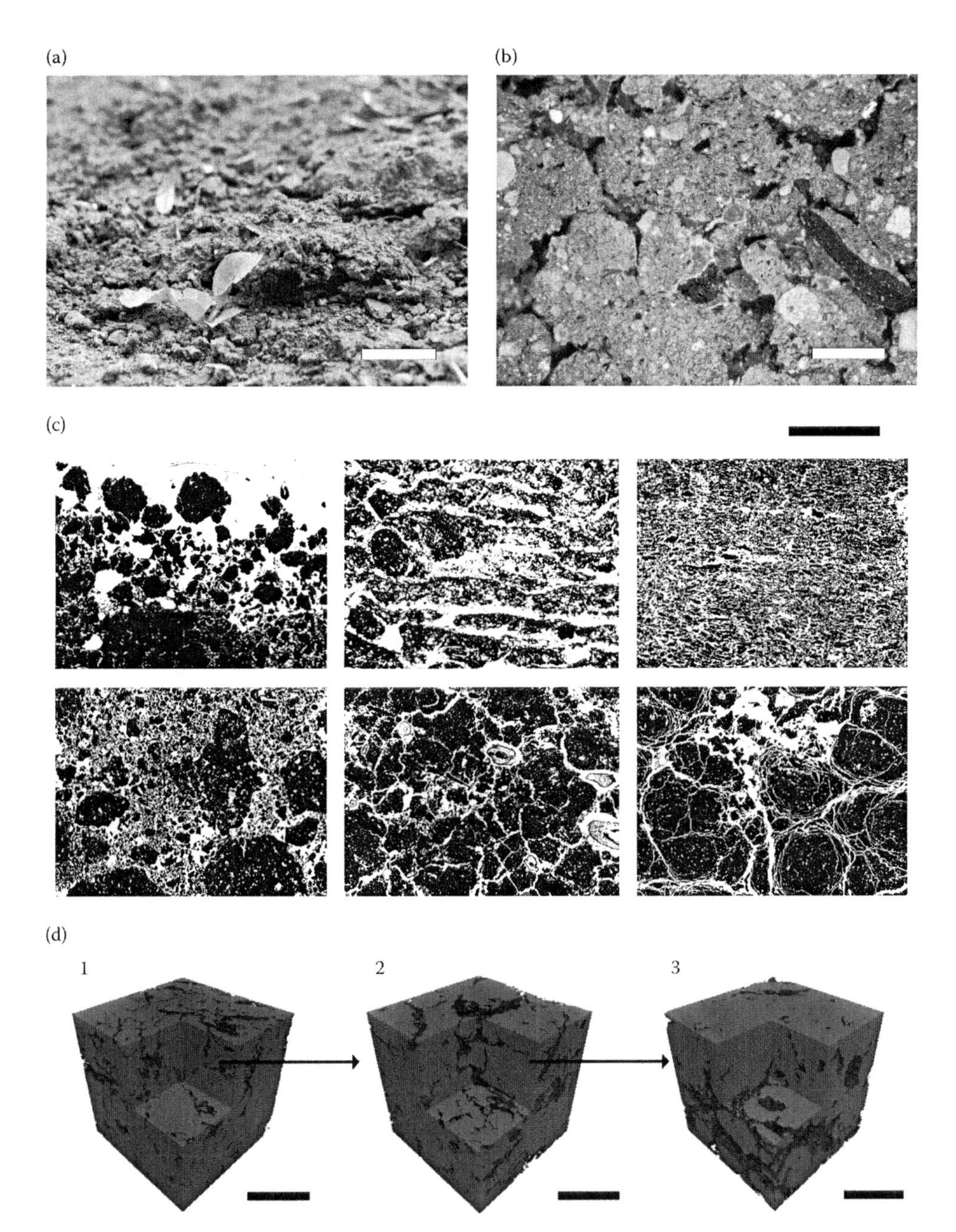

FIGURE 29.1 **(See color insert.)** The soil stage across scales and dimensions. (a) The familiar scale to humans. Scale bar = 10 cm. (b) Soil aggregates and pore network visualized in cut-face of a resin-embedded undisturbed soil. Scale bar = 5 mm. (c) Contrasting pore network morphologies as visualized in two dimensions via thin sections of resin-embedded soils. Scale bar = 10 mm. (d) Three-dimensional reconstruction of soil pore networks at three successive spatial scales derived from x-ray CT scans of a clay loam beneath a grass racecourse. Cube 1: scale bar = 10 mm, pores greater than 50 μm resolved; Cube 2 zoomed-in image of the corner zone of Cube 1, scale bar = 5 mm, pores greater than 25 μm resolved; Cube 3 zoomed-in image of the corner zone of Cube 2, scale bar = 2.5 mm, pores greater than 12.5 μm resolved. ((a), (b) From Karl Ritz; (c) From Young, I.M. and Ritz, K. 2005. *Biological Diversity and Function in Soils*, ed. R.D. Bardgett, M.B. Usher, and D.W. Hopkins, pp. 31–43. Cambridge: Cambridge University Press; adapted from Fitzpatrick, E.A. 1993. *Soil Microscopy and Micromorphology*. Chichester, UK: John Wiley; (d) From Craig Sturrock, CPIB, Hounsfield Facility, University of Nottingham.)

29.3.1 Prokaryotes

The prokaryotes are distinguished from all other life forms in that their genetic material is not confined to a nucleus, but occurs in a free form in their cells.[3] The two forms of prokaryote are bacteria and archaea. They are the smallest independent life forms, which occur in soil, with each individual cell typically being of the order of 1 μm (10^{-6} m) in size, with the shapes of such cells occurring as spheroids, rods, spiral, helical, or S-shaped forms. Archaea are much less abundant than bacteria in soils, and until recently it was considered that they only occurred in extreme environments. However, recent evidence arising from the revolution in environmental genetics has revealed that archaea are essentially ubiquitous, although their precise functional roles still remain elusive. Reproduction occurs by a process of binary fission, where the cells simply divide into two, and thence by a continual doubling-series until resources become limiting or they are consumed by other organisms. This process tends to result in close-packed aggregations of cells, or colonies, the size of which is generally governed by the amount of available food. In most soils, such substrate is usually very limited, hence belowground bacterial colonies tend to be small (Figure 29.2a). Exceptions are where nutrient resources are more abundant such as in the vicinity of plant roots (the "rhizosphere"), or where nutrient-rich organic material such as fresh plant litter, dung, or cadavers are present, or the soil has passed through the gut of a worm or other soil animal (Figure 29.2b). The majority of soil prokaryotes are heterotrophs and hence gain their energy from such fixed (organic) sources of carbon (C), but a wide range of autotrophs also occur, who fix CO_2 using energy sources derived from light (photosynthesis), or via other inorganic chemical reactions. Most colonies are bound to soil surfaces, held there by the sticky mucilages, which also serve to bind the constituent cells together. Being predominantly attached to pore walls, such colonies must then rely on their delivery of substrate via diffusion in soil water, and such delivery is then regulated by the nature of the soil pore network. Actively mobile cell forms also exist, where each cell is propelled through water films via oscillations of hair-like protrusions called flagellae. Taking our filmic perspective, there are both heroes and villains[4] among the prokaryotes. The hugely diverse and versatile range of types involved in the delivery of almost all of the key soil functions are the heroes—they are ubiquitous extras in the soil plot, taking part in the fixation of carbon, and the cycling of virtually all nutrient elements, including decomposition, nitrogen (N) fixation, and phosphorus (P) solubilization. Many form symbiotic associations, including mutualistic relationships based on N-fixation, for example, with plants (e.g., *Rhizobium* with leguminous plants), ferns (*Azolla:Nostoc* rafts in rice paddies), or fungi (lichens). The villains include a wide range of plant and animal pathogens that can be phenomenally destructive, especially when offered monotonous contiguous areas of genetically uniform hosts—leading to easy pickings. Variants of the bacterium *Pseudomonas syringae* have an extraordinarily wide plant host range and can be very virulent, and both tetanus and anthrax are notorious soil-dwelling animal pathogens.

29.3.2 Fungi

The most familiar forms of fungi to most people are mushrooms and toadstools, but these are merely the reproductive structures of one particular group of this hugely diverse group of organisms. Fungi occupy a distinct taxonomic kingdom, separate from prokaryotes, plants, and animals. They are ubiquitous in soils and can constitute the larger part of the biomass belowground, particularly in soils high in organic matter. There are two basic growth forms of fungi, single-celled yeasts, and a filamentous form called a hypha that grows by apical extension. Hyphae then branch periodically to form space-filling networks called mycelia (Figure 29.2c), or indeed more complex structures such as mushrooms. Fungi are all heterotrophs, acquiring their nutrition by exuding enzymes from hyphal tips into the substrata through which they are growing, and absorbing the resultant nutrients from the digested soup that arises. This filamentous growth form is highly suited to explore complex pore networks such as those prevailing in soil, enabling the fungus to explore the belowground labyrinth for food resources

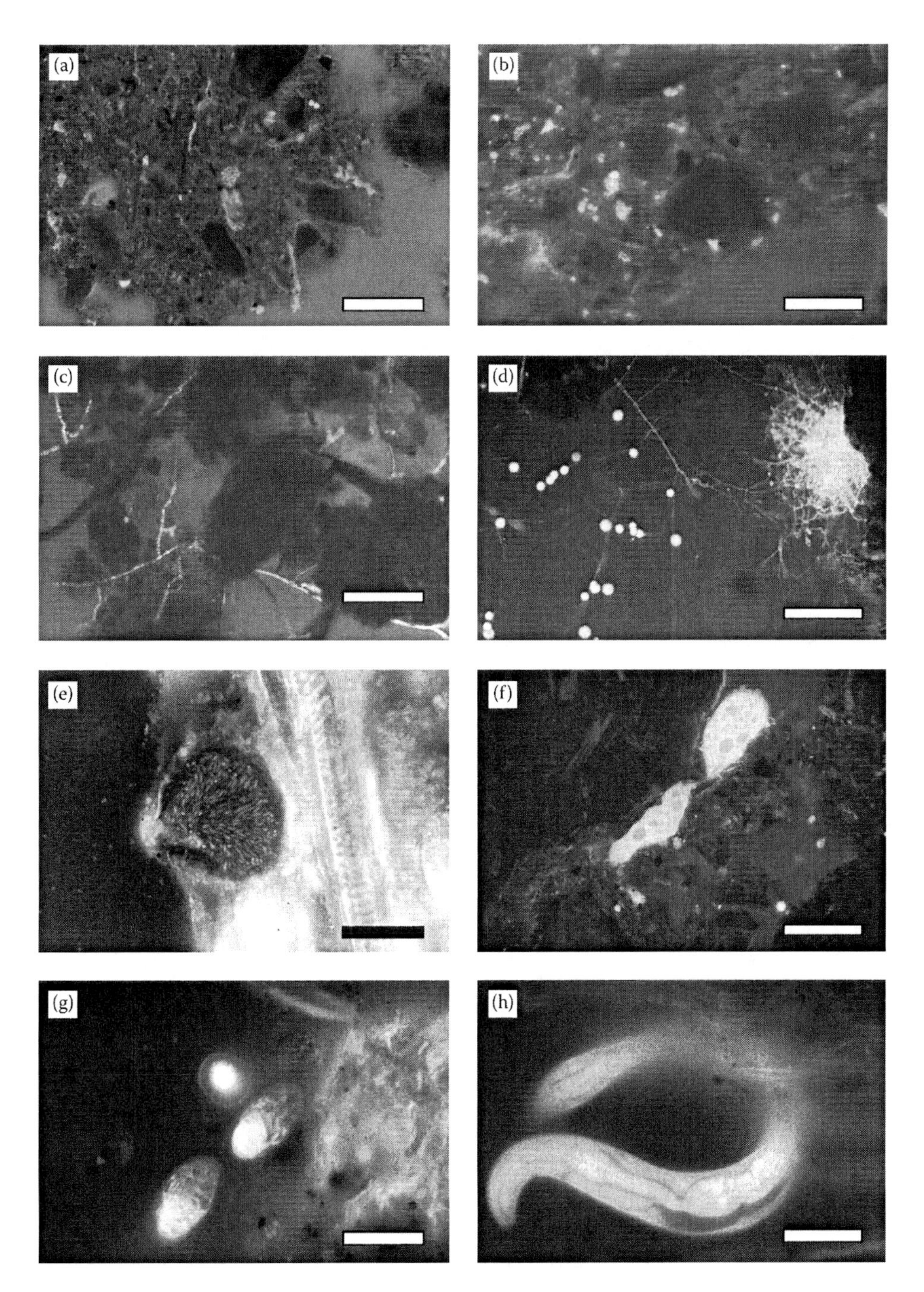

FIGURE 29.2 (**See color insert.**) Soil biota visualized in biological thin sections of undisturbed arable soils. (a) Bacterial microcolony in topsoil zone of arable soil. (b) Bacterial colonies in earthworm cast. (c) Mycelium of *Rhizoctonia solani* foraging through pore network. (d) Unidentified fungus bridging pore in exploratory mode and proliferating in exploitative mode on localized substrate; the bright spots are spore clusters. (e) Fungal perithecium (flask-shaped fruiting body containing spores) embedded in decomposing root, with opening from which spores are extruded aligned with lumen of pore. (f) Naked amoeba passing through soil pore neck. (g) Testate amoebae; three individuals in vicinity of bacterial colonies located on pore wall surface. (h) Bacterial-feeding nematode. Scale bars (a)–(g) = 20 μm; (h) = 100 μm mm. (From Ritz, K. 2011. *Architecture and Biology of Soils*, pp. 1–12. Wallingford, UK: CABI.)

by branching sparsely and then exploiting such resources encountered by branching profusely (Figure 29.2d). Soil fungi are apparently very sensitive to the spatial organization of their environments, for example, some species locate their spore-bearing structures in a manner that will optimize likely spore dispersal pathways (Figure 29.2d and e). The shape of many mushroom caps is adapted to optimize the patterns of air-flow around the cap to enhance carriage of spores dropping down from the gills into the wider local atmosphere. Mycelia are indeterminate, which means the elongation and branching process can continue more or less indefinitely and thus can grow to enormous dimensions. There are many documented instances of some mycelia in forests being several hectares in extent and weighing several megatonnes, making such fungi by far the largest organisms, in terms of both mass and extent, on the planet. Like the prokaryotes, they are involved in all aspects of the soil plot. They decompose organic matter with aplomb, producing enzymes capable of very efficient degradation of some of the more recalcitrant woody material that plants produce. While this is important for ecological function, it can be hugely destructive where structural timbers of buildings are attacked, for example, by dry rot (*Serpula lachrymans*), which is essentially a woodland fungus flexing its enzymes in what it detects as fair game—coarse woody debris connected to a soil-like (plaster and mortar) substratum. Fungi also form mutualistic relationships, called mycorrhizas, with the vast majority of the plant kingdom. Specialized symbiotic fungal hyphae penetrate plant roots, and then exploratory mycelia grow through the soil using carbon derived from their host plants. The fungi then absorb water and nutrients, particularly phosphorus, from the soil and transmit such materials back to the plant, thus enhancing growth of their partners. Mycorrhizas are so ubiquitous that in most soil-grown plants, roots are rarely simply roots in the botanical textbook sense but nearly always hooked-up to a fungal network.[5] A diverse range of fungi are also antagonistic to most other life forms, with the plant pathogens ranking as some supreme villains. Rapacious plant root pathogens include "take-all" of cereals and the Honey-fungus (*Armillaria*) root rot of trees. There are many relationships between fungi and soil-dwelling insects. Some species of ant essentially farm fungi in their nests, in highly regulated fungal-based composting systems that provide food for their colonies.

29.3.3 Protozoa

Protozoa are single-celled microbes typically less than 1 mm in size. They are mobile, but only in water films, and so their movement is constrained by the presence of liquid water. Soil-inhabiting protozoa occur in three morphological forms. Flagellates possess limited numbers of whip-like flagellae similar to (but larger than) those of bacteria, which are used for locomotion. Ciliates possess numerous short, hair-like cilia that cover the cell surface, and locomotion is achieved by wave-like pulses of these cilia passing synchronously across the cell. Amoebae are naked cells that can deform their shape to "ooze" through water and across surfaces by extruding protuberances in the direction of movement and contracting the cell surface in the hindmost regions. They can then also pass through soil pores (Figure 29.2f). Some amoebae form silicaceous "tests" (Figure 29.2g), in which the organism resides when not grazing, which are bottle-like structures created from numerous microscopic plates (Figure 29.3). A further class of common soil-inhabiting protozoan organisms is the mycotozoa or slime molds (their taxonomic provenance is rather academic; they are a relatively unknown group, but are ubiquitous and beguilingly diverse). Much of the time, these occur as single-celled amoeboid cells in the soil, but on occasion, they form remarkable structures called plasmodia, which are huge masses of millions of swarmer cells in the form of an often brightly colored slime, which moves by a scaled-up oozing process. This structure then also passes through the soil matrix and litter surfaces and consumes organic matter and microbial prey in its path. Some plasmodia can be a square meter or more in dimension, and form spectacular sheets of brightly colored slime draped across the forest floor and climbing up tree trunks (Figure 29.4). These plasmodia are then real-life versions of an often-revisited horror film device, "The Blob," but in this case from inner space! Protozoa play a key role in nutrient cycling in soils. They consume bacterial and fungal cells, essentially grazing across bacterial colonies, capturing free-living cells

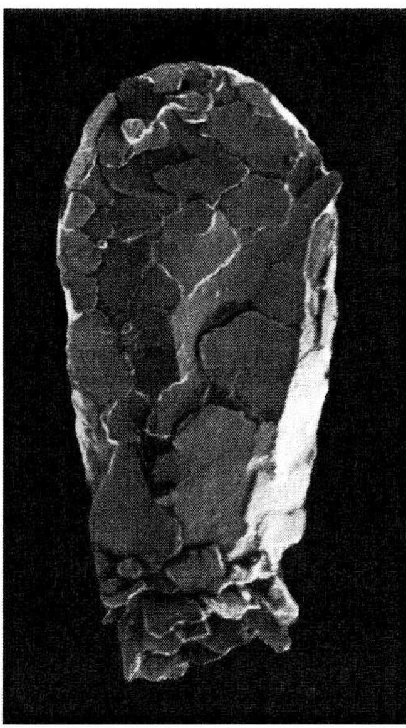

FIGURE 29.3 **(See color insert.)** Scanning electron micrographs of testate soil amoebae showing siliceous plate-based construction of the test. Scale bar = 10 μm. (From Wilhelm Foissner, Universität Salzburg.)

in the soil water phases, or burning holes in fungal hyphal walls with enzymes and then consuming the contents. Other types browse and consume particulate organic material. Potential food sources, whether other biota or organic matter, can be protected from attack or consumption by protozoa if it is located in pores smaller than the body size of the forager or predator, or in pores where the entry points to them are too small to allow entry. It is a moot point the extent to which such size protection operates for amoeboid protozoa, which can clearly "ooze" through constrictions, presumably only constrained by whether their nuclei will fit through the gap (Figure 29.2f).[6] Following all of these forms of consumption, the excretion products are often rich in forms of nitrogen, which are available to plants and other soil microbes.

29.3.4 Nematodes

So far, the cast has involved single-celled organisms. Nematodes, the next in our cast list, are multicellular (i.e., composed of coherent aggregations of cells) worm-like organisms, and as such, have considerably more complex anatomies than the microbes, including a mouth, an anus, a very basic nervous system, and various relatively simple organs. They are motile, and like protozoa, confined to water for their locomotion, essentially by a process of wriggling. Soil nematodes are of the order of millimeters in size and are ubiquitous, occurring in all soil habitats from rainforests to the dry valleys of Antarctica. They have a more or less common body plan but their feeding habits range very widely, in essence there being nematode species' adapted to consume anything that is organic, and/or alive. They achieve this via specialized mouthparts adapted to their respective primary food sources, such as: siphons to suck up particulate organic matter, bacteria (Figure 29.2h), or protozoa; hollow needles to suck the contents out of plant roots or fungal hyphae; sharp, tooth-like structures to penetrate the body walls of other nematodes or fauna; and graters to enable penetration of plant tissue. Nematodes, like protozoa, play a key role in nutrient cycling in that they excrete nutrients following ingestion and digestion of their food sources. Plant pathogenic forms are also pernicious in that they can cause massive yield reductions in many crops, and can reach such levels of infestation in some agricultural soils that certain crops cannot be grown.

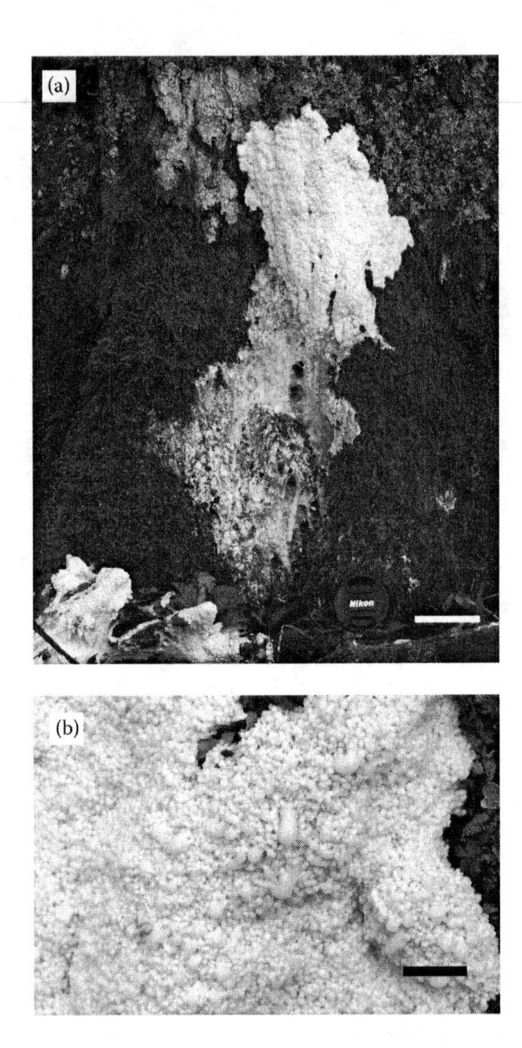

FIGURE 29.4 *It Came from Inner Space*. Plasmodium of slime mold in Scottish woodland, ascending a tree trunk. (a) Distant view. The plasmodium extends 1 m further out of the frame of the photograph to the left. Scale bar = 10 cm. (b) Detailed view of plasmodium. Scale bar = 10 mm. (From Karl Ritz.)

29.3.5 Worms

There are two main forms of worms (in taxonomic terms, annelids) common in soils, the enchytraeids and the earthworms. Enchytraeids are typically millimeters to centimeters in size and nonpigmented. They feed on organic matter and thus act as decomposers, and are most abundant in soil zones that contain relatively large quantities of organic matter. Earthworms are larger soil annelids and can range from centimeters to meters in length. They create burrows by ingesting soil via their mouths and extruding it via the anus. The worm castings that they excrete are either deposited on the soil surface for some species or back-filled in the burrows for those types that prevail in the soil fabric. Large quantities of soil can be processed in this manner by worms, and thus huge masses of soil brought to the soil surface over time. Darwin calculated that worms can turn over the top 15 cm of soil in 10–20 years, by measuring the rate at which stones on the surface became buried. Earthworms are perhaps the iconic soil organism, and the one that most people are aware of. Alas, in films, worms are often favored as a means of invoking horror.[7]

29.3.6 Arthropods

Arthropods are defined as animals without backbones and a rigid external skeleton (exoskeleton). This is a catch-all term, which in the soil context includes insects (e.g., collembola (springtails), ants, termites, beetles), arachnids (mites and spiders), myriapods (millipedes and centipedes), and isopods (woodlice). There is a vast range of soil arthropoda, most of which are of the order of a millimeter to a centimeter or so in size, due to the physical constraints imposed by the soil matrix. These organisms are essentially all about "legs and mouthparts"—they are mobile, and forage through the soil matrix (and across the soil surface) seeking their food resources. As is the case for nematodes, anything organic in the soil context, dead or alive, is potentially consumable by one or many species of arthropod. And the critter will accordingly be adapted via its legs (fast for predator, stable for herbivore) and mouthparts (chewing, sucking, snipping, grabbing, tearing, shredding, etc.). Mites in particular cover the gamut of feeding habits and are concomitantly diverse in their form (Figure 29.5). All these fauna collectively form an efficient cohort of ecosystem engineers, comminuting their substrate, excreting fecal pellets that act as organic-rich substrate for a myriad of microbes, and burrowing and moving soil material that restructures their collective habitat.

29.3.7 Viruses

Viruses are in the twilight zone of soil biota as strictly they are not organisms as such, since they are incapable of independent reproduction, but are infectious agents that can only replicate via

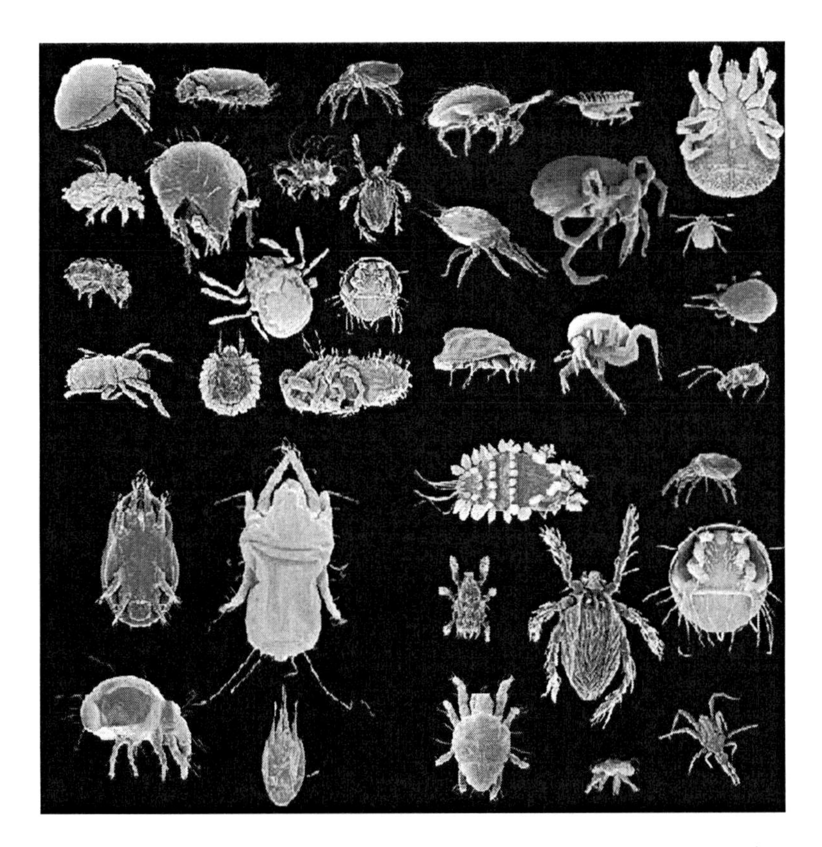

FIGURE 29.5 Montage depicting diversity of some soil mites. Scales variable—most mites of the order 0.5–2 mm body length. (From David Walter, University of the Sunshine Coast, Australia.)

interactions with their hosts. Viruses are known to be highly abundant and play important roles in regulating biotic communities in aquatic systems, but very little is known about the nature, extent, and actions of viruses in soil systems, with most emphasis (and then very modest) on viral pathology of plants. It is certain that there are viral interactions occurring within the majority of the soil biota, and that viral ecology will be strongly affected by interactions with the soil matrix, particularly with respect to adsorption phenomena to soil minerals. But soil virology is essentially an unexplored belowground frontier. If viruses are involved in significant regulation of belowground communities, this is a potentially serious knowledge gap.

29.4 THE PLOT

So much for the cast—an almost inconceivably large group of actors of stultifying diversity, range of sizes, origin, background, and form. If we freeze-frame almost any soil in time, the mis-en-scène will reveal the biota distributed throughout the soil set, largely in relation to their physical size, the recent locations of water throughout the soil pore network, and the recent or likely distribution of their food sources. The small proportion that are not in long-term inactive phases ("quiescent"—or more euphemistically stated in acting circles "at rest") will be requiring to locate a source of energy relatively promptly or then also shut down until resources become available. If we now cue "action," then the biological engine of the earth will start up, commence work, and a panoply of interactions between the biota and their underground environment will commence, delivering the range of ecosystem goods and services that make the terrestrial system function.

The Millennium Ecosystem Assessment (2005) has provided a useful framework for the four key scenes to the plot, and as in any effective script there are links, connections, and feedbacks between the different phases.

29.4.1 Scene 1: Supporting Services

29.4.1.1 Soil Formation

Soils are formed over geological time by a range of physical and chemical processes acting on the parent material and involving small- and large-scale movement and transformation of mineral constituents by heating, cooling, freezing, thawing, wetting, and drying, and by a range of biochemical processes, mediated by the soil biota, including transformation of mineral constituents, for example, the etching and (literally) tunneling of minerals by fungi, and the weathering of rocks by lichens. Another critical biologically mediated function is that of the incorporation and transformation of organic compounds and matter into the regolith. This is fundamentally underpinned by primary production.

29.4.1.2 Primary Production

The big green autotrophs—on land, predominantly the vascular plants—drive the terrestrial system by capturing the energy in sunlight and delivering a proportion of it belowground as root exudates, sloughed root cells, and eventually root debris. Litter deposited on the root surface is incorporated into the soil fabric by the faunal engineers, and this organic material provides energy for the soil biota, fuelling the myriad of processes that they carry out. During these transformations, organic matter becomes physically protected by being sequestered into tiny pores beyond the reach of organisms or even enzymes, ultimately at the molecular scale as highly recalcitrant films on mineral particles. Huge random polymers ("humus") are also formed by condensation reactions, which are resistant (but not immune) to further decomposition. These processes serve to establish a pool of latent energy or slow-release power, which ensure that the biological engine will likely have some means to prevail in the longer term, by keeping it ticking over, or idling. The nutrients that plants require to grow are in turn made available by nutrient cycling.

29.4.1.3 Nutrient Cycling

Life, or rather the biochemicals that underpin it, requires a suite of elements as well as the carbon backbone of all living matter. A relatively small proportion of all extant elements are involved here and there is an approximate hierarchy of requirement classified as major, macro-, and micronutrients elements,[8] but the key is that they must all be provisioned appropriately in both space and time, via dynamic processes of delivery and transformation. In soil systems, all such nutrient elements are then "cycled" by both abiotic and biotic processes, whereby they are transformed from one chemical form to another and transferred between compartments, usually involving associated transfers of energy. The soil biota are implicated to a greater or lesser extent in cycling all nutrient elements. Such processes can involve a "loss" from the soil system as a result of biotic activity, for example, in gaseous forms (carbon dioxide, methane, nitrous oxide, dinitrogen). A full description of biologically mediated nutrient cycling in soils would be truly epic, and is even beyond the scope of a chapter that contains the term in its title. Aspiring directors are referred to the Further Reading section!

29.4.2 Scene 2: Provisioning Services

Provisioning services provided by soils are recognized as relating to food, fiber, and biomaterial (including fuel) production, manifestation of habitats and refugia, and generation and sustenance of biodiversity. As rehearsed above, the soil biota are implicated in plant production via many strands of supporting services, but there is rather a circular aspect to our plot here since the soil biota also directly contribute to provisioning biodiversity, to a massive extent in the belowground compartment, as well as providing the underpinning basis for aboveground biodiversity. The plot then thickens in that the soil biota themselves create and condition soil structure and hence the habitat and associated refugia in which they themselves reside. Biotic roles in soil structural genesis are many and varied. Soil constituents are glued together via sticky biochemicals produced by microbes to enable colony cohesion and adhesion to pore surfaces, and by earthworms and plant roots to lubricate passage through the soil matrix. The surfaces of aggregates and pore walls are also coated by hydrophobic materials exuded by many microbes to protect their cells from desiccation. Fungi in particular do this, since the surface-area-to-volume ratio of their mycelia poses a particular demand in this respect. Fungal hyphae and plant roots also serve to bind the fabric of the soil together via the filamentous and branching nature of their mycelial and root networks, notably across a range of size scales. The faunal engineers also serve to structure and restructure the soil matrix via burrowing, consumption, and excretion of soil particles, and remolding using mouthparts. In one sense, organisms build their habitat, and this raises a more appropriate concept of soil "architecture," since this admits the interactions that occur between life belowground and their living space. The next twist in this aspect of the plot is that the extreme levels of biodiversity prevailing in soils arise as a result of the multiscale characteristics of this architecture. A consequence of the belowground labyrinth is that there is essentially an indefinite number of refugia for soil organisms, microbes in particular, to reside within and be physically protected from predation. But it also results in an extreme form of spatial isolation, which tends to result in the evolution of genetically distinct forms, which then prevail. This is a form of 3-D island biogeography, akin to the processes first recognized by Darwin in the Galapagos. Zones of soil that are relatively proximal in unconstrained space may be orders of magnitude apart in literal space when the soil pore network is taken into account.

29.4.3 Scene 3: Regulating Services

Regulating services are essentially the "governors" of ecological systems, and for soils include water storage quality, erosion control, and atmospheric gas regulation. Water storage is underwritten by the soil pore network, governed by biotic action as discussed above. Water quality is mediated via the action of the soil biota in terms of regulation of nutrient cycling, degradation of pollutants, attenuation of pathogens (via predation), and once again, via soil structural integrity, which can affect the amount of particulate material deposited into water courses, particularly via processes of erosion. The latter is affected by the stability of both soil aggregates and the entire soil surface. Soil communities growing in this critical interface between the atmosphere and the soil matrix are very different from those in the subsurface, often dominated by microbial photoautotrophs, and their roles in stabilizing and governing erosion processes are remarkably poorly understood. The soil biological engine mediates atmospheric gas regulation mainly by nutrient cycling phenomena; for example, the following gases are produced in the soil via sequential transformation processes associated with the decomposition of organic matter and then released into the atmosphere: CO_2 from respiration, methane from anaerobic processes, and nitrous oxides from denitrification. Such gases can also be absorbed into the soil system via various processes of biologically mediated fixation. A further important regulatory function, which involves a form of self-reference or circularity, is that of biotic regulation, whereby certain components of the soil biota regulate others via processes of predation and consumption. This connects to nutrient cycling provisioning, but is of significance in other terms, for example, attenuation of pathogens and modulation of invasive organisms.

29.4.4 Scene 4: Cultural Services

While the previous three scenes need not necessarily involve a human audience (they have been occurring on the planet since the onset of soil formation in geological time), cultural services are gratuitously related to our perception of the biosphere and indeed beyond. Thus, habitat provision, leading to the landscapes that humans find desirable in cognitive terms are provided by the soils that support them, and thence largely by the soil biota that underpin them as explored in the previous scenes. Soils play a significant role in the preservation of cultural heritage, principally via archaeological routes. Biotic roles here can again be those of villain, in that many organic archaeological artifacts (wood, clothing, bones, teeth, etc.) may represent energy-containing substrate for hungry soil organisms, and be decayed or destroyed via decomposition processes, particularly in aerobic soils. Some soil fungi play a direct cultural, or more specifically culinary, role in many societies. The Perigord Truffle is a renowned epicurean delight, and ranks among the most expensive wild food available.

In cognitive terms we then have a paradox, in that the soil biota are for the most part out of sight and out of direct cognition, but the roles they play, and the fascinating and tangled plots they weave, *all the time*, are of huge significance to humankind.

29.5 COMING SOON

Soil biology as a discipline has been studied for a century or so, but it is clear that there remains a vast amount that remains to be understood—and discovered—about life in the underworld. The main constraints relate to the extraordinary biodiversity that prevails in soils, their opacity (literally and conceptually), and techniques to deal with these issues. There is also an extensive and pervasive underappreciation of the role of soil biota by scientists, policy makers, and public alike, which is, to an extent, understandable given the invisibility of what goes on belowground.

This notwithstanding, recent advances in our understanding of the biological engine of the earth suggest what may be coming soon.

29.5.1 Biological Indicators of Soil Health

Soil health is a term much used, abused, debated, and professed on by many involved with studying, managing, exploiting, and protecting soils. The simplest definition is arguably that a healthy soil is one that is fit for purpose. This allows that soils serve many purposes and hence the definition must relate to the context and perspective under consideration. More sophisticated definitions involve notions of resilience (ability to recover from perturbation), sustainability (delivery of functions into the future), and multifunctionality. Effective and interpretable indicators of soil health are therefore often required. Given the pivotal role that the soil biota play in driving and governing so many soil functions, it is rational to consider that life belowground will reflect both the current status of the soil as an integrated system and the capacity of the system to function—possibly in future terms as well. Hence, the biota should be able to serve as an indicator, or even provide a means of quantifying, soil health in both a general and specific manner, and this has been the topic of much research by soil ecologists. But one of the consequences of both the multiplicity of function and enormous biodiversity belowground is that there are apparently no straightforward relationships between biotic structure or properties and function (general or specific), resilience, sustainability, or whatever. There is a wealth of context dependency and idiosyncrasy, which frustrates all involved. There is, however, often a straightforward mapping of what a soil biologist will attest as a "powerful bioindicator of soil health" and whatever particular facet of soil biology or biochemistry has preoccupied them for much of their careers. In some respects, this is rational too, over and above the psychology that is involved, since as rehearsed above there is a large diversity in all biotic groups or soil biochemistry that has developed in the context of any particular circumstance, and there is bound to be some form of relationship between them. But it is naive to expect that one facet of the biota will serve as a universal interpretable indicator fit for all purposes, and as such, multivariate (sometimes called multiphasic) approaches are likely to be more effective, notwithstanding the fact that their interpretation may be more challenging. The big question is what level of abstraction of the massive construct of the soil biota is most appropriate to allow for *interpretable* indicators of the state of a soil system, in contemporary and future terms? This issue preoccupies scientists and policy makers alike, and will continue to do so, since its resolution is challenged by the complexity of soil systems.

29.5.2 Biotechnological Potential and Informatics

There are a wide range of biotechnological products of great significance derived from soil organisms, including antibiotics, crop protection products, and single-cell protein sources. Examples include some of the earliest antibiotics such as streptomycin (from the bacterium *Streptomyces griseus*), the fungicide azoxystrobin (from the fungus *Strobilurus tenacellus*), the antimicrobial compound lumbricin (from the earthworm *Lumbricus rubellus*), and the mycoprotein Quorn® (from the fungus *Fusarium venenatum*). To date, the production of these compounds relies on the ability to grow such organisms in large-scale culture systems. The first glimpses of the true genetic diversity of soil microbes became apparent in the 1980s, as advances in molecular biology, and particularly the ability to analyze genetic sequences, began to be applied to soil systems. Subsequent application of these techniques as they become more resolute, rapid, and larger-scale persistently reveal that soil genetic diversity is extraordinary and extreme. Initially, what almost every DNA sequence revealed had not been hitherto seen, even from the same soil sample. Gradually, as billions of sequences are accrued and collated on databases, some modest similarities are being mapped, but still relatively few.

A big surprise was that it became apparent that the vast majority of bacteria in soils are not revealed by techniques based on cultivation on nutrient media, which soil microbiologists had so assiduously been applying for nigh on a century, and which underpin the industrial-scale biotechnological utilization of specific microbes. Thus, there are thousands of bacteria and archaea in soils, which have never been cultivated, nor their actions in the soil or *in vitro*, determined. The new and next generations of ultra-high-throughput techniques for genetic and other molecular biochemical analyses (the slew of so-called "-omics" techniques, such as proteomics [proteins], metabolomics [metabolites], and doubtless more to come) will result in insights leading to significant new biochemical understanding, and thence new biotechnological concepts and applications. These techniques are leading to hitherto unattainable levels of resolution in measurement, and will drive the development of new ways to analyze and visualize highly complex and massive data sets. While it is likely that new products will be identified within this vast morass of unknown microbes, and subsequently exploited, the potential goes far beyond this. This may include a move away from the one-organism-one-product concept, since soil ecology particularly demonstrates the impact that the *community context* has upon the behavior of both individual organisms and the functions of the system as a whole. Such phenomena are at the core of the "probiotic" concepts becoming more applied in medical scenarios, and have potential in the management of soil biota as well. This will influence concepts in industrial biotechnology, including beyond the fermentation tank and at the farm scale. Most farmers are essentially applied soil ecologists, and increasingly, sophisticated biotechnologists. More effective management of the soil biota will lead to more effective production systems, and the challenge is to do this in a sustainable manner. Understanding how the soil biological engine functions will underpin this. We know the basics—that this has to relate to the provision of energy to the biota, and the organization of their habitat. Future approaches must then consider the system-level context (given the diversity), and this will arise from development and application of environmental informatics approaches.

29.5.3 Live Set: Do Not Disturb

That soils are opaque to visible light severely constrains our ability to study what goes on in them, and certainly in relation to the location, movement, and action of the biota. In filmic terms, soil scientists are confronted with the script term *fade to black* almost by default. It is also understood that soils function by virtue of their spatial organization and the integrity of such arrangements, and thus techniques that involve disintegrating them in order to make observations and measurements are prone to be of limited effectiveness. Noninvasive techniques, such as x-ray and gamma-ray tomography, offer a means to visualize the internal structure of soils, and have been successfully applied to reveal the larger soil biota such as fauna and plant roots (Figure 29.6). Microbes are much more challenging because of the high spatial resolution required, and the difficulty of attaining any contrast in the signals between the organisms and their background, which is particularly "noisy" in soils due to the complexity of soil structure. Nonetheless, tomographic resolution is increasing with ongoing development, and techniques will likely emerge that will enable microbes to be visualized, including those based on pattern-based image analysis of morphology in three dimensions, such as has been pioneered for root systems. Optical probes based on laser spectroscopy are also being developed that can enable

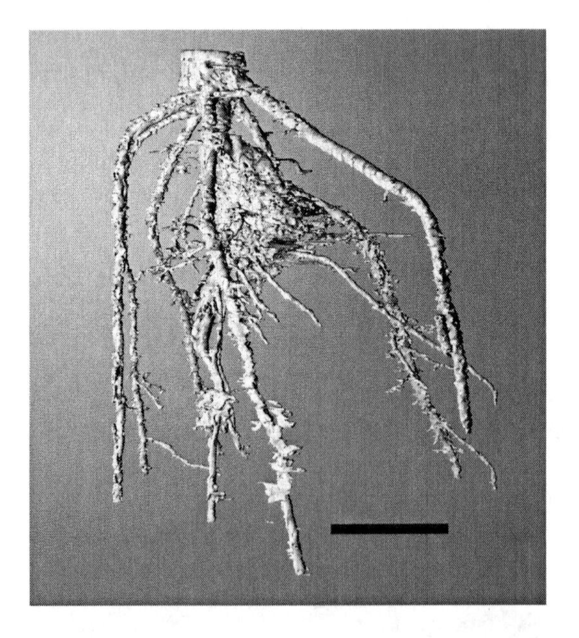

FIGURE 29.6 **(See color insert.)** Computer-assisted tomographic image of soil-grown maize root system. Scale bar = 15 mm. (From Stefan Mairhofer, CPIB, Hounsfield Facility, University of Nottingham.)

the *in situ* (and hence mappable) noninvasive measurement of a range of chemical properties, including gases and pH. In time, these will likely include specific biochemicals, and thence to organisms.

Another means to noninvasively measure the activity of soil biota, which has long been applied, is via the production of CO_2 or consumption of O_2 during respiration. This works since the gases diffuse out of the soil and can thus be detected without intrusion into the system. However, organisms also produce a wide range of volatile organic compounds (VOCs) that similarly diffuse from soils and can be characterized by spectrometry. The profiles of VOCs from both individual organisms and whole soil communities are often complex, and can apparently relate to the physiological state of the biota and environmental circumstances such that the "smell" of soil could be an effective means to noninvasively determine the state and action of the biota.

29.6 THE WRAP

Life on Earth thus clearly relies upon life in earth. Soils are fundamentally living systems, and those who are required, tasked, or committed to managing soils need to take this into account. Prior to the onset of industrial agriculture, this occurred via experientially based soil management practices, by which civilizations developed their agriculture, which of necessity had to rely solely on biological systems to deliver the requisite soil functions. Industrial agriculture short circuits many of the biologically delivered functions by a process of substitution, largely based on fossil-fuel-derived energy. This can be very effective purely in terms of delivering yield (i.e., the production function), but there is no such thing as a free lunch and the follow-on effects in terms of ecosystem integrity and resilience are of significance and consequence for sustainability. The challenge is to generate and maintain such productivity while also balancing the delivery of the other range of ecosystem goods and services we require. The soil biota undoubtedly have a principal role to play in this respect. The full story of life in earth has so far only been glimpsed, but is beginning to be understood. The potential for the biological engine to deliver the work required for sustainable civilization is unquestionably there. How to realize this will require new and imaginative solutions, new ways of looking at things, and most challenging of all, some quite astonishingly sound direction. *Dear Mr Tarantino....*

NOTES

1. More literal relationships between film and soil have been explored by Landa (2010). A few examples are given in this chapter, and below, for filmic analogies between soil biota and their natural history and functions. Readers are invited to consider—and spot—others. The author would be pleased to receive suggestions.
2. The concept of surface area in soils is less straightforward that at first might appear, because it is a scale-dependent measure. The apparent area depends on the size of the ruler you use to measure it... or indeed how small you are. This area would relate to that available to the smallest soil organisms, the bacteria.
3. All other life forms carry their genetic material in membrane-bound organelles within each of their cells called nuclei.
4. This is of course a perspective entirely from a human standpoint. In ecological terms, good and evil are redundant concepts.
5. This networking theme is in some sense explored in the film *Avatar* (2009).
6. Predator avoidance by such size exclusion mechanisms was a trick the kids invoked in *Jurassic Park* (1993); amoeboid-like circumvention of this device was awesomely shown in *Terminator 2* (1991), where a liquefied predatory robot simply oozed through the opening in that favorite Hollywood device, the elevator hatch.
7. Examples include *Squirm* (1976), *The Seven Percent Solution* (1976), *Dune* (1984), *The Lair of the White Worm* (1988), *Tremors* (1990), and many others with worm-like Hollywood clichéd monsters emerging from the soil underfoot.
8. For the record, these are generally considered to be: major nutrients—carbon, hydrogen, oxygen; macronutrients—nitrogen, phosphorus, sulfur, potassium, sodium, magnesium, chlorine, calcium, silicon;

and micronutrients—boron, manganese, iron, cobalt, copper, zinc, molybdenum, iodine, selenium. As for most aspects of biology, the details belie these generalizations and variation in dependency varies between plants and animals, and indeed within these groups. But the intriguing point is that these "chemicals of life" represent a tiny proportion of the periodic table.

FURTHER READING

Bardgett, R.D., M.B. Usher, and D.W. Hopkins. 2005. *Biological Diversity and Function in Soils*. Cambridge: Cambridge University Press.

Clothier, B. and M.B. Kirkham. 2014. Soil: Natural capital supplying valuable ecosystem services. In *The Soil Underfoot*, ed. G.J. Churchman, and E.R. Landa. pp. 135–149, Boca Raton: CRC Press.

Darwin, C. 1883. *The Formation of Vegetable Mould, Through the Action of Worms*. London: John Murray.

Fitzpatrick, E.A. 1993. *Soil Microscopy and Micromorphology*. Chichester, UK: John Wiley.

Jeffrey, S.R., C. Gardi, A. Jones, L. Montanarella, L. Marmo, L. Miko, K. Ritz, G. Peres, J. Rombke, and W.H. Van der Putten. 2010. *Soil Biodiversity Atlas of Europe*. Commission, Office for Official Publications of the European Communities: Luxembourg.

Landa, E.R. 2010. In a supporting role: Soil and cinema. In *Soil and Culture*, eds. E. R. Landa and C. Feller, pp. 83–105. Dordrecht: Springer.

Marschner, P. and Z. Rengel. 2007. *Nutrient Cycling in Terrestrial Ecosystems*. Berlin: Springer.

Millennium Ecosystem Assessment. 2005. *Ecosystems and Human Well-Being: Synthesis*. Washington, D.C.: Island Press.

Ritz, K. 2011. Views of the underworld: *In situ* visualisation of soil biota. In *Architecture and Biology of Soils*, eds. K. Ritz and I. M. Young, pp. 1–12. Wallingford, UK: CABI.

Ritz, K. and I.M. Young. 2011. *Architecture and Biology of Soils*. Wallingford, UK: CABI.

Stevenson, F.J. 1986. *Cycles in Soil: Carbon, Nitrogen, Phosphorus, Sulfur, Micronutrients*. Chichester, UK: John Wiley.

Turbé, A., A. de Toni, P. Benito, P. Lavelle, N. Ruiz, W.H. Van der Putten, E. Labouze, and S. Mudgal. 2010. *Soil Biodiversity: Functions, Threats and Tools for Policy Makers*. Bio Intelligence Service, IRD, and NIOO, Report for European Commission (DG Environment), Brussels.

Wall, D.H., R.D. Bardgett, V.M. Behan-Pelletier, J.E. Herrick, T.H. Jones, K. Ritz, J. Six, D.R. Strong, and W.H. Van der Putten. 2012. *Soil Ecology and Ecosystem Services*. Oxford: Oxford University Press.

Young, I.M. and K. Ritz. 2005. The habitat of soil microbes. In *Biological Diversity and Function in Soils*, ed. R.D. Bardgett, M.B. Usher, and D.W. Hopkins, pp. 31–43. Cambridge: Cambridge University Press.

Young, I.M. and J.W. Crawford. 2014. Soil biophysics—The challenges. In *The Soil Underfoot*, eds. G.J. Churchman and E.R. Landa. pp. 371–377, Boca Raton: CRC Press.

30 Sustaining "The Genius of Soils"

Garrison Sposito[*]

CONTENTS

> Now give we place to the genius of soils, the strength of each, its hue, its native power for bearing.
>
> **—Publius Vergilius Maro, 29 BCE**

30.1 INTRODUCTION

The lines quoted above from a prose translation of the second book of Virgil's *Georgics* (Fairclough and Goold 1999, 149) draw attention to the friability, the color, and the inherent fertility of soils, three key properties constituting the "genius" of natural porous media found at the land surface. This iconic Latin poem about farming goes on to contextualize these three properties in a very practical manner for a variety of agricultural crops ("the denser better for Ceres, the lighter for Bacchus") in what the poet-translator David Ferry has called "one of the great songs, perhaps the greatest we have, of human accomplishment in the difficult circumstances of the way things are" (Ferry 2005, xiv). While Virgil's *Georgics* celebrate the unique characteristics of soils in nature, the poem does not underestimate the arduous task of cultivating them for human use (Ferry 2005, xiv):

> The farmer works the soil with his curved plow, This is the work he does, and it sustains His country, and his family, and his cattle, His worthy bullocks and his herd of cows. No rest from this.

Writing in a special issue of *Science* magazine with the engaging title, "Soils—the Final Frontier," McNeill and Winiwarter (2004) outlined a history of soil cultivation over millennial timescales in a narrative that also underscores the arduousness of the task and its imminent precariousness, their proposition being that "civilization, with its systemic links between cities and hinterlands, over

[*] Department of Environmental Science, Policy and Management, University of California at Berkeley, Mulford Hall MC 3114, Berkeley, CA 94720-3114, USA; Email: gsposito@berkeley.edu

the past 5000 years has posed an ongoing challenge for farmers trying to maintain soil fertility" (McNeill and Winiwarter 2004, 1627). They conclude their survey with the exhortation that we all should do better to pay more attention to soils and our use of them (McNeill and Winiwarter 2004, 1629): "Soil ecosystems remain firmly, but uncharismatically, at the foundations of human life. The intensity and scale of modern soil use and abuse suggest there is much yet to be discovered about soils and their relations with people. Equally, current behavior implies that there is much that is already known that is not yet converted into prevailing practices. Soil ecosystems are probably the least understood of nature's panoply of ecosystems and increasingly among the most degraded." Note the use of the adverb "uncharismatically" to give emphasis in the first sentence. No doubt, this is what Leonardo da Vinci had in mind in creating his famous epigram from which the present volume takes its title. Whatever inspired Virgil to write so extensively and beautifully about soils in the first century BCE had faded away by the 1500s CE, such that Leonardo could remark ironically concerning the typical human preoccupation with charismatic features of the material world, illustrating once again, the Aesopian moral, that familiarity breeds contempt.

The luxury of taking soil for granted may carry an increasingly higher price tag, however, because of the looming issue of human population growth, also the topic of a recent special issue of *Science* magazine, this one titled provocatively, and somewhat ominously, by the single word, "Population." Notwithstanding a consistent global decline in human fertility during the past 40 years, current estimates of where the world human population will be at mid-century exceed 9 billion (Roberts 2011; Tilman et al. 2011), with about 80% of the global populace expected to reside in Asia (56%) and Africa (24%). This mid-century estimate represents an increase of the current world population by about 30%, but corresponding estimates of the percentage increase in crop production that will be required to meet the resulting world food demand (e.g., for cereals) are more than twice as large (de Fraiture et al. 2009; Hanjra and Qureshi 2010; Gregory and George 2011; Tilman et al. 2011). Since food demand convolutes both population size and personal income, increases in the latter alone can shift consumption trends toward higher-caloric intake and a more diversified diet, thus increasing food demand. Moreover, the large estimated percentage increase in crop production, for reasons to be made transparent in the next section, is expected to come mainly from increasing the crop yield per hectare cultivated—crop intensification—and not from simply increasing the amount of global land area converted into agricultural use. This is a daunting challenge, one that cannot be met in the absence of a deep understanding and appreciation of the very properties of soils that Virgil was praising in his mellifluous poem.

30.2 PROLEGOMENA

One effective way to frame a challenging problem is first to go through a preliminary exercise of considering the constraints that any attempt to address the problem has to respect. In the context of increasing food production, recent detailed studies of the patterns of global land and water use, along with their ecological impacts, lead to three such constraints that in all likelihood will limit how crop production increases of the magnitude noted in the introduction can be achieved. One of them is about land use, another is about water use, and the third one is about the way most crops are raised.

30.2.1 Land Conversion for Crop Cultivation Is Nearing Its Planetary Limit

Agricultural land, defined here as both cropland and pastureland, now occupies about 38% of the ice-free area of the global terrestrial surface, this occupation having been achieved at the expense of losing 70% of the grassland, 50% of the savanna (i.e., grassland dappled with trees), 45% of the temperate deciduous forests, and 27% of the tropical forests that were present before the invention of agriculture (Foley et al. 2011). The ecological consequences for biodiversity, as well as for water, air, and soil resources, stemming from this dramatic change in land use, particularly, the change that has occurred during the past five decades, are clearly outlined by Ellis (2011) and carefully tallied by Foley et al. (2005, 570–571), who suggest that "modern agricultural land-use practices may be trading short-term increases in food production for long-term losses in ecosystem services," the latter being defined as the flows of matter and energy from ecosystems that benefit humans (Dominati et al. 2010).

In particular, the ecological impacts of croplands, now occupying 12% of the ice-free global terrestrial surface area (Foley et al. 2011), are dramatic [e.g., more atmospheric nitrogen is now converted for agricultural purposes into reactive and, therefore, potentially pollutant, forms of nitrogen than in all natural processes combined (Rockström et al. 2009a)]. Rockström et al. (2009a,b) have concluded from their survey of such impacts that a "planetary boundary" must be imposed on the future expansion of croplands, eventually to limit them to occupy not more than 15% of the ice-free global terrestrial surface. In proposing this limit, Rockström et al. (2009a, 17) worried out loud that humankind "may be reaching a point where further agricultural land expansion at a global scale may seriously threaten biodiversity and undermine regulatory capacities of the Earth System." This same concern was echoed by Barnosky et al. (2012, 57), who, among other things, called for "increasing the efficiency of existing means of food production and distribution instead of converting new areas," while warning that a planetary-scale "critical transition" of global ecosystems may be in the offing.

In a recent comment on the proposed planetary boundaries, Running (2012) suggests that global net primary production (i.e., the biomass, aboveground and below, which is made available annually to consumers by photosynthetic plants) has remained effectively constant despite major land-use changes over recent decades. Noting that humans now appropriate about 38% of the global net primary production for themselves, he goes on to estimate that all but 10% of the remaining 62% is not realistically harvestable. This residual 10% is thus set as a limit on any additional human co-opting of global net primary production.

Taking a less dystopian perspective, Smith et al. (2010) performed a careful survey of global land-use trends over the past 50 years, noting that, on average, only about one-fifth of the food production increase achieved during that time was directly attributable to the expansion of harvested area. Noting the dual importance of population growth and changes in diet when assessing food demand, and alluding to a possible decrease in the quality of nonagricultural land brought into future crop production, they predict from their extensive compilation of model forecasts of land-use change under a variety of competitive drivers and pressures that crop extensification (i.e., the transformation of nonagricultural ecosystems into cropland) will increase the global harvested area only by about 20% at mid-century, which is rather close to the limit imposed by the "planetary boundary" of Rockström et al. (2009a,b). Smith et al. (2010, 2955) conclude that, given "increasing competition for land to deliver non-food ecosystem services, it is clear that per-area agricultural productivity needs to be maintained where it is already close to optimal, or increased in the large proportion of the world where it is suboptimal."

30.2.2 Blue Water Use by Croplands Is Also Nearing Its Planetary Limit

Blue water (Falkenmark and Rockström 2004, 2006; Rockström et al. 2009c) refers to the water flowing in streams and rivers, stored in lakes and reservoirs, or pumped from groundwater, the water we humans use for drinking, bathing, cooling, and diluting our pollutants, and the water we share with other forms of life in aquatic and terrestrial ecosystems, who use it for exactly the same purposes. Thus, it is blue water that supports irrigated agriculture, which accounts for 90% of the global consumptive use of this resource. ["Consumptive use" describes blue water withdrawn, used, and subsequently evaporated into the atmosphere without being returned to the blue water flow system; it is also termed the "blue water footprint" (Mekonnen and Hoekstra 2011; Hoekstra and Mekonnen 2012).] This is a very large fraction of the total consumptive use of blue water, but irrigated croplands do produce nearly 40% of the global food supply (Rost et al. 2008; Hanjra and Qureshi 2010).

Determining how much blue water should be allocated to the nonhuman biosphere is an abiding dilemma (Hanjra and Qureshi 2010; Strzepek and Boehlert 2010). Hoekstra et al. (2012) have added to the debate recently by adopting a precautionary principle that sets 20% of a natural blue water flow as the upper limit of human consumptive use of that flow, where "natural" is defined as the sum of the observed flow plus the consumptive use by humans that has reduced the flow below its "natural" level. Thus, a consumptive use that does not leave at least 80% of a natural blue water flow available for reuse is deemed to pose a serious risk to the health of any ecosystem served by the flow. Note that more than 20% of a natural blue water flow might be withdrawn; the precautionary limit applies only to the consumptive use of the flow, the blue water footprint. Rockström et al. (2012) highlight the importance of this consideration by pointing out that current blue water withdrawals have caused about one-fourth of the perennial rivers in the world no longer to reach the ocean or, if they do, to reach it only some of the time. In addition, some of the major inland lakes on the planet are drying out. These sobering omens led Rockström et al. (2009a,b) to propose another "planetary boundary" at which future blue water consumptive use is limited to 50–60% more than the current amount. But this limit may prove to be too optimistic when human uses of blue water that compete with agriculture are projected in detail (Molden 2009; Strzepek and Boehlert 2010; Rockström et al. 2012).

In light of these concerns, and building on the precautionary principle described above, Hoekstra et al. (2012) have very recently offered a new quantitative indicator of blue water scarcity to supplant the previous widely applied indices (e.g., the ratio of blue water withdrawal to renewable supply) by the one that can be measured accurately on various temporal and spatial scales. Instead of withdrawal, they propose using the blue water footprint, and instead of renewable supply, they propose using 20% of the natural blue water flow, the nominal human consumptive use they assume can exist without major ecological impact. When the ratio of these two quantities, evaluated for a chosen spatial scale (say, that of a river basin) over a chosen time period and expressed as a percent, is less than 100%, blue water scarcity is deemed to be "low," but when the ratio exceeds 200%, blue water scarcity is deemed to be "severe," meaning that more than 40% of a natural blue water flow is already being consumed by humans (Hoekstra et al. 2012).

Hoekstra et al. (2012) determined blue water scarcity for more than 400 river basins worldwide using monthly hydrologic data. These river basins represented about two-thirds of the global population and three-quarters of the irrigated regions of the world. The blue water footprint in the basins comprised agricultural, domestic, and industrial consumption, with agriculture turning out to account for 92% of the global footprint. In about half of the river basins examined, representing 2.7 billion people, blue water scarcity was found to be "severe" during at least 1 month of every year. In about one out of 12 of the basins, representing one-half billion people, blue water scarcity was "severe" for at least 6 months of every year. Twelve of the river basins investigated experienced "severe" blue water scarcity all year long. Hoekstra et al. (2012, 7) concluded that "our results underline the critical nature of water shortages around the world." From this understated portent, we may conclude that the notion of meeting global food demand in 2050 simply by a major expansion of irrigation systems seems an unlikely prospect.

30.2.3 Most of the Water Consumed Globally by Croplands Is Green Water

Green water (Falkenmark and Rockström 2004, 2006; Rockström et al. 2009c) refers to the water that accumulates in soil as a result of the infiltration of rainfall. Thus, it is water available to plant roots and to the soil biota (microbes, earthworms, soil insects, etc.) because it did not run off into streams and rivers or percolate down into an aquifer to replenish groundwater. The flows of green water are simple: either transpiration—movement into plant roots and then upward in the plant to exit through its leaves—or evaporation directly from the land surface, with both pathways returning green water to the atmosphere as water vapor. Thus, green water moves from soil into the atmosphere, whereas blue water moves in channels at the land surface or in the pores of a groundwater aquifer (Rockström et al. 2009c).

The global flow of green water accounts for about 65% of the total global flow of any water, green or blue (Rost et al. 2008; Rockström et al. 2009a). There are also "virtual flows" of green and blue water, in the form of the water stored in the food commodities that are traded among nations.

Green water accounts for a whopping 85–88% of these virtual flows (Konar et al. 2012; Hoekstra and Mekonnen 2012), which, as is the case for blue water resources, are increasingly subjected to pressures that will almost certainly drive up competition for them among nations (Suweis et al. 2013). In comparing green and blue water flows, it is important to note that only about 30% of the global blue water flow is estimated to be accessible to human appropriation (Rockström et al. 2009a), while nearly all of the global green water flow in principle could be accessible to human use. Nonagricultural ecosystems currently consume about three-fourths of the global green water flow, with the remaining one-fourth going through agricultural ecosystems partitioned into equal shares for croplands and pasturelands (Rost et al. 2008).

Mekonnen and Hoekstra (2011) [see also Hoekstra and Mekonnen (2012)] have very recently performed an exhaustive, country-by-country, crop-by-crop assessment of the green and blue water footprints of croplands, afterward comparing their detailed results with a half-dozen similar but perhaps less complete earlier tallies of the consumptive use of water by crops. Their data, which improve on but do not change the implications of the previous work, show that nearly 90% of the water consumed by croplands worldwide is green water. Even for irrigated croplands, the global green water footprint is actually larger than the blue water footprint (Rost et al. 2008; Mekonnen and Hoekstra 2011). One implication of this not-widely-appreciated fact, as noted by Rockström et al. (2009c) and emphasized by Mekonnen and Hoekstra (2011), is that an opportunity exists for meeting the global food demand developing over the next 40 years by optimizing green water use on extant croplands without further taxing the blue water resources of the world in a major way. Thus, although we may be running out of options for the amount of land that can be cleared for crops and the amount of blue water that can be appropriated for irrigation, we are not necessarily up against the same kind of constraint when it comes to green water.

It is pertinent at this juncture to revisit the passage from the historical survey by McNeill and Winiwarter (2004) quoted in the introduction. For example, their use of the adverb "uncharismatically" now must be extended to include green water. Indeed, where do we find the great poems inspired by green water? Where do we go to view the classic paintings or the scenic photographs or the riveting films celebrating the place of green water in our culture? Charismatic blue water is the polar bear of the hydrologic cycle, while green water is its Cinderella, quietly doing the world's work sustaining global ecosystems without fanfare—or respect. This is the deeper meaning of that last sentence in the quoted passage by McNeill and Winiwarter (2004). But there is something else just as important: their use of the noun "soil ecosystems." Yes, soils are ecosystems, in the very same sense as forests are ecosystems and savannas are ecosystems or, for that matter, in the same sense as oceans are ecosystems. Soils are ecosystems with the same kinds of intricate complexities and critical services that we customarily attribute to other terrestrial ecosystems. It is just that most people do not think or talk about them that way.

30.3 WHAT WE TALK ABOUT WHEN WE TALK ABOUT SOIL

Soil scientists, on the other hand, do think and talk about soil ecosystems, and recently, a number of them have banded together in an organized effort calling for a systematic research agenda that would place soils on the same conceptual footing as above-ground ecosystems (Robinson et al. 2012, 2013). Fortunately, an excellent review of soil characteristics already exists that is entirely consistent with this emerging perspective, while being fully cognizant of green water as soil water (Palm et al. 2007). The framework proposed in this review for an understanding of soil ecosystems is based on the premise that "...the natural capital of soils that underlies ecosystem services is primarily determined by three core soil properties: texture, mineralogy, and soil organic matter.... Secondary soil properties, such as aggregation, bulk density, nutrient content, and pH are determined by the combination of these core soil properties" (Palm et al. 2007, 101). [The term "natural capital" used by Palm et al. (2007), may be defined as the stock of assets that permits soils to function beneficially for humans in terrestrial ecosystems (Dominati et al. 2010).]

Texture refers to the percentage distribution of sand-, silt-, and clay-sized particles in a soil [sand, 2.0–0.05 mm; silt, 0.05–0.002 mm; and clay, <0.002 mm diameter (Schaetzl and Anderson 2005)]. For example, a soil with 40% sand, 40% silt, and 20% clay is said to have a loam texture, and conversely, a soil with 20% sand, 40% silt, and 40% clay has a clay texture. Texture determines the nature of the soil pore space and the propensity for aggregate formation by soil particles, thereby affecting green water availability, soil erodibility and concomitant blue water flow, deep percolation and concomitant solute leaching, and soil aeration. It also determines the physical qualities of the microhabitat offered by soils to the soil biota, as well as the storage and movement of the green water upon which the soil biota and all higher plants growing in soil depend. Texture mediates the uptake of dissolved nutrients, the fate of contaminants, and the effects of floods.

Mineralogy refers to the variety of primary minerals (i.e., those inherited from the parent material out of which a soil has formed) and secondary minerals (i.e., those formed during the weathering of primary minerals) in soils. Primary minerals directly contribute to soil nutrient capital (i.e., the stock of essential plant nutrients made available to the soil biota and plant roots through weathering), while secondary minerals mediate nutrient cycling (Schaetzl and Anderson 2005; Sposito 2008). Thus, mineralogy plays a major role in the capacity to provide nutrient elements for uptake by the soil biota and higher plants and to retain them against leaching loss. It also determines the potential for indigenous metal toxicity (e.g., aluminum toxicity resulting from the weathering of aluminum-containing minerals) and the ability of soil to attenuate exogenous contaminants. Mineralogy has also long been recognized for its major role in stabilizing organic matter in soils (Mortland 1970; Allison 2006; Schmidt et al. 2011), thereby influencing the provision of nonmetal nutrients, such as carbon, nitrogen, and sulfur, to the soil biota and plant roots.

Palm et al. (2007) present a global map of areas covered by soils with low nutrient capital, identified by having a primary mineral content of <10% in their sand and silt fractions. Most of these areas are in the Global South, particularly the tropics, but the southeastern United States along with parts of Canada and northern Europe are also included. Irrespective of climate, the soils in these areas are inherently limited for supporting croplands, a prospect underscored by comparing the map provided by Palm et al. (2007) with another map shown by West et al. (2010) that delineates the "carbon debt" incurred by land conversion into crop cultivation. [The "carbon debt" of a cropland is defined as the ratio of the change in carbon stock resulting from land conversion (carbon stock in cropland minus that in the prior natural vegetation) to the annual crop yield, with the carbon stock expressed in units of metric tons of carbon per hectare.] Notably, the areas of highest "carbon debt" are in the tropics and overall, the global map presented by West et al. (2010) closely resembles that presented by Palm et al. (2007). West et al. (2010, 19645) concluded from their analysis that, "particularly in the tropics, emphasis should be placed on increasing yields on existing croplands rather than clearing new lands," that is, crop intensification should drive the future of food production.

The carbon-containing compounds in soil that are not either minerals or carbon dioxide, are collectively termed "soil organic matter" or "humus." They vary enormously in provenance and molecular complexity. Soil organic matter comprises molecules produced to sustain the life cycles of the soil biota and higher plants, along with a diverse assembly of solid materials, not usually identifiable as individual molecules, which are complex residual products of carbon cycling by the soil biota, plant roots, and fires. Soil organic matter is the primary contributor of nonmetal chemical elements to soil nutrient capital and it is the most important repository of protons and electrons engaging in the chemical reactions that play dominant roles in the cycling of indigenous elements and the detoxification of contaminants (Sposito 2008). It is also by far the largest repository of carbon in the part of the carbon cycle that circulates on submillennial timescales (Sposito 2008).

As noted above, most soils are composed of weathered solid matter inherited from the rocks or deposits on which they form. Texture and mineralogy, as well as the critical nutrient cycling these two fundamental properties control, thus, are soil attributes inherited from the underlying lithosphere. In contrast, soil organic matter is derived from continuous inputs and subsequent decomposition of organic compounds by the soil biota, and this process supplies the bulk of the carbon, nitrogen, sulfur, and other nonmetal nutrients that differentiate fertile soils from weathered rock. These three soil characteristics identified by Palm et al. (2007) can be grouped metaphorically as typifying the "genetics" (texture and mineralogy) and "diet" (organic matter input) of a soil ecosystem. In the absence of disturbance (e.g., erosion), they help determine the health of soils, just as genetics and diet help determine the health of human beings in the absence of physical injury. In this metaphorical context, a healthy soil ecosystem maintains high natural capital conditioned on its "genetics" and "diet": texture, mineralogy, and organic matter. Accordingly, one could designate areas having soils with low nutrient capital, such as those mapped by Palm et al. (2007), as regions with unhealthy soils by virtue of their "genetics."

But this is not the complete picture: the three fundamental soil characteristics are necessary but not sufficient determinants of soil health, a point also acknowledged by Palm et al. (2007). What must be added is some form of catalysis to drive the processes keeping soils healthy. Again, we can resort to a comparison with human health, borrowing a metaphor from very recent studies showing the importance to the latter of the human microbiome, the community of microbial organisms, mainly bacteria, which live in and on humans (Pollan 2013). (See also Gina Kolata, "In good health? Thank your 100 trillion bacteria," *New York Times*, June 13, 2012.) These microbes have coevolved with humans, changing with their diet and influencing the ways in which they function, whether healthy or sick (Ley et al. 2008; Spor et al. 2011; Pflughoeft and Versalovic 2012). In the same way, the soil microbiome, the community of microbial organisms, mainly bacteria, which inhabit soil ecosystems, change with their "diet" and influence the ways in which they function, healthy or not (Wardle et al. 2004; Bardgett and Wardle 2010; Wall 2012). Like the human microbiome, soil microbiomes are dominated by a few major taxonomic groups of bacteria [in fact, almost the same ones (Pflughoeft and Versalovic 2012; Fierer et al. 2012)], and like the familiar terrestrial biomes—forests, grasslands, and deserts—soil microbiomes are strongly affected by climate (Fierer et al. 2012). One kilogram of healthy soil can serve as the habitat for up to a trillion bacteria (Kirstin and Miranda 2013). These microorganisms, which are major contributors to global biodiversity, catalyze important soil physicochemical processes, especially those occurring within a few millimeters of plant roots, a zone known as the rhizosphere (Bardgett and Wardle 2010; Kristin and Miranda 2013). And, it is through the tiny rhizosphere that as much water flows annually as passes out of all the rivers of the world into the oceans.

30.4 FRAMING THE CHALLENGE

A very useful way of thinking about how to optimize green water use for crop intensification was put forth by Rockström and Falkenmark (2000), and then further elaborated by Falkenmark and Rockström (2004) and Rockström et al. (2007). Their strategy, developed in light of the established proportionality between crop biomass and transpiration, is based on increasing both green water availability and the productive flow of green water. By "green water availability" they mean the percentage of the nominal transpiration requirement for maximal yield of a crop that is actually available in a soil during a growing season. Evidently, the upper limit of this quantity is equal to 100 times the ratio of precipitation to the transpired water requirement (both expressed in the same units), whereas the lower limit is set by the green water content below which crop failure will occur regardless of any other conditions that may obtain, this lower limit also being expressed as a percentage of the transpired-water requirement (Rockström and Falkenmark 2000). Green water availability is dependent on both weather and climate, as well as on the hydrologic behavior of soil (e.g., its propensity for generating either lateral or vertical blue water flows and its water-holding capacity). Thus, green water availability is small during periods of low rainfall (or yearlong in arid climates) and it is reduced when rainfall is significantly converted into blue water running into streams and rivers, or when deep percolation of soil water into groundwater is significant.

The second variable introduced by Rockström and Falkenmark (2000), the "productive flow of green water," refers to the percentage of green water flow that occurs through transpiration, as opposed to occurring through evaporation. The percentage of green water flow that is deemed productive thus will intrinsically depend on crop characteristics (e.g., rooting structure and the nature of the canopy), but it will also be contingent on the efficacy of the rhizosphere in promoting the flow of green water into plant roots.

Considering these two "master variables" quantitatively, one finds that a chosen crop yield can be achieved either by a high-enough green water availability in spite of a low productive flow of green water or by a high-enough productive flow of green water despite a low green water availability (Rockström and Falkenmark 2000; Falkenmark and Rockström 2004). Figure 30.1 illustrates this compensating relationship through an example typical of maize grown in the semiarid regions of sub-Saharan Africa (Rockström and Falkenmark 2000; Rockström et al. 2007). For this situation, the upper limit of green water availability ("Rain deficiency" in Figure 30.1) is near 90%, as set by the average growing-season rainfall and the green water requirement of maize for maximum yield, while the lower limit where crop failure would occur is near 10%. The upper limit of the productive flow of green water is 85%, as constrained by unavoidable soil evaporation occurring at the beginning of the growing season. Under typical on-farm conditions, because of combined rainfall losses to blue water flow as runoff into streams or deep percolation into groundwater, the green water availability is under 40% (lowest dotted horizontal line in Figure 30.1) and only 30% of the green water flow is productive (i.e., transpired by the crop, vertical dotted line in Figure 30.1), leading to a large "Green water loss" and consequently a very low average maize yield of 1 metric ton per hectare. However, for the same 40% green water availability, if 85% of the green water flow were to become productive, the yield could triple (intersection of the dotted horizontal line with the

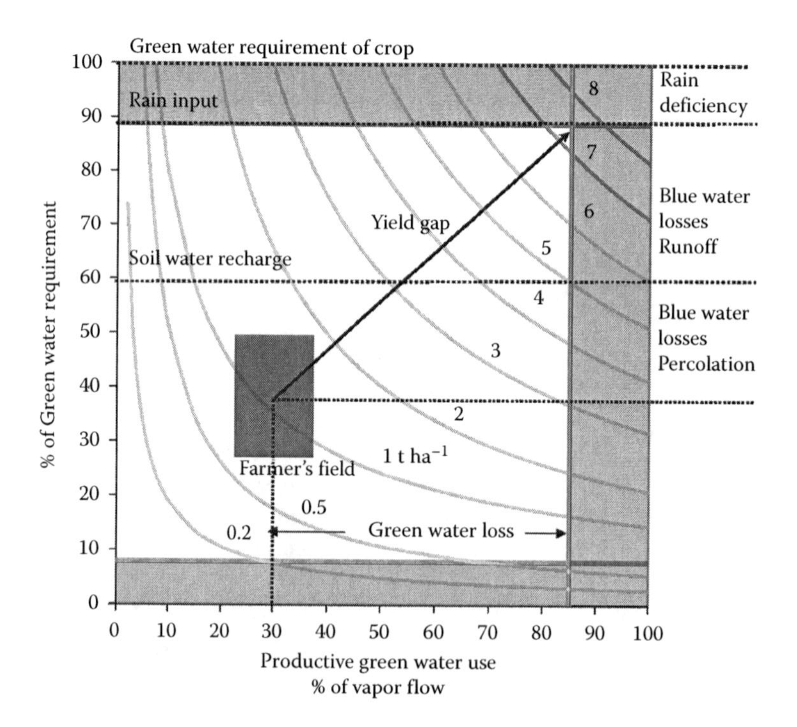

FIGURE 30.1 Maize yield (values labeling gray curves, in units of metric tons per hectare) achieved with differing green water availability (vertical axis) and productive flow of green water (horizontal axis). The dark gray rectangle indicates typical ranges of the two independent variables for farms in the semiarid regions of sub-Saharan Africa, which achieve an average yield of 1 t/ha. See the text for additional details. (Reprinted with permission from Rockström, J., M. Lannerstad, and M. Falkenmark. 2007. *Proceedings of the National Academy of Sciences* 104:6253–6260. Copyright 2007, National Academy of Sciences, USA.)

gray vertical line in Figure 30.1), but, if the green water availability could also be increased, say, to reach 60%, the yield could quintuple (intersection of the dotted horizontal line labeled "Soil water recharge" with the gray vertical line in Figure 30.1). On the other hand, yield would not improve, even if 85% of the green water flow could be made productive, were the green water availability to drop, say, to 20% from the effects of a severe drought (intersection of the 1 metric ton per hectare curve with the gray vertical line in Figure 30.1).

Rockström et al. (2007, 6254) infer from this kind of strategizing that "there are no hydrological limitations to attain a doubling of yield levels," at least with respect to maize grown in semiarid sub-Saharan Africa, a point recently echoed by Foley et al. (2011) in their assessment of the prospects for increasing global crop yields in response to population growth. Instead, the limitation on maize yields in semiarid sub-Saharan Africa mainly appears to be one of the limiting plant nutrient availability (Foley et al. 2011), which, if remedied, would naturally increase the productive flow of green water as a result of increasing crop biomass. Precisely the same point was made by Sánchez (2010) in his closely reasoned critique of the Green Revolution as applied to sub-Saharan Africa, building on an earlier perspective developed by Sánchez and Swaminathan (2005, 443) for mitigating malnutrition in that part of the world: "In Africa, soil nutrients and water management are the major limiting factors, not improved crop varieties." They go on to outline a strategy for increasing crop yield that utilizes the metaphor of soil health, with nutrient-depleted soils accordingly deemed unhealthy, noting that these soils are also typically low in organic matter, that is, they are also unhealthy by virtue of their poor "diet."

This prediction was tested by the initiation of the Millennium Villages Project in 2005 (Sánchez et al. 2007, 2009) as part of an effort to achieve the United Nations' Millennium Development Goals announced at a summit of world leaders in 2000. Nziguheba et al. (2010) have reviewed the Millennium Villages Project in great detail while surveying the first 3 or 4 years of results and then offering a post-factum analysis. The interventions employed may be summarized by the following bulleted list:

- Eighty villages in 10 sub-Saharan countries grouped into 14 clusters of major agroecological zones and farming systems located in "hunger hotspots" (i.e., more than one-fifth of the children less than 5 years old are malnourished)
- Science-based community-led interventions to improve soil health, human development, and social, financial, and physical infrastructures
- Increasing soil nutrient capital with subsidized mineral fertilizers (urea for nitrogen and diammonium phosphate for phosphorus; occasionally, potassium and sulfur amendments are also provided)
- Providing improved seeds, also subsidized, either hybrids or open-pollinated cultivars
- Improving agronomic practices, including early planting, spacing, seed and fertilizer placement, harvest and storage methods, and small-scale water management

Three years after these interventions began, maize yields were more than doubled across the project sites, reaching an average 4.4 metric tons per hectare, up from an average of 1.2 metric tons per hectare and well above the goal of 3.0 metric tons per hectare (Nziguheba et al. 2010).

As noted by Sánchez (2010, 300), these encouraging results reflect the expected increase in the productive flow of green water arising from larger crop biomass, as well as a synergistic crop canopy effect: "At current African yields, about two-thirds of soil moisture is lost via soil evaporation, leaving only one-third of the captured rainfall available for plants. But when cereal yields rise from one to three tons per hectare, the crop canopy closes and the balance flips over: only about a third is lost by soil evaporation, and two-thirds is funneled through the plants as transpiration." In reference to Figure 30.1, this effect can be seen by imagining a vertical line marking 67% productive green water flow intersecting the lowest horizontal dotted line, implying a yield around 2.5 metric tons per hectare. A positive correlation between maize yield and rainfall variation was also observed during the first 3 years of the project (Nziguheba et al. 2010), indicating that not only was the productive flow of green water increased by nutrient-capital interventions and the synergistic canopy effect, but green water availability was also increased from 40% to about 50% by natural rainfall variation.

We note in passing some obvious improvements in the interventions summarized above that must be achieved to make them sustainable (Sánchez et al. 2009; Nziguheba et al. 2010): Use of synthetic fertilizers should be replaced by the use of nonsynthetic fertilizers and legumes; subsidies for fertilizers and seeds should be replaced by credit programs; and subsistence monocultures such as maize should be diversified to include high-value crops. But the overall take-home message from this encouraging example is that science-based interventions directed at improving the productive flow of green water increase crop productivity by restoring "the genius of soil."

30.5 ADDRESSING THE COMPLEXITY

The synergistic crop canopy effect that played an important role in promoting the remarkable maize yield increase achieved during the early years of the Millennium Villages Project is a simple example of what is termed in ecology as a "plant–soil feedback" (Ehrenfeld et al. 2005; Bardgett and Wardle 2010). In the language of systems analysis, the output of an enhanced transpiration process—increased crop canopy size—induced a feedback—shading of the soil—that affected the input to the transpiration process, that is, productive green water flow, by amplifying it. Generally, a plant–soil feedback is a change in soil conditions induced by plants; this change in soil conditions then leads to a further change in the plants. The feedback can affect either the biotic or the abiotic properties of soil, as well as its biological, chemical, or physical processes (Ehrenfeld et al. 2005).

If the plant-induced change in soil acts to amplify a given soil input to the plant, the plant–soil feedback is termed positive, but if it acts to diminish the soil input to the plant, it is termed negative. In the case of the maize canopy example, the feedback was positive, since the productive flow of green water was increased as a result of the decrease in soil evaporation induced by a larger maize canopy. In contrast, suppose that the growth of a more extensive root system resulting from improved soil nutrient status had encouraged root parasites and pathogens to become more active in the soil. This plant–soil feedback, shifting the composition of the soil microbiome to increase its population of pathogens, is commonly observed in terrestrial ecosystems (Kulmatiski et al. 2008, 2012; Bardgett and Wardle 2010). It is, of course, negative because it would lead to a decrease in plant biomass. However, if an enhanced root system instead was to promote symbiotic relationships between plant roots and soil microbes that enhanced nutrient uptake, the plant–soil feedback would be deemed positive (Schnitzer et al. 2011).

Recent reviews of laboratory, field, and modeling research on plant–soil feedback, including meta-analyses that survey the results of previous studies (Ehrenfeld et al. 2005; Kulmatiski et al. 2008, 2012; Brinkman et al. 2010; Bardgett and Wardle 2010; Bardgett 2011; Wall 2012) consistently point to the complexity associated with attempts to unravel feedback mechanisms or even assess whether a plant–soil feedback is positive or negative, the problematic partly arising from the intricacy of root architecture (Pierret et al. 2007) and the extraordinary diversity of the soil microbiome, with a multitude of interactions possible among its species and between its species and plant roots (Bardgett and Wardle 2010; Kristin and Miranda 2013). In light of this complexity, it is not surprising to learn that some studies conducted in the past are now seen with more critical eyes as flawed in the design, such that their results may have limited value. The recent meta-analysis of Brinkman et al. (2010, 1072), concluding with the sobering proviso that "this does not mean that all previous plant–soil feedback studies should be considered unreliable," acknowledges that the extant literature does tend to support the hypothesis that plant–soil feedback has strong effects on plant communities and that the sign of the feedback is often negative. A recent study by Kulmatiski et al. (2012) intriguingly suggests that, for plant species to thrive in monocultures, typical of croplands, plant–soil feedbacks must always be positive, not negative. Clearly, there is an opportunity here to address plant–soil feedback as part of a strategy to increase green water availability and productive flow in croplands (Bakker et al. 2012).

As noted above, green water flow necessarily must go through the rhizosphere to be productive. By definition, the rhizosphere is significantly affected by plant roots, naturally leading to the question of whether plant–soil feedbacks exist in it that would directly impact green water availability as well as productive green water flow. Plant roots continually exude into the rhizosphere complex mixtures of carbon compounds, collectively termed mucilage. One abiotic result of this feedback—besides the evident biotic result of stimulating the soil microbiome (Bakker et al. 2012; Kirstin and Miranda 2013)—is to increase the water-holding capacity of soil (Carminati et al. 2010, 2011). This in turn implies an increase in the availability of green water because its loss to deep percolation should then decrease. Evidence also exists supporting the notion that the incorporation of mucilage into the rhizosphere causes it to have different water-conducting properties from nonrhizosphere soil (Carminati et al. 2011), with the overall outcome of this difference being that "the rhizosphere acts as a buffer that softens the hydraulic stress experienced by roots in soils, and favors water availability to roots during drought" (Carminati et al. 2010, 174). This plant–soil feedback, although originating at very small spatial scales, may also be manifest at much larger spatial scales in driving ecosystem response to drought (Bengough 2012; Marasco et al. 2012).

30.6 SUSTAINING "THE GENIUS OF SOILS"

Michael Pollan succinctly defined the often-invoked environmentalist term "sustainable" as a characteristic of any human activity that does not undermine the conditions required for its existence (Michael Pollan, personal communication 2009). (See also Michael Pollan, "Our decrepit food factories," *New York Times*, December 16, 2007.) In the vernacular of systems analysis, a human activity is sustainable if it does not cause negative feedback. Thus, in the present context, any agricultural practice dedicated to increasing global food production by at least 65–70% during the next 40 years can be judged sustainable only if it does not undermine the natural capital of the soils and the integrity of the water resources used to grow crops. Adding considerable gravity to these criteria is the fact that the fundamental science necessary for accurately assessing and monitoring the natural capital of soils and the integrity of water resources has not yet been fully developed. [Consider the paucity of knowledge about the composition and workings of the soil microbiome (Bardgett and Wardle 2010; Kirstin and Miranda 2013) or the high degree of uncertainty that attends the estimate of 20% as the fraction of any natural blue water flow that can be consumed without ecological damage (Hoekstra et al. 2012).]

Knowing that cropland expansion and increasing the withdrawals of blue water for irrigation are not major options for meeting future demands on global croplands places useful constraints on what priorities should obtain as the necessary fundamental science is developed over the coming decades. Knowing further that a major goal is instead to optimize the availability and productive flow of green water not only helps to focus research priorities, but also highlights the importance of methodologies that, unshackled by adherence to conventional disciplines, fully respect the complexity of soil ecosystems, which "remain firmly, but uncharismatically, at the foundations of human life" (McNeill and Winiwarter 2004, 1629). Perhaps uncharismatic today, but not 20 centuries ago when Virgil sang the prescient lines (Ferry 2005, 65):

> But if the soil breathes steaming vapors out,
> Drinks moisture in, and when it wants to, breathes
> The moisture out again, and if it's always
> Green with the greenness of its grasses and
> Never corrodes the blade of the plow with rust,
> Then that's the place to drape your flourishing vines
> Upon your elms, the place that will produce
> Rich olive oil, the place (as the tilling will show)
> That makes the plowing easy for the beasts
> Because the soil is easy for the plow.

ACKNOWLEDGMENTS

Gratitude is expressed to my Berkeley colleagues, Ronald Amundson and Michael Pollan, and to Pedro Sánchez, Columbia University, for reviewing this chapter in draft form and for many helpful discussions about soils, farming systems, and food security. The preparation of this chapter was in part supported by funds allocated to the author as the Betty and Isaac Barshad Chair in Soil Science, administered through the College of Natural Resources, University of California at Berkeley.

REFERENCES

Allison, S. D. 2006. Brown ground: A soil carbon analogue for the green world hypothesis. *American Naturalist* 167: 619–627.

Bakker, M. G., D. K. Manter, A. M. Sheflin, T. L. Weir, and J. M. Vivanco. 2012. Harnessing the rhizosphere microbiome through plant breeding and agricultural management. *Plant and Soil* 360: 1–13.

Bardgett, R. D. 2011. Plant–soil interactions in a changing world. *F1000 Biology Reports* 3: 16 (doi:10.3410/B3-16).

Bardgett, R. D. and D. A. Wardle. 2010. *Aboveground–Belowground Linkages*. New York: Oxford University Press.

Barnosky, A. D., E. A. Hadley, J. Bascompte et al. 2012. Approaching a state shift in earth's biosphere. *Nature* 486: 52–58.

Bengough, A. G. 2012. Water dynamics of the root zone: Rhizosphere biophysics and its control on soil hydrology. *Vadose Zone Journal*, doi:10.2136/vzj2011.0111.

Brinkman, E. P., W. H. van der Putten, E.-J. Bakker, and K. J. F. Verhoeven. 2010. Plant–soil feedback: Experimental approaches, statistical analyses and ecological interpretations. *Journal of Ecology* 98: 1063–1073.

Carminati, A., A. B. Moradi, D. Vetterlein et al. 2010. Dynamics of soil water content in the rhizosphere. *Plant and Soil* 332: 163–176.

Carminati, A., C. L. Schneider, A. B. Moradi et al. 2011. How the rhizosphere may favor water availability to roots. *Vadose Zone Journal* 10: 988–998.

Dominati, E., M. Patterson, and A. Mackay. 2010. A framework for classifying and quantifying the natural capital and ecosystems services of soils. *Ecological Economics* 69: 1858–1868.

Ehrenfeld, J. G., B. Ravit, and K. Elgersma. 2005. Feedback in the plant–soil system. *Annual Review of Environment and Resources* 30: 75–115.

Ellis, E. C. 2011. Anthropogenic transformation of the terrestrial biosphere. *Philosophical Transactions of the Royal Society A* 369: 1010–1035.

Fairclough, H. R. and G. P. Goold (trans.). 1999. *Virgil: Eclogues, Georgics, Aeneid 1–6*. Cambridge: Harvard University Press.

Falkenmark, M. and J. Rockström. 2004. *Balancing Water for Humans and Nature*. London: Earthscan.

Falkenmark, M. and J. Rockström. 2006. The new blue and green water paradigm. *Journal of Water Resources Planning and Management* 132: 129–132.

Ferry, D. (trans.) 2005. *The Georgics of Virgil*. New York: Farrar, Straus and Giroux.

Fierer, N., J. W. Leff, B. J. Adams et al. 2012. Cross-biome metagenomic analyses of soil microbial communities and their functional attributes. *Proceedings of the National Academy of Sciences* 109: 21390–21395.

Foley, J. A., R. DeFries, G. P. Asner et al. 2005. Global consequences of land use. *Science* 309: 570–574.

Foley, J. A., N. Ramankurty, K. A. Brauman et al. 2011. Solutions for a cultivated planet. *Nature* 478: 337–342.

de Fraiture, C., L. Karlberg, and J. Rockström. 2009. Can rainfed agriculture feed the world? In *Rainfed Agriculture: Unlocking the Potential*, eds. S. P. Wani, J. Rockström and T. Oweis, pp. 124–132. Cambridge: CAB International.

Gregory, P. J. and T. S. George. 2011. Feeding nine billion: The challenge to sustainable crop production. *Journal of Experimental Botany* 62: 5233–5239.

Hanjra, M. A. and M. E. Qureshi. 2010. Global water crisis and future food security in an era of climate change. *Food Policy* 35: 365–377.

Hoekstra, A. Y., M. M. Mekonnen, A. K. Chapagain, R. E. Mathews, and B. D. Richter. 2012. Global monthly water scarcity: Blue water footprints versus blue water availability. *PLoS ONE* 7(2):e32688, doi:10.1371/journal.pone.0032688.

Hoekstra, A. Y. and M. M. Mekonnen. 2012. The water footprint of humanity. *Proceedings of the National Academy of Sciences* 109: 3232–3237.

Konar, M., C. Dalin, N. Hanasaki, A. Rinaldo, and I. Rodríguez-Iturbe. 2012. Temporal dynamics of blue and green virtual water trade networks. *Water Resources Research* 48: W07509, doi:10.1029/2012WR011959.
Kristin, A. and H. Miranda. 2013. The root microbiota—A fingerprint in the soil? *Plant and Soil* 370: 671–686.
Kulmatiski, A., K. H. Beard, J. R. Stevens, and S. M. Cobbold. 2008. Plant–soil feedbacks: A meta-analytical review. *Ecology Letters* 11: 980–992.
Kulmatiski, A., K. H. Beard, and J. Heavilin. 2012. Plant–soil feedbacks provide an additional explanation for diversity–productivity relationships. *Proceedings of the Royal Society B* 279: 3020–3026.
Ley, R. E., C. A. Lozupone, M. Hamady, R. Knight, and J. I. Gordon. 2008. Worlds within worlds: Evolution of the vertebrate gut microbiota. *Nature Reviews in Microbiology* 6: 776–788.
Marasco, R., E. Rolli, B. Ettoumi et al. 2012. A drought resistance-promoting microbiome is selected by root system under desert farming. *PloS ONE* 7(10): e48479, doi:10.1371/journal.pone.0048479.
McNeill, J. R. and V. Winiwarter. 2004. Breaking the sod: Humankind, history, and soil. *Science* 304: 1627–1629.
Mekonnen, M. M. and A. Y. Hoekstra. 2011. The green, blue and grey water footprint of crops and derived crop products. *Hydrology and Earth System Sciences* 15: 1577–1600.
Molden, D. 2009. The devil is in the detail. *Nature Reports Climate Change* 3: 116–117.
Mortland, M. M. 1970. Clay–organic complexes and interactions. *Advances in Agronomy* 22: 75–117.
Nziguheba, G., C. A. Palm, T. Berhe et al. 2010. The African green revolution: Results from the millennium villages project. *Advances in Agronomy* 109: 75–115.
Palm, C., P. Sánchez, S. Ahamed, and A. Awiti. 2007. Soils: A contemporary perspective. *Annual Review of Environment and Resources* 32: 99–129.
Pierret, A., C. Doussan, Y. Capowiez, F. Bastardie, and L. Pagès. 2007. Root functional architecture: A framework for modeling the interplay between roots and soil. *Vadose Zone Journal* 6: 269–281.
Pflughoeft, K. J. and J. Versalovic. 2012. Human microbiome in health and disease. *Annual Review of Pathology: Mechanisms of Disease* 7: 99–122.
Pollan, M. 2013. Some of my best friends are bacteria. *New York Times Magazine*, May 19, 2013, 36–43, 50, 58–59.
Roberts, L. 2011. Nine billion? *Science* 333: 540–543.
Robinson, D. A., B. A. Emmett, B. Reynolds et al. 2012. Soil natural capital and ecosystem service delivery in a world of global change. *Issues in Environmental Science and Technology* 35: 41–68.
Robinson, D. A., N. Hockley, D. M. Cooper et al. 2013. Natural capital and ecosystem services, developing an appropriate soils framework as a basis for evaluation. *Soil Biology and Biochemistry* 57: 1023–1033.
Rockström, J. and M. Falkenmark. 2000. Semiarid crop production from a hydrological perspective: Gap between potential and actual yields. *Critical Reviews in Plant Science* 19: 319–346.
Rockström, J., M. Lannerstad, and M. Falkenmark. 2007. Assessing the water challenge of a new green revolution in developing countries. *Proceedings of the National Academy of Sciences* 104: 6253–6260.
Rockström, J., W. Steffen, K. Noone et al. 2009a. Planetary boundaries: Exploring the safe operating space for humanity. *Ecology and Society* 14(2): Art. 32.
Rockström, J., W. Steffen, K. Noone et al. 2009b. A safe operating space for humanity. *Nature* 461: 472–475.
Rockström, J., M. Falkenmark, L. Karlberg, H. Hoff, S. Rost, and D. Gerten. 2009c. Future water availability for global food production: The potential of green water for increasing resilience to global change. *Water Resources Research* 45: W00A12, doi:10.1029/20007WR006767.
Rockström, J., M. Falkenmark, M. Lannerstad, and L. Karlberg. 2012. The planetary water drama: Dual task of feeding humanity and curbing climate change. *Geophysical Research Letters* 39: L15401, doi:10.1029/2012GL051688.
Rost, S., D. Gerten, A. Bondeau, W. Lucht, J. Rohwer, and S. Schaphoff. 2008. Agricultural green and blue water consumption and its influence on the global water system. *Water Resources Research* 44: W09405, doi:10.1029/2007WR006331.
Running, S. W. 2012. A measurable planetary boundary for the biosphere. *Science* 337: 1458–1459.
Sánchez, P. 2010. Tripling crop yields in tropical Africa. *Nature Geoscience* 3: 299–300.
Sánchez, P. and M. S. Swaminathan. 2005. Hunger in Africa: The link between unhealthy people and unhealthy soils. *Lancet* 365: 442–444.
Sánchez, P., G. L. Denning, and G. Nziguheba. 2009. The African green revolution moves forward. *Food Security* 1: 37–44.
Sánchez, P., C. Palm, J. Sachs et al. 2007. The African millennium villages. *Proceedings of the National Academy of Sciences* 104: 16775–16780.
Schaetzl, R. and S. Anderson. 2005. *Soils: Genesis and Geomorphology*. Cambridge: Cambridge University Press.
Schmidt, M. W. I., M. S. Torn, S. Abiven et al. 2011. Persistence of soil organic matter as an ecosystem property. *Nature* 478: 49–56.

Schnitzer, S. A., J. N. Klironomos, J. HilleRisLambers et al. 2011. Soil microbes drive the classic plant diversity–productivity patterns. *Ecology* 92: 296–303.
Smith, P., P. J. Gregory, D. van Vuuren et al. 2010. Competition for land. *Philosophical Transactions of the Royal Society B* 365: 2941–2957.
Spor, A., O. Koren, and R. Ley. 2011. Unraveling the effects of the environment and host genotype on the gut microbiome. *Nature Reviews in Microbiology* 9: 279–290.
Sposito, G. 2008. *The Chemistry of Soils.* 2nd edition. New York: Oxford University Press.
Strzepek, K. and B. Boehlert. 2010. Competition for water for the food system. *Philosophical Transactions of the Royal Society B* 365: 2927–2940.
Suweis, S., A. Rinaldo, A. Maritan, and P. D'Odorico. 2013. Water-controlled wealth of nations. *Proceedings of the National Academy of Sciences* 110: 4230–4233.
Tilman, D., C. Balzer, J. Hill, and B. L. Befort. 2011. Global food demand and the sustainable intensification of agriculture. *Proceedings of the National Academy of Sciences* 108: 20260–20264.
Wall, D. H. (ed.) 2012. *Soil Ecology and Ecosystem Services.* Oxford: Oxford University Press.
Wardle, D. A., R. D. Bardgett, J. N. Klironomos, H. Sätäla, W. H. van der Putten, and D. H. Wall. 2004. Ecological linkages between aboveground and belowground biota. *Science* 304: 1629–1633.
West, P. C., H. K. Gibbs, C. Monfreda et al. 2010. Trading carbon for food: Global comparison of carbon stocks vs. crop yields on agricultural land. *Proceedings of the National Academy of Sciences* 107: 19645–19648.

Index

C

T

U

V

W

X

Y